Praise for *Darwin's Sacre*

New York Times "Editors'

"Rich in detail, remarkably readable and engaging . . . *Sacred Cause* is the finest birthday tribute to Charles Darwin in many years."
Thomas Hayden, *Washington Post Book World*

"A candidate for the most insightful, and perhaps the most radical [Darwin bicentennial publication], is *Darwin's Sacred Cause*. In this thorough and highly researched, yet readable and even entertaining book . . . Adrian Desmond and James Moore seek to humanize the father of evolutionary theory. His drive to uncover a single common ancestor for humans, they say, was prompted squarely by his opposition to the great social evil of his day: slavery. . . . The authentic Darwin, the authors conclude, is 'a man more sympathetic than creationists find acceptable, more morally committed than scientists would allow.'"
Christian Science Monitor

"Arresting. . . . The case they make is rich and intricate."
New York Times Book Review

"[I]f you read only one book to honor the bicentennial, make it *Darwin's Sacred Cause*. This first-rate work breaks new ground and persuasively locates the inspiration for Darwin's theory in his abolitionism."
M. G. Lord, *Los Angeles Times*

"A new reading of Darwin's life and scientific work." *Nature*

"A masterful achievement. . . . [*Darwin's Sacred Cause*] establishes the range and reach of early to mid-nineteenth-century scientific concerns about slavery and the races of man. It makes the general theory of evolution an outgrowth not just of a special theory of human evolution but of moral sensibilities and political agenda. It leaves in tatters a notion of a protected apolitical and amoral domain inhabited by Darwin's science."
Steven Shapin, *London Review of Books*

"Particularly admirable are the detailed analyses of the racial attitudes of numerous intellectuals and scientists. . . . [The authors] have, with great thoroughness, displayed the variety of ideological and scientific positions on slavery during the first half of the 19th century . . . a book of deep scholarship, which considerably expands our appreciation of Darwin's accomplishment."

Robert J. Richards, *American Scientist*

"Of all the scores of books that greeted Darwin's double anniversary last year, none matters more than this. Desmond and Moore . . . now in this triumph of re-interpretation plant the hatred of slavery—and of the science that sought to justify it—at the heart of Darwin's career."

Boyd Tonkin, *The Independent*

"A first-rate read that transforms and refreshes even the most familiar parts of Darwin's story. Surprises abound as slavery slips from the background to assume a leading role. . . . Here is Darwin's achievement recast, magnificently, for a world aware as never before of the enormity of our slaving past but also the distance travelled since. Masterpieces come no more timely."

Gregory Radick, *Times Higher Education Supplement*

"Indispensible reading for those seeking to understand Darwin's work on human evolution . . . and bound to generate considerable debate."

Isis

"Adrian Desmond and James Moore offer a new vision of the architect of evolution by natural selection. . . . The Darwin who emerges from this meticulous analysis is profoundly humanitarian, despising slavery because he abhorred cruelty to any creature."

Philip Ball, *Observer*

DARWIN'S SACRED CAUSE

RACE, SLAVERY, AND THE QUEST FOR HUMAN ORIGINS

Adrian Desmond
&
James Moore

THE UNIVERSITY OF CHICAGO PRESS

The University of Chicago Press, Chicago, 60637

First published in the United Kingdom by the Penguin Group in 2009

University of Chicago Press edition 2011

Printed in the United States of America

20 19 18 17 16 15 14 13 12 11 1 2 3 4 5

ISBN-13: 978-0-226-14451-1 (paper)
ISBN-10: 0-226-14451-8 (paper)

Library of Congress Cataloging-in-Publication Data

Desmond, Adrian J., 1947–
Darwin's sacred cause : race, slavery, and the quest for human origins / Adrian Desmond and James Moore.
p. cm.
Includes bibliographical references and index.
ISBN-13: 978-0-226-14451-1 (pbk. : alk. paper)
ISBN-10: 0-226-14451-8 (pbk. : alk. paper) 1. Darwin, Charles, 1809–1882—Ethics. 2. Darwin, Charles, 1809–1882—Political and social views. 3. Human evolution—Philosophy. 4. Slavery—Philosophy. 5. Slavery—Moral and ethical aspects. I. Moore, James R. (James Richard), 1947– II. Title.
GN281.4.D475 2011
306.3'6208996017521—dc22
2010046170

♾ The paper used in this publication meets the minimum requirements of the American National Standard for Information Sciences—Permanence of Paper for Printed Library Materials, ANSI Z39.48-1992.

I was quite delighted . . . to hear of all your varied accomplishments and knowledge, and of your higher attributes in the sacred cause of humanity.

Darwin, writing in 1859 to the naturalist/anti-slavery activist Richard Hill, the first gentleman 'of colour' in the Jamaican magistracy, assigned to adjudicate between former slave-holders and slaves

Contents

Illustrations

Acknowledgements

Trekking in the hinterlands of Darwin's world, we have amassed a large number of debts. Many individuals, each with their own expertise in wide-ranging fields, have shared all manner of esoteric knowledge with us – the sort that can often be gained nowhere else.

For answering specific queries, sometimes at short notice, we thank colleagues at universities, research libraries, historical societies, museums, professional bodies, heritage sites and scholarly projects around the world: Stephen Alter, Patrick Armstrong, Rich Bellon, Robert Bernasconi, Daniel Brass, Nick Cooke, Martin Crawford, David Dabydeen, John W. de Gruchy, Jeremy Dibbell, Mario di Gregorio, Richard Drayton, Martin Fitzpatrick, Sheila Hannon, Keith Hart, Uwe Hossfeld, Karl Jacoby, Peter McGrath, Chris Mills, Richard Milner, Duncan Porter, Greg Radick, Tori Reeve, Peter Rhodes, Nigel Rigby, Kiri Ross-Jones, Nicolaas Rupke, Matthew Scarborough, Lester Stephens, Keith Thomson, David Turley, Gene Waddell, Sarah Walpole, James Walvin, R. K. Webb and Leonard Wilson.

Without the help of dedicated library and archival staff we would still be searching for vital documents. In particular, for making materials available to us, we are grateful to the libraries of the American Philosophical Society (Valerie-Anne Lutz van Ammers), Christ's College (Candace Guite) and Corpus Christi College (Gill Cannell), Cambridge; Dartmouth College Library (Sarah Hartwell, Rauner Special Collections); Edinburgh University Library; John Murray Archives; John Rylands University Library, Manchester (Les Gray); Keele University Library (Helen Burton); National Library of Jamaica (Nicole Bryan); National Library of Scotland (Anna Hatzidaki, Robbie Mitchell); Parliamentary Archives, House of Lords Record Office (Mari Takayanagi); Smith College Library (Susan Boone); Southampton Reference Library

(Vicky Green); Suffolk County Record Office, Ipswich (Pauline Taylor); University College London Library; and Waring Library, Medical University of South Carolina (Kay Carter).

We wish to express special gratitude to William Darwin for permission to publish extracts of Darwin's letters and manuscripts and to the Syndics of Cambridge University Library for allowing us to quote from unpublished materials in the Charles Darwin Collections and in other manuscript deposits.

For permission to study, and in some instances quote from, documents in their collections, we also thank: A. K. Bell Library, Perth, Scotland; American Philosophical Society (Samuel George Morton Papers); Birmingham University Library (Harriet Martineau Papers and Church Missionary Society Unofficial Papers); Cambridgeshire Archives Service; Dartmouth College Library (Ticknor Autograph Collection); Down House and English Heritage (*Beagle* Field Notebooks); Durham University Library (Political and Public Papers of 2nd Earl Grey); Edinburgh University Library; Ernst-Haeckel-Haus, Friedrich-Schiller-Universität Jena (Darwin–Haeckel Correspondence); Gray Herbarium Library, Harvard University (Asa Gray Papers); Houghton Library, Harvard University (Louis Agassiz Papers, Charles Eliot Norton Papers, Charles Sumner Correspondence); Imperial College of Science and Technology (Thomas Henry Huxley Archives); Jesus College, Cambridge (Arthur Gray Notes); Mitchell Library, Sydney (Philip Gidley King the Younger Journal, Autobiography and Reminiscences); National Archives, Kew (Logs of HMS *Beagle* and *Samarang*); National Library of Jamaica (Feurtado Manuscript); Natural History Museum, London (Richard Owen Correspondence and Alfred Russel Wallace Family Papers); National Library of Scotland; Shropshire Archives (Katherine Plymley Diaries); The Trustees of the Royal Botanic Gardens, Kew (Asa Gray Letters); Royal College of Surgeons of England; Royal Society of London (FitzRoy–Herschel Letters); Suffolk County Record Office, Ipswich; UK Hydrographic Office, Taunton (FitzRoy–Beaufort Correspondence); Wedgwood Museum Trust, Barlaston, Staffordshire, for permission to quote from materials in the Wedgwood Archive; Zoological Society of London.

A handful of colleagues went the second mile, conjuring arcana on demand: special thanks to Andrew Berry at Harvard University, Tim Birkhead and Ricarda Kather at the University of Sheffield, Helen Burton at Keele University Library, Lisa DeCesare at the Gray Herbarium, Harvard University, Rachel Mumba at Durham University and Vanessa Salter at the Wilberforce House Museum. Gwen Hochman at Harvard Law School went the third mile, traipsing around Boston. Nor can we forget our English and American editors, Stuart Proffitt, Amanda Cook and Jane Birdsell, who kept us firmly targeted.

Writing on *Darwin's Sacred Cause* began in early 2007 while Moore was a fellow of the Institute of Advanced Study at Durham University. He is especially grateful to Ash Amin, the Executive Director, for timely discussions about the book; and to Maurice Tucker, Master of University College, and members of the Senior Common Room, for Castle's good fellowship. Members of Moore's spring 2005 graduate seminar, 'Darwin, Sex and Race', in the Department of the History of Science at Harvard University were a tremendous stimulus: Topé Fadiran, Adam Green, Max Hunter, Sarah Legrand, John Mathew, Aaron Mauck, Matt Moon, Mac Runyan, Alex Wellerstein and Nasser Zakariya.

Our personal debts are many. Ralph Colp Jr, David Livingstone, Mark Noll, Bob Richards and Nicolaas Rupke gave us early sight of their latest fine books. In Cambridge, Nick Gill, Boyd Hilton, Simon Keynes, John Parker and Simon Schaffer generously shared their encyclopaedic knowledge (respectively) of Darwin's manuscripts, Georgian politics, the *Beagle* voyage, Victorian botany and everything else. Editors at the Darwin Correspondence project, especially Samantha Evans, Shelley Innes, Alison Pearn and Paul White, let us interrupt them from time to time. Tony Lentin and Sheila Thorpe in England, Gordon Moore in the United States and Maggie Fankboner in Canada cheered *Darwin's Sacred Cause* to the finish line. John Greene, Randal Keynes and David Kohn saw us through. Such debts can only be repaid in kind. Warmest thanks to all.

Introduction: Unshackling Creation

Global brands don't come much bigger than Charles Darwin. He is the grizzled grandfather peering from book jackets and billboards, from textbooks and TV – the sage on greeting cards, postage stamps and commemorative coins. Darwin's head on British £10 notes radiates imperturbability, mocking those who would doubt his science. Hallow him or hoot at him, Darwin cannot be ignored. Atheists trumpet his 'atheism', liberals his 'liberalism', scientists his Darwinism, and fundamentalists expend great energy denouncing the lot. All agree, however, that for better or worse Darwin's epoch-making book *On the Origin of Species* transformed the way we see ourselves on the planet.

How did a modest member of Victorian England's minor gentry become a twenty-first-century icon? Celebrities today are famous for being famous, but Darwin's defenders have a different explanation.

To them Darwin changed the world because he was a tough-minded scientist doing good empirical science. As a young man, he exploited a great research opportunity aboard HMS *Beagle*. He was shrewd beyond his years, driven by a love of truth. Sailing around the world, he collected exotic facts and specimens – most notably on the Galapagos islands – and followed the evidence to its conclusion, to evolution. With infinite patience, through grave illness heroically borne, he came up with 'the single best idea anyone has ever had' and published it in 1859 in the *Origin of Species*. This was a 'dangerous idea' – evolution by 'natural selection' – an idea fatal to God and creationism equally, even if Darwin had candy-coated this evolutionary pill with creation-talk to make it more palatable. Evolution annihilated Adam; it put apes in our family tree, as Darwin explained in 1871 when he at last applied evolution to humans in *The Descent of Man*. Secluded on his country estate, publishing book after ground-breaking book, Darwin cut the figure of a detached, objective researcher, the model of the successful scientist. And so he won his crown.

The most that can be said for this caricature is the number of people who credit it. Not only evolutionists and secularists, but many creationists and fundamentalists see Darwin's claim to fame – or infamy – in his single-minded pursuit of science. Doggedly, some say obstinately, he devoted his life to evolution. A zeal for scientific knowledge consumed him, keeping him on target to overthrow God and bestialize humanity. Brilliantly, or wickedly, Darwin globalized himself. By following science and renouncing religion, he launched the modern secular world.

This isn't just simplistic; most of it is plain wrong. Human evolution wasn't his last piece in the evolution jigsaw; it was the *first*. From the very outset Darwin concerned himself with the unity of humankind. This notion of 'brotherhood' grounded his evolutionary enterprise. It was there in his first musings on evolution in 1837.

Today we are beset by polemics of every stripe, comic attempts to pummel Darwin into this shape or that, to convict or acquit Darwin of beliefs – atrocities even – associated with his name. (A recent title about German history says it all: *From Darwin to Hitler*.)

To reverse Marx's dictum for a moment: the point is not to change Darwin, the point is to understand him. Darwin was neither saint nor satan. Looked at in his own day, he was complex, sometimes even contradictory, never quite as one imagined, but vastly more interesting and informative. And the real story behind his journey to evolution – *human* evolution – is much richer than anyone realizes. It is a story we have been piecing together for years, trying to grasp what could have made this gentle naturalist such an anomaly in his age, and so determined in the face of overwhelming odds.

Darwin was the most gentlemanly gentleman anyone had ever met. He was diffident, afraid to ruffle feathers, at home with the conservative Anglican dons, wanting only his own quiet vicarage lifestyle, away from urban tumult and religious harangues. The dons he emulated detested a bestial human evolution – so hysterically that the Cambridge cleric who taught him geology talked of stamping with 'an iron heel upon the head of the filthy abortion' to 'put an end to its crawlings'. Yet even as Darwin listened to his mentors after the *Beagle* voyage, he was privately musing on a missing 'monkey-man', our ancestor.[1] How could this be? What drove him to deny the cherished tenets of his privileged Christian society? Surely it had to be some overwhelming impulse that outweighed all others, pitching Darwin into words and deeds that frightened even him.

People have been looking in the wrong place for the answer. With the opening up of Darwin's treasure trove of unpublished papers over the past generation, the clues began to appear – some were even there in his most famously little-read book *The Descent of Man, and Selection in Relation to Sex* (to give its full and telling title). But this was the end point of Darwin's journey, and its contents have so confused readers that they even now call it 'two books' on different subjects, sex and ancestry, and so miss the point entirely.[2] Darwin's *human* project remains obscure. But it is foundational, and without understanding it, we cannot understand why Darwin came to evolution at all.

'Roots' is where we begin, *Roots* as in Alex Haley's stirring historical novel – the enslavement of black Africans that so outraged Darwin's generation. It was Darwin's starting point too, his abhorrence of racial servitude and brutality, his hatred of the slavers' desire, as he jotted, to 'make the black man [an]other kind', sub-human, a beast to be chained.[3] Roots were what Darwin's human project was all about. And to understand why he started thinking about the roots – the origin – of black and white races, we have to appreciate his moral anchorage in the noontide of the British anti-slavery movement. It is the key to explain why such a gentleman of wealth and standing should risk all to develop his bestial 'monkey-man' image of our ancestry in the first place.

Always retiring, often unwell, Darwin never threw himself into abolitionist rallies and petitions (as his relatives did). While activists proclaimed a 'crusade' (his word) against slavery,[4] he subverted it with his science. Where slave-masters bestialized blacks, Darwin's starting point was the abolitionist belief in blood kinship, a 'common descent'. Adamic unity and the brotherhood of man were axiomatic in the anti-slavery tracts that he and his family devoured and distributed. It implied a single origin for black and white, a shared ancestry. And this was *the* unique feature of Darwin's peculiar brand of evolution. Life itself was made up of countless trillions of sibling 'common descents', not only black and white, but among all races, all species, through all time, all joined up in bloodlines back to a common ancestor.

It was in Darwin's most generous, relativistic phase, at the height of

Britain's radical political period in the late 1830s, when the slaves in the colonies were finally freed, that he extended the kinship to all groaning, degraded, disparaged races of animals. He saw them sharing our own deep ancestry; 'we may all be netted together', we may all feel a common pain, he jotted in a notebook.[5] He had saved the blacks, stopped the slaves being seen as some 'other kind'. But by embracing the whole of creation – breaking life's shackles and allowing it too to evolve, as black and white men had done from a joint ancestor – he ironically opened himself to vilification by the Christian world whose belief in racial brotherhood he shared. A major criticism of the *Origin of Species* (particularly during the American Civil War) was that Darwin had now bestialized the *white* man, by contaminating his ancestral blood. Darwin had upturned the racist logic, only to 'brutalize' his own Anglo-Saxon kind (as it was said), uniting them, not only with black people, but with black apes.

Here was Darwin at his most paradoxical. And in this impasse lies the moral of our story, literally. Rather than seeing 'the facts' force evolution on Darwin (other circumnavigating naturalists had seen similar phenomena all over the globe), we find a *moral* passion firing his evolutionary work. He was quite unlike the modern 'disinterested' scientist who is supposed (supposed, mark you) to derive theories from 'the facts' and only then allow the moral consequences to be drawn. Equally, he was the reverse of the fundamentalists' parody, which makes his enterprise anti-God, inhuman and immoral. We show the *humanitarian* roots that nourished Darwin's most controversial and contested work on human ancestry. The ensuing picture is, as a result, dramatically different from previous ones, revealing a man more sympathetic than creationists find acceptable, more morally committed than scientists would allow.

Our reconstruction of Darwin's trajectory – lighting the path he trod after returning with the *Beagle*, through his private notes and drafts to the *Origin* itself – finally explains some of his otherwise anomalous statements. Reading the greatest one-origin-for-all-the-races work, by an anti-slavery advocate (James Cowles Prichard's *Researches into the Physical History of Mankind*), Darwin scribbled, 'How like my Book all this will be.'[6] 'My Book' was of course what became the *Origin of Species*. And, as this suggests, it too was meant to discuss mankind. The irony is that the *Origin* ultimately said next to nothing about human origins. Explanations of the human races and of an ape ancestry were dropped at the last moment.

Why, in the book that critics knew was *really* about mankind, had Darwin decided to say next to nothing on the subject? And why was he forced to come clean twelve years later and write the *Descent of Man*, and then incongruously fill the book with butterflies and pigeons and 'sexual selection'? Sexual selection is critical to our reading of Darwin's morally-fired 'human brotherhood' approach; it was central and critical, too, in Darwin's answer to the American South's (and London's) pro-slavery pundits who proclaimed the black and white races as separate species. Understand Darwin's strategy, and the oddities of his books and their anomalous timing make sense.

Today Darwin is *the* 'scientist' to reckon with. His theories about people and society are debated more widely than ever. The media buzzes with stories about evolutionary psychology, sociobiology and eugenics, about gender, race and sex differences and the possibility of improving human nature.

Some choose to emphasize the darker side to Darwin's evolution, notoriously hinted at in the *Origin*'s subtitle: *The Preservation of Favoured Races in the Struggle for Life*. From Albania to Alabama, from Russia to Rwanda, Darwin's theories have been used to justify racial conflict and ethnic cleansing. Perpetrators of the worst atrocities have seen themselves as 'favoured races' surviving Darwin's bloody 'struggle'. And so an ocean of ink has been spilt to prove that none of this was in Darwin's works, nor was it a logical consequence. Darwin's was a pure, untainted science.

We aren't out to prove the uncorrupted purity of Darwin's corpus, or indeed deify his corpse. Nor do we celebrate any unsavoury consequences of his work, or coerce him into siding with religious groups or atheists. We undercut all of these contrasting attempts to hijack Darwin for today's ends. The real problem is that no one understands Darwin's core project, the nucleus of his most inflammatory research. No one has appreciated the source of that moral fire that fuelled his strange, out-of-character obsession with human origins. Understand that and Darwin can be radically reassessed.

In sounding the depths of Darwin's anti-slavery we have exploited a wealth of unpublished family letters and a massive amount of manuscript material. We use Darwin's notes, cryptic marginalia (where key clues lie) and even ships' logs and lists of books read by Darwin. His published notebooks and correspondence (some 15,000 letters are now known) are an invaluable source: sixteen of the projected thirty-two volumes of *The Correspondence of Charles Darwin* have already been published by an international team whose deciphering and transcribing can only be called heroic. Add to this the extraordinary growth of historical studies on transatlantic race, racism and slavery, and we are equipped to connect Darwin for the first time with the most forceful moral movement of his age.

The discovery and recovery of Darwin's letters is still something of a rolling revolution. Even as we write new ones are turning up – not least, from the son of the most famous 'immediatist' abolitionist in the world, the American William Lloyd Garrison. It confirms what we had come to suspect, that Darwin was an admirer of the most uncompromising, non-violent Christian leader in the anti-slavery movement. Garrison was, in Darwin's words, 'a man to be for ever revered'. Darwin was overjoyed on hearing that the blistering anti-slavery passage in his *Beagle* journal had been read to the elderly Garrison, whose son reported to Darwin how it shed 'a new and welcome light on your character as a philanthropist'. To think, Darwin replied, that a man 'whom I honor from the bottom of my soul, should have heard and approved of the few words which I wrote many years ago on Slavery'.[7] This correspondence shows how much is still to be learned about Darwin.

Not only is the evolutionary upshot of his hatred of slavery unknown, Darwin's humanitarian imperative itself has never been brought adequately to the fore.[8] We try to show how it locks him into the context of nineteenth-century abolitionism, and how it speaks directly to our post-colonial age, with its hatred of ethnic cleansing and apartheid. Ours is a book about a caring, compassionate man who was affected for life by the scream of a tortured slave.

Finally, a word about terminology. Although 'scientific racism' is said to start about 1860 – taking over from an earlier xenophobia – we believe that such a hard-and-fast line is problematic. If 'racism' is taken to mean categorizing difference in order to denigrate, control or even enslave, then its scientific components and rationale can be traced much earlier. It is generally unknown that American slavery-justifying race-agitators were actually booming their hatreds as early as 1841 at the British Association for the Advancement of Science. Even before that they were linking subjugation to the anatomical 'inferiority' of blacks. These ideologies and slaving justifications bathed the cultural debates embroiling Darwin throughout the 1840s and 1850s, and we think that 'racism' is an acceptable historical term in Darwin's context.

We must add to this an apology for our ethnological *faux pas*. All the actors in our story – Darwin included – held derogatory views of other peoples to some extent, and used terms to match. While we try at times to mention the modern self-appellations of peoples, it would be confusing to mix Darwin's and modern ones, and so we have stuck to contemporary usages in most places. That goes for inclusive language also. We use a mix of contemporary and modern terms for *Homo sapiens*: 'man', 'mankind', 'humankind', 'humanity' and 'human beings'. Of course we regard such terms, along with 'savages', 'Kaffirs', 'Bantu', 'Hottentot' and so on, together with 'race' and ethnic labels generally, as historically constructed, but it would be tedious to place them in inverted commas every time; so also 'Negro', a respectful period-term for black people, and 'mulatto', a negative term.

This, then, is the untold story of how Darwin's abhorrence of slavery led to our modern understanding of evolution.

1
The Intimate 'Blackamoor'

No 'evil more monstrous has ever existed upon earth'. So said the leading anti-slavery campaigner Thomas Clarkson on celebrating the end of the slave trade. Clarkson was supported and part-financed by Charles Darwin's grandfather, the master potter Josiah Wedgwood. But the words could equally have been Darwin's – or those of his other grandfather, the libertine, poet and Enlightenment evolutionist Erasmus Darwin. For all of them slavery was a depravity to make one's 'blood boil', in Charles Darwin's words, a sin requiring expiation: 'to think that we Englishmen and our American descendants . . . have been and are so guilty'. The trade – transporting snatched Africans to labour till their deaths in the fields and factories of the New World and elsewhere – was outlawed in the British dominions in 1807. That was two years before Darwin was born, and he would grow up awaiting abolition of the damnable slavery system itself in the British colonies (which came in 1833).

Even then Charles Darwin continued to see the worst excesses of slavery for himself on the *Beagle* voyage (1831–6) and he was revolted by its 'heart-sickening atrocities'.[1] Slavery, justified by the planters' belief that black slaves were a separately created animal species, was the immoral blot on his youthful landscape and a spur to his emancipist study of origins – evolution, we call it today. The enormity of the crime in the eyes of the Darwins and their Wedgwood cousins was understandable: the African slave abductions had resulted in probably the largest forced migration of humans in history.

The mass action against slavery between the 1780s and 1830s engendered a new feeling of patriotic pride in British liberty after the loss of the American colonies. The campaign made anti-slavery 'unprecedentedly popular' in Darwin's formative years.[2] He was not alone in growing up in such a humanitarian environment; he was not alone in sharing its goals. But Darwin's undermining of slavery was a unique scientific response and it would shape the modern world.

Charles Darwin's family engagement with abolitionism began with his grandfathers, the doctor, philanderer, poet and prodigiously fat Erasmus Darwin on the one side, and the stern Unitarian and industrial potter Josiah Wedgwood on the other. These men would meet on full-moon-lit nights, with other prime movers who would power a technological revolution, in an informal Lunar Society of Birmingham.

Birmingham was a proud manufacturing town, full of self-made industrialists, its iron foundries exporting goods through Liverpool docks. But some trade items were sinister. In Erasmus's day almost 200 British ships were plying the slave trade between Africa and Jamaica, more than half out of Liverpool. These ships alone transported 30,000 slaves a year. The city had grown fat on the trade in flesh and was 'so risen in opulence and importance' as to be inured to the immorality of it all. The tall ships heading back to Africa carried 'a cheap sort of fire-arms from Birmingham, Sheffield, and other places', as well as powder, bullets and iron bars, all made in the Birmingham area, and all used to barter for the slaves. The *local* trade was partly funding slavery. There was a growing awareness of the fact: in 1788 the ex-slave Olaudah Equiano had passed through Birmingham on a propagandist anti-slavery tour and been well received. There was plenty of call on the new industrial furnaces, with Jamaica's slaves being punished by having 'iron-collars [fastened] round their necks, connected with each other by a chain'.[3] Erasmus Darwin was livid in 1789 on learning of the destination of these foundry products. Already planning ways to get Parliament to stop the trade, he wrote to Josiah Wedgwood: 'I have just heard that there are muzzles or gags made at Birmingham for the slaves in our islands.' One of these instruments, scarcely suitable for beasts, could be 'exhibited by a speaker in the House of Commons' during a coming debate. Or what about a specimen of the 'long whips, or wire tails', used in the West Indies? When it came to swaying a debate, 'an instrument of torture of our own manufacture would have a greater effect, I dare say'.[4] Wedgwood canvassed the likely impact with his London friends.

All emancipation stirred Erasmus Darwin. In the 1780s, when Quakers and then Anglicans began organizing against the trade, Erasmus lent them his devastating pen. Like everyone, he stood aghast at the *Zong* slave-ship atrocity, in which 133 sickly blacks were flung overboard so that the owners could claim the insurance on their lost 'property'. Fellow Lunar members were equally affected by the Africans' plight, not least Darwin's best friend and eccentric in benevolence, the Rousseauian Thomas Day (who so cared for animals that he refused to break horses). He versified in *The Dying Negro* on a runaway slave who chose suicide rather than be separated from his white lover.[5] Verse best served the cause, and in mastering the art, Erasmus himself mastered his passions. He could celebrate 'the loves of the plants' in his bucolic *Botanic Garden*, but then out of the blue, the sunny lines flashed incandescent, a lightning rod for his wrath:

> E'en now in Afric's groves with hideous yell
> Fierce SLAVERY stalks, and slips the dogs of hell;
> From vale to vale the gathering cries rebound,
> And sable nations tremble at the sound! –
> – YE BANDS OF SENATORS! Whose suffrage sways
> Britannia's realms, whom either Ind obeys;
> Who right the injured, and reward the brave,
> Stretch your strong arm, for ye have power to save!
> . . . hear this truth sublime,
> 'HE, WHO ALLOWS OPPRESSION, SHARES THE CRIME.'[6]

No Doctor Pangloss wrote those lines, or

> The whip, the sting, the spur, the fiery brand,
> And, cursed Slavery! thy iron hand . . .[7]

The revelations of slave misery and brutality were such as no enlightened Europe should endure.

And Erasmus was a child of the Enlightenment. A fervid republican, he flag-waved for the American and French revolutions. He invented machines, predicted the future and wrote poetry copiously. All progress for him was linked to an upward-sweeping, sex-driven material evolution, as set out in his vast bio-medical treatise on 'the laws of life', *Zoonomia*. That evolutionary vision might have been buried in the book but for Erasmus's popular poetry. Verse made the doctor's love of sex and progress a political force. 'Darwinianism', it was dubbed – to versify in the manner of Erasmus Darwin. In the long years of Tory crackdown in Britain following the French Revolution, his works were slammed as atheistic and subversive. Even young Charles Darwin at Edinburgh University in 1825–7 would read diatribes on his grandfather's 'unbounded extravagance', which 'benighted, bewildered, and confounded' readers with its near blasphemous idolatry of the material world.[8] And it would provide a salutary tale, reminding the grandson of the need for circumspection.

Old Erasmus loathed cruelty, whether to man or beast. All creatures in his evolving world were sensitive, suffering and deserving of respect, even the humblest. Never holier-than-thou, he delighted in being lowlier-than-thou. *Zoonomia* trumped the Bible's 'Go to the ant, thou sluggard' with 'Go, proud reasoner, and call the worms thy sister!' With mind reaching so low in the scale of nature and morality held so high in this vaunted 'age of reason', nothing would keep Erasmus from fighting against cruelty to any beast, or any human with skin-deep difference. To achieve the greatest health and happiness there should be 'no slavery . . . no despotism'. In his botanical *Phytologia*, on the subject of West Indies' sugar cane, he burst out, 'Great God of Justice! grant that it may soon be cultivated only by the hands of freedom, and may thence give happiness to the labourer, as well as to the merchant and consumer.'[9]

His fellow members of the Lunar Society concurred. Skilled 'mechanicks' and 'chymists', manufacturers and doctors, poets, even the radical Unitarian minister Joseph Priestley: the Lunaticks some called them, and they were indeed mad about science, with nine of the twelve being elected Fellows of the Royal Society. Iron, coal and steam were elemental to them, fly-wheel revolutions as vital as political revolutions. Their beam engines and great factories seemed to be driving forward the progress that Erasmus saw running up through nature, with no end in sight. Science-based enterprise would elevate and emancipate humanity as surely as good men acting together would stamp out the barbaric slave trade.

Most were strong Dissenters (they stood outside the state-established Church of England and suffered discrimination as a result – thus many were political and moral reformers). The Lunaticks held a progressive, unconventional faith like Erasmus's. One was his patient Josiah Wedgwood I, Charles Darwin's maternal grandfather, renowned for his fashionable tableware and vases. Prosperity was no hedge against mortality for the master potter – ailments plagued the Wedgwood family. Josiah's smallpox-affected right leg was amputated below the knee, a terrifying ordeal without anaesthetic. Josiah insisted on watching the operation. Afterwards the anniversary became 'St Amputation Day', but he endured shooting pains for the rest of his life from a phantom limb and ill-fitting wooden prostheses.[10] Josiah and Erasmus both knew suffering.

In his potting sheds at 'Etruria' in Staffordshire, Wedgwood sought to make 'such *machines* of the *men* as cannot err', and a timepiece ran the shifts clockwork fashion. His religious universe ran in a similar way. The family were 'rational Dissenters', discarding the Trinity and Jesus's divinity as corruptions of early Christianity. In their creedless Unitarianism – taught rigorously by Wedgwood's factory chemist Priestley – God's world was like a self-perfecting engine, with each person improved by following the perfect man, Jesus. Salvation was open to all, without regard for rank, ritual or race. Women found the levelling ethos empowering, and the next generation of Wedgwood wives and daughters would contribute more than their equal share to the anti-slavery cause.

Such radicals did not fare well in the rumour-ridden aftermath of the French Revolution. In July 1791 Priestley's chapel, house and laboratory were gutted by a reactionary mob crying, 'No philosophers – Church and King for ever!' Other Lunar men, terrorized, armed themselves and their factories.[11] Darwin and Wedgwood became more circumspect after Priestley fled to America. The next generation made its peace with the Church, baptizing their children (thus Charles Darwin was christened an Anglican). A Wedgwood grandson even became the family's vicar and young Charles Darwin, after failing at medicine, would be sent to Cambridge University to prepare for ordination.

Josiah and Erasmus had teamed up to fight the slave trade, the corpulent doctor with his sharpened pen, poised like a spiky buttress beside the peg-leg potter, whose flair for merchandising and London showroom gave him metropolitan connections. Parliamentary petitions against the slave trade had sprung up in 1788, over a hundred of them countrywide, the people's voice in an age when few had the vote. Erasmus sent Josiah's on to the Lunatick 'Birmingham F.R.S.s' for their signatures and forwarded it on for more to 'Dr Darwin of Shrewsbury', Erasmus's son Robert, Charles Darwin's father. But the sugar merchants and planters' lobby was powerful, and Parliament only agreed to regulate conditions on the slave ships. Even so, as William Wilberforce, the leading parliamentary spokesman for abolition, prepared to open the first debate in the House of Commons, a shocking broadsheet appeared on the streets. It showed the section of a loaded slaver with 482 black bodies packed below decks, the legal limit being an appalling 454.[12] All parties redoubled their efforts to counter this black agony brought on by white greed.

The anti-slavery side of manufacturing fought back. In London, the Society for Effecting the Abolition of the Slave Trade, founded in 1787, needed an official seal. Wedgwood, one of the first on its Committee, produced the image: a black man on one knee, shackled hand and foot, with eyes and hands pointing heavenward, pleading 'Am I not a Man and a Brother?' The words might have been his but the kneeling figure resonated as a piece of Christian art. A seal and woodcut were made, and Etruria cast a small oval medallion, with the slave shown in relief, 'in his own native colour'. It stood out in the pottery's range of collectable cameos. But it was not for sale. Wedgwood produced thousands at his own expense for distribution by the Society and its friends. In 1789 Benjamin Franklin, president of Philadelphia's abolition society, was sent a consignment. Batches were continually fired to keep pace with parliamentary debates, and inevitably the tiny ovals became a must-have solidarity accessory, a sort of poppy or yellow ribbon of its day.[13] Gentlemen mounted it on snuff-boxes, women in hair-pins, bracelets and pendants. Commercial variants came out, with coat-buttons, shirt-pins, medals and even mugs showing, sometimes ignominiously, 'the poor fetter'd SLAVE on bended knee / From Britain's sons imploring to be free'.

Darwin printed those lines beside a woodcut in his *Botanic Garden*, with a note explaining the cameo was 'of Mr. Wedgwood's manufacture'. He had 'distributed many hundreds, to excite the humane to attend to and to assist in the abolition of the detestable traffic in human creatures'. Darwin continued, imploring:

> Hear, oh, BRITANNIA! potent Queen of isles,
> On whom fair Art, and meek Religion smiles,
> Now AFRIC'S coasts thy craftier sons invade,
> And Theft and Murder take the garb of Trade!
> – The SLAVE, in chains, on supplicating knee,
> Spreads his wide arms, and lifts his eyes to Thee;

With hunger pale, with wounds and toil oppress'd,
'ARE WE NOT BRETHREN?' sorrow choaks the rest;
– AIR! Bear to heaven upon thy azure flood
Their innocent cries! – EARTH! Cover not their blood![14]

In London, Josiah shared his business skills with the Committee. At home he distributed tracts and held 'country meetings' to rouse the gentry. Were they to know a 'hundredth part of what has come to my knowledge of the accumulated distress brought upon millions of our fellow creatures by this inhuman traffic', they would rise up in protest. He helped former slaves, notably Olaudah Equiano. An Igbo, kidnapped in Nigeria as a child, Equiano had endured every trial and was now tramping the country, proselytizing and publicizing his autobiography. Bristol lay on his itinerary, but it was a dangerous slave-sugar port, and he turned to Josiah for help. He feared being press-ganged 'on the account of my Publick spirit to put an end to the accursed practice of slavery' and ending up back in chains. Josiah stood ready to have his London business manager intercede with the Admiralty, should the need arise.[15]

The frail, earnest Wilberforce became abolition's most effective parliamentary vote-winner. He kept Wedgwood abreast of evidence submitted to the slave-trade Select Committee, set up by the government as a delaying tactic. After Erasmus Darwin and Josiah Wedgwood heard of the 'muzzles' being forged in Birmingham, Wilberforce's motion to ban slave imports into the West Indies was defeated: the Tory evangelical Wilberforce went to the Whig Unitarian's pottery works to talk strategy:

. . . to Wedgwood's – Etruria, got there to dinner – three sons and three daughters, and Mrs. W. – a fine, sensible, spirited family, intelligent and manly in behaviour, – situation good – house rather grand . . . Discussed all evening.[16]

Morals, manners and money spanned the ideological divide, as they would always do in British anti-slavery. The children noted by Wilberforce in his diary were Charles Darwin's future aunts and uncles – and Susannah, who would be his mother.

The Spirit rather than steam-power animated the 'Saints', another London lobby looking to Wilberforce as its parliamentary voice. So called for being righteous, the Saints hated slavery as the deepest sin, which, without repentance, would surely bring down God's wrath on the nation. Some sat on the London abolition Committee. The families intermarried and a few put down roots in Clapham, a village south of London: hence they became more prosaically the 'Clapham Sect'. As a young man, Charles Darwin had relatives in the village who still preserved its ethos.

Clapham's was a rarefied air, far from the factory din. Most of the Saints had inherited wealth: the Member of Parliament Henry Thornton massively so, with Bank of England directors in the family. At his mansion, Battersea Rise, the Saints met in the library to pray and plan for a worldwide moral revolution. Wilberforce lived here before moving into 'Broomfield' next door. His own neighbour in 'Glenelg' was another MP, Charles Grant, director of the East India Company. These men ran banks and businesses and practised law, and favoured Cambridge for their sons: Charles Darwin, raised in similar privilege, would join them at the University.

The Saints were Anglican evangelicals, living in hope of eternal salvation more than material progress. Their religious life centred on the parish church. Here the moral revolution planned at Battersea Rise acquired God's conquering grace and would stretch even to 'the uttermost parts of the earth'. With five Saints Members of Parliament they added political clout to the power of grace. Their Church Missionary Society and the British and Foreign Bible Society became world brands, but the Abolition Society was the jewel in Clapham's crown. Numerous other societies were set up to regenerate morals and manners, from 'bettering the condition of the poor' to the 'Society for the Prevention of Cruelty to Animals'. The latter sanctified an 'extension of humanity to the brute creation', and Darwin, fifteen when the SPCA was formed, would have seen it that way. He could be left white with rage at the sight of a horse being whipped, or later in life prosecute a local for mistreating sheep.[17] He shared the Society's core values.

So a coalition of disparate groups formed around a slave crying, 'Am I not a Man and a Brother?' The man who acted as a lynchpin holding these parties together was another guest at Wedgwood's Etruria. Thomas Clarkson was a grim, intense, tall and burly investigator and propagandist. He trudged from port to port collating evidence from slave-ship sailors, merchants and customs officials to build damning reports on the brutality of the slave trade. He was supported through thick and thin by Wedgwood, who provided a Midlands base and finances. No one did more to shape the culture of anti-slavery into which Charles Darwin was born than Clarkson.

Wilberforce may have taken up abolition during a talk with prime minister William Pitt at his home on the Kentish North Downs in May 1787. As legend has it, seated 'in the open air at the root of an old tree at Holwood just above the steep descent into the vale of Keston', Wilberforce heard God's call to fight the slave trade (the spot would lie within strolling distance of Down House where Charles Darwin lived half a century later). In truth, his action had followed months of arm-twisting by Clarkson, then a young activist sitting with Wedgwood on the Abolition Society's Committee.

Educated at St John's College, Cambridge, Clarkson overlapped with Wilberforce and other future Saints. At Magdalene College, too, a set of earnest evangelicals thrived under the mastership of the Revd Peter Peckard, a reforming Whig who admired their sincerity almost as much as he hated the slave trade. In these colleges abolition first took root in the university, feeding into Clapham and then the wider movement.[18] The defining moment for Clarkson came in 1784 when he heard Peckard thunder against the slave trade as a 'Sin against the light of Nature, and the accumulated evidence of divine Revelation', from the towering twenty-foot University pulpit in Great St Mary's.

Peckard ended with a prayer 'for our brethren in the West Indies', which sparked a debate about whether black men really were brothers of Cambridge dons. Peckard answered in a pamphlet, *Am I Not a Man? and a Brother?*, stamped with a kneeling slave and published to coincide with an anti-slave trade petition from the university's Senate. He scorned claims that 'the native inhabitants of Africa are not of the Human Species; that they are Animals of an inferior class; or if they have any relationship to the human race, they are some spurious brood'. And were the offspring of black and white an 'unnatural mixture', as planters alleged? No. Peckard, a friend and patron of Equiano and his English wife and children, had observed the descendants of 'a Negroe married to a white woman' and found them just like himself. 'Black and White men, though different in *Sort* are the same in *Kind*, and consequently . . . Negroes are Men.'[19]

It was a direct rebuttal of the pernicious planter lore. Jamaican slavery apologists even in the 1770s were making 'Negroes' a different *species* of human. More, they saw black–white crosses leading to infertile hybrids – setting hares running that Charles Darwin would himself chase decades later. Out of this planter literature also came the notion of the 'contamination' of racially pure white blood – something again that would surface in the American South's propaganda when Charles Darwin was working on his evolutionary theory. Species separation was arguably a common planter position, founded on observations of a demoralized, uneducated, mentally stifled slave population. Visitors were horrified to find the idea of separate species a common opinion still in Jamaica in the 1820s, and in Brazil (where Charles Darwin's hands-on experience lay) the black was regarded 'as merely an intermediate step between man and the brute creation'.[20]

Not merely a different species, then, but one so 'lowly' as to touch the beasts. 'Negroes' were 'the vilest of the human kind, to which they have little ... pretension of resemblance'. For many planters it was a convenient post hoc justification for black bondage. Africans were the 'parent of everything that is monstrous in nature', and their women are as 'libidinous and shameless as monkies'. From that it was a short step to the calumny that black women take 'these animals frequently to their embrace' – so close were Negro and monkey species that even sexual unions were common. Destitute of morality, incapable of civilization, black people were hardly above the ape themselves. The chimpanzee was little known; a black person, it seems, not much more.[21] The upshot was so much mud-slinging about the unlikelihood that 'an oran-outang husband would be any dishonour to an Hottentot female'.

Peckard had implacably opposed all of this. Such planter propaganda was also anathema to Wedgwood and Clarkson. They expected it to incur the 'heaviest judgment of Almighty God, who made of one blood all the sons of men'. Clarkson had taken up the cause in 1785 when Peckard, vice-chancellor of the university, set an essay competition on the subject 'Whether it is lawful to enslave the unconsenting'. Most classically educated undergraduates focused on the ancient world; Clarkson tackled the current Atlantic trade. He uncovered conditions so abominable that he gave up all thoughts of the pulpit to devote his life to fighting the abduction of Africans. On the slave ships, the chained blacks, with sixteen inches allotted to each, 400 to 500 per ship, were packed even on jutting shelving, there to endure the Atlantic crossing on one pint of water a day and two meals of yams or beans. On good voyages 'only' 5 per cent died. Reports of captured slavers told of the wretches being mostly unable to stand, with some 'completely paralysed'. A naval chase might result in worse atrocities. In Charles Darwin's student days slavers could still jettison chained blacks overboard and claim loss of property on the insurance, or drop them over 'wedged up in casks' to force the pursuing cruisers to stop and rescue the slaves, allowing the slavers to slip away.[22]

Copies of Clarkson's published *Essay on the Slavery and Commerce of the Human Species* (1786), which won the vice-chancellor's prize, went to MPs, including Wilberforce. Out of this lobby came the nucleus of the London Committee of the Abolition Society, with Clarkson as its roving ambassador. From London to Birmingham, Manchester, Staffordshire and Shrewsbury he travelled, then to the slave-trade ports of Bristol and Liverpool (where he was assaulted on a quayside). Clarkson seemed to be everywhere, riding 35,000 miles collecting facts before a breakdown – and lack of cash – made him pause.

Wedgwood money refloated him. After the Saints set up the Sierra Leone Company to create a bridgehead in West Africa for liberated slaves, Wedgwood was one of those Clarkson approached. Shares in the company, £50 each, had to be kept from the slave merchants. Men like Wedgwood would put the good of Africa above 'Commercial Profit' and introduce 'Light & Happiness into a Country, where the Mind was kept in Darkness & the Body nourished only for European Chains'.[23] In Staffordshire to obtain the money in 1791, Clarkson found the Wedgwoods already using East Indian sugar rather than slave sugar from the West Indies, supporting the consumer boycott then in full swing. If petitions against the slave trade could not force Parliament's hand, lost tax revenue surely would. Clarkson had tens of thousands of sugar-abstinence pamphlets distributed, with Josiah paying for 2,000, each again emblazoned with 'Am I not a Man and a Brother?' Wedgwood pledged 'to procure petitions' from surrounding towns and he sent thirty guineas to cover expenses before the forthcoming parliamentary debate.[24] From this point on there would be continual Wedgwood financing for the closure of the slave trade.

Even after Clarkson had fallen out with the Clapham Saints over Sierra Leone, Wedgwood was there. Even after visiting revolutionary Paris and coming away, as Clarkson told Wedgwood, entertaining 'Notions of Liberty & having totally left the Church', and when others would no longer 'subscribe a single Shilling' to a Jacobin and leveller,[25] Charles Darwin's Unitarian grandfather stood firm. He still reimbursed Clarkson's running costs.

It was thanks to such men, then, that the trading in slaves was finally outlawed in the British dominions in 1807, just before Charles Darwin was born. Some have seen it as a system in decline anyway, with the costly tariffs propping up West Indian sugar imports over those from the free-labour East Indies (which the Wedgwoods were buying) making abolition economically attractive. The original words of the abolition bill talked of the trading in Africans as 'contrary to the principles of justice, humanity, and sound policy', but the words 'justice' and 'humanity' were dropped on its third reading in the House of Commons. And yet justice and humanity were crucial to Darwin/Wedgwood feeling and rhetoric, no less than to the country at large. The oppressed were portrayed in radical literature as 'black Englishmen' who were being crushed and exploited; the Dissenters of nearby Derby (grandfather Erasmus's home) called slavery 'a System full of Wickedness, hateful to God'.[26] For Darwin's family the system was cruel and ending it was a moral imperative, not an economic necessity.

Charles Darwin was the second son of Dr Robert Darwin and Josiah Wedgwood's daughter Susannah. He was born in 1809 at The Mount, the three-storey mansion that his father (known sternly as 'the Doctor') had built on the outskirts of Shrewsbury, high on a bluff above the River Severn. The building symbolized the Doctor's own self-confidence. The Mount stood for solid security, like a bank – and, indeed, Dr Darwin was a noted financier and trustee of local charities. Even with a thriving practice, most of his income came from shrewd investments, including mortgages to the gentry.

His mother dying in 1817, Charles Darwin was effectively brought up by his older sisters (Marianne, Caroline and Susan; he also had a younger sister, Catherine). Along with a love of liberty, the girls taught him respect for life and sympathy for God's creatures. This caring education would continue when young Darwin visited his Wedgwood cousins: here he would learn, when fishing with worms, how to kill them first in brine to spare them the suffering of being spitted. Compassion and anti-cruelty were paramount in the family.

Josiah Wedgwood II ('Uncle Jos') – Susannah's brother – had followed his father at the factory and on to the London abolition Committee. He knew Wilberforce and Clarkson from their visits to Etruria, and they valued his youthful commitment as much as his wealth. Clarkson himself publicized Jos's election to the Committee. This was a dozen years before even Wilberforce's friends from Clapham had joined. 'The time is fast approaching', Uncle Jos was assured by his father, 'when [the slave trade] & the other abuses of government will be destroyed, & justice . . . will be admitted as the ruling principle of legislation & politics'.[27]

With a family of his own and a mortgage from Dr Darwin, Jos had bought a wooded estate in Staffordshire ten miles from the factory and thirty from Shrewsbury. The Elizabethan house, Maer Hall, near Newcastle-under-Lyme, had a fine library stocked with the latest novels, divinity, science, agriculture and travel books, suitable for an industrialist's country retreat. It was still being decorated when, after eleven abolition bills lost in fifteen years, on 25 March 1807 the trade in black lives was finally made illegal throughout the British colonies. Maer Hall, freshly painted, its grounds restored, matched the buoyant mood. Jos's youngest daughter, Emma, was born there in early 1808, with Dr Darwin attending. A golden age lay ahead for the family: forty years memorable for their 'peace, hospitality, and cordiality' as much as their energy and social commitment, an era in which Wedgwoods and Darwins together would see slave-owning itself abolished.

Jos's four daughters were all older than Charles Darwin, all humanitarians, all devoted to the same causes. After Susannah's death, the Darwins and their cousins half lived together, filling the void. Staying with the Wedgwood girls at Maer was also a way of avoiding the Doctor, who was increasingly overweight and overbearing. 'Charley' spent many of his happiest days at Maer, fishing and hunting, or browsing Uncle Jos's books. 'Life there was perfectly free', he remembered.[28] Charles would end up marrying Emma. He would also end up understandably worrying about the effects of in-breeding, in families no less than in evolution.

Humanitarian efforts were central to their lives. At Maer, the women supported asylums, benefit clubs, infirmaries, reading rooms, Sunday schools, anti-cruelty societies, 'temperance teas' and more. Jos's sister Sarah ('Aunt Sarah'), living nearby, was the soul of generosity. A rich rentier, she was staunchly anti-slavery. No more urgent concern was expressed in the vast correspondence of these Wedgwood women than for the emancipation of black slaves.

Although the trade had been outlawed in 1807, merchants and sea captains were willing to flout the law for the lucrative prize. And other countries – despite pressure from Britain and harassment from the Navy – remained engaged. But the big desideratum now was not so much the end of *trade* as of slavery itself. The West Indies cane fields were still worked by black slaves, even if planters could not import more. In the century to 1810, about 1.75 million people had been abducted from Africa and sold in the British colonies. Some 750,000 were alive when slavery was abolished in the 1830s. These broken black bodies under the lash had been destined to serve the sugar tooth of the nation – until the nation, with its lobbies of activists, finally baulked. Darwin's young world, then, was buoyed by this huge humanitarian upswelling in the country – the whole family (and most notably his sisters and Wedgwood cousins) were swept along.

Young Darwin had other connections to anti-slavery in Shrewsbury. Longnor Hall, just south of town, was the seat of the Revd Joseph Plymley, archdeacon of Shropshire, a 'kind and liberal' man, with a 'compassionate tenderness for the sufferings of all his fellow-creatures'. Plymley became Clarkson's mainstay; he helped administer his expenses and the pair made Shrewsbury the abolition headquarters in the Midlands. Plymley's son Panton had been reduced to tears on reading Clarkson's *Essay*, and he came of age fighting for abolition.[29] At Longnor, anti-slavery was as much part of the children's education as in the Wedgwood and Darwin households. Panton became Shrewsbury's Whig MP in Darwin's youth and sat as a county magistrate with Dr Darwin. The intimate details of many a family's account books as well as their bodies were familiar to the Doctor. After the Plymleys changed their name to Corbett when the archdeacon inherited estates from an uncle (who had himself raised the county's first anti-slave-trade petition), Dr Darwin's accounts showed a Corbett loan for £2,000, and he would later confide details of the Corbett medical history to Charles Darwin, then studying the evolution of the mind.[30]

As a 'corresponding member' of the London Committee, Archdeacon Plymley – as he then was – anchored one end of an anti-slavery axis stretching from Shrewsbury to the Wedgwoods in Staffordshire. Within this catchment area and beyond the activists distributed 'short works of authority amongst the people' and got up petitions. They remained implacable while 'the national character is stigmatized by injustice & murder'. So it was as Darwin grew up, and in his teens he became close to the archdeacon and his family while hunting on their Longnor estate.[31]

Liverpool was the port closest to Darwin's Shrewsbury home: an exotic place still showing relics of the trade, with African produce, from gum rubber, Gambian cotton, peppers and musks, to Calabar mahogany and gaily coloured cloths. In 1818 the nine-year-old first visited the port, with his brother Erasmus Alvey (his only brother, five years older). Here their mother's family had business interests and anti-slavery friends, and along the docks they probably first encountered black people, stranded by the trade. Even twenty years later Darwin still had a 'vague memory of ships' from this trip.[32]

By now there were a third of a million slaves in Jamaica alone. The anti-slavery tracts painted a terrifying picture for a boy, of the 'legal constitution of slavery ... written in characters of blood, and hung round all those attributes of cruelty and revenge which jealousy, contempt, and terror could suggest'. The quarterly *Edinburgh Review*, favoured by Darwins and Wedgwoods, damned this 'atrocious crime' and called for 'every Englishman who loves his country [to] dedicate his whole life, and every faculty of his soul, to efface this foul stain from its character'. The pamphlets retailed harrowing accounts of the brutality and relentless drudgery suffered by cane-cutting slaves. Any education was frowned upon by the masters. With marriage outlawed and planters debauched, 'concubinage' was presented as the norm, with owners taking their pick of the black women.[33] Behind the outrage lay barely concealed concerns about miscegenation and the fate of the mixed-race children, the 'mulattos'.

It was the cruelty, though, that generated the national anger. Certainly it was what outraged Darwin. With the start of the *Anti-Slavery Monthly Reporter* in 1825, the sickening minutiae of slave life was picked over in gory detail, making it widely known. Men and women worked regimented in gangs under the whip in these 'factories in the field'. They were driven like cattle in a team at planting and cutting time; any slacking resulted in both men and women being pinned to the ground, stripped and flogged. Missionaries would tell tear-jerking stories of a multiply-whipped old slave, 'his posteriors ... completely exposed, much lacerated, and bleeding dreadfully', the man suffering, barely able to walk and shortly to die. Others told of maggots infesting wounds and pregnant women accused of some infraction being forced as a punishment to work in the fields.[34]

With the Wedgwoods a conduit for tracts, and his sisters fierce proselytizers, the sensitive boy in Shrewsbury in the early 1820s would have read of the 'husbands' and 'wives' from rival work gangs being wrenched apart and their children sold off; or of planters using threats to sell them as a spur to recalcitrant 'niggas'. Darwin was later (in Brazil and after) deeply disturbed by the break-up of families and the sale of children – 'picture to yourself', he would say, the fear 'of your wife and your little children . . . being torn from you and sold like beasts to the first bidder!' He must have been primed to expect this. He always felt for the children, who were worked as water-carriers in the fields or in menial jobs around the house. Again it would show in his outrage in Brazil on seeing 'a little boy, six or seven years old' struck repeatedly with a horse-whip for handing him a tarnished glass of water.[35] His tract-upbringing could only have strengthened his conviction that this iniquitous system ran on cruelty.

Shrewsbury petitioned Parliament to end slavery in 1824 as Darwin faced his last year in school. The family also started donating to the new and strategically understated 'Society for the Mitigation and Gradual Abolition of Slavery' – known as the Anti-Slavery Society – run by Wilberforce and Thomas Fowell Buxton (the philanthropist also famous for his charitable work among distressed weavers and his criticisms of prison discipline: from 1818 he was an MP and Wilberforce's humanitarian successor in the House of Commons). In the end, the theme common to all of this anti-slavery activity was summed up by the Unitarian words (and Darwin's mother and Wedgwood cousins were Unitarians) prefacing anti-slavery resolutions: the 'Universal Father [has] made of one blood all nations'.[36]

But petitions and resolutions were no substitute for meeting an ex-slave. As Charles Darwin prepared for Edinburgh University late in 1825 that prospect lay in store.

'Susan is hotter than ever about Slavery,' sister Catherine reported as Darwin suffered his first Scottish frost. After years of Susan's dinner-table talk of anti-slavery Charley was suddenly confronted with the black-slave predicament. Surprisingly his experience came early, in his first university year, a time often dismissed by historians as barren. The wide-eyed freshman Charles Darwin might prove a wastrel in medicine (as his father feared), but the first year was not wasted. Indeed, it likely helped condition his life's work on the deepest social – and scientific – issues.

There are only four surviving letters by Darwin from his time at Edinburgh University (1825–7), and these only offer snap impressions from the early days. The boy – sixteen still when he matriculated – reported back on his professors, some of them friends of his father to be greeted with letters of introduction. But that didn't stop him retailing his frustrations with the lectures.

Darwin certainly intended to study hard. He read voraciously to start, and that would become a lifelong habit. His library cards record subjects ranging from the viscera to insects, philosophy to optics. Some have seen his dedication signalled by his failure to borrow a single novel. In fact, since he reported home within weeks of term's start, 'I have been most shockingly idle, actually reading two novels at once', he had clearly brought them in his trunk or was buying them.[37] The rot may already have set in by then, January 1826, as suggested by his attitude to Professor Andrew Duncan Jr.

Duncan was full of the latest Continental thought. Here was the doctor famed for his extraction of cinchona from Peruvian bark – the source of anti-malarial quinine – who could thus boast of saving thousands of colonial lives. Even if the medical school was in decline, it still offered the best education for doctors and remained the largest exporter of medical graduates to the four corners of the world. A generation had paid their half-crowns to hear Duncan's lectures on medicines and sample his concoctions of plant extracts.[38] But to Darwin it was all doses and prescriptions and deadly boring. He did struggle for a while. He bought Duncan's textbook, *The Edinburgh New Dispensatory*, and scribbled notes on the contents of about sixty medicines. But the youngster was itchy. To sister Caroline he complained of 'a long stupid lecture from Duncan':

> But as you know nothing either of the Lecture or Lecturers, I will give you a short account of them. – D^r^. Duncan is so very learned that his wisdom has left no room for his sense, & he lectures, as I have already said, on the Materia Medica, which cannot be translated into any word expressive enough of its stupidity.

And so the complaints went on. The fact is he was young, away from home, dislocated, and it left these lectures 'on a winter's morning ... something fearful to remember'.[39]

The winters *were* miserably cold. That January was vicious: the Tweed froze, allowing laden carts across. The lochs iced over. Most years were as bad, with November pea-soupers followed by severe snowfalls, when postal riders had to abandon their horses. Foreigners walking to the shore feared their 'ears and nose ... would drop off', and it was nothing for six-foot drifts to plague walkers into March. So when Darwin complained about wasting 'a whole, cold, breakfastless hour on the properties of rhubarb!', one senses his blue toes.[40]

He lost interest in medicine. But historians have been wrong to write the year off – or have been looking in the wrong place for its importance. It did not lie in the lecture theatre or Infirmary. It lay in a room in the University Museum, where a pleasant black face could be seen daily.

Darwin told sister Susan on 29 January 1826 that 'I am going to learn to stuff birds, from a blackamoor I believe an old servant of D^{r}. Duncan', meaning the professor's father, past eighty and still teaching physic. Darwin wasn't sure of the outcome, clearly, because he added that the arrangement had the benefit 'of cheapness, if it has nothing else, as he only charges one guinea, for an hour every day for two months'. Evidently the sixteen-going-on-seventeen-year-old saw nothing untoward in paying money to apprentice himself to a Negro, and the forty or so hour-long sessions which he had with the 'blackamoor' through that frosty winter clearly made an impact: as an old man he recalled 'a negro lived in Edinburgh, who had travelled with Waterton and gained his livelihood by stuffing birds, which he did excellently; he gave me lessons for payment, and I used often to sit with him, for he was a very pleasant and intelligent man'. That statement in itself is surprising, lacking so much of the racist hauteur that characterized British society from mid-century.

Darwin's teacher 'John' (we think his name was) had arrived in Glasgow from Guyana in 1817 with his Scots 'master' Charles Edmonstone. It had been a one-way trip. John could not go back to the West Indies, because returning freedmen no less than escapees risked re-enslavement: his master being at home in Dunbarton, John would simply have lacked protection. So he settled in Scotland. Glasgow was an industrial town built partly on plantation sugar, and where the 'West India party is nowhere stronger'. Perhaps that made it less congenial, or lucrative, for he had migrated over to cosmopolitan Edinburgh by at least 1823. Here he was living at 37 Lothian Street, close to the museum and a few doors down from Darwin's digs.[41] And perhaps this Guyanese ex-slave had received a bit more education beyond bird-stuffing (the plantocracy's resistance to educating their slaves would not have deterred Edmonstone's close friend, the South American explorer Charles Waterton, with whom John travelled through the jungle). Self-employed, trading on his association with the celebrated Waterton, John was given space in the University Museum, where his craft knowledge came at a guinea a term. Here he taught genteel shooters such as Darwin how to stuff their trophies.

It was not uncommon to see men like John on Edinburgh's streets. Given the huge Scottish planter community and the exodus from the plantations as a result of the abolitionist agitation, with estate managers returning, black servants in train, we might have expected it.

But Americans were aghast at what they saw. With black males outnumbering black females by four to one in the British seaports, the men took companionship where they could. The Revd John Bachman, a visiting Charleston naturalist, though a passionate defender of human racial unity, would report seeing 'well dressed young white men and women walking arm in arm with negroes in the streets of Edinburgh'. He was dumbstruck that 'however revolting this sight was to our American feelings, yet it did not appear to be regarded with the same repugnance by the communities in Europe.' Other Yankee men of science had been equally repelled. Benjamin Silliman (a geologist whose Yale *Journal* Darwin would come to admire) had been flabbergasted two decades earlier. Seeing black or Hindu men with white women 'struck me rather unpleasantly':

> It would seem that the prejudice against colour is less strong in England than in America; for, the few negroes found in this country, are in a condition much superior to that of their countrymen any where else. A black footman is considered as a great acquisition, and consequently, negro servants are sought for and caressed.

He had reported, surprisingly, that an 'ill dressed or starving negro is never seen in England'. It wasn't something to be praised, but stomached;

and reasoned with, for how could the British be so peculiar? 'As there are no slaves in England, perhaps the English have not learned to regard negroes as a degraded class of men, as we do in the United States'.

There's the rub, slavery. It made the liaisons of black men with 'white girls of the lower orders of society' intolerable in American eyes. And stunningly incomprehensible, so that Silliman had to gawp and report for incredulous Yankee readers: 'A few days since I met in Oxford street a well dressed white girl who was of a ruddy complexion, and even handsome, walking arm in arm, and conversing very sociably, with a negro man, who was as well dressed as she, and so black that his skin had a kind of ebony lustre'. What was outrageous to Americans held no qualms for Darwin, and one begins to sense how John may have found the British people at their most tolerant. By 1826 the slave-emancipation programme in Britain had so heightened the humanitarian conscience that black acceptance seemed assured.[42]

The abolitionist sentiment was stronger in Edinburgh than Glasgow. An Edinburgh branch of the Anti-Slavery Society had sprung up in 1824, and by 1826 it was taking more from sale of publications than any other outside London. But hecklers would still disrupt its meeting on 1 February 1826, when the eloquent advocate Henry Cockburn was damning the outrages perpetrated in Britain's sugar-growing islands as worse than 'all the horrors of the dark fancy of Dante'.[43]

Even if Scotland wasn't at the forefront of the abolitionist movement, Darwin's sisters kept him on track. Caroline on 27 February relayed her surprise to Charles 'that your friend Archdeacon Corbet has not called an antislavery meeting' in Shrewsbury, as part of the same countrywide response as Cockburn's rally; 'he has always done so before'. With the 'great stir in England at this moment', the family was responding. Uncle Jos was pouring money into the African Institution, which aimed to develop the repatriated slave colony of Sierra Leone into a Christianizing and civilizing seed point on the dark continent.

Not that such sacred subjects were resistant to the boys' devilish pranks, with Charles probably egged on by his older and more outgoing brother Erasmus, also in Edinburgh for a year, finishing his medical studies begun at Cambridge. When Caroline complained in 1826 that, at a dinner party, one of Erasmus's friends 'evidently rather laught at the suffrings of the poor slaves' and took the Tory rag *John Bull*'s line supporting 'the slave holder's part', the boys sent their own copies of

John Bull home. Scurrility was *John Bull*'s forte, reform its *bête noire*. It admitted no tampering with Britain's ancient law, nor Catholic Emancipation (currently much agitated), which would 'open a door to let in the monstrous POPERY'; nor would it appease the abolitionists, who would let in the monstrous black man. To *John Bull* free black labour was a joke. It lampooned the returns from the repatriated Negroes of Sierra Leone (from ostrich feathers – 2 ounces – to coffee – 13,889 pounds) as 'HUMBUG'. The 19 March 1826 *John Bull* sent by Charley suggested only the most dependable West Indies slaves should be freed, and even then it did not expect 'favourable consequences'. The 2 April number was worse, for it included a report which extolled 'a happy and contented [slave] population'. And the West Indies lobby, it said, should be proud of the fact. The Darwin girls were furious. Catherine and Caroline sternly reported that Susan was enraged to see the 'odious' 'John Bull pretending that the Slaves lead a life of Comfort and Happiness'.[44]

The sisters instilled the notion that anti-slavery, like religion, was a grave concern. The humanitarian ethos of the age was fragile, at times shallow, and allowed of no faltering or frivolity. The heckling, the *John Bulls*, the Commons planter lobby were a cancerous moral growth whose torture-justifying iniquities demanded a leaden response.

Some, like Charles Lamb, in his *Essays of Elia* (1820), saw in the black faces of the streets so many benign 'images of God in ebony', even though 'I should not like to associate with them – to share my meals and my good-nights with them'. No such qualms for young Charley. Darwin's sisters had done their work well: not only did he find John 'pleasant' and was prepared to sit with him hour after hour, but he became, in his own words, an 'intimate'.[45]

In the 1820s blacks were being affected as much as whites by industrialization. The paternal ethos having been eroded, black coachmen and turbaned boys had ceased to be pampered pets of courtly circles. Old avenues of assimilation were closing as new ones opened in an increasingly insecure world of wage earners. They now sought jobs in the expanding cities, merging into 'lower white society'. They might have been crossing-sweepers, coachmen or servants (as formerly the case of this 'blackamoor', according to Darwin). Blacks were not only on Edinburgh's streets, but in the Regius Museum. The wave of abolitionist feeling brought the conviction that they could become not only gentlemen, but professional gentlemen. They encountered fewer obstacles than a later generation. They were appearing in the law, churches and medicine, and many were directed into colonial service, largely in West Africa, Sierra Leone particularly, where slaves were being repatriated. The Jamaican half-caste Staff Surgeon William Ferguson of the Army Medical Department, whose father was a white Scot and mother a black African, and who had studied at Edinburgh, would be lieutenant-governor of the colony by 1841.[46]

Darwin's teacher 'John', when he belonged to Charles Edmonstone, would have been one of the 400–500 slaves who worked his plantation at Mibiri Creek, up the Demerara River from Stabroek (Georgetown) in Guyana. Edmonstone's friend, the traveller and animal stuffer Charles Waterton, mentioned John in his freshly published *Wanderings in South America* (1825), and this association with the celebrity of the moment gave John his cachet. Neither of them was exactly conventional: Edmonstone's wife was half-Arawak Indian, half-Scot, and soon Squire Waterton would arrange to marry their daughter (then fourteen, although he decently waited a couple of years). Waterton was a Catholic and many of the Arawaks in Guyana had already converted to Catholicism. As Darwin was sitting with John, the last section of Waterton's nine-foot-high 'Keep Out' park wall (costing £10,000) was being completed around his Yorkshire estate, Walton Hall, which he would share with his wife's two 'mulatto' sisters.

One of John's claims to fame, then, as Darwin noted, was that he had travelled through the South American rainforest with the man of the moment. Waterton's nonchalantly sensational accounts of riding caimans (to obtain undamaged skins) and wrestling boas were amplified in the reviews as Darwin started his medical studies.[47] Accusations flew thick and fast about his exaggerations, but the men accompanying him swore to their truth, and John would have told Darwin as much. The heart-stopping escapades even ran in the serious papers, with the wag Sydney Smith dilating on caiman-wrestling in the *Edinburgh Review*. (The well-connected Darwin knew all about Smith, a friend to the Wedgwoods, who referred to him as 'The Cid'.) Waterton, wrote Smith, was 'a Roman Catholic gentleman of Yorkshire of good fortune' who showed his 'unconquerable aversion to Piccadilly' by preferring life with forest Indians. His wit worked ceaselessly on this 'gentleman's delight' in scenes where 'every trait of civilized life is completely and effectually banished'. Darwin's sympathy would have been with Waterton. Waterton's stories were so well known that, by April 1826, the *British Critic* refused to repeat them because they had run in 'most of the newspapers'.[48] So Waterton was the most talked-about traveller of the moment, the wilds of Guyana the most familiar terra incognita. Waterton had cleverly constructed literary adventures to stagger belief out of his transcendent experiences in forest glades.

Say what you would, Waterton had penetrated the Guyanese jungle like no Englishman before, walking barefoot and bareheaded with his Indians and slaves (there would be no convoys of backpacking porters carrying the white man and his essential home furniture, as in later Victorian safaris). Even the discomforts, the chiggers, ticks and flies, were, as Smith noted, a slap at civilization, a sort of 'Keep Out of the Jungle' notice to those not prepared to suffer. Smith duly drew the moral: 'Nature . . . seems to be gathering all her entomological hosts to eat you . . . out of your coat, waistcoat, and breeches. Such are the tropics. All this reconciles us to our dews, fogs, vapours, and drizzle . . . to our old, British, constitutional coughs, sore throats, and swelled faces'. Darwin must have understood the beguiling contrast in the icy depths of winter.

Waterton preferred 'the freedom of the savage', at least to the industrial squalor at home. As a Catholic, denied equal rights with Protestants until 1829, he had little class adhesion. He spurned the society that rejected him, preferring the jungle parasites to those of his own class in England, whom he saw crushing the 'hordes of enslaved, oppressed' and angry factory workers.[49]

Certainly Waterton's sympathy for 'savage' life contrasted to the stock travellers' tales. Even missionaries could paint stark portraits of 'the vices of savage life', the worst showing Africa as 'one uniform scene of plunder and ferocious broil', where 'robbery and murder' were 'a daily occupation'. Such impressions justified the corralling of heathen tribes where they could be instructed in the principles of 'religion and social order'. This was far from Waterton's image. But then he was very different. Portrayed as a man out of time by later biographers, an 'eccentric', Waterton was a man reacting to his time. He defied convention, cropped his hair (and was said to have 'looked like a man recently discharged from prison'). Just as the Jacobite Catholic fostered the social exclusion by marrying an Arawak princess's daughter and building a huge wall around his stately home, so his taxidermy would poke fun at the museum men. Waterton's famous 'Nondescript' hoax – the hind-quarters of a monkey stuffed, some said, to resemble a particular tax-man's face – mocked the zoologists whose life was spent naming creatures without comprehension. He took strength from his jungle-bred knowledge. It gave him his rival authority. He refused to donate to the national collections and he used his own park as an alternative route to social visibility. Sydney Smith thought the 'Nondescript' an abuse of Waterton's taxidermic powers, but even he managed to identify the backside-face as 'a Master in Chancery – whom we have often seen backing in the House of Commons'. It meant that taxidermy's profile in spring 1826 had also never been higher.[50]

Between John and the Squire, young Darwin was initiated into a different, welcoming account of jungle life. The theatrics of Waterton's South American wanderings were retailed endlessly at the precise moment (February–April 1826) when Darwin was sitting with his erst-while companion. We can only guess what Darwin learned in the forty-odd hours spent with John, but preservation must have accompanied talk of Waterton and the rainforest – the lure of the exotic to contrast with the snow and frost.

Presumably the talk turned to a slave's life. One assumes that John had been a house slave on the Mibiri plantation. Charles Edmonstone was one of the colony's oldest residents and a 'Protector' (a sort of liaison officer/pacifier) of the Carib and Arawak Indians. Possibly, like Waterton, the slaver found slavery an iniquity. Waterton claimed that 'he whose heart is not of iron can never wish to be able to defend it: while he heaves a sigh for the poor negro in captivity'. Yet Waterton himself was a slave-owner, and the operative point for most planters was the lucrative return. Demerara's 73,000 slaves produced 44 million pounds of sugar annually, besides the rum, coffee and cotton. The records show Edmonstone, appointed captain in a Demerary militia battalion in 1804, repeatedly leading sorties to capture fugitive slaves or attack rebel-Negroes' camps, with a result that an indebted Governor relieved him of paying any tax. Even Waterton's own slaves periodically escaped and were retaken.[51] These men were themselves trapped in an iniquitous system.

John was probably a Christian. At least, the indigenous religions of these slaves from West Africa had been crushed. But most planters had resisted slave conversions to Christianity as dangerous. Despite the home government's insistence on Christian instruction, the Demerara Governor had threatened to deport the Revd John Smith of the London Missionary Society if he even taught the slaves to read.

Smith had become a cause célèbre the year before Darwin entered Edinburgh, and the family was fuming at his treatment. In 1823 a slave uprising had led to Smith's imprisonment on the planters' trumped-up charges. (He had preached 'Christian liberty' but had actually urged the ringleaders *not* to rebel.) Sentenced to hang, he had died of pneumonia while incarcerated. The House of Commons debated a censure motion on the Governor in 1824 and the family was privy to the strategizing beforehand, as Cabinet members consulted with a Darwin in-law – indeed a man much admired by Darwin himself – the reforming Whig MP Sir James Mackintosh. (His wife Kitty Allen was sister to Jos's wife Bessy Allen, Darwin's 'Aunt Bessy'. The Allen sisters, including the piquant and clever Fanny Allen, in their intimate letters to the Wedgwoods, would record the social world in which Charles Darwin grew up.) To Mackintosh, Wilberforce had 'conferred on the world' the 'greatest' of all possible benefits. After abolition, Mackintosh believed, 'hundreds and thousands' would be led 'to attack all the forms of corruption and cruelty that scourge mankind'.[52]

Mackintosh was a philanthropist to a fault. 'He could not hate', laughed Sydney Smith. 'The gall-bladder was omitted in his composition'. Only the Indian administrators he upset as a Bombay judge by refusing to pass death sentences might disagree, or the planters' lobby in the Commons. In May 1824 Fanny Allen had been in the House to witness the censure motion on the Demerara Governor for this 'monstrous violation of justice'. Speaking was the Whig barrister Henry Brougham: he was a man of encyclopaedic learning – or to his foes 'encyclopaedic ignorance' – who would champion parliamentary reform, educational openings (he helped found London University) and, here, slave emancipation. He was another on the abolition Committee with Uncle Jos, and closely linked to Mackintosh. Brougham's speech was the 'most incomparable thing I ever heard', raved Fanny. 'I could have screamed or jumped with delight'.[53]

The Smith débâcle stoked the family's anger. Like the rest of the nation, they now knew that the planters allowed only such religion or education as promoted docility. But John's lot on a Demerara estate must have been much better. Indeed he must have been a favourite to have been freed and brought across the Atlantic.

For Darwin, primed by his sisters on the barbarity of slavery, his hours with John were curiously timely: here was Waterton's companion from a country that had just witnessed a slave revolt. Darwin presumably learned of Demerara slavery from the inside. Perhaps, too, he saw through Waterton's spectacles a friendlier image of wild peoples. Then there was the lure of the exotic. In the Guyanese forests walked peoples in their 'rudest state', Indians with boar-teeth necklaces, their bodies coloured with scented red paint. While Darwin's brother Erasmus sent reports from Glasgow's Hunterian Museum on their dazzling stuffed hummingbirds, John probably relayed his own hummingbird impressions to Darwin: of 'the glittering mantle' glimpsed as it darted 'through the air almost as quick as thought! . . . now a ruby – now a topaz – now an emerald'. Dozens flitted through the jungle, as Waterton's bird-stuffer knew well, for 'Nature', said his old master, 'has not known where to stop in forming new species and painting her requisite shades'.[54]

These weeks with John confirmed Darwin's belief that black and white men possessed the same essential humanity. Late in life, he would say that the races he encountered during his voyage on the *Beagle*, no less than the 'negro with whom I happened once to be intimate', obviously shared so 'many little traits of character' with himself. It showed 'how similar their minds were to ours'. Coming from a family which had campaigned to emancipate Britain's colonial slaves, and obeying the abolitionist command to honour black people as 'equal humans', young Darwin had been happy to be taught by a 'full-blooded negro'.[55]

1. The original anti-slavery medallion, cast in 1787 by Darwin's grandfather Josiah Wedgwood I. The black-on yellow jasperware bas-reli was manufactured by the thousand and distributed to promote slave-trade abolition.

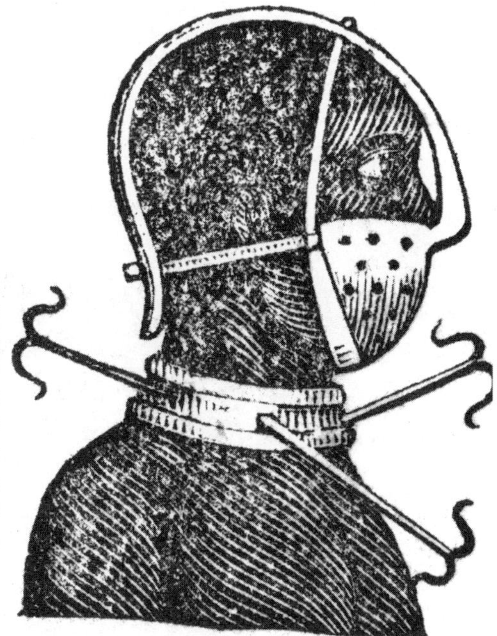

2. Darwin's other grandfather, Erasmus Darwin, proposed to shock the House of Commons by exhibiting an 'instrument of torture' manufactured in Birmingham. He had in mind this sort of contraption: an iron mask with a pronged punishment collar, which prevented a slave from eating, talking or lying down.

3. The young Charles Darwin.

4. Thomas Clarkson, architect of the fight against the slave trade, who enlisted William Wilberforce to wage the campaign in Parliament. Darwin's grandfather Josiah Wedgwood I was one of Clarkson's chief allies.

me again with him; we dined
at Mrs Peeles, who was still
alone; she appeared in very good
spirits, & said she was very well
last Sunday, but still better today,
she lamented she could not ac-
=company us to Church, having
made a promise to Dolly before
she set out that she would
not venture to Church.—
12th dined at the Hall, met
Joseph, Messrs Glover, Hale,

Charles Darwin, Henry Johnson,
& Mr Harrison, Watie's Curate
at Corley.—
13th The Mrs Panton's arrived
at the Hall.—
14th I dined at the Hall.—
15th The Mrs Panton, Josepha,
Matilda & Mildred dined at the
Hall.— Joseph & Robert joined
us in the evenn.—
16th I dined at the Hall.—
17th Robert & Mildred dined
with us.—

5. (*Above*) In his teens Darwin hunted and dined with Clarkson's main Midlands lieutenant, Archdeacon Joseph Corbett, at Longnor Hall, near Darwin's home in Shrewsbury. This entry in a diary kept by Corbett's sister shows Darwin and a fellow medical student, Henry Johnson, among a Longnor party on 12 September 1825, just before the pair went up to Edinburgh University.

6. (*Left*) The Revd Joseph Corbett, Archdeacon of Shropshire, the county's leading abolitionist, and a loyal friend and supporter of Thomas Clarkson.

7. Edinburgh University Museum in the 1820s, where a young Darwin spent many hours in 1826 being taught to preserve birds by a freed slave from Guyana.

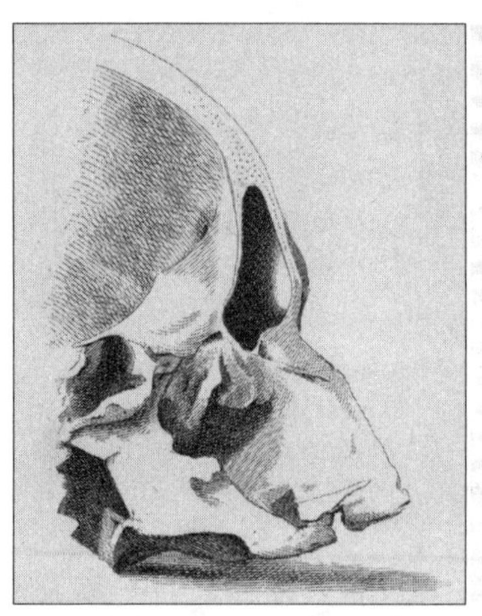

8. Darwin's anatomy professor at Edinburgh vehemently opposed the phrenologists. Darwin studied dissections of the bones of the frontal sinus of the skull (here in cross-section, the dark cavity above and behind the bridge of the nose), which proved that the 'bumps' on the brain in this region could not be 'felt' by phrenologists and so invalidated their system.

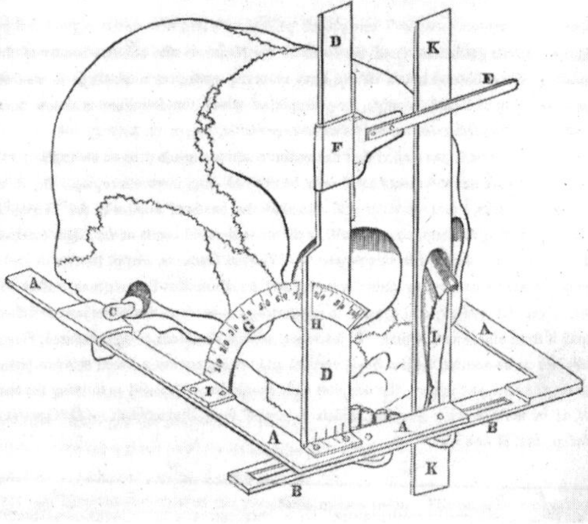

9. The skull-measuring instruments used by phrenologists in the Edinburgh of Darwin's day to characterize individual differences were quickly adapted and augmented by anthropologists to distinguish and rank whole races. This one measured the angle of the face and jutting jaws – the closer to perpendicular the face, the 'higher' the race.

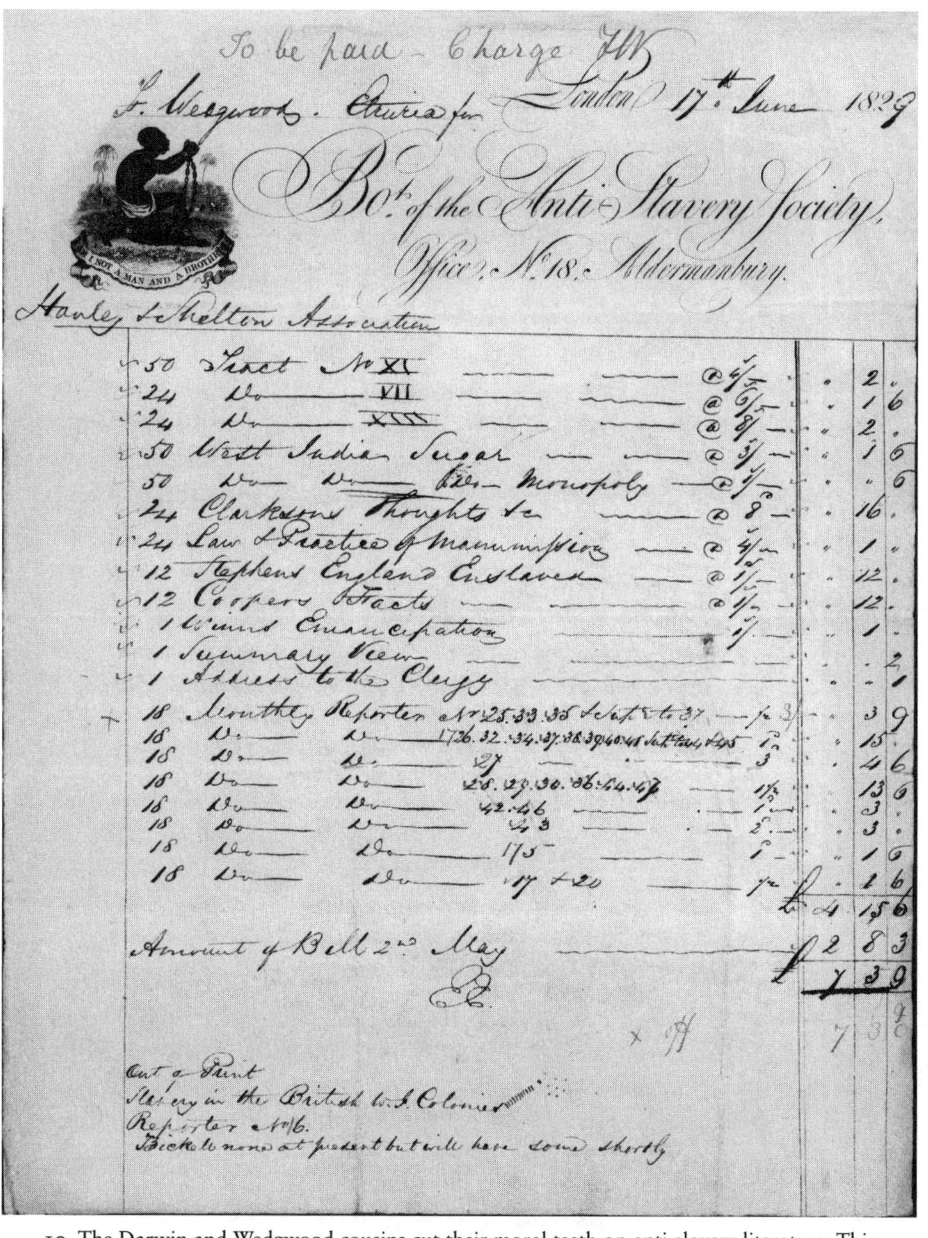

To be paid – Charge FW

F. Wedgwood. Etruria for London 17th June 1829

Bot. of the Anti-Slavery Society, Office, No. 18, Aldermanbury.

I NOT A MAN AND A BROTH

Hanley & Shelton Association

50 Tract No XI @ 4/- . 2 .
24 Do VII @ 6/- . 1 6
24 Do XIII @ 8/- . 2 .
50 West India Sugar @ 3/- . 1 6
50 Do Do Free Monopoly @ 1/- . . 6
24 Clarksons Thoughts &c @ 8 . 16 .
24 Law & Practice of Manumission @ 4/- . 1 .
12 Stephens England Enslaved @ 1/- . 12 .
12 Coopers Facts @ 1/- . 12 .
1 Winns Emancipation 1/- . 1 .
1 Summary View . . 2
1 Address to the Clergy . . 1
18 Monthly Reporter No 25.33.35 & Sup. to 37 pr 3 . 3 9
18 Do Do 17.26.32.34.37.38.39.40.41 Sup. to 44 & 45 1 . 15 .
18 Do Do 27 3 . 4 6
18 Do Do 28.29.30.36.44.47 1½ . 13 6
18 Do Do 42.46 1 . 3 .
18 Do Do 43 2 . 3 .
18 Do Do 175 1 . 1 6
18 Do Do 17 & 20 pr . 1 6
£4 15 6
Amount of Bill 2nd May £2 8 3
£7 3 9

7 3 9

Out of Print
Slavery in the British W. I. Colonies
Reporter No 16.
Tickets none at present but will have some shortly

10. The Darwin and Wedgwood cousins cut their moral teeth on anti-slavery literature. This is an invoice to Darwin's cousin Francis Wedgwood for propagandist tracts to be distributed by the Wedgwood family's Hanley and Shelton Anti-Slavery Society: from shilling pamphlets on the West India sugar boycott to Thomas Clarkson's *Thoughts on Improving the Condition of the Slaves in the British Colonies*, James Stephen's *England Enslaved by Her Own Colonies* and T. S. Winn's book of 'practical advice' to slaveholders, *Emancipation*. Of course the *Anti-Slavery Monthly Reporter* was *de rigueur*.

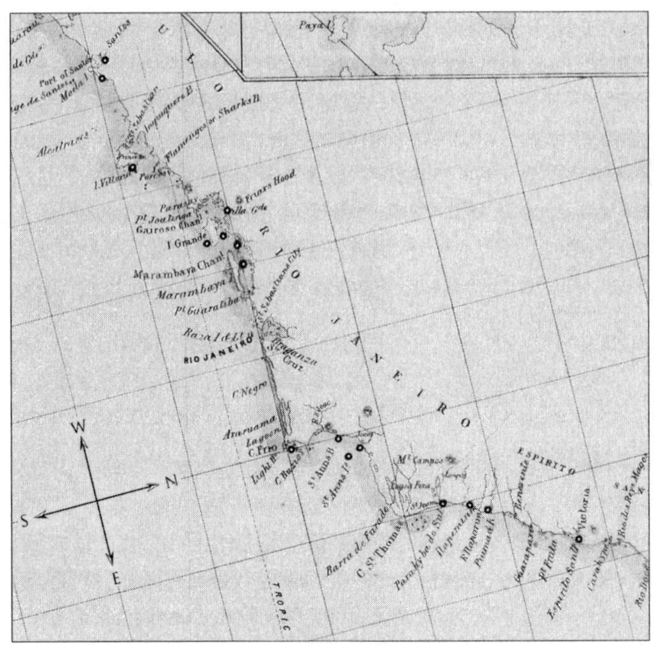

11. On the *Beagle* voyage Darwin found evidence of the slave trade in almost every South American port. The dark spots on this map of Brazil's central slaving coast between Santos (Sâo Paulo) and Victoria (Vitória), some 200 miles either side of Rio de Janeiro, show the known sites where slavers landed their human cargoes and fitted-out for return trips to Africa. (© National Maritime Museum, Greenwich, London)

12. At Bahia in Brazil, Darwin heard first-hand accounts of the hellish conditions endured by slaves on the voyage from Africa. This is the Bahia-bound slaver *Veloz*, intercepted in 1829 by a Royal Navy frigate. She carried 517 Africans (the 55 dead having already been thrown overboard). They were branded, like sheep, with their owners' marks, and each was allotted two square feet of space under a deck scorched by the equatorial sun.

13. Darwin saw the whip in action and heard the screams. 'Domestic punishments' in Brazil, as pictured from life by the anti-slavery German artist J. M. Rugendas, whose engravings Darwin knew: a child weeps as its mother's palm is paddled, a young woman awaits the whip (*centre*) while the colonial family (*right*) amuses itself.

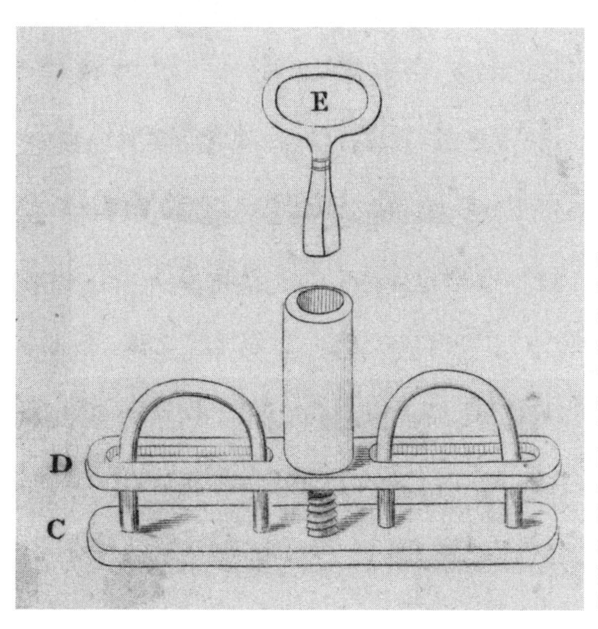

14 Thumbscrew, the sort used on female slaves by the 'old lady' who lived opposite Darwin near Rio de Janeiro. The thumbs were put through the two circular holes at the top. Turning a key (E) raised the bar (D), causing acute pain. Continued turning forced blood to ooze from the thumb tips. Removing the key left the tortured slave locked in agony.

15. Darwin combatted the 'pluralists', who saw each human race as a separate species with its own unrelated bloodline. For them there was no 'common descent' of black and white races. This detail from an 1827 map of the 'human genus' in Bory de Saint-Vincent's natural history dictionary shows the separate species of humans believed to exist in Patagonia (dark area) and Tierra del Fuego (light area beneath). Darwin had his own copy of the dictionary on the *Beagle*, but he and the captain, Robert FitzRoy, knew Bory was wrong – the Patagonians and the Fuegians were blood relations.

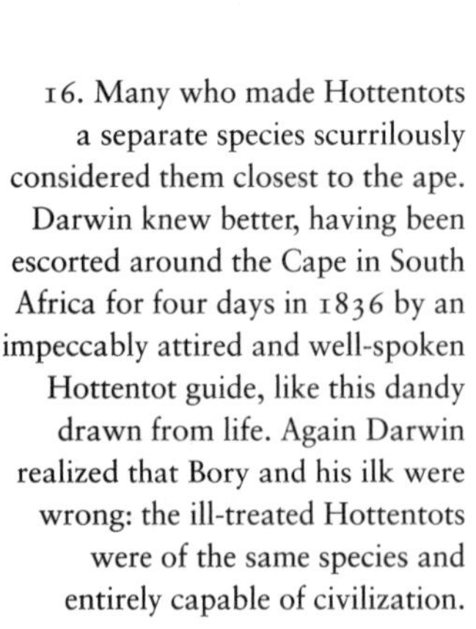
16. Many who made Hottentots a separate species scurrilously considered them closest to the ape. Darwin knew better, having been escorted around the Cape in South Africa for four days in 1836 by an impeccably attired and well-spoken Hottentot guide, like this dandy drawn from life. Again Darwin realized that Bory and his ilk were wrong: the ill-treated Hottentots were of the same species and entirely capable of civilization.

17. Darwin and his first-born, William, in an 1842 daguerreotype: a quarter-century later, William's jesting over a racist atrocity in Jamaica would propel his father towards publishing on human origins.

18. The dandy Robert Knox – a teacher in Edinburgh while Darwin was in residence – became a travelling showman after the Burke and Hare murder scandal. For him, the human races comprised many species and competition among them was the driving force of history. Like a growing number, he denied any 'common descent'. In London in 1847, he exhibited this Hottentot family, whom the press likened to beasts awaiting extinction.

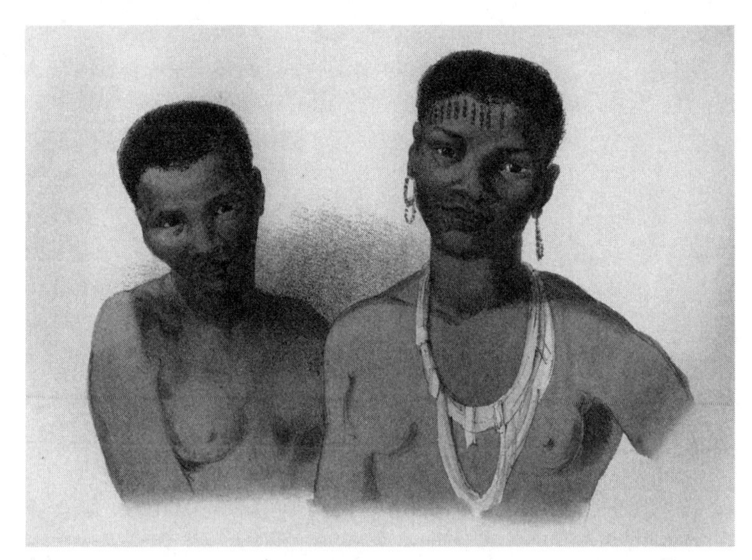

19. James Cowles Prichard was the foremost defender of the unity of the human species. His monumental *Researches* on mankind was closely studied by Darwin. Not only did Prichard consider Hottentots the same species; he bucked the racist trend by picturing their womenfolk in his book with 'pleasing and even handsome features'.

20. Louis Agassiz, the charming and debonair Harvard zoology professor, was so viscerally revolted by black people that he too came to see the races as separate, unrelated and immutable, and thus became Darwin's most formidable opponent outside Britain in the 1850s.

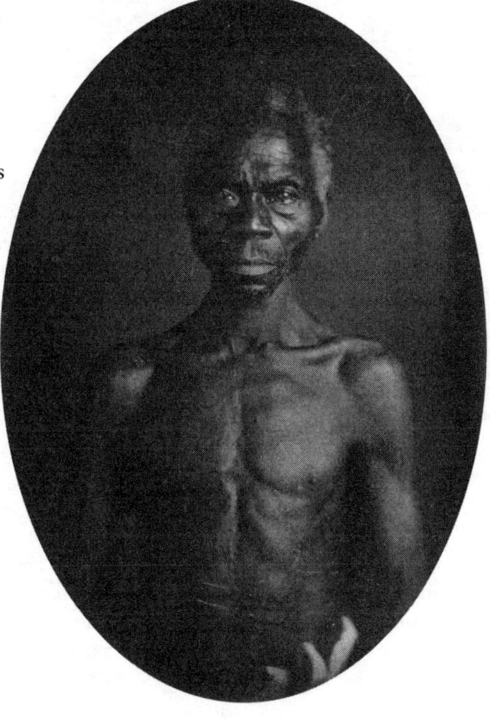

21. Agassiz commissioned this daguerreotype as part of a series taken in 1850 on a plantation near Columbia, South Carolina, to show the distinctive physical traits of 'pure' Africans. It was labelled 'Renty' of the Congo tribe, and the grim gaze speaks volumes. Ironically, this is the earliest known image of a named slave. (© 2008 Harvard University, Peabody Museum Photo 35-5-10/53037)

22. The European, the American and the Negro (*left to right*) were among Agassiz's eight primordial human species, each created in a different geographic zone with its own peculiar fauna. In this crude illustration of Agassiz's view of the inhabitants of four of the world's zones, from the 'pluralist' bible *Types of Mankind*, the Hottentot (*far right*) is also shown as a separate species.

24. (*Right*) Darwin's genealogy of the 'eleven chief races' of fancy pigeons. The ancestral rock dove is at the top. He explained that the breeds in italics are those 'which have undergone the greatest amount of modification. The lengths of the dotted lines rudely represent the degree of distinctness of each breed from the parent-stock, and the names placed under each other in the columns show the more or less closely connecting links. The distances of the dotted lines from each other approximately represent the amount of difference between the several breeds.'

23. Examples of the fancy pigeon races kept by Darwin: foreground a carrier (*left*) and pouter (*centre*), behind them jacobins with their ruffs (*left*) and mottled (*right*), and a tumbler at the back on the ledge. By proving that these were all descended from an ancestral rock dove, Darwin removed a major obstacle to the belief that the races of mankind, with their own varying physiques, hair types and skins, could have descended from a common human parent.

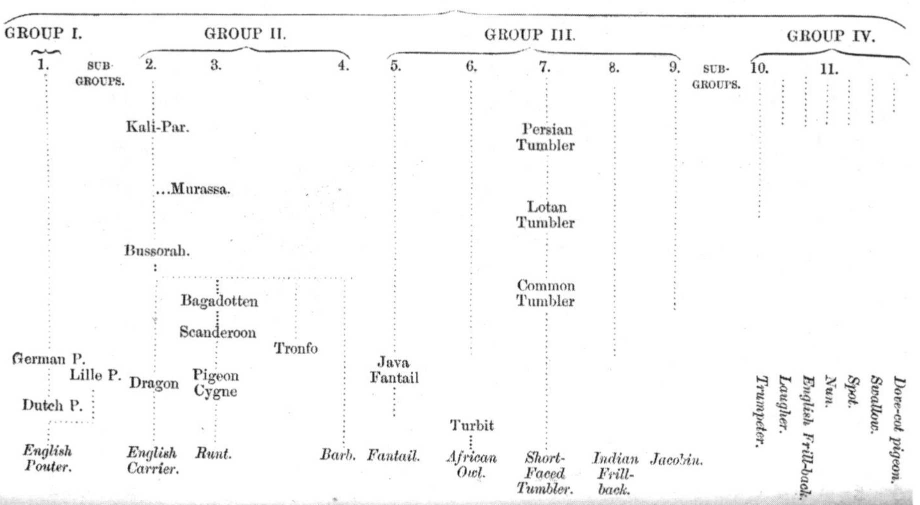

25. (*Left*) Asa Gray, Harvard botany professor, anti-slavery patriot and Darwin's outstanding defender in the United States.

26. (*Below*) Gray's residence, the Garden House, Cambridge, Massachusetts, where Darwin's ideas on evolution were premiered in America. On 12 May 1859, at a meeting here of the Cambridge Scientific Club, with Agassiz present, Gray argued on Darwin's authority that 'what are termed closely related species may in many cases be lineal descendants from a pristine stock, just as domesticated races are'.

27. Charles Darwin, facing a war over slavery and a fight for 'common descent'. His son William took the photograph on 11 April 1861; the American Civil War broke out the next day.

28. (*Above*) Richard Hill, anti-slavery activist and Jamaica's first coloured stipendiary magistrate: Darwin praised his work for the 'sacred cause of humanity'. (By courtesy of the National Library of Jamaica)

29. (*Left*) Charles Lyell, Darwin's life-long mentor, in 1863. Lyell's failure that year to endorse natural selection and human evolution in his *Antiquity of Man* drove Darwin to despair.

30. The 'Wilberforce Oak', where in 1787 William Wilberforce was said to have heard God's call to fight the slave trade, stood about a mile from Down House. Shown here in 1873 is the Revd Samuel Crowther (*centre*), the first native African to become an Anglican bishop, with colleagues from the Church Missionary Society. The oval commemorative bench in the background to the left of the trunk stands on the public footpath where Darwin would pass en route to collect plants on Keston Common.

considered the majority 'black or brown population' a drawback to shipping, where whites would have been more expeditious and trustworthy in trade.[10]

Crossing the equator to Bahia in Brazil, she anchored on 28 February 1832 in All Saints Bay amid a forest of masts. At the mouth of the bay stood the provincial capital, Salvador, population 65,000, a third of African descent. The tropical jungle here was no more exotic than the human jungle. Some 41 per cent of all Africans had ultimately been shipped to Brazil. The wharves and streets in Bahia showed the dregs, but also the unexpected: turbaned, bright-shawled Muslim slaves who were sophisticated, Arabic-writing literates and 'vastly superior to most of their masters'. Everyone but slaves owned a slave, even women, priests and free blacks. But times were hard. Slave imports had slowed to a trickle since the trade had been outlawed by Brazil two years earlier, and for months all new arrivals had been rightfully free. Without fresh labour, the sugar plantations would fail, the economy collapse; and as scarcity drove up prices, slave merchants took ever greater risks. Near the *Beagle* a Spanish schooner, the *Segunda Tentativa*, lay brazenly at anchor, fitted out for Africa. She and dozens like her were depositing their slave cargoes along the porous coast, while steam-powered cane factories worked blacks harder, giving the phrase 'dark Satanic mills' an awful new sense.[11]

Darwin had only read about such a world. Now as half-naked slaves bore FitzRoy in a 'sedan chair' from the crowded 'lower town' to the salubrious neighbourhood above, Darwin paced Salvador's sweltering streets, facing flesh-and-blood realities. He saw all the labour being 'done by the black men', who stagger under 'heavy burthens, beat time & cheer themselves' by a song. The 'excellent manners of the Negroes' astonished him, as did the perfect courtesy of black waitresses and the delight of the children to whom he showed his pistol and compass. He went armed, but found it unnecessary: most of the slaves seemed happier than he had imagined. Yet he knew there would be 'many terrible exceptions'. In the equatorial midday, his eyes had not adjusted to the shadows.

To others the evil lurked out of sight. With smuggling still endemic, blacks could be crowded into warehouses like cattle. Cruelty stemmed from individual depravity. Travellers would hear stories of masters threatening to boil blacks to death, or reports of suicides among slaves as the last act of resistance. Others saw torture instruments, gags, shackles and pronged collars.[12] Darwin would change his tune.

Back on the *Beagle*, he had time to reflect as sailors took their lashes for neglect of duty and drunkenness. Uncle Jos had been right: a captain could be like a slave-master. One, Darwin heard, even presumed to have ecclesiastical powers, making a bishop of his chaplain so he could consecrate a cemetery. FitzRoy had served under this captain on the frigate *Thetis*, a fast scout and dispatch vessel, which had been under orders to intercept slavers in South American waters – even if FitzRoy, little concerned with slavery, had described their work as 'going from Port to Port . . . always busy, yet I should be puzzled to say about what'. Game-shooting had preoccupied him, and his chances of promotion.[13] After the *Thetis*, FitzRoy had become flag lieutenant to the rear admiral at Rio de Janeiro, replacing Charles Paget, who took command of the small but resilient teak-built East Indiaman HMS *Samarang*, which, by chance, lay in All Saints Bay with the *Beagle*. Still in their twenties, the young captains were old friends, and Paget often came aboard.

At this time one incident had a profound impact on Darwin. FitzRoy had thought him a 'very pleasant messmate', but the captain was a powder keg with a short fuse. Came the day and there was a terrible explosion over slavery. FitzRoy recounted a conversation proving the benevolence of the system. A local slave-owner, asking his slaves in the captain's presence if they wished to be freed, had received the inevitable answer – no. Anyone schooled in anti-slavery had heard this one a thousand times. Darwin, tetchy, forgetting Henslow's advice to 'bridle your tongue when i[t] burns', indeed, forgetting whom he was speaking to altogether, asked whether the captain thought 'the answers of slaves in the presence of their master . . . worth anything'. Darwin's response came straight from the family's propagandist tracts and was probably more rooted in historic outrage than personal vindictiveness. But he was walking in a minefield unawares. FitzRoy's rejoinder was furious, as he accused Darwin of doubting *his* word rather than the slaves'. For such insolence, men could be thrown off a ship, officers court-martialled. FitzRoy nearly threatened as much by declaring that they could no longer mess together – the very reason Darwin was on board.[14] From now on, Darwin's acceptance of anti-slavery would no longer be passive and covert.

Darwin was left messing in the gun-room, taking moral support from the voluble Second Lieutenant Bartholomew Sulivan, respected for his 'simple and manly piety' and abhorrence of the slavers. But FitzRoy's storms blew over fast, and Darwin was readmitted to the captain's table. For Darwin there was even an impromptu vindication. The *Samarang* had just spent six months patrolling the coast between Rio and Pernambuco, drilling, firing practice rounds and pursuing suspect ships. She had boarded a Brazilian brig and found no slaves, but a French warrant to detain vessels flying the tricolour would keep her vigilant.[15] The *Samarang*'s Captain Paget knew these waters; so did his father, a rear admiral (and now a reform MP for Caernarfon, having snatched the seat from a rotten squirearchy in the May 1831 election), who had captured Spanish shipping here. No one could gainsay Captain Paget when on one of his visits he opened up about the slave trade.

Paget reeled off a string of facts 'so revolting' that, had Darwin been told them in England, he would have put them down to a fanatic's hyperbole. He jotted in his journal:

> The extent to which the trade is carried on; the ferocity with which it is defended; the respectable (!) people who are concerned in it are far from exaggerated at home ... [I]t is utterly false (as Cap[tain] Paget satisfactorily proved) that any [slaves], even the very best treated, do not wish to return to their countries. – 'If I could but see my father & my two sisters once again, I should be happy. I never can forget them.' Such was the expression of one of these people, who are ranked by the polished savages in England as hardly their brethren, even in God's eyes.

From the mouths of slaves, FitzRoy had been answered. Happy in bondage? He who fights slavery, Darwin confided to his journal, is helping to relieve 'miseries perhaps even greater than he imagines'.[16]

He found another ally in the great biogeographer Alexander von Humboldt, whose *Personal Narrative* of South American travels had been a parting present from Henslow. Humboldt had ploughed the field that Darwin now picked over (and, in time, the *Beagle* would sail up the west coast of South America in the cold 'Humboldt current', named in his honour). At the turn of the century Humboldt's self-financed five years in the closed Spanish-controlled areas of Central and South America had seen him travel 6,000 miles overland. To Humboldt was owed the grand idea of tying the fauna and flora to the geography of each region. Twenty years were spent writing up his research. At Bahia Darwin was slowly digesting Humboldt's majestic seven tomes, revelling, thinking that, 'like another Sun' Humboldt 'illumines everything I behold'.

On a rolling ship, pages were crudely cut and passages wildly scored as Darwin reached Humboldt's indictment of New World bondage. 'What a melancholy spectacle is that of christian and civilized nations, discussing which of them has caused the fewest Africans to perish in three centuries, by reducing them to slavery!' he thundered. So powerful was his chapter on Cuba, hub of the West Indian trade, that it was published separately, a *Political Essay* quantifying slave mortality with German precision and attacking the planters' morals. The slave who 'has a hut and a family' in the southern United States might be 'less miserable than he who is purchased' in Cuba, but Humboldt refused to extol such lessened misery. For this, his essay was censored in America, the anti-slavery passages cut. He also discussed slave sugar production in the 'English islands' of the Caribbean. Humboldt demanded that anti-trafficking laws be strictly enforced, punishments be inflicted, 'mixed tribunals' be set up between nations, and the 'right of search [be] exercised with equitable reciprocity'.[17] At Rio de Janeiro Darwin would see just such a regime in force.

The Sugar Loaf, Pâo de Açúcar, stands sentinel at the entrance to Rio harbour, beckoning weary crews to safety and pleasures ashore. In this vast fortified bay in the 1820s was the South American Station, the Royal Navy's headquarters in the South Atlantic. A dozen vessels mounting over 300 guns sailed out from here. On 4 April the tiny ten-gun *Beagle* slid in among them as conspicuously as possible, setting every yard of canvas as she passed the Admiral's flagship: the third-rate colossus HMS *Warspite*, one of the old 74-gun ships, with huge firepower on two gun decks, and a full complement of 600 men. Darwin perched on the deck and was even told to hold 'a main royal sail in each hand and a topmast studding-sail tack in his teeth' as the ship shortened sail. He laughed afterwards on hearing that his role in the manoeuvre had been redundant, a joke played by the crew.[18]

On the 'Brazils station', resources were overstretched because of the necessity of showing the flag on both sides of the continent. Britain's interests here were commercial and strategic rather than territorial. Free trade was the aim, but in Brazil's young monarchy and in the republics of Argentina and Uruguay revolts were common, which unsettled the

British merchants. The Navy's task was to restore confidence, to make 'peace for the pursuit of profit'. Political instability and, at sea, piracy were the enemies, though bad weather and worse charts didn't help, as the fate of the *Thetis* showed. The squadron transported bullion and banknotes for creditors. From Rio they were shipped to Portsmouth and thence to the Bank of England. The *Thetis*, laden with gold and silver, had struck the cliffs of Cabo Frio island in fog and broken up, losing $800,000 in treasure and twenty-eight lives. Salvage operations had just finished as the *Beagle* passed the site,[19] a grim reminder that surveying was about life and death as well as dollars.

As in Bahia, criminal penalties and compulsory inspections were driving the slave trade underground, but in the vicinity of Rio the smuggling took place on a vast scale. Countless Africans were landed weekly in hidden bays along thousands of miles of coast. The slavers returned to port innocently carrying only ballast while the captives, suitably disguised and often with official connivance, were herded to market to be auctioned 'legally' alongside Brazil-born Creoles. Prices had risen sharply thanks to the 'spectacular expansion' of coffee production, which cost lives. On the *fazendas* (large farms), punishing work schedules, brutal discipline and disease pushed slave mortality above that of livestock. Out of 100 slaves, 25 might still be working after three years.[20] So fresh labour was constantly in demand, and, while the *Beagle* was in South American waters, the illegal African trade grew at an alarming rate.

Darwin rode inland to see these *fazendas* with a Scottish 'Slave-Merchant' and a 'black boy as guide'. He found a simple patriarchal way of life. The surrounding forest with its ferns and lianas filled him with 'wonder, astonishment & sublime devotion'. Before dawn one day, lost in the 'solemn stillness', he heard 'the morning hymn raised on high' by the blacks as they began work. Some slaves arranged their houses 'like the Hottentots', trying to 'persuade themselves that they are in the land of their Fathers'. 'Miserably overworked & badly clothed', was Darwin's view of these expendable people. Sad stories came to light, of the old woman runaway who 'dashed herself to pieces from the summit of the mountain' rather than be recaptured. Darwin visited the site; he knew that 'in a Roman matron this would have been called the noble love of freedom: in a poor negress it is mere brutal obstinacy.'[21] Half-understood scenes were recorded in disbelief: of the Irish planter quarrelling violently with his English manager, who kept a pet 'illegitimate mulatto child'. Shouting something about 'pistols', the planter threatened to sell the child, and then 'all the women & children', dragging them 'from their husbands' to the vile Valongo slave market in Rio. And this, Darwin noted, was a comparatively humane man. 'Against such facts how weak are the arguments of those who maintain that slavery is a tolerable evil!'[22]

He returned to the *Beagle*, skirting riots in the town, and on 25 April moved his collecting kit into a cottage on Botafogo Bay, fringed by a beach of 'dazzling whiteness' behind the Sugar Loaf. Augustus Earle, the precocious Royal Academy-trained artist and the *Beagle*'s draughtsman (whom Darwin frankly found 'eccentric') joined him, as well as young King, the midshipman, indulging his flair for natural history. He and Darwin netted butterflies and caught beetles by 'hanging an umbrella upside down in the bushes and shaking the branches'. When King returned to duties, Darwin collected corallines along the coast and hiked to the Botanic Garden and the town.

Twice he climbed the 2,000-foot Corcovado, notorious as a refuge for runaway slaves. On the granite dome (where 'Christ the Redeemer' stands today, arms outstretched) he met a gang of slave-hunters, 'villainous looking ruffians, armed to the teeth', bounty hunters paid per capita, dead or alive: in the former case, Darwin noted, 'they only bring down the ears'. Lucky escapees, he knew, often found employment, even in their masters' neighbourhoods. 'If they will thus work when there is danger, surely they likewise would when that was removed.'[23]

Anti-abolitionists had claimed that blacks were sluggards, an accusation which had infuriated the anti-slavery lobby. Darwin's proselytizing sisters would have had none of it. Nor Darwin. But the stereotype was entrenched. Ever since the great classifier Linnaeus stamped *Homo afer* as cunning, lazy and lustful the mud had stuck, dirtying all dark peoples. The concept of 'race' was being fashioned as industrialization was throwing up an urban underclass, to be judged by the new thrift-and-industry yardstick of 'worth'. Now blacks found themselves being tarred with the same brush as British indigents: they were feckless. Even if abolitionists could never surmount the prejudice, the missionaries would issue a rather damning counter-claim, that at least the 'Christian negroes are industrious'.

Darwin would continue to hear this 'slothful' calumny for decades, not least from books about the 'gol-durned lazy niggers' in the American South. 'Indolence' was a byword among the slave-owning friends of fellow Edinburgh-man Samuel Morton (one of whom was indicted in Georgia for 'cruel treatment of my fat, lazy, rollicking Sambos'). In these Slave States the vilification would become complete, as 'niggers' became 'Mere animals', who, without the whip, 'will lie down and bask in the sun'. Only the lash could convert 'the lazy, murdering, thieving, fetish worshipping African' into a productive worker. The abolitionists might not be able to defeat the muck-raking defamation, but Darwin refused to accept this slur. As he put it in Brazil, 'What will not interest or blind prejudice assert, when defending its unjust power?'[24]

Darwin's 'commonplace Journal' recording all of this was shipped home in instalments, along with reassurances about himself – yes, he and the captain were getting along, although he had felt the effects of FitzRoy's 'vanity and petulance'.

FitzRoy was Darwin's entrée into Rio society. In these months Darwin was the only member of the *Beagle*'s company invited to dine with admirals, chargés d'affaires and the good and great. The high life here could be as richly various as the insect low-life he snared, with 'good houses, in beautiful situations', and sea views unsurpassed anywhere in the world. These airy villas raised FitzRoy's noble mind above the city's 'almost naked negroes' and 'offensive sights and smells'. King recalled a mansion on the road to Botafogo, where the merchant Mr Young's slaves were 'all dressed in the purest white'.

> We generally sat down twelve to dinner with as many butlers to attend as the dining room opened out with french windows on a garden in which slaves with watering pots kept watering the flowers and cooling the air which blew into the room fragrant with the odours of the roses and the freshness of the moistened soil. The guests were mostly English and American officers with occasionally a Minister of State . . . After a long sitting at dessert and wine two or three carriages would drive up to the door and everyone went off to the Opera . . .

Such scenes would have been familiar to Darwin. The British had long been 'the most visible and wealthiest of the foreigners' here. And he enjoyed rubbing shoulders with the top brass and hearing their yarns. Rear Admiral Sir Thomas Baker was a veteran of the Napoleonic Wars and merchant convoys to the West Indies.[25] His flag captain Charles

Talbot had, the year before, rescued Brazil's abdicating emperor Dom Pedro I and brought his household and goods safely to Europe in the *Warspite.*

For all his excesses, Dom Pedro had proclaimed the end of the slave trade, making Brazil's 1826 abolition treaty with Britain fully enforceable. This was a special concern of another dining companion, Arthur Ingram Aston, secretary of the legation and chargé d'affaires, 'one of the very few gentlemen' Darwin met in Rio (meaning, someone of standing of whom he approved). Their first evening passed so pleasantly, almost like 'a Cambridge party', that more were arranged. Aston, an Oxford MA, had taken over from the minister who negotiated the abolition treaty. For the previous three years he had been the senior official representing Britain's interests in halting the slave trade. London was 'increasingly annoyed at the Brazilians' ability to find new loopholes to continue the trade whenever the Royal Navy and Foreign Office closed old ones'. Aston's job was to convey that annoyance forcefully but politely. So effective were he and his successors that the British legation 'virtually assumed the role of an abolitionist society in Brazil'.[26]

Aston's advisers on the trade sat on the British–Brazilian 'mixed commission' set up by the abolition treaty to oversee the disposal of seized slavers and their cargoes. That year, 1832, the British commissioners formally reported that no slave-vessel was 'brought into this port for adjudication', but they knew the contraband was being landed along the coast. So did Darwin: slaves were brought ashore at Botafogo Bay while he stayed there – shamelessly, within sight of a civil prison for blacks. He understood that a British official with a 'large salary' lived at Botafogo, whose job was to prevent these landings, and he had heard mutterings about him among 'the lower English'. 'The Anti-Slavery people ought to question about his office', Darwin told the family. He probably made the point to Aston. Sister Susan replied that she would report the 'ill conduct' to their well-connected aunt Sarah Wedgwood at Maer,[27] 'who will I daresay take notice of it'.

Topic of the hour at consular parties was the Brazilian decree of 12 April that all new slaves from Africa should be shipped back at the traders' expense. Emancipation wasn't the motive, it was fear of 'Africanization' – 'piling barrels of black gunpowder into the Brazilian mine', as whites put it, leading to an explosion from below. Darwin almost wished for the bang, 'for Brazil to follow the example of Hayti', where, as every abolitionist knew, Toussaint l'Ouverture had led a revolution of ex-slaves to create the first independent republic ruled by blacks. 'Considering the enormous healthy looking black population, it will be wonderful if . . . it does not take place' in Brazil, he told the family. But white fears of dominance by a 'rude and stupid race' ran deep.[28]

Aston and the commissioners actually fought to stop the repatriation of blacks, liaising with the Whig Foreign Secretary Palmerston, Darwin's old Cambridge MP. British policy was avowedly humanitarian. Slaves suffered unimaginably in the hellish bowels of ships sailing from Africa. But at least, as Palmerston put it, 'the hope of profit . . . affords some motive to the Trader to preserve the lives of his Cargo; on the voyage back to Africa, that slender restraint upon ill-usage would be removed, and the pecuniary interest of the Trader would tell indeed in the contrary direction'. In plain words that meant death by dumping at sea.

What FitzRoy thought of the consular company is moot. He declared that Darwin was the life and soul of parties, a 'regular trump', and it was perhaps after a particularly 'merry' dinner, on the way to a piano recital, that Aston remarked to Darwin that FitzRoy's fine manners recalled those of his uncle Castlereagh. Political lines were drawn as the 'varsity' men stood together. Darwin wrote to his 'chief Lord of the Admiralty', Henslow, that however well he and the captain were getting on, it was not at the expense of his 'Whig principles'. 'I would not be a Tory, if it was merely on account of their cold hearts about that scandal to Christian Nations, Slavery'.[29]

For the next year and a half, the *Beagle*'s base was the River Plate separating the volatile republics of Argentina and Uruguay. 'Anything must be better than this detestable Rio Plata', Darwin grumbled. 'I would much sooner live in a coalbarge on the Cam'. Twice he sailed south with the *Beagle* to Tierra del Fuego and the Falkland Islands, and he took every chance meanwhile to travel inland in search of evidence for how the continent had been formed.

On arrival at Montevideo in July 1832, the *Beagle* came under the protection of HMS *Druid*, a new fifth-rate frigate not long back from delivering $2 million of West Indian sugar profits to the Bank of England. En route to Rio, Captain Gawen William Hamilton had boarded a schooner outside Bahia, the *Destimida*, and found her carrying five blacks, said to be crew, and taking on water at two feet an hour. Nor were her papers in better order, and the master, from the Floridas in North America, was a well-known slave-trader. The *Druid*'s men searched high and low for contraband and had all but given up when 'an officer pushed his sword into the bung-hole' of a water-butt.[30] A cry went up, the cask was opened and out crawled five Africans. Forty more were found jammed into the crevices between the water casks under the false deck. Captain Hamilton seized the ship and cargo and hauled them before the mixed commission in Rio. The judges unanimously agreed to emancipate the fifty slaves, who were given 'certificates of liberty'.[31]

Buenos Aires and Montevideo had been Britain's last attempted conquests in South America. A quarter-century before, warships had landed thousands of troops in the River Plate. Both capitals had fallen in street-to-street fighting, with hundreds killed, only to be retaken within eighteen months by Creole soldiers backed by former slaves, who forced the British out of the river. It had been Castlereagh, as secretary for war, who had drawn the lesson that conquering this continent 'against the temper of its population' was a 'hopeless task'. Since then a Royal Navy captain would send a landing party ashore 'on only the most desperate of occasions'.

Castlereagh's nephew now faced one. Montevideo, chief slave port for the southern continent, had routinely imported Africans by gross weight in 'tonnes' like any commodity, or as 'pieces' to be sold for profit like working animals. But in recent years, as whites fought for independence, the slaves had plotted revenge. Offers of manumission induced some to become soldiers, but many freed slaves became mercenaries acting in their own interests. In August, one such group of 250 'mutinous negro soldiers' stormed the Montevideo prison, armed the black inmates and then seized the Citadel and munitions dump near the harbour. Desperate officials begged the *Beagle* for help and FitzRoy went ashore to meet them with the British consul-general Thomas Samuel Hood.[32]

No doubt lives and property were at stake. While an American man-of-war sent in sailors to occupy the Custom House, FitzRoy landed his fifty finest, 'armed with Muskets, Cutlasses & Pistols', Darwin among them. They marched to the central fort and seat of government, under orders not to touch the trigger unless threatened. 'A most unpleasant job', the captain admitted, like 'treading upon cracked ice', as they waited anxiously all night for reinforcements to secure the Citadel. Darwin got such a bad headache that he returned to the *Beagle* at sunset, before the bloodshed, which he dreaded. The next day, with everyone back on board, 'volleys of Musketry' were heard as the rebels drove out the military governor. Refugees from both sides pleaded with Fitzroy 'to give them shelter', which he refused. 'It has all ended in smoke', Darwin wrote home. Constitutional government was restored only after the president Don Fructuoso Rivera thundered in with '1,800 wild Gaucho' and Indian cavalry. Darwin was left wondering 'whether Despotism is not better than such . . . anarchy', even if it was at the black rebels' expense.[33]

To hasten the *Beagle*'s survey, FitzRoy bought a schooner to act as a 'tender' (a supply vessel which could sail close inshore). He named her *Adventure*. She had been owned by a sealer, though, as Darwin knew, 'Sealer, Slaver & Pirate are all of a trade'. Truly, for one of her mates was recognized as a pirate who had boarded HMS *Redpole* (a vulnerable brig-sloop like the *Beagle*). The *Redpole*'s commander had been shot, the crew forced to walk the plank and everyone else murdered, including eleven English passengers. Although the *Black Joke* had caught up with the buccaneers, one had escaped, and here he was, in July 1833, standing in irons before FitzRoy.

The slave trade was relentless; just how much so was brought home to Darwin at the River Plate. The *Destimida*, the slaver caught with blacks in casks, had arrived here *yet again* to be fitted with chains and branding irons, ready for a new run. Other ships were also clearly fitting out for slaving. Hood remonstrated with the Uruguayan foreign minister that the traffic was in 'direct and open infraction' of the country's constitution. The minister conceded that his government had approved the import of '2,000 negro *colonists*' – four laden ships' worth – 'which he considered a fair and legitimate trade'.[34] The traffic had been covertly re-sanctioned, through bribery. The anticipated slave landings gave FitzRoy the perfect excuse to justify buying the *Adventure*. Whatever his real feelings, he told the Admiralty:

If other trades fail, when I return to old England . . . I am thinking of raising a crusade against the slavers! Think of *Monte Video* having sent out *four slavers*!!! Liberal and enlightened Republicans – and their Prime Minister 'Vasquez' has been bribed by *30,000* dollars to wink at the violation of their *adored constitution*!! The '*Adventure*' will make a good Privateer!![35]

FitzRoy was capitalizing on events. He was using the anti-slavery brief, not the survey, to justify purchasing the *Beagle*'s sister ship. The arch-Tory was even declaring himself ready to join a 'crusade' against slavers if it brought down 'Liberal and enlightened Republicans' anywhere. But the ruse failed and he was not reimbursed. Nor would the *Adventure* stray from the *Beagle* to carry out anti-slaving patrols or rival Ramsay's *Black Joke*.

As the *Beagle* prepared to sail from the Plate at the end of 1833, word came that the slave 'colonists' had begun to land. The irony was that the *Beagle*'s old tender, purchased during the previous voyage, the *Adelaide*, itself a former slaver (and last seen as a tender to the flagship at Rio), had been sold and had *returned* to slaving. It was she who now entered Rio harbour after reputedly having landed 'near 200' Africans somewhere along the coast.[36] Nothing better illustrates the centrality of slaving concerns in these waters than that the *Beagle*'s own former tender was a once and future slaver.

Now more than ever anti-slavery advocates were needed, here and at home. The family's letters told Darwin that Sir James Mackintosh, bulwark of Commons anti-slavery, had died. There passed a giant who had helped fashion his Whig world, indeed a man whose humane commitments paralleled the liberal Darwin's so precisely: freedom from slavery, animal cruelty and abuse of all sorts. The families were tied as Mackintosh's daughter Fanny married Maer's most eligible, Hensleigh Wedgwood, and brought him to live in the 'Circle of Clapham'. Henslow had safely received the natural history specimens that Darwin had shipped home. And Darwin's old professor was hearing all about the slave trade from Captain Ramsay of the *Black Joke*, who was staying with him in Cambridge before sailing for the West Indies.

FitzRoy had been so anxious about the fierce Reform Bill politics that he had chased a Falmouth packet to get the latest news. At Rio they heard that the bill had passed both houses. The country had been 'on the very verge of Revolution',[37] but now a reformed Parliament was sitting. The Tories had been crushed, and, best of all, Uncle Jos had been elected a Whig MP for Stoke on Trent.

At Shrewsbury and Maer the family devoured Darwin's journal, and after hearing at Montevideo of Jos's 'fine Majority' he sent the next instalment. 'Our new member', sister Caroline called her uncle. Jos was staying with Fanny and Hensleigh and bussing up from Clapham to the Commons as Mackintosh had done. The debates were fatiguing, most speeches dull, but Jos remained 'a staunch supporter' of Earl Grey's Whig ministry, which the family thought quite 'perfect for they let no abuses remain'. Wedgwood assured them that 'some strong measure in favour of Emancipation of the Slaves' would be carried in the 1833 session.[38]

That was what Darwin, cooped up all those months with FitzRoy, biting his tongue, wanted to hear. As he caught up with the news, his confidence rose, and he wrote home:

> I have watched how steadily the general feeling, as shown at elections, has been rising against Slavery. – What a proud thing for England, if she is the first Europaean nation which utterly abolishes it. – I was told before leaving England, that after living in Slave countries: all my opinions would be altered; the only alteration I am aware of is forming a much higher estimate of the Negros character. – it is impossible to see a negro & not feel kindly towards him . . .

His true colours were unfurled to a liberal Cambridge friend who thought the 'Tories (poor souls!) . . . gone past recovery'. Darwin talked of blasting 'that monstrous stain on our boasted liberty, Colonial Slavery'. He had seen enough of it to be 'thoroughly disgusted with the lies & nonsense' retailed in England.

Jos's Whigs in power did their job. On the day Parliament finally freed Britain's 800,000 colonial slaves, 28 August 1833, Darwin was ending a wild gallop over the Argentine pampas. Months later, the news reached him at sea off the coast of Chile. 'You will rejoice as much as we do', said his sister Susan.[39]

Darwin spent more time hiking and hacking than on board. Being on terra firma cured his seasickness, and forests and mountains, coasts and islands, held riches beyond any he dredged from the deep. He shipped back huge collections of insects, birds, mammals and fossil skulls. His vision of South American geology – from the pampas to the Andes – owed everything to one audacious young hammerer battering an interesting path for himself in London: Charles Lyell.

Lyell, an urbane Oxford-educated barrister, had found geology more congenial than the law. He was a well-travelled, reform-minded sort. His ambitiously titled *Principles of Geology* (three volumes, 1830–33) not only tried to 'free the science from Moses' – throw out the Flood – but to cast most rival geologists adrift too. *Principles* proposed that mountain ranges were built, not by cataclysmic upheavals as most thought, but by tiny, incremental steps. Indeed all past events could be explained by modern, environmentally observable causes. Peaks rose through earthquake or volcano, and degraded through rain or frost. The past had not been another country at all; the ancient world no apocalyptic scene of catastrophic violence. An anti-revolutionist in a different way, FitzRoy had given Darwin the first volume before leaving England; Lyell's second came at Montevideo and the third reached him in the Falklands. Poring over them, Darwin learnt to look at the old world through Lyell's new eyes, and by 1834 his conversion was complete. Lyell's was almost an 'evolutionary' remodelling of the landscape through each geological epoch. Everything Darwin saw, even the Concepción earthquake in 1835 that raised the seabed and stranded mussels, seemed to confirm it.[40] The Andes were not testimony to a sudden paroxysm of the crust, but a reminder of untold aeons of gentler uplift. Everything Darwin saw fitted into Lyell's vision of earth history.

Principles wasn't just the latest geology text: it put flesh and bones to the theology taught at Cambridge. Humans to Lyell were latecomers in the timeless flux of life, divinely appointed as the only species moral and progressive. They were of one family: 'the varieties of form, colour, and [bodily] organization' of the different races are 'perfectly consistent with the generally received opinion, that all the individuals of the species have originated from a single pair', Lyell declared, though the 'pair' was metaphoric. More than Moses was going overboard this time, for Lyell personally jettisoned Adam and Eve as well. He asserted that the human races are like those in any animal species. They all swirl round a 'common standard', all slightly different; and they intermarry freely, with the unions of the remotest varieties as 'fruitful' as 'those of the same tribe'. The mixed offspring are not odd sterile 'mules' – so their parents were not separate species. Their adaptiveness to 'every variety of situation and climate' has allowed 'the great human family' to extend 'over the habitable globe'.

As the *Beagle* sailed south, this racial panoply began to unfold. It chimed with Darwin's religious beliefs, which were still so orthodox that the officers once laughed at his priggishness in quoting the Bible as an 'unanswerable' moral authority. Darwin admired the 'fervor' of Catholics on seeing a 'Spanish lady' kneeling alongside 'her black servant'. Though on the eve of the *Beagle*'s first passage to Tierra del Fuego, it was, of course, to an Anglican chaplain he went to receive Holy Communion, with a shipmate from the *Druid*. The 'Philosopher' (as the crew called him) still sought salvation. At home he hadn't been above some banter about 'fighting with those d—— Cannibals', and even shooting 'the King of the Cannibals Islands' (a popular song in the 1820s), which rather worried Hensleigh and Fanny.[41] It was all bravado. Face to face with strangers, Darwin tried to see them sympathetically, in their environmental totality, as Humboldt had viewed New World slaves.

Slavery was a brutal fact of the voyage, people bought and sold, used and abused like beasts. His sensitivity to the whip might have stemmed from his reading of anti-slavery tracts, but here he would see its effect: he stayed in one house where a young mulatto was beaten 'daily and hourly', he said, 'enough to break the spirit of the lowest animal'. And that was the point. Bestialization was implicit in the system; it was as though the whip hands were attempting to break humans the way prehistoric peoples had broken horses during their 'domestication'. Then again, American slave-masters were said to have no more fear of 'rebellion amongst their full-blooded slaves than they do of rebellion amongst their cows and horses. That was because the tranquillity of Negoes in their approach to civilization resembled the content of domestic animals.' In the 'breaking' of animals originated the yokes, leg-irons, chains, lashes and branding irons so familiar to the slave-master: these instruments usually adorned the overseer's walls as a permanent threat. Slaves had been reduced to childlike, cattle-like dependency, with a result that 'Sambo' was rendered a broken brute in the planter literature. It would start Darwin thinking about the way masters tried to distance themselves from slaves, tried to make them appear as soulless beasts of burden.

Only a half of Brazil's population were black slaves, a quarter were white Creole and European, and the rest mestizos of every shade – *mulato*, *mameluco*, *caboclo*, *cafuzo*, *zambo*, *cabujo* – all permutations of the offspring from African, white and Native Indian unions. Darwin had been sensitive to the diversity since seeing the bright mulatto children at Porto Praya. He noticed how ornamental scars preserved African tribal identities among the slaves, even though they were forbidden to speak their own languages. South of the River Plate, the picture changed. On the plains of Patagonia and in the cordilleras of the Andes, black faces were rare. Slavery did not thrive where tractable white or native labour was plentiful. Yet life here could be as hard, and stories of man-eating primitives prepared Darwin for his encounters with 'curious tribes'.[42]

Nevertheless, these Patagonian plains were witnessing a genocidal conflict, open and violent – and on Darwin's two shore excursions he saw its bloody upshot. The Indians were being annihilated. FitzRoy blamed the 'war of extermination' on '*independent* Creoles'. He reported back to Beaufort sarcastically that their '*Revolution* (what a *glorious sound*)' had provoked the hostility of the Indians, who preferred Spanish rule. Darwin's concern was more ethical (even though, beneath it, lurked a more sinister contemporary understanding of civilization, which measured 'progress' by means of productivity, in which the Indians scored badly): plainly, a 'villainous Banditti-like army' was cleansing the pampas, slaughtering Indians to make way for livestock and ranchers. 'The War is carried on in the most barbarous manner. The Indians torture all their prisoners & the Spaniards shoot theirs.'[43]

Darwin met the *commandante* in charge, General Juan Manuel de Rosas, a 'perfect Gaucho', grave, unsmiling, himself a torturer. Like other genocidal despots, this future dictator of Argentina perhaps had a soft spot for nature, if not naturalists. Darwin was given safe passage deep inland along a string of the *Rosistas* stations. And it was from His Excellency's vast estancia bordering the Rio Salado near Buenos Aires that part of a tessellated carapace of the giant fossil armadillo *Glyptodon* was shipped back to the London College of Surgeons. Darwin himself found similar remnants nearby on the Rio Salado. So perhaps, after Darwin had (as he recorded in his journal) delicately commented to the strongman that the genocide on the plains seemed 'rather inhuman',[44] the safely extinct megafauna gave them some common talking ground.

'Inhuman' was a ridiculous understatement. According to Rosas, Indians were pests to be eradicated, like rats. Darwin was appalled. Throats slit, prisoners shot, all women 'above twenty years old' butchered lest they 'breed', children sold for the price of a horse – such 'shocking barbarity' in a so-called 'Christian, civilized' land would bring a pyrrhic victory. The gauchos saw it as 'the justest war, because it is against Barbarians'. Darwin knew that the extermination of every 'wild Indian in the Pampas North of the Rio Negro' would leave the country 'in the hands of white Gaucho savages instead of copper-coloured Indians', the former being as 'inferior in . . . moral virtue' as they are 'a little superior in civilization.'[45] A civilized race was not necessarily a more moral one.

Comparisons of colours and morals and physiognomies became inevitable. They were not always happy; he interpreted the sights as much through the prejudicial stereotypes of his own age as anything, but they sowed the seeds of later thought. Most of the vicious gauchos were of mixed race, 'between Negro, Indian & Spaniard'; many had 'moustachios & long black hair curling down their necks' and wore ugly expressions, as 'men of such origin' usually did. In the town of Patagones Darwin found fewer mixes, and more pure-blooded Indians and Spaniards. The Indians intrigued him. One group, allied with Rosas, was a 'tall exceedingly fine race', some of them 'very fair' but with a 'countenance, rendered hideous by the cold' and hunger, like that of the shorter Fuegians further south. Then there were the young women, with their glistening eyes, coarse 'bright & black hair', elegant feet and limbs. They might even be 'beautiful', he thought, and one, he imagined, 'flirted' with him. But 'like the wives of all Savages', they were 'useful slaves', leaving the men to fight, hunt and feed themselves ravenously. When crouched round a fire, 'gnawing bones of beef', they 'half recalled wild beasts'. The meat diet must have enabled them, 'like other carnivorous animals', to withstand sustained hunger and exposure.

Just how tough they were Darwin would himself find out on a six-week trek across the plains. Four 'strange beings' joined him: a 'fine young Negro', a 'half Indian & Negro', and two 'quite non descripts, one an old Chilian miner of the color of mahogany, & the other partly a mulatto'. 'Mongrels', he dubbed the pair, 'with such detestable expressions I never saw'. They survived on ostrich, armadillo and fresh foetal puma, camping under the stars or sheltering at militia outposts. At one, Darwin's host was 'a Negro lieutenant born in Africa' who kept a corral and a neat 'little room for strangers'. Nowhere did he meet 'a more obliging man than this Negro', which made it 'the more painful . . . that he would not sit down and eat with us.'[46] Racial protocols and customs met every exigency; no outpost was exempt.

*

The lower the latitude, the more agreeable Darwin and FitzRoy seemed to find each other. The poop cabin, with its chart table and chairs, measured about nine feet by twelve, and the captain's mess eight by ten, close quarters for an arch-Whig and a Tory aristocrat. But Darwin's shore-going helped ease the tension, and he would come back to spend a 'whole day . . . relating my adventures & all anecdotes about Indians to the Captain'. Darwin wasn't just a 'good pedestrian' and 'good horseman'; to FitzRoy, Darwin had become a 'sensible, shrewd and sterling fellow'. He feels 'at home . . . and makes every one his friend', the Admiralty heard.

Despite the politics and row about slavery, Darwin saw eye-to-eye with FitzRoy on most things, not least the primitive tribes they met. All belonged to Lyell's 'great human family' and were, as the captain put it, 'of one blood'. Climate and diet had shaped their bodies, and habit their mental faculties; not that there had been much but superficial change anyway. FitzRoy saw 'far less difference between most nations, or tribes', than between the individuals within each.[47] His theology would harden, but for the moment FitzRoy's and Darwin's beliefs about humans were practically at one: the nations comprised a single 'human race'.

Harder-headed savants might disagree, and the poop cabin shelves sagged under the weight of contradictions. Here sat Darwin's seventeen volumes of the great natural history *Dictionnaire*, co-edited by Jean-Baptiste Bory de Saint-Vincent. Bory was an old Napoleonic officer turned radical transmutationist. He operated in a dissident, republican, anti-Catholic Paris. Here the levelling milieu was congenial to his science: he insisted that vanity drove humans to elevate themselves, while leaving apes allied to the 'stupid brutes'. The entry on 'Man' was his doing, and it appalled FitzRoy.[48] Bory marched apes up through time to become humans, savage and civilized, all fifteen species of them. Adam he made simply the Jews' 'first man'; and there must have been as many Adams as human species. All of Bory's human species were aboriginally distinct; that is why he made them 'species'. His world map showing their 'original distribution' in a blaze of colour looked like a stained glass window, which was all the religion FitzRoy saw in it. He *knew* this was wrong factually as well as theologically. The Patagonians and Fuegians were *not* separate species – he had lived among them. And the Fuegians were *not*, as Bory thought, like black Ethiopians and Australians either; he had three of them on board to prove it. All the races intermarry and reproduce, French 'false philosophers' notwithstanding. It was arrant nonsense to suppose the 'separate beginnings of savage races, at different times, and in different places'.[49]

Those who believed in the separate creation or emergence of each human race or species were 'pluralists'. For them the various human species were not blood-kin at all. Each species in its geographical home had a separate bloodline back to the beginning, which never connected to any other species. There was no *common* ancestor for all the races. Some American writers were already arguing that the 'origin of the different races of men' was the most intriguing subject in natural history. A few laughed Moses out of court and dismissed as flippant talk of climate turning one race into another. These pluralists had aborigines first appearing (not as ancestral pairs, which was zoologically daft, but as viable populations) adapted to the spot where they are now found. So black and white had separate ancestries and differed more from each other than one species of dog did from another. With increasing agitation over American slavery, pluralism was a perfect legitimating philosophy. Books were already denying that the separate races or species were equal or 'sprang from the same primitive root'. Slave and master were thus unrelated, which made the planters' actions towards their 'inferior' captives easier to justify.

To FitzRoy and Darwin such heresies trampled on their terrain. They were out in South America, studying the natives and the power of civilization and 'domestication'. Others saw Fuegians and all Indians as unprogressive, like animals, 'motionless; fixed to a spot . . . each generation pursuing the same time-beaten track'. These lowly peoples, whose 'moral deficiencies' left them unable to advance, were destined never to build a city or raise an Indian Shakespeare. 'Savagism' was their natural state, 'essential to their existence', just as civilization was to the Caucasian. They could not 'flourish in a domesticated state'; it wasn't in their constitution, and every civilizing scheme involving cruel attempts to take them out of the jungle would 'necessarily fail' and lead to their extinction. But Darwin had been immensely influenced by London's effects on the Fuegians aboard the *Beagle*, with their fine clothes and Court manners. To him Bory's pluralism was a living, refutable issue. He and FitzRoy were in the nether regions, testing its claims and finding them wanting. Darwin made his own comparisons between the Fuegians and Patagonians. French savants had 'separated these two classes of

Indians' in 'defining the primary races of man', but 'I cannot think this is correct', Darwin noted in August 1833, showing his first interest in racial origins.[50]

So contact with 'pluralism' first forced Darwin to think through the philanthropic image of unity, of shared blood, and what it meant for human relationships. A few months later, during the *Beagle*'s last sojourn at the tip of the continent, he tested his conclusion among the tribes living along the Straits of Magellan.

The Patagonians to the north were tall, six feet and more, but then so was Darwin, and he looked them in the eye and found it 'impossible not to like' them, 'they were so thoroughily good-humoured & unsuspecting'. He noted their aptitude for languages, Spanish and English, which would 'greatly contribute to their civilization or demorilization', or both, for they usually went 'hand in hand'. While Darwin warmed to 'our friends the Indians', FitzRoy's blood ran cold. Shorter than Darwin, he felt uneasy when surrounded by 'two hundred men and women . . . all between five feet ten inches and six feet six inches – very stout – & large limbed, with large features, & deep sonorous voices'. They seemed 'extremely bold, and confident', although fond of talking with strangers. The captain remained wary. His responses too were shaped by preconceptions and a pointed literature – particularly the eighteenth-century *Description of Patagonia* by a slave-ship's surgeon and Jesuit convert. Gregarious the Patagonians might be, but they had 'superstitious minds' and lacked 'moral restraints'. Beneath their agreeable veneer, FitzRoy was sure these humans 'disgraced' themselves by 'the worst barbarity'.[51]

While FitzRoy and Darwin saw eye-to-eye on human origins, how they judged human nature differed radically. Darwin at Edinburgh had witnessed phrenology's attempt to clamp leg-irons on human potential, and at Cambridge any vestigial belief he had had in phrenology had been 'battered down' by Mackintosh. FitzRoy's experience had been the reverse. He was now convinced that the head and features were the key to the human mind. Sea captains – on whose character judgements the crew's survival depended – made good converts to phrenology. Even convict ships leaving for Botany Bay had their potential troublemakers among the convicts phrenologically identified and segregated. But phrenology had gone further, to make these skull-shapes racial markers, allowing the primitives to be ranked by degrees of inferiority. FitzRoy wanted these phrenological facts recorded. A pair of Patagonian skulls robbed from a grave on his previous South American voyage had been passed to the College of Surgeons in London. The living descendants of the grave occupants now confirmed his diagnosis: 'simplicity and shrewdness, daring and timidity' – a singularly 'wild look which is never seen in civilized man' characterized Patagonian faces.[52]

On that previous voyage, south of the Magellan Straits, during the scuffle and seizure of the Fuegians, a musket shot had felled a stone-throwing young native. An autopsy had been performed on the spot and the corpse carefully measured. These remains too had gone to the College of Surgeons: FitzRoy had deposited the skull, bones and 'prepared skin of the Head', with a description of the 'Phrenological organs'. For him the evidence pointed to savagery, superstition, even cannibalism: 'extremely small, low forehead . . . prominent brow small eyes . . . wide cheek-bones; wide and open nostrils; large mouth . . . thick lips', and so on. From this he imagined he could see how transformed their civilized and Christianized compatriots on board were. The returning captives, with features 'much improved' by a Church Missionary Society education, were new men; they were now physiognomically separated from their wild Fuegian brethren. Unlike other phrenologists, FitzRoy saw education actually alter features. It could throw off chains, raise 'savages'. As the soul was saved so the face was rendered serene. Skull-bending by brain use was heady nonsense to Darwin. But the captain's 'faith in Bumpology' aside, they agreed: their Fuegian passengers were human equals changed from 'savages . . . as far as habits go' into 'complete & voluntary Europaeans'.[53] Maybe not quite voluntary, but 'civilized' all the same, showing the 'progressive' possibilities inherent in all peoples.

Neither FitzRoy's words nor the 'civilized' Fuegian faces could have prepared Darwin for the shock of seeing their wild brethren. Altogether he spent about a month in Fuegian waters between 1832 and 1834, and his contacts there were the most disturbing of his entire voyage.[54]

Shore-going was worrying, destabilizing, frightening and fascinating. There he stood, an impeccably correct young gentleman from the closed and clothed culture of Cambridge, surrounded by men and women of his own age, many nearly naked, their faces painted, 'hair entangled', red skin 'filthy & greasy'. Women too; this was possibly the first time that he had seen such female flesh, at least off the dissecting slab. It must have been an extraordinary sensation, the sort to cause swooning at home – not a neatly turned ankle, but a less-than-neatly-turned near-naked torso. To him they were humans in the raw, jabbering and gesticulating, looking scarcely like 'earthly inhabitants'. Yet they were all too earthly; they seemed more like animals than English people. He had seen people treated like animals, but never *living* like them, and he kept coming back to the thought, unable to shake it off. Without fit clothes or proper homes, the Fuegians wandered about foraging and slept 'coiled up' on the soggy earth. When threatened, they fought as if by instinct, with courage 'like that of a wild beast'. Surely 'no lower grade of man could be found'. Darwin too was grading peoples, but not anatomically, like the phrenologists and pluralists. His was the conventional sliding scale of plastic qualities, behaviour and morality, with their technological and civilizational consequences. And chance, adaptation and terrain played a key role in their development.

While Fitzroy peered at heads and the officers made monkey faces (drawing a Fuegian's 'still more hideous grimaces'), Darwin observed their hosts' humanity. 'More amusing than any Monkeys' they might be, but they were also, he scrawled in a notebook in 1833, 'innocent *naked* most miserable very wet'. A group came running so fast to greet them that 'their noses were bleeding, & they talked with such rapidity that their mouths frothed'. Yet these were 'quiet people' on the whole. They sat in a row, naked, 'watching & begging for everything'. Families arrived. He huddled around a 'blazing fire' with one, trying to teach them choruses. Poignantly, he watched as one of the abducted Fuegians was reunited with his 'mother, brother, & uncle' after three years; they told him that his father had died, only to learn that it had already been revealed to him in a dream. Then he lit a small fire and, looking 'very grave and mysterious', watched the smoke rise. Rowdy neighbours turned up. A gullible Darwin believed the joke that some were 'bold Cannabals' who ate their elderly women in times of want. But for all that, he was still shocked to think of firing on 'such naked miserable creatures', whatever the justification.[55]

He believed that the Fuegians were 'essentially the same' as himself, 'fellow creatures' of one God, yet what a chasm separated them! 'Bumpology' had nothing to do with it. On first seeing the wild Fuegians, he compared the gulf to the difference between a 'wild & domesticated animal'. Two years later, in 1834, deeper reflection had stretched it wider. The Cambridge graduate, barely twenty-five years old, pondered nature's power of 'improvement'. It could turn an 'unbroken' man into the greatest of intellectuals; it could stretch a savage, if not to the stars, at least into a star-gazer like 'Sir Isaac Newton'. Such was its potency.

Why did men exist so 'high' and so 'low'? Had the Fuegians been created here and remained in the same state ever since? No, Bory was wrong, these people had relatives in the 'fine regions of the North'. They must have come down and become 'fitted' to this stormy, cold environment. The abducted Fuegians, throwing off their clothes and returning to their old 'savage' ways, proved he was right. But if such changes could take place, maybe all humans had adapted, improved: staying wild here, becoming domesticated there. 'One's mind hurries back over past centuries, & . . . asks could our progenitors be such as these?' It was hardly an original thought, but for Darwin it would be formative. The captain saw the Fuegians as 'satires upon mankind'. The 'Philosopher' considered it less derogatory to believe that 'the difference of savage and civilized man' is only that of 'a wild and tame animal'. Interestingly it was in Tierra del Fuego, perplexed and troubled by an alien race, that Darwin decided to spend his life studying natural science.[56]

'Whence have these people come?' The question he first asked in Tierra del Fuego kept returning. Sailing up the Chilean coast in 1834, he looked for ancestral relationships. Language was no help: 'everything I have seen convinces me of the close connection of the different tribes, who yet speak quite distinct languages'. And then another 'complete puzzle', the absence of Indians in the Chonos archipelago. With food abundant, why was no one there to eat it? Given the pampas genocide and expectations of 'the final extermination of the Indian race in S. America', he first presumed these Indians extinct. But on Chiloé island further north he found 'little copper-colored men of mixed blood', said to be the descendants of Indians who had been shipped in by the Spaniards to be 'slaves to their Christian teachers'. After hearing stories about missionaries tempting tribes with gifts to leave their homes, the answer came to him: the Chonos Indians had decamped, not died out. All the continent's aborigines must have similarly shuffled themselves and dispersed.

The Chiloéans taught him something else about origins, or rather their lice did. The 'disgusting vermin' plagued everyone there. Darwin assumed from their large size that they had arrived with the Chonos

Indians from the south, where the tall Patagonians had the same parasite. He collected specimens to compare with English lice, said to be softer and smaller, and then he heard from an English whaler's surgeon that the 'blacker' ones infesting dark-skinned Sandwich islanders died promptly when they crawled on to British sailors. Different lice adapted to live on different races – the possibility was intriguing. 'Man springing from one stock according his *varieties* having different parasites – ', Darwin began a note, meaning that with human races diverging from a common ancestor, perhaps their lice did too. And then he drifted off, musing, 'It leads one into many reflections.'[57]

Human diversity blossomed as he re-entered the tropics. In northern Chile the Indians had a 'slightly different physiognomy' – 'more swarthy, their cheek bones . . . very prominent', facial expressions 'generally grave & even austere', their hair 'not so straight & in greater profusion'. He had picked up FitzRoy's language but none of his brain science. Even so, Darwin could be as censorious. At Lima in Peru they both noticed the rich racial mix. FitzRoy counted 'at least twenty-three distinct varieties', which Darwin recorded as 'every imaginable shade of mixture between Europoean, Negro & Indian blood'.

At last the *Beagle* began her long homeward voyage, crossing the Pacific first to the 'frying hot' Galapagos Islands in 1835. Here scalding lava whirls looked like they had just issued hissing from a vent, recalling the cinder-strewn wasteland around Wolverhampton's furnaces. New land, this was, fresh from the sea, part colonized; each island, oddly, with its own marginally different variety of giant tortoise. And the inquisitive, fast-running mockingbirds, too, seemed to vary through the desolate chain. More specimens for the hold, as the *Beagle* sailed on through the South Seas and Indian Ocean, bringing Darwin into contact with more ethnic groups in thirteen months than during the previous four years. As he chased the racial rainbow back to England, his journal recorded each hue with mounting astonishment.

Darwin's journal contained little of that derogatory racial judgement that became widespread in the ethnological literature in later decades. The Tahitians with their 'naked tattooed bodies' and mild expressions, were the 'finest men' he ever saw: 'very tall, broad-shouldered, athletic, with . . . limbs well proportioned' and 'the dexterity of Amphibious animals in the water'. 'A white man bathing along side a Tahitian' looked like 'a plant bleached by the gardeners art' next to one 'growing in . . . open fields'. The analogy he had used in Tierra del Fuego was taking root: civilized people are simply domesticated varieties of the species. And civilized domesticity came by degrees. So the Maori, though evidently from 'the same family' as the Tahitian, still had an eye full of 'cunning & ferocity', revealing his savagery, while his cousin was a 'civilized man' by comparison.[58]

The Fuegians FitzRoy had taken to London had been 'civilized', so much so that they had been presented at Court ('paraded' might be a better word). But once back in their own terrain they had cast off their European clothes and run wild. The young missionary dumped with them, Richard Matthews, lacking any native help, had had to be rescued. A bemused Darwin had always thought Matthews odd and devoid of energy, but then he was pathetically and literally out of his element: still in his teens, he had been left in a hut at the end of the world, with tea trays, soup tureens, a vegetable patch and instructions to found an Anglican mission! The holy folly lasted a fortnight, and then FitzRoy brought him back aboard, planning to drop the boy with his brother, a missionary in New Zealand.

Like his Parliamentary failure at Ipswich, the Fuegian fiasco launched the captain on a mission of atonement. Tahiti was a paradise synonymous with sexual licence, celebrated as such by Erasmus Darwin poetically, and by sailors more physically. Crossing the Pacific, Darwin prepared by reading the Congregationalist minister William Ellis's 'most interesting' multi-volume *Polynesian Researches*. It was a typically rose-coloured account, as Darwin knew, where others on his shelf were critical, but authoritative because the humble missionary had learned to speak Tahitian. Sandwiched between the Christian philanthropic wrappers praising the moral progress and eradication of 'delusive . . . idolatries' was a huge repository of local lore, customs, ethnology and natural history. Such reading would become one of Darwin's lifelong loves.

During the *Beagle*'s ten-day stay in Tahiti Darwin helped FitzRoy take moral soundings among the natives to gauge this 'progress'. Over Christmas 1835 they did the same at the Bay of Islands in New Zealand. This beautiful land-locked archipelago lured sailors and whalers from all over the Pacific with its cheap drink and sex. A few years earlier one visitor had been the ship's former artist Augustus Earle, Darwin's house-mate at Botafogo.[59] He had left the *Beagle* in Brazil, but his Bohemian 'eccentricity' continued to plague Darwin and FitzRoy. Earle's *Narrative of a Nine Months' Residence in New Zealand* – a mission-bashing account of his previous sojourn at the Bay of Islands, published in 1832 – had caught up with the ship, aggravating the officers and giving FitzRoy a new chance to redeem himself.

An inveterate voyager, Earle had been blown many places by chance. So, on a voyage from Rio to Calcutta via the Cape, he had become stranded on the desolate isles of Tristan da Cunha, to be saved by a passing ship destined for New Zealand. Like flotsam he had ended up at the Bay of Islands in 1827. His book indelicately described the flourishing sex trade he found; how the sailors would greet the boatloads of 'sweethearts' rowing out to them. Afterwards, these native girls disembarked with a 'tinge of vulgarity . . . their beautiful forms hid under old greasy red or checked shirts'. He portrayed the prostitution as a moral trade-off, stemming from the disappearance of female infanticide. Now, with these 'fine young women' receiving presents from European crews, the families were naturally 'anxious to cherish and protect their infant girls' as potential income. '[I]f one sin has been, to a certain degree, encouraged, a much greater one has been annihilated.'

The Church Missionary Society would tolerate neither. They had denounced the depraved whalers and Earle alike. The book was his revenge. It lampooned the gospel-touting prigs in their cosy homes, ringed by a stockade 'to keep out the "pagan" savages'. The evangelists had perplexed the natives with 'abstruse' religious doctrines and 'absurd opinions' rather than teaching them the practicalities for a harsh life: 'how to weld a piece of iron, or to make a nail'. One saving grace Earle did concede: the missionaries had spared a poor slave girl from human sacrifice. On anti-slavery they could at least agree.

He recognized as clearly as FitzRoy and Darwin the 'hideous shape' of local slavery. People were seized in war, when the chiefs took the pretty women as wives and 'greatly prized' the children. Bondage was for life, release only by death, which often was not long coming. Darwin met a chief who had 'hung one of his wives & a slave for adultery', imagining he was following 'the English method' of punishment. Earle listed other atrocities, sickening ones, such as the slaughtered teenage slave he found being roasted by her owner. FitzRoy trekked to the spot with a CMS missionary, who told him the chief was now supplying 'his female vassals' to whaling ships, which was better than eating them. FitzRoy and Darwin had seen the sex slavery for themselves:[60] entering the bay, the *Beagle* had passed three whalers with 'many of these women' on board.

Slavery was endemic. Darwin even had to let a slave carry his small bundle, such was the supposed 'indignity' of doing it himself. On these wild Antipodean islands, beyond the reach of lawful diplomacy, missionaries alone were able to 'redeem' those in sin, by teaching them basic skills, Christianity and cricket. FitzRoy and Darwin, believing this, observed the mission closely for ten days. Earle had perversely, it seemed, disparaged or denied these good works, taking vengeance after the missionaries denounced his own sexual exploitation of slaves.[61]

Darwin's sisters doubted there could be anything good in Earle's book after it was panned in the *Edinburgh Review* and religious press. Darwin was on God's side. He told the girls (in a letter from the Bay of Islands, carried by one of the whalers) that the missionaries treated their traducer with 'far more civility, than his open licentiousness' deserved. He cut to the Christian heart of the matter in a letter to Henslow: Earle's type sneered at mission work because they are 'not very anxious to find the Natives moral & intelligent beings'.

Darwin's own attitude to indigenes was complex, but he did have faith in the morality and intelligence of many tribespeople. His was a standard philanthropic view of native abilities: it required only a condescending hand to start their 'rise'. This mix of *noblesse oblige* and humanitarianism grew out of attitudes to his own culture, divided into 'higher' and 'lower' orders. He thought the former on board were those 'whose opinions are worth anything'. Yet even that was cross-cut by his sympathy for the underdog, whether slaves, Maoris or dogs, literally. Sympathy was the first step to questioning the very concept of the 'lowly' in creation.

When Christmas came to this Antipodean 'little England', Darwin and FitzRoy attended a service conducted by the missionaries in English and Maori and had dinner with their families. FitzRoy passed the hat and £15 was raised for the tiny church being built amid the 'filthy hovels' in Kororareka, 'the very strong-hold of vice' at the Bay of Islands. By New Year's Eve, back on the high seas en route to Australia, the captain had decided to record his moral soundings. He and Darwin would write up their observations, and publish an article defending the missions from the likes of Earle.[62]

When it came to writing their missionary article Darwin drew support from Sir James Mackintosh, Cousin Hensleigh Wedgwood's father-in-law. Darwin dug out his copy of Mackintosh's *History of England* and noted passages praising early Christian missionaries for lifting the nation's 'faithless and ruthless' barbarian ancestors towards civilization. Darwin's copy of Humboldt yielded support for missionaries in the Americas, and FitzRoy put these extracts at the head of the article. Even members of a 'thorough-bred savage' race, such as his Fuegians, said FitzRoy, could be transformed into 'well behaved, civilized people' (he neglected to mention his own captives' recidivism).

Selections from 'our journals' followed next in the article, the captain having approved of Darwin's sentiments. The Tahitians' church attendance, Bible reading and simple prayers had impressed Darwin deeply. Alcohol was proscribed (there was no word of Darwin's passing his hip flask around, causing the natives to murmur 'Missionary' under their breath); 'profligacy' was abolished and 'licentiousness' much reduced, from which Darwin concluded that the missionaries' critics were sexually frustrated. Glum at finding their paradise lost, they despised 'a morality which they do not wish to practice'. FitzRoy agreed. At the Bay of Islands, the missionaries were peacemakers and temperance preachers to a society corrupted by guns, spirits and prostitution. Lotus-eaters like Earle naturally objected to those who sought 'to check, or expose, the impropriety of their own hitherto unrestrained immorality'. If evangelical teaching failed to reform such men, at least it 'acted like an enchanter's wand' on the natives, miraculous in its 'moral effect'.[63] For a moment, Darwin sounded like Charles Simeon.

In Australia, Darwin saw the Aborigines as 'harmless' savages, wanderers like the Fuegians, with equally 'abominably filthy' bodies, though at King George's Sound in March 1836 their genial high spirits and wild rituals delighted him. But there was a tension on the continent, and its irony and injustice were not lost on the visitor. Tasmania, founded as a penal colony, now kept its Aborigines 'as prisoners'. The two thousand remaining had been terrorized and herded towards the Tasman Peninsula. It must have reminded Darwin of the pampas clearances. He didn't mention the role of missionaries, such as the heavy-set Methodist and former bricklayer George Robinson, who had gone into the bush to bring in the last Tasmanian hold-outs. For this he was awarded £8,000 in cash and land grants, and made commandant of the 'Christianizing concentration camp' on Flinders Island which held the last tattered remnants of Tasmania's indigenous peoples.[64] Perhaps Darwin was not aware of this 'good' missionary work, although that seems surprising. But he did find the 'cruel' removals tragic, because 'without doubt the misconduct of the Whites' had led to the Aboriginal hostilities.

Darwin's first contacts were always revealing. In South Africa he found the racial situation even more explosive. The so-called 'Sixth Kaffir War' had ended a few months earlier. Now for the first time he would meet 'wild' black Africans and surprisingly civilized ones in their homeland. What is extraordinary, again, about his encounter with the Cape's 'Hottentot' people (today's Khoikhoi) was his lack of conscious disparagement, despite a widespread derogatory literature which depicted them as brigands and butchers.

Just how derogatory was illustrated by the Cape expert, Edinburgh-trained Dr Andrew Smith. He was army surgeon to Robert Knox's old 72nd Regiment, and as knowledgeable as Knox on ethnology. (Smith would become one of Darwin's prime sources as he began his evolutionary explorations.) To him the 'Bushmen' were notorious for their 'universally outrageous conduct'. They were cruel, 'thieves', 'deeply versed in deceit, and treacherous in the extreme', their aggressive tendencies manifesting in the 'depredations and murders they had committed on the colonists'.[65] Smith's list of atrocities was compiled using a typical civilizer's yardstick, with notches cut according to the 'savage's' observation of Cape law, missionary morality, obedience and servility. The Boers had already tried to annihilate these peoples, and many still welcomed their demise; but not before a number of Hottentot 'Venus' corpses had been taxidermically stuffed, to illustrate the 'peculiarities attributed to these nymphs by travellers'.[66] One such grotesquerie could be seen in Cape Town as Darwin passed through, a reminder of the barbarian underbelly of civilized morality.

Darwin showed no such disparaging attitude to the Hottentots; quite the reverse. He judged them as the 'ill treated' party. The Boers had considered these 'savages' patently inferior and resented the English for releasing them from slavery. Hence Darwin's note: 'Dutch hospitable but not like the English, Emancipation not popular to any people . . .' 'Hottentot' was itself a derogatory Dutch word meaning 'stutterer', on account of their curious clicking language. This language and their diminutive stature would have piqued Darwin's interest. In Cape Town in June 1836, he hired 'a young Hottentot groom as a guide' and set off to explore. For four days the two were on the road together, riding through the desolate scrub in the mountains to the east. They hardly saw a soul and were thrown into each other's company. Again, Darwin's experience was one-to-one, as it had been with the Fuegians and his 'blackamoor' at Edinburgh. His guide spoke perfect English '& was most tidily drest; he wore a long coat, beaver hat, & white gloves!'[67] The perfect diminutive gentleman – and that was the point. Darwin's encounter was with what he would call a 'tame' or zoologically 'domesticated' Hottentot: domesticated not in the pejorative sense, but like white men themselves. It demonstrated again the pliancy of the human constitution in any of its forms.

Yet these small pale people, 'like partially bleached negroes', were one of the most vilified races. It was nothing to hear of them 'at the very nadir of human degeneration', closing the gap with chimpanzees. They had 'singularly formed heads & faces', Darwin observed. As such, phrenologists would place these 'miserable representatives of humanity' halfway between men and monkeys, and even Knox's Edinburgh friend Morton thought they were 'the nearest approximation to the lower animals'. Darwin stuck by his belief in the unity of the human races and attributed to his impeccably attired companion the same aptitude for civilization as he would a white man. Darwin might have treated such people as he would the 'lower orders' at home, but these personal contacts could only have firmed his resolve against an increasingly vicious craniology. As a sympathetic writer said, 'When we know the Hottentot better, we shall despise him less'.[68] Darwin, it seems, did.

In Cape Town he met Andrew Smith himself, just back from an expedition to the tropical Limpopo River region. It had been a typical trading-and-Christianizing trek, taking twenty Hottentot scouts and trackers. Commodity brokers were continually searching out new raw materials, and Smith had been financed by merchants to get a more accurate knowledge of the tribes, geography and 'natural productions' upcountry. Smith had returned armed with enough information for a four-volume zoology book. One can see why he would become Darwin's prime source on the Cape tribes and fauna. 'Pleasant', Darwin considered him, and the expedition 'most interesting'. The two men took 'long geological rambles', and Darwin visited Smith's museum. What did he see? Everyone was drawn to the trophies Smith had brought back. But a more sinister exhibit was also a 'must', at least for the troops – the stuffed naked woman. The 'Hottentot Venus' was described as 'Smith's', so it was probably in the museum.[69] What would it have said to Darwin? Black people and animals treated with the same barbarism, fit to be stuffed?

Another who had seen the grim exhibit was the astronomer Sir John Herschel, son of William Herschel, the discoverer of Uranus. Sir John was at the Cape to observe Halley's comet and map the southern skies. Darwin was in awe of the man and even before his four-day trip had introduced himself.[70] Herschel was a Whig after the Darwins' heart, all for progressive governance running with mechanical efficiency, like the cosmos itself. This meant aborigines pursuing what Herschel considered more 'rational', controllable, agricultural lifestyles. It was important in this time of heightened tensions after the Kaffir (Xhosa) war to get race relations stabilized. Like many English liberals, he saw agriculture as a civilizing activity that could bring native wage earners into the economic calculus. This put him, like Darwin, on the side of the mission liberals – in fact on the side of the most liberal in the mission camp: the Scottish General Secretary of the (Dissenters') London Missionary Society, the Revd John Philip, 'a goodly portly man', in Herschel's words. Herschel backed Philip and his 'compassionate approach' to race relations, even if the 'Boers hate him cordially' for promoting an 'ungodly equality' between the races.[71]

Missionaries were at the centre of colonial English society, although Darwin too noted the hostility they encountered. The FitzRoy–Darwin article began by deploring this misplaced 'feeling against the Missionaries' here, as in New Zealand. The ill will was hardly surprising. They functioned as civil authorities in tribal regions, negotiating disputes, and as honest brokers with the colonial government (or not quite so honest if the Boers were to be believed). But Philip's beliefs were extreme, and paralleled Darwin's. Philip championed the fundamental equality of all humans. All were capable of salvation, all had the same aptitude: the distinctions were more a matter of social deprivation than innate disability.[72] He also thought (as Darwin had in Patagonia) that the natives were often enslaved by men their moral inferiors.

Philip had been instrumental in gaining the free Hottentots greater liberties, notably in getting rid of the pass system which restricted their movement. And he had brought the issue up in London before the anti-slavery groups. This was known to the Wedgwoods, who naturally owned a copy of Philip's propagandizing *Researches in South Africa*. Then came slave emancipation in the Cape in December 1834, which had infuriated many colonists, as Darwin well knew: the freeing of 36,000 slaves had led to labour shortages for the farmers, while the lifting of travel restrictions had encouraged population displacements and inevitable conflict. The new governor, Sir Benjamin D'Urban, had moved into and annexed the Xhosa lands after the Kaffir War to produce a buffer zone against these 'treacherous and irreclaimable savages'. An angry Philip had appealed to the Colonial Secretary in London, Lord Glenelg of Clapham (Charles Grant, Fanny and Hensleigh Wedgwood's neighbour), who had forced D'Urban – six months before Darwin arrived – to pull back his frontier line. The Boers' loss of land and slaves – and 'the intolerable doctrine that heathen blacks and Christian whites should be treated on a footing of equality'[73] – was the breaking point that finally triggered their 'Great Trek' of 1836–7. Some 7,000 Dutch settlers crossed the Orange River, out of the colony, to found what would become the 'Orange Free State'. One can sense why the evangelical philanthropists were so 'cordially' hated in expansionist and commercial quarters.

In the Cape the philanthropists were opposed by racial phrenologists such as Smith who denigrated blacks. The *South African Quarterly Journal* (with Smith as editor) used racial phrenology to support 'pessimistic portraits of the African character'. Phrenologists were already shipping Hottentot heads to Edinburgh colleagues by the 1820s (the accompanying notes could be chilling: 'shot by some Boors . . . in the act of stealing cattle – and being left till the vultures and hyenas had picked his bones' clean). These racial phrenologists urged an end to conciliatory policies and a return to land seizures. Darwin was being pushed further away from a science which fixed and separated racial temperaments, with its disparaging views of blacks. His heritage and Hottentot experiences put him on the mission side, with Herschel, who refused to see blacks as innately inferior or to condone the theft of their lands.[74] Herschel couldn't even bear to sit down at the same table with D'Urban. So the *Beagle* naturalist found his idol *in person* happily siding with the angels.

When Darwin dined with Herschel, they possibly discussed more than

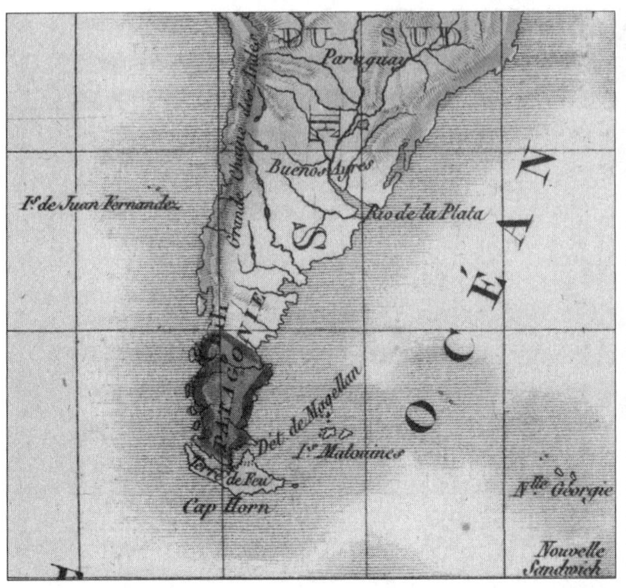

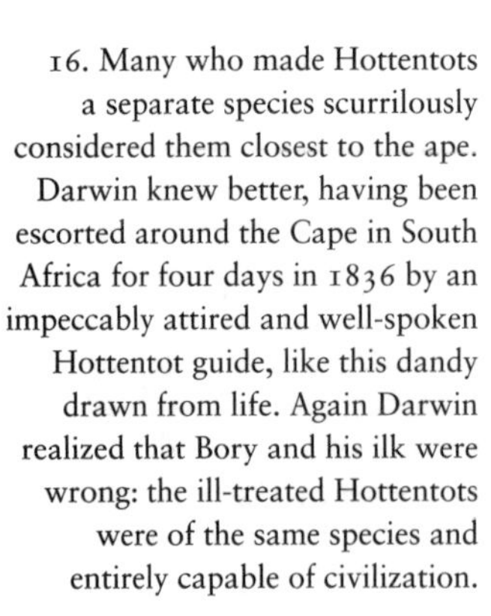

D'Urban's drubbing or Hottentot exploitation. They might even have mooted Lyell's *Principles of Geology*. Herschel, looking to a rationally efficient mechanics of creation, thought that, along with slowly changing landscapes, Lyell should have tackled the simultaneously changing species. The cause of life's fossil succession was to Herschel the 'mystery of mysteries'. It was the great problem for the future. The astronomer's rational cosmos sat ill with the seemingly serendipitous appearance of species, a sort of Creative chaos, one dropped here, another there. If landforms were changed by climatic forces like those today, shouldn't species be considered as produced by the physiological forces we know today? He was not presciently promoting what would later be called 'evolution'. He did not have a bestial transmutation in mind – that was only one of a number of solutions to the problem.[75] But he was attempting to lift the injunction on studies of the potential causes of the succession of species.

On the high seas again Darwin and FitzRoy finished their article defending the missionaries, and the accompanying letter to Herschel at the Cape was dated 'at sea, 28 June 1836'. The manuscript was timely, and it demanded safe-keeping. Two copies were sent off in separate ships, in case of loss. Herschel dutifully placed the article in the *South African Christian Recorder*, a new missionary monthly edited by Philip, and sent copies of the newspaper back to FitzRoy and Darwin.[76] It was Darwin's first published article.

Sailing north across the Atlantic, they landed in Brazil one last time in August 1836 to enable the fastidious FitzRoy to recheck his longitude readings. In Bahia the tropical luxuriance re-enchanted Darwin, but for all of the jungle's 'glories', this paradise still had its rotten human core.

Slave trading was still actively pursued, perhaps even more so because (as the British Consul put it) of the connivance and 'shameless venality of the lower magistracy'. Out of the mouldy core had burst the maggots. A slave revolution had erupted in January 1835, leaving signs everywhere: bizarrely, even the 'finest mango trees' that Darwin remembered from 1832 had gone, destroyed during the riots. It had been the continent's 'most effective urban slave rebellion ever' and showed at least that the blacks possessed military skills.[77] African-born Muslim slaves, armed with swords and Koranic verses, had attacked Bahia's capital San Salvador. Soldiers and civilians had shot and killed some seventy of them, while 500 were sentenced to prison, whipping, deportation or death. It spurred a renewed slavery debate in the Rio parliament. How much Darwin took in during five days in the town is unclear, but news of the uprising surely reminded him of the Haiti-style revolution that he had predicted for this slave country.

Of all the sights and sounds of five years' circumnavigation, one would haunt him. Pernambuco was the *Beagle*'s last port of call on the continent. It was mid-August, the rainy season was ending. The town was 'filthy' and 'disgusting'; the houses stood 'tall & gloomy' in the stench. The minority Europeans looked like 'foreigners' in a sea of 'black or . . . dusky colour' skins. Months earlier, 300 more Africans had been landed south of town and marched to the plantations. And shiploads of frightened, chained men and women were even now on the high seas. Planters were grouping to protect their 'barbarous contraband', leaving the British Consul pleading for more naval patrols.

Somewhere in this repellent part of Brazil, probably in the old town of Olinda, which he reached by paddling through 'putrid exhalations' from mangrove swamps, Darwin heard a scream when passing a house. Abuse he had witnessed; he had seen the instruments of torture – shackles, thumbscrews, whips – but he had surprisingly never witnessed pain deliberately inflicted. Behind walls, invisible, unreachable but horribly real, came 'the most pitiable moans' and shrieks. Sensitive, squeamish, Darwin was tortured himself. There could be no rushing to the police, as at Cambridge; no fleeing to the street, as at Edinburgh. There was no one else to help, yet to intervene was impossible. Anger and frustration left him feeling 'as powerless as a child' – powerless to help himself or aid an innocent; perhaps, like the child in Edinburgh under the knife, strapped down, screaming. The emotion would haunt him for life. A 'distant scream'[78] would always bring back memories of that tortured slave.

It was a grim note to end an extraordinary journey. Cloistered Cambridge might have given Darwin his moral anchorage, but he had left England rudderless. He might have imbibed his sisters' anti-slavery ethos, devoured Shrewsbury and Maer's tracts, but their dry teachings were so much print pathos, to be stowed as ethical baggage. For the young Darwin to feel as well as understand took tropical pain, real shackled legs and sweat-soaked backs broken in cane fields. Throughout the voyage, Darwin had never ceased tacking into slavery's wind, sailing against an evil South American gale. He was returning bronzed, salted, and born again.

The island and continent hopping had left him world-weary and wiser. Never in five years had he been able to escape slavery, now seen for what it was, a global empire of evil requiring a global remedy. Never would he be inured to those wretched slaving vessels anchored, 'as commonly is the case', in each port. Yet how proud he was of knowing that stranded 'negroes liberated from slave ships', like those at the military outpost on Ascension Island, were 'paid and victualled' by His Majesty's government. How often, too, Darwin must have seen amiable John Edmonstone, his ex-slave bird-stuffer, in these oppressed peoples. Every cured skin sent home must have reminded him, and every lined black face.

Thus it was to the end, in July 1836, on the South Atlantic rock called St Helena where Napoleon had been banished and buried. Here was another 'curious little world within itself', littered with imported species and human flotsam from four continents. The fortress was an allegory of his personal voyage, even of his black experience generally. Here was John again, many Johns: stoic, good-natured, interesting. Among St Helena's stranded 'lower orders' was a poor community of 'emancipated slaves . . . blessed with freedom . . . which I believe they fully value'. It was valued as much on a windy prison-island as in frosty Edinburgh. Darwin's guide was an elderly man who had mastered the jagged volcanic terrain as a youthful goatherd. He was 'of a race many times mixed', his skin 'dusky' but pale, his face pleasant. On their long daily walks, Darwin marvelled at how this 'very civil, quiet old man' would talk with equanimity 'of the times when he was a slave'.[79]

After the voyage, FitzRoy and Darwin, captain and companion, would go their separate ways in every sense. FitzRoy's theology would harden and become more literalist. In a few years he would claim that Noah's sons, Shem, Ham and Japheth, had fathered the white, black and red races respectively. Since the Flood their offspring had produced 'every variety' of complexion and hair by intermarriage. He would even accept that God's curse on Ham's 'Negro' son Canaan, making him a 'servant of servants', was the origin of black bondage. Such views grew increasingly common as demands peaked for the abolition of American slavery. By 1838 religious tracts were already depicting the whole of Africa filled with Ham's cursed descendants. In the Southern Slave States, particularly, God's favours and curses were seen as very unequal, and Noah's descendants were ranked accordingly.[80] Pro-slavery ideology was served as well by the curse of Ham as by a pluralist craniology. But Darwin's views would tend in the opposite direction. He would take the 'common ancestry' approach which he and FitzRoy shared on the *Beagle* to its ultimate conclusion.

Oppressed people claimed his concern at the end of the voyage as at the start. Ever-present was his desire to alleviate the suffering, abolish the evil, undermine the institution – to share in 'the glory of having exerted himself' in undercutting slavery.

He had not been able to rush into a stranger's house to end the torture. But there were subtler ways to intervene. For five years he had endured seasickness and loneliness; now he knew the reward would come as he put his experiences to use.

> No doubt it is a high satisfaction to behold various countries, and the many races of Mankind, but the pleasures gained at the time do not counterbalance the evils. It is necessary to look forward to a harvest, however distant it may be, when some fruit will be reaped, some good effected.

Darwin had already shipped crates of specimens back to Henslow. And five more boxes on board were ready to go. Among the myriad rocks, bones, skins, plants and fossils, there was not a single human skull.[81]

The young Darwin would have no truck with craniology, no sympathy with its emphasis on the separateness and ranking of races. No skull collecting would mark his science. He would find a very different way of approaching black and white, slave and free.

5

Common Descent: From the Father of Man to the Father of All Mammals

If the well-salted supernumerary wasn't leaning towards evolution when he stepped ashore, the landlubber was fully committed within months. But Darwin's image of a transforming nature was highly peculiar. Indeed it was probably unique.

His Edinburgh teacher, the radical Robert Grant, like Lamarck's disciple, the anti-Church Bory, envisaged nature's transformation in what today seems an alien way. Humans had monkey ancestors, and through them a bloodline all the way back to fishes, true enough. But it was, by and large, a single straight line. For these evolutionists, today's monkeys do not share in our ancestry. There was no forking of our line to produce them (or vice versa). They had their *own* parallel pedigree all the way back. Their lineage has so far only risen up as far as monkeys, but these living monkeys will be tomorrow's humans. Bizarrely, for evolutionists before Darwin nature was composed of so many parallel lines, all progressing upwards and passing through the same stages: some had got as far as fish, some monkeys, and one all the way to humans.[1] Bory even split up the human races and gave *each* its own bloodline back to some spontaneously generated germ in ancient times. So black people, with their own chimp ancestry, had yet to step up to the top white rung on their ladder. Whites had already passed through the chimp and black stages, and had reached the apotheosis. This is the crux: *living black and white peoples were unrelated*, they shared no *common* ancestor.

Darwin's image was completely different. Maer's anti-slavery axis and Cambridge's religious anchorage had committed him to the human races as a brotherhood. Like siblings they shared a common parentage: the races were united by blood. The metaphor Darwin visualized near the beginning of his evolutionary journey was of a genealogical 'tree': many branches meeting in the past in a joint ancestor.

At the outset in his 'evolution notebooks' – a series of small leather-bound pads in which, from 1837, he jotted thoughts on the origin and affiliation of life – Darwin tried to visualize this human pedigree extended to all animals and plants. His notes were telegraphic, breathless, bitty and often cryptic, with misspellings common, but he kept coming back to the human analogy:

> My theory [he wrote in May 1838] explains that *family* likeness [in animals], which as in absolute human family is undescribable, yet holds *good*, so does it in real classification [. . .] I cannot help thinking good analogy might be traced between relationship of all men now living & the classification of animals —[.][2]

So Darwin's evolutionary image was of a *converging* bloodline. His heresy was to extend human racial relationships to all the branches of creation, and to push the trunk deep into the geological soil. Hardly had he begun speculating than he called the couple of Jurassic shrew-sized mammals known from Oxford's slate the greatest-grandparents of all the diverse living families, from humans to hedgehogs, bats to Bactrian camels:

> we have one Marsupial animal [ran one of his earliest notes, in summer 1837] . . . the father of all Mammalia in ages long gone past. . & still more so known with fishes & reptiles.—[3]

Those last words mean that we could push further back in vertebrate ancestry still, to join all the reptiles and mammals by searching out their common fish ancestor.

Darwin had taken his commitment of racial 'common descent' to the ultimate extreme. But always he was looking backwards from the human races, always looking to fathers with their common linking features, then to forefathers:

> The *races* of men differ chiefly in colour, form of head & features (hence intellect?) & what kinds of intellect) quantity & kind of hair forms of legs— hence the father of man kind probably possessed a structure in these points[.]

And from this 'father of man' he quickly jumped to a deeper ancestry before coming up again:

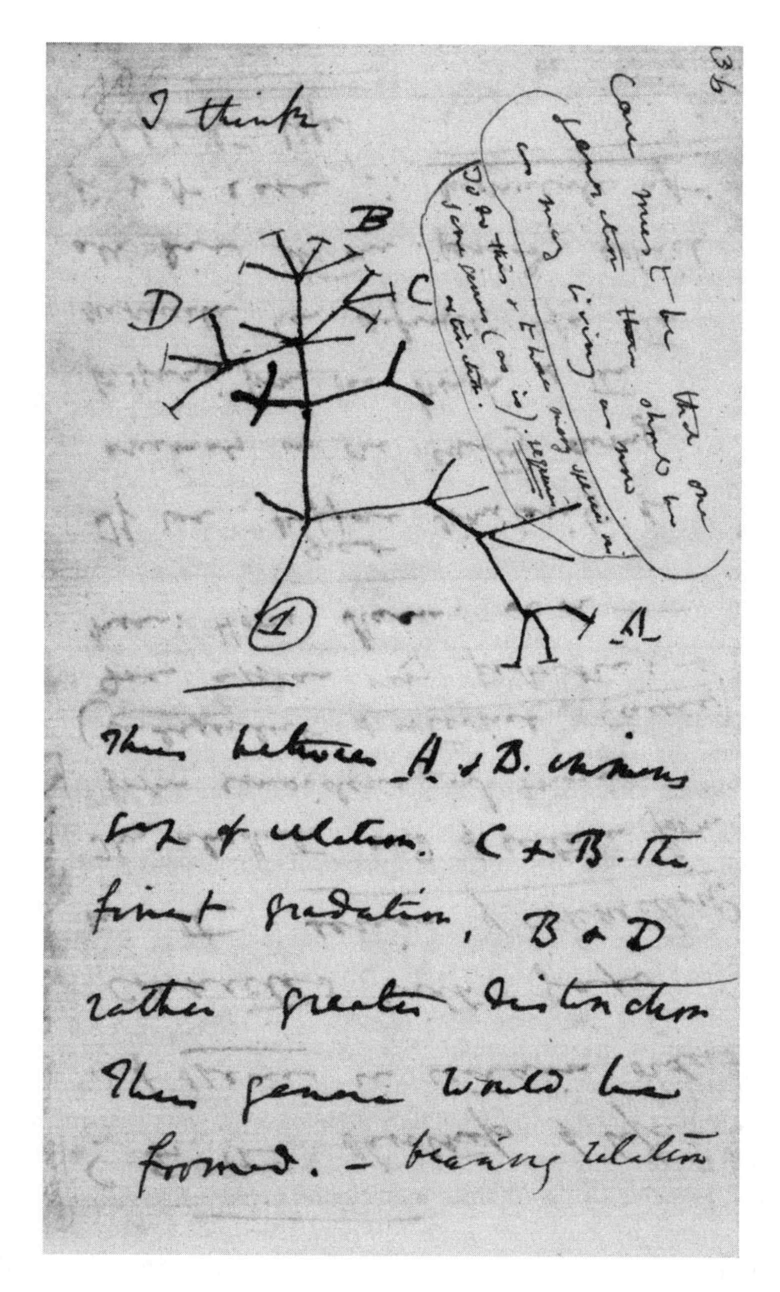

Given his heritage of anti-slavery and human brotherhood, and his shock at seeing slavery in the raw during the *Beagle* voyage, it is no surprise that Darwin, on arriving home, used the human genealogical image to model the 'common descent' of all life. This diagram in his first evolution notebook, *c.* July 1837, shows the pedigree concept firmly lodged in his mind.

Now we might expect that animal half way between man & monkey, would have differed in hair colour [. . .] form of head & features;. but likewise in length of extremities, how are races in This respect [. . .] (Negro or father of negro probably was first black at base of nails & over white of eyes,—[.][4]

These snippets of thought from late May 1838 show us tantalizing glimpses into Darwin's route to 'common descent', a route he might have begun taking during the last stages of the voyage. They also illustrate Darwin's abolitionist debt.

Two savages, [are so far apart that they resemble] two species [. . .] civilized Man, May exclaim with Christian we are all Brothers in spirit— all children of one father.— yet differences carried a long way.[5]

The races are related yet different, so the question for the future was, how did this racial difference come about? Solve that, and the whole evolutionary problem could be cracked.

The speed with which Darwin adopted evolution hints at a gestation on the voyage. Those shocking events around the world – the brutality, black slavery, segregation and degradation – tested his moral resilience on racial unity. The sensitive humanitarian had plainly been upset, and the barbarity of slavery brought his barely visible abolitionism into sudden sharp relief. It demanded a new commitment, the sort he had been unable to give till now.

His encounters with Fuegians in cravats, and Hottentots in white gloves, had proved human cultural adaptability as no anti-slavery tract could. Nothing could have better pointed up the pliancy of race. It justified the abolitionist faith in blacks being able to pull themselves up by their bootstraps. The impeccably mannered Hottentot was living proof of the evil of considering 'wild' humans grovelling beasts. These racial encounters sharpened Darwin's sense of injustice – the tortured slaves, genocide of the Pampas Indians, the dragnet round-ups of the Tasmanians, the stuffing of Hottentot skins. The result was an astonishing outpouring early in his notebooks which reveals the anguish behind his evolutionary venture.

Here is his remarkable breathless dash, jotted in February 1838 – a stream of consciousness that stretched from the aggrieved races through the denigration of all animals, to its elegant evolutionary solution:

Animals— whom we have made our slaves we do not like to consider our equals.— Do not slave holders wish to make the black man other kind? Animals with affections, imitation, fear. pain. sorrow for the dead.— respect

---

We have no more reason to expect [to be able to find] the father of man kind. than [to find the father of the extinct South American llama-like] Macrauchenia yet he may be found [. . .] if we choose to let conjecture run wild then animals our fellow brethren in pain, disease death & suffering & famine; our slaves in the most laborious work, our companion in our amusements, they may partake, from our origin in one common ancestor we may be all netted together.—[6]

Darwin faced the Brazilian prejudice as his grandfather had tackled the West Indian one: slave-masters relegating slaves to beasts of burden. Of course slavers were making black men an 'other kind', one that was 'brutish, ignorant, idle, crafty, treacherous, bloody, thievish, mistrustful, and superstitious': sable-skinned and woolly-haired 'like the bestial fleece', itself infested by peculiar black lice.[7] Darwin was fighting the defamation at root.

But animals too were being degraded, for they 'know the crys of pain, as well as we'. The thrust of Darwin's work on evolution, which 'netted together' all suffering creation, human and animal, was now set against the hauteur of Cambridge's clerical professors with their 'godlike' image of man. In his inscrutable (and very private) way, Darwin summoned all his *Beagle* anger at the denigration he had witnessed:

Has not the white Man, who has debased his Nature & violates every best instinctive feeling by making slave of his fellow black, often wished to consider him as other animal— it is the way of mankind. & I believe those who soar above Such prejudices, yet have justly excalted nature of man. like to think his origin godlike, at least every nation has. done so as. yet.—[8]

After all he had seen, Darwin loathed the cosmic arrogance that could lead to these views. He castigated the devout dons, who thought themselves above prejudice against blacks, yet who separated 'godlike' humans from a bestial nature – the great Revd William Whewell, for instance, who despised slavery but whose whole philosophy depended on humans having godlike ideas rooted in an immortal soul.

These were harsh, levelling criticisms, written in a political climate of harsh, levelling violence and intellectual attacks on the arrogant power of kings, priests and aristocrats. The years 1838–42 were disturbing and destabilizing for Britain, as demands for the recognition of lower-class political rights (which had been ignored in the reforms of the early 1830s) peaked with riots and uprisings.[9]

Darwin was jotting his speculations, too, as the slaves were ending their required 'apprenticeship' period. (Although technically slavery had been outlawed in 1833–4, the slaves had been forced into 'apprenticeships' – ostensibly to prepare them for freedom, but the reality was a further four to six years of forced labour on the colonial plantations.) And this reminds us that there was another side to Cambridge that he *was* endorsing. He stood ever more resolutely on the moral foundation provided by his family and his Cambridge teachers. He never lost faith in the brotherhood of races. That humans *were* divided into races he barely questioned. But the time was coming to investigate why they had separated and how they had formed. In his first evolution notebook he jotted 'man has not had time to form good species', meaning humans had not been around long enough for the races to diverge into distinct populations that cannot cross-breed. And everything suggested the 'probability of [humans] starting from one point'. The babble of tongues – and he had heard enough in five years to understand why the Tower of Babel drove humans apart – offered a proof that the races hadn't been separately created by God. Had they all appeared at once, 'created as now', languages 'would surely have been more homogenious', enabling them to communicate.[10] Evolution suggested that as tribes diverged, so had their tongues.

At the end of these initial months ashore, Darwin – like a good Edinburgh materialist – had extended the racial common bloodline to such heretical lengths that caution and some secrecy was needed for the next twenty years.

'Why, the shape of his head is quite altered'! Dr Darwin exclaimed as his son walked into The Mount on 5 October 1836. It was his first sight of Charles after the voyage, and the young man had shed so many pounds that he looked top-heavy. Mental weight he had also put on. They could tell from his journal; this had arrived in instalments to be read aloud after dinner and then sent on to the Wedgwoods at Maer.[11]

Unlike FitzRoy's Fuegians, whose faces had changed under civilization, Darwin's hadn't degenerated in the wilds. The phrenological remark was a joke, perhaps like the huge engraving of George IV that Darwin was startled to find gracing his father's sitting room, 'for the honor & glory of the family', he laughed to FitzRoy afterwards. The family's liberal conscience, Darwin's pole star during the voyage, still shone bright, and at home his 'radical sisters' helped him catch up. They talked of the 'great deed' of West Indian emancipation, even if the slaves had to endure apprenticeships. And like other 'immediatists', the sisters were outraged by the '20 million [pound] compensation' paid to slave-owners. Still, they rejoiced at Darwin's admiration for the blacks and blanched at the 'murderous war' against the Argentine Indians, shocked that 'anything so wicked ... as the conduct of General Rosas' could happen 'at the present day'. They had been dying to hear more since the last portion of his journal had arrived from South America. Now he could regale them in person, about the Antipodes, Africa and beyond, with missionaries to the fore. His and FitzRoy's article hadn't reached England, but Darwin's summary had the sisters in 'an uproar' over the 'shabby' government dealings and Xhosa policy at the Cape. At least the Whigs were back in office in Britain after an 'odious' Tory interlude, and Darwin assured the 'sisterhood' that his five years cooped up with FitzRoy hadn't dented his principles. They were 'as firmly fixed and as wisely founded as ever'.[12]

Maer expected him, so a weekend was devoted to Uncle Jos and the Wedgwood cousins. Jos, 'heartily tired' of Parliamentary routine and the commute, had not stood for re-election. It was enough that he had voted for emancipation; his political mission had ended as his father's had, in his sixty-fifth year. The Anti-Slavery Society in Hanley had continued to issue declarations and addresses.[13] When their petitions went up in smoke as fire gutted the Houses of Parliament in October 1834, it was like an Old Testament burnt-offering for the nation's sin.

At Maer the globe-trotter was the centre of attention, the men extracting geological lectures, while Bessy and the sisters hoped that '*the lion* Charles' (as she called him) would humour Aunt Sarah. Since emancipation, Sarah had opened her ample purse to the new American Anti-Slavery Society. This seemed in a 'more hopeless state' than its British counterpart had been at the start. In America there were worse cases of cruelty to confront and stronger 'prejudices against the coloured population'. Everyone knew what Sarah wanted to talk about. Usually she exasperated guests; they were pinned down, made to feel 'tied to a stake' as she compelled them to talk. But Charles was full of 'just the subjects' that pleased her; so absorbed was she that when the carriage arrived to take her home, she sent it away. Maer got its American news through the anti-slavery press, and the latest had shocked everyone: a blow-by-blow account of 'the trial & execution of 10 men, 5 black & 5 white by Lynch law'. Apparently 'the negroes were examined by flogging before they were hanged'![14] With emotions running high, Darwin undoubtedly described comparable horrors from around the world.

Cambridge turned out to be 'not half so merry' as before. Old friends had gone, Christ's College was full of ghosts. Indeed, while Darwin was at Maer, a pall fell upon the university with the death of Charles Simeon, who for half a century had put the spiritual force into anti-slavery. Town and gown thronged the funeral in King's College Chapel and Darwin's old professors were overcome by emotion. That week a lioness in a visiting menagerie gave birth to cubs, which an undergraduate christened 'Whewell', 'Sedgwick' and 'Simeon', showing who ruled Cambridge hearts and minds. Darwin could tell: even among the 'young lordlings', whose high life he resented, many more now went to church on Sunday. Henslow's family felt like 'nearest relations', but Darwin turned down their offer of a room and decamped to digs for a few months to sort his *Beagle* specimens and revise his journal for publication. He dined out on stories about Tierra del Fuego, probably twitting the dons about meeting that 'miserable' being, a 'wild man'. But Cambridge was a far cry from seafaring, and the Philosopher soon realized that his science thrived on hurly-burly. There was no place for natural history like London, even if there was no nature in the 'odious dirty smokey' capital.[15] London was the imperium of science.

In March 1837 Darwin took lodgings next door to his brother Erasmus's townhouse in Great Marlborough Street. Here he unloaded his crates of rocks, skins and pickled specimens. The family fortune enabled Darwin to escape all financial fears. He had no need to work at all (Erasmus never did). But toil he would for life, beginning now, at his science. The Geological Society was a stroll away, as was the stuffed Piccadilly museum belonging to London Zoo. So too the museum of the College of Surgeons, just rebuilt to lavish proportions with gigantic South American armadillos and ground sloths as their star attraction.

All were pivotal points on Darwin's intellectual landscape, and staffed by the experts who now received his *Beagle* haul.

Darwin's name preceded him. Henslow had already shown off his letters and specimens, but no one recognized Darwin's rising star more than the cultured Charles Lyell, whose *Principles of Geology* had been so influential. They became close. A sort of urban Henslow, Lyell took over as Darwin's patron. He entered 'heart & soul' into Darwin's career and proved to be a religious kindred spirit. As a Whig and (at this stage still) a liberal Anglican, he shared the Darwin–Wedgwoods' faith in progressive freedom: for him emancipating lives, minds and the marketplace would achieve the greatest good. (In this respect, Henslow's evangelical humanitarianism was more circumscribed.) Wealthy, from a landed family, Lyell had a superior air, and in *Principles* had acknowledged 'some connexion between an elevated and capacious forehead in certain races and a large development of the intellectual faculties'. He was status-conscious, in science and out, but despite assuming the vast superiority of the whites he still believed in one human family and called all men brothers. He could also admit his own faults. Hence, with 'a pang', he gave up his theory that coral islands had formed on the tops of rising volcanoes.[16] His new protégé had convinced him of the reverse: that reefs formed as rings around submerging mountains. The student was revising the teacher's textbook. It would be a lifelong trend.

Lyell had an eager Darwin elected to his own club, the exclusive Athenaeum, that 'microcosm' of 'ruling-class egalitarianism'. Here he mixed with the greats in science and literature. He might have been elected as one of the 'forty thieves' (that is, without ballot), but the club made him feel 'like a Lord', while dinner at two shillings and ninepence didn't force him to sell the family silver. Then, with Lyell's help again, came his fellowship of the Geological Society, set up in grand public premises in Somerset House on the Strand. In its way this was an even more exclusive club: the Geological kept reporters out, fearing its discussions involving deep time could offend religious sensibilities. Success in the capital called for self-discipline, and Lyell's tips were good. 'Don't accept any official scientific place' – too time-consuming; take a leaf out of Herschel's book: he was productive at the Cape, rather than 'President of the Royal Society, which he so narrowly escaped being'. Darwin was badgered into becoming secretary of the Geological in 1838, but otherwise followed Lyell's advice.

The two clearly sparked off each other. A chance remark of Darwin's would elicit Lyell's considered response, which 'often made me see more clearly', Darwin recalled. Lyell built up Darwin's confidence and launched him into an unequalled spate of writing. In two years Darwin had published a half-dozen major scientific papers, had seen his *Beagle* journal into press, landed a £1,000 government grant to describe his zoological specimens, and half finished a book on South American geology. For his efforts, a couple of years later, on 24 January 1839, shortly before his thirtieth birthday, he was elected Fellow of the Royal Society.[17]

Through that period, unbeknown to Lyell or any scientific colleague, Darwin had been surreptitiously filling his evolution notebooks. These provided a record of his extraordinary intellectual feat. The mental turmoil he suffered is barely graspable, and on reading his spidery scrawl even now one is staggered at his daring. 'Once grant that species [. . .] may pass into each other [. . .] & whole fabric totters & falls'. His demolition work was knocking down some imposing Creative edifices. It took gall to turn the old world to rubble. A young man's bravado fired these jottings, as he ran full-tilt against the combined might of his Cambridge dons, indeed the conservative world: 'the fabric falls! But Man— — wonderful Man [. . .] he is not a deity, his end under present form will come [. . .] he is no exception—'.

As Darwin effectively bestialized man (in the best sense) he also began to recoil inwards, to start his withdrawal from a potentially hostile society, to protect the thing most cherished by a gentleman, his character. He knew he was treading on hot coals. Just look at his burning conclusions: 'love of the deity [is the] effect of [the brain's] organization. oh you Materialist!— Read Barclay on organization!!' That reminder of official Edinburgh's censorious attitude to such heresies said it all: Barclay was the extra-mural anatomy teacher who had damned all talk of atoms being self-powered or the brain's structure being the sole explanation of thought. Yet Darwin persisted, mimicking the provocative jibes that radical reductionists were being vilified for. 'Why is thought. being a secretion of brain, more wonderful than gravity a property of matter? It is our arrogance, it [is] our admiration of ourselves.' This was atheist street-slang, the patter of radical tracts levelling an overbearing Church and acquisitive aristocracy, and whether Darwin voiced such thoughts to the family is moot. He did record a note which suggests that one family member doubted this denial of any ghost in the machine: 'Hensleigh says to say. *Brain* per se thinks is nonsense; yet who will venture to say germ within egg, cannot think' – but in what context Hensleigh said this we don't know.

Anyway, these years saw the beginnings of Darwin's nervy, stomach-churning bodily reaction. Illness was to plague him until he put the human question aside again, thirty-five years later.[18] He was now on a much longer voyage, and sailing into very dangerous waters.

After years on the *Beagle* his habit of quiet note-taking had become second nature. The fifteen pocketbooks he filled during the voyage contained the raw material of his journal. Now the new series on evolution worked through more heretical themes. In his first notebook, on pages 3 and 4, Darwin looked at the one issue that had been critical to his understanding of human potential: domestication. How did wild humans become and *stay* civilized? His Fuegians had reverted. Perhaps you could not teach old dogs new tricks. Perhaps the youngsters were more malleable; or, as he jotted critically,

> the young of living beings, become permanently changed or subject to variety, according to circumstance [. . .] child of savage [might move the tribe towards civilization] not civilized man [who, like the Fuegians, reverted to savagery] [. . .] There may be unknown difficulty with *full grown* individual with fixed organization thus being modified,— therefore generation to adapt & alter the race to *changing* world.—

In short, perhaps the children of 'savages' were more culturally malleable. Or the adults, civilized and thus altered by changing conditions, produced children already acculturated, because somehow the changes had affected the adults' reproductive system, enabling them to pass on the changes. Thus at the outset Darwin was testing his understanding of domestication to explain how humans could be altered to keep pace with conditions. It was as if civilization was the goal, to be built on the fine 'instincts of wisdom [and] virtue', and that an aptitude for it had already become hereditary in the white races.[19]

At this early stage, Darwin also started questioning the notion of 'progress', that staple of the age. He asked himself: 'Each species changes. does it progress. Man gains ideas.' The human parallel here was with the cumulative knowledge produced by civilization

('Man gains ideas'). But if that was the yardstick, then in what sense did animals and plants 'progress'? Was it built into the evolutionary system? Again and again he toyed with the knotty problem of evolutionary improvement, but quickly jettisoned the 'high' and 'low' ruler and substituted a more spreading, adaptational, everything-has-its-own-yardstick approach. Humans weren't the be-all and end-all:

> It is absurd to talk of one animal being higher than another.— *We* consider those, where the cerebral structure/intellectual faculties most developed, as highest.— A bee doubtless would when the instincts were.—[20]

This startlingly relativist admission flew in the face of conventional wisdom. But 'common descent' writ large had always implied a bushy, tree-like spread, as every family genealogist knew. Darwin had simply made the terminal twigs adapt to very different circumstances. The bees had gone their own way. The yardstick was how well they had adapted – not how closely they approached humans. Even the materialistic radical Bory had positioned mankind at the apex of creation, allowing white to look down on black, black on all those below, right down to the bees. Darwin was beginning to think the unthinkable, that there was no 'up' and 'down' at all. These kingly concepts did an injustice to life's adaptation to niches.

Another theme started immediately was racial interbreeding and its upshot. 'Dr. [Andrew] Smith says he is certain that when White Men & Hottentots or Negros cross at C. of Good. Hope the children cannot be made intermediate, the first children partake more of the mother, the later ones of the father'. This too would play and play as Darwin tried to sort out what really resulted from intermarriages and whether the resulting mixtures fed evolution or, quite the reverse, stopped racial divergence from occurring (because of traits back-blending). What obsessed him was the mechanism which kept the races apart and pushed them further away. 'No doubt', he wrote, 'wild men do not cross readily, distinctness of tribes in T. del Fuego. the existence of whiter tribes in centre of S. America shows this.' With races only incipient species, how far would 'black & white' have to depart before they 'keep to their type' and refuse to intermarry or have sterile offspring?[21] These were tantalizing glimpses; trains of thought for the future.

As he contemplated Africa, the missionary magazine containing his article finally arrived, having been posted by Herschel from the Cape.

Coincidentally, Darwin had just been discussing with sister Caroline a long reflective letter from Herschel to Lyell about his *Principles of Geology*, which Lyell had passed on. This broached the 'lapse of years since the first man made his wonderful appearance'. Going beyond Egyptian and Chinese records, Herschel saw it as a vast period, extending far into remote antiquity, when the languages 'separated from one stock'.[22]

Herschel's letter drew a further lesson from language. He called words 'battered relics of past ages', full of meaning, like rocks, those remnants of former worlds, to the geologist. Darwin was now viewing living *species* that way: as end products, carrying information about their ancestral route. They too had appeared *naturally*. Herschel too thought that 'the replacement of extinct species by others . . . would be found to be a natural in contradistinction to a miraculous process'. This succession of species, in time, through the rocks, or across space (for example the Patagonian plains), had haunted Darwin for months. He saw 'no more wonder in extinction of species than of [an] individual', so why shouldn't the replacement of species be as natural?

Darwin recalled missionary reports of 'strange contagious disorders' spread by healthy Europeans to 'natives of distant climes', killing off 'the Aborigines in both Americas – Cape of Good Hope – Australia & Polynesia'. Could these *individual* deaths lead to the disappearance of whole races and the emergence of new ones? He sent such a string of abstruse queries to his father via sister Caroline (can contagion be caused by puncturing the skin? could an experiment be set up to see whether puncturing the hide of a dead dog transfers some contagion to a living one?) that 'the Governor will think I am gone mad'. Something in the 'first mingling' of races seemed to prove fatal.[23] Maybe it was what wiped out the ancient megafauna of South America – those giant armadillos and ground sloths he had shipped home.

A short way into his first notebook and Darwin was still envisaging how the racial 'common descent' and extinction imagery reinforced the idea of a 'tree' of life. Go back to that father of all mammals, the little Jurassic insect-eater: very few of his descendants will be alive. Such were the vagaries of existence, survival was very chancy. Many branches of the tree would have been wiped out by accidents, extreme climate change, and so forth.

In same manner, if we take a man from. any large family of 12 brothers & sisters [. . .] it will be chances against any one of them having progeny living ten thousand years hence [. . .] so that by tracing back. the fathers would be reduced to small percentage.—[24]

Human and animal pedigrees were on a continuum. But humans provided the cosy family example: families rising and sinking; some dying out, others forming great dynasties. So the bushy tree of life, based on 'common descent', would have many dead ends. Human families could be wiped out because of

dislike to marriage, heredetary disease, effects of contagions & accidents [. . .] one man killing another.— So is it with *varying* races of man [. . .] the whole races act towards each other, and are acted on, just like the two fine families no doubt a different set of causes must act in the two case, May this not be extended to all animals [. . .]

The homely example was human, easily graspable (he was thinking as much of presentation, if he published), but the image extended from families to races to all animals. It was the same process. The more family branches wiped out, the greater the gap between the surviving groups. On the larger picture this explains the 'great gap between birds & mammalia'.[25] The intermediates had died out; their branches had terminated in the distant past.

Though Darwin could now explain these great gaps between, say, mammals and birds, when it came to humans and apes he was not averse to squeezing them closer. He knew this was where he would meet the stiffest resistence. After visiting London Zoo's orang-utan he sympathetically talked of its 'expressive whine' and 'intelligence when spoken [to]; as if it understood every word', and 'its affection'. He contrasted this with the cruel Fuegian stories of cannibalism and continued, 'look at savage, roasting his parent, naked, artless, not improving yet improvable'. Again, down went the gauntlet: how dare man 'boast of his proud preeminence'![26]

Humans were not the sole source of insights into transmutation, but part and parcel of Darwin's project. Yet racial denigration symbolized by slavery did provide an emotional powerhouse to drive him on.

When he returned from the voyage, there were probably as many blacks in London as Englishmen in Africa. Some too were making good, and better than Hottentot grooms. Best of all was that University College London could award a slave's son from Trinidad, John Carr, its inaugural prize in English Law in 1839 (just as Darwin was closing his main evolution notebooks) – and have no less a person than Lord Brougham present it. Blacks and Bengalis being treated equally was one fact of British life that visiting Americans found repugnant. Many were staggered that half-caste Bengalis were 'received into society and take the rank of their [white] fathers', and repelled to see 'in the streets of London genteel young ladies, born in England, walking with their half-brothers, or more commonly with their nephews, born in India'. Some admitted to embarrassment, because it threw America's own intolerance into relief, in the land of the supposedly 'free'. Attitudes were beginning to harden in Britain, too, but medicine still attracted many foreigners at University College (a stone's throw from Darwin's future house on Upper Gower Street). Here were graduating Bengali students. Robert Grant, Darwin's old dissident mentor, who had come down to take the Chair of Comparative Anatomy, awarded his gold medal to one in 1846.[27]

In an age of laissez-faire, many saw black bondage as 'the symbol of all the forces which thwarted individual liberty in their own society'. At the same time one anatomist, Friedrich Tiedemann from Heidelberg, when visiting Britain to study skulls, skeletons and racial artifacts in 1836, had disproved the 'dastardly allegation' advanced by defenders of the 'abominable traffic' in slaves, 'that the negro race is altogether a degraded and inferior section of the human family'. Tiedemann, like Darwin, was predisposed towards equality, and presumably in London to obtain a better hearing for his findings from British humanitarians. His statistical study proved 'that the brain of the Negro is as large as that of the European'. This caused a furore in the racial-phrenology camp, which was trading on a more debasing assumption. They labelled it 'prejudice'. Some criminals too had large brains, they said rather callously (but totally typically). And then they switched their target to prove that it was not overall brain size that counted anyway, but the development of the organs of 'intellect' and capacities for civilization.[28]

Since arriving in London, Darwin had continued to look for arcane proofs of the unity of the human races. He had hopes that head lice – like those he saw on Chiloé island, which he knew differed on white and black men – would make a 'good argument for origin of man one', but the lice would prove perennially troublesome. Still, racial unity was his starting point for explaining the common descent of all life using a pedigree approach. With humans, to understand the relations between cousins, you need to look to their common grandparent. So with 'pig & tapir'; it is no good looking for a living link, but for 'some common progenitor'. Then came the search for clues to how the human races had actually started. Darwin hunted down stories of hairy men, albinos and the afflicted being banished, and speculated that these could found new colonies (all straws were grasped in the early days). More likely, 'In first settling a country.— people very apt to be split up into many isolated races', when geography could put up physical barriers to back-blending.[29] If geography didn't, it was the races' own 'repugnance' to one another, something he had seen enough of around the world.

Fanny and Hensleigh Wedgwood continued to entertain in old Clapham style. Even if crusading evangelicalism was no longer headquartered here, Clapham's country air beckoned the faithful, the mansions beside the Common a reminder of glory days gone by. The parish church still rang with Wesley's hymns and the charities worked diligently, while the children, friends and in-laws of the old Saints kept Clapham a byword for moral reform and anti-slavery. But the momentum of anti-slavery now rested as much with Unitarians, scientific Whigs and radical free-thinkers, men and women who shared the Saints' moral seriousness if not their theology or wealth.[30]

Darwin's brother Erasmus drove from the West End to Clapham in his smart cabriolet. Erasmus, cultured and clever, was strangely languid, with a 'sarcastic' expression that could terrify, though it was probably caused by too much opium. Fanny Wedgwood indulged him, and he would linger at Clapham for weeks while Hensleigh was absorbed in compiling a dictionary. Erasmus's love for Fanny was an open secret, like his devotion to her children. 'Papa has long been alarmed for the consequences', Darwin had heard from sister Catherine, '& expects to see an action in the Papers.' Darwin came down to Clapham himself and teased Erasmus by letting 4-year-old Julia, 'his greatest of darlings', sit on his knee. Fanny and Hensleigh looked over his *Beagle* journal, and the whole family was 'very anxious' for their verdict. They thought Charles's prose too interesting to criticize; what most pleased them were his accounts of missionaries in Tahiti and New Zealand.[31]

Darwin dined with activists in London. Harriet Martineau, self-reliant and strong-willed, a cloth manufacturer's daughter who made her way by her prolific journalism, met Erasmus first. A Unitarian, a free-trader and rather radical, Martineau knew all the political worthies. Just back from America, she was single, deaf and indomitable, with strong views on the dissolubility of marriage. Her relationship with Erasmus was well developed by the time Darwin saw them together. He was astonished at Erasmus's driving her about, murmuring about being '(to use his own expression) . . . not much better than her "nigger" '. Fanny would see Erasmus signal 'the Cab is waiting' and Harriet jump up, ready to be taken home, which made them 'very much married'. Fanny and Erasmus knew it was only a 'comfortable relationship', like her own marriage, 'excusing him' from 'reading her books'.[32] It didn't excuse Darwin. He now not only had the most detailed castigatory travelogues of Southern slavery – Martineau's three volumes of *Society in America*, and three more of *Retrospect of Western Travel*, all published within two years of Darwin's return – but their author mooching around the dinner table. He would read them all. But there was no rush while she was a fixture at Erasmus's brilliant parties next door and stood ready to compare 'our methods of writing'.[33]

Already a literary lioness for her Whig government propaganda, Martineau was instilling the spirit of American abolitionism into a British anti-slavery movement, flagging after its achievement in emancipating the slaves in its colonies. She wanted the movement now to target the American South with its thriving plantations and internal slave trade. Martineau's two years of fact-finding was intended to measure American society against the nation's founding beliefs, but she was never a neutral observer. Even before returning home she had come out for immediate and complete emancipation without compensation for slave-owners. Any Darwin or Wedgwood woman visiting America might have had the same experience. (Aunt Sarah Wedgwood, articulate and self-opinionated, was in many respects a middle-aged Martineau with money.) They all shared the same radical Unitarian–humanitarian heritage, to which Harriet added the moral obligation to speak out.

Martineau's stories were standard fare at dinner parties. Having arrived in America amid rising anti-abolitionist violence, she had dared to speak at Boston's 'female' branch of the American Anti-Slavery Society as angry protesters stoned the building. She stood up for what she called the 'holy cause', and William Lloyd Garrison, the ultra-abolitionists' driving force, publicized her words in his inflammatory rag the *Liberator*. Travelling with her ear trumpet through the South, she denounced slavery as an 'utter abomination' and 'inconsistent with the law of God'. For abusing Southern hospitality the slave-holders hated her. Newspapers invited her back so they could cut out her tongue. In Charleston, South Carolina, where she saw a woman sold with her children in the slave market, they called her a secret 'incendiary', and she learnt of plans for her lynching. The prospect galvanized her: after 'witnessing & being implicated in the perils & struggles of the abolitionists', she wrote *Society in America* in the white-hot hope of mobilizing a moral army to free the blacks.[34] This was Darwin's frequent dining companion as he penned his own incendiary racial-evolution notes.

While Yankees joked about her book being 'placed on the *Index Expurgatorius* of the South', Maer devoured every word, and Cousin Emma Wedgwood knew the Darwin sisters would 'really like Miss Martineau'. The women found her 'uncommonly acute', and not just as an observer. She was a consummate literary strategist, weaving anti-slavery through her chapters to make it 'impossible for the Americans' to remove it from their edition. Some family members thought she was diluting the message by also rubbing in the 'sufferings of woman' in general. To Aunt Fanny Allen, this simply detracted from her 'noble true & powerful' remarks about 'the real sufferers, the slaves', though she still believed the book would do 'infinite good'. It was 'impossible not to catch some of her hopefulness on Slavery'.[35]

Martineau was an acquired taste, but the 'wonderful woman' had a knack for attracting 'geniuses', Darwin marvelled – Whig grandees, *Edinburgh Review*ers, freethinking professors, Erasmus who would send her a single rose from time to time. Erasmus saw her soft side; so did Charles, and he had to admit she was 'not a complete Amazonian'. 'Thinking too much' exhausted her, as it did him as he filled his notebooks.[36] More to the point, she shared his first-hand experience of slavery as no one else in their circle. They had both witnessed brutality, felt threats of violence, heard shrieks of pain. They had seen black men treated as animals by white men behaving as animals. The sight of so much New World slavery, far from dulling their senses, had quickened their conscience, and in 1838 Darwin took notes on yet another new title by Martineau, *How to Observe*.

She had sketched out the book en route to America, aiming to distil what the 'philosophical traveller' needed to know about how the moral sense manifested in different peoples. Darwin compared it with a book on ethical philosophy by Sir James Mackintosh and found a common ground: besides '*some* universal feelings of right & wrong', humans have a 'moral sense' that varies even from race to race, Darwin noted, just as breeds of dogs have 'different instincts'. Moral feelings are as 'natural' to people as herding instincts are to deer. Yet however fixed mankind's 'conscience or instinct' might seem to be, it can be changed and improved.[37] It was more grist to Darwin's mill as he ground on, trying to grasp how savages became sophisticates.

The American South for Martineau was an incongruous mix of politeness and injustice, patrician slavery without guilt. On the plantations charity and barbarity were bedfellows, yet many whites considered slave life idyllic. It was blindness due to ignorance, and her remedy was education: teach white Southerners the nation's libertarian principles and slavery would cease. Darwin sympathized but his approach would undercut the very wellsprings of the 'domestic institution' (a common euphemism in America for slavery): the belief that slaves were another kind, or could be treated as such. His grandfathers had supported the American Revolution. The enlightened Dissent so prominent among the Founding Fathers had been bred into him, and with it the same 'overruling commitment' to anti-slavery that shaped Martineau's mission.[38] But the brutality Darwin had witnessed on the *Beagle* voyage saw his new dedication manifest itself very differently: it led him to forge a common evolutionary bond. He was uniting the races at the most fundamental level.

Or rather, he had progressed to investigating what had caused them to diverge. How does one species split into races? National characters were different; physiques varied; 'Kaffirs' stood head and shoulders above a 'Hottentot'. 'The American [Indian] in Brazil is under same conditions as Negro on the other side of the Atlantic. Why then is he so different'; 'look at them both savage—look at them both semi-civilized'. They were distinct, yet lived under the same conditions. What could have prised their common ancestor into two such peoples?

His omnivorous reading threw up all manner of leads. Colonial surgeons provided one clue that Darwin chased on an initial bet. An old soldier who had fought Napoleon and been dispatched to the West Indies, and who made tropical diseases his speciality, raised the Negro's resistance to 'endemic fevers' (which rendered him so useful to the army): something to do with the skin's texture, Darwin hazarded. Fascinated, he noted 'that the ideosyncracy of the Negro (& partly Mulatto) prevents his taking any form of Malaria— adaptation & species-like'. Was there something in the African constitution which made it resistant? Such an 'adaptation' would give the Negro a 'species-like' distinction.[39] Comparative immunity was something Darwin could study as one possible cause of the races moving apart.

He delved deeper into human prehistory. The medical failure looked to the tell-tale signs of anatomical relics: 'The rudiment of a *tail*, shows man was originally *quadruped*.— . Hairy.— could move his ears'. It was proof that monkeys lay back along the trail. In the most shocking of revelations, one he kept secret from the world for three decades, Darwin even envisaged the existence of a midway 'monkey-man'. Fossil monkeys had been unknown till this point, but, opportunely, a swathe now turned up: a femur in India's sub-Himalayan mountains, then a perfect skull; while news of Brazilian finds began to trickle through, then word of French and Greek discoveries. Within months around 1837 there were fossil monkeys known on three continents. These proved one thing: any prediction of the impossibility of even more disturbing creatures from the lack of fossils was 'valueless'.[40] Monkeys now, monkey-men next.

Which group of monkeys lurked in our past was too soon to say. But Darwin had the quirkiest way with such things. He noted how male baboons 'knew women'. Andrew Smith had often seen them 'try to pull up petticoats' and 'clasp them round [the] waist & look in their faces', making noises that signified 'recognition with pleasure'. 'These facts may, be turned to ridicule, or may be thought disgusting', Darwin jotted, but to a 'philosophic naturalist' they were 'pregnant with interest'. Oddly the American monkeys at London Zoo showed 'no desire for women', but the Senegalese were attracted, suggesting, he laughed, that 'the monkeys understand the affinities of man, better than the boasted philosopher' who denies his own ancestry.[41] Humans had descended from Old World primates – the monkeys knew it instinctively!

In July 1838 Darwin took his human exploration a stage further. He opened a new notebook to examine the deeper moral and metaphysical implications of human evolution. Insanity, emotion, memory, facial expression, hereditary dispositions, materialism (his own included), the meaning of free will. Nothing was excluded. It was intractable terrain, more often than not attacked the way he knew best, anthropomorphically. Evolution gave Darwin a handle: 'Experience shows the problem of the mind cannot be solved by attacking the citadel itself.—the mind is function of body.—we must bring some *stable* foundation to argue from.' That was mental evolution, making human morals and mores the best part of brute instinct. The upshot was of course that our instincts of revenge, anger and such are modified relics and, like the tail, 'slowly vanishing' – remnants of our quarrelsome time in the trees. 'Our descent, then, is the origin of our evil passions!!—The Devil under form of Baboon is our grandfather!' The breathlessness was exhilarating: 'Origin of man now proved', he scribbled.[42]

By June 1838, Hensleigh and Fanny Wedgwood had just moved in with Erasmus, next door to Darwin. Hensleigh had lost his job as a police magistrate. His scruples about Christ's precept not to swear had got the better of him, so he could no longer administer and swear oaths. Maer disagreed with his exegesis but respected his 'great sacrifice . . . to a Christian principle', though it meant that Clapham had to be given up as an economy measure. They had thought of America, but another job offer (to be Registrar of London cabs) allowed Hensleigh to stay. It meant Darwin saw more of him, and conscientiousness – one topic of Darwin's new notebook – became an inevitable talking point. 'Hensleigh says the love of the deity & thought of him or eternity, only difference between the mind of man & animals.' Darwin doubted it; as far as he could tell it was rather 'faint in a Fuegian or Australian!' And as always there was his solution – 'why not gradation.— no greater difficulty for Deity to choose, when [man's moral consciousness is] perfect enough for future state [of evolution], [. . . than] when good enough for Heaven or bad enough for Hell'.[43] There was a creative scale even on finer moral issues, perhaps culturally related, with no one-size-morality-fits-all.

Erasmus by this time was on growing terms with the sour Scottish essayist Thomas Carlyle, a rather discordant figure in Darwin's dining circle. Carlyle had latched on to Erasmus as a kindred cosmopolitan, 'an Italian, German travelling University sort of man . . . very polite, good, quiet'. Erasmus introduced him to the others, and Carlyle found the Wedgwoods endearing but Martineau less so. Carlyle loved to talk and once silenced a party by lecturing them on the virtues of silence, so a noisy chatterbox wasn't welcomed. Martineau, he griped, kept boasting and insisting on 'your waving banners too'.[44]

Her banners were not Carlyle's. Martineau's crusading anti-slavery was just too much, and her books full 'of illustrious obscure mortals whom she produces on you, of Preachers, Pamphleteers, Antislavers' – 'absolutely more than enough'. Her Dissenting optimism and faith in progress – her bedrock belief in material improvement through moral education – he dismissed. While she went on in 'her deft Unitarian-Poetic way' about slavery, society and God, he was resolving morals into raw intuitions, spinning rhetorical cobwebs from his alienated Scots-Calvinist consciousness. He scorned Priestley, chief among Unitarians, and Darwin's freethinking grandfather Erasmus, for dismissing spiritual explanations. He mocked the French who taught 'that "as the liver secretes bile, so does the brain secrete thought"', retorting with the reductio ad absurdum, so 'Poetry and Religion . . . are "a product of the smaller intestines!"'

Darwin the secret materialist – happy to have brains secrete even religious notions as physiological by-products – was on his own private collision course with Carlyle. But it was not as a materialist that he would find the Scot so objectionable. What improved the world in Carlyle's mind was native genius, the 'heroic' Anglo-Saxon racial soul. Here Darwin was meeting the 'sage' who symbolized the racialist manifest destiny: the supreme white races, straining at the bit, ready for empire. 'Must the indomitable millions, full of old Saxon energy and fire', asked Carlyle, 'lie cooped-up in this Western Nook, choking one another . . . while a whole fertile untenanted Earth, desolate for want of the ploughshare, cries: Come and till me, come and reap me?' Human destiny lay in the hands of this race and its heroes. America, too, was '*All-Saxondom*', he told his critical admirer Ralph Waldo Emerson in Massachusetts, and 'we will right cheerfully go thither to hold such festival'.[45] Carlyle's vision of a racialized white America was Martineau's *bête noire*.

Even Erasmus, a fellow dyspeptic, had trouble understanding this maverick in their midst, so alien was his creed from the Darwins'. Carlyle's sneering at science and dismissing its material explanations as 'arrogance' annoyed Darwin. But he had the upper hand. Little did Carlyle know what he was up to: the Scot would have vented sulphurous fumes on seeing the black and white races fraternally united, and exploded on hearing of our bad habits being traced to baboon rages. As for thoughts being called secretions of the brain, like bile from the gall bladder – that would have left him bilious himself. It is difficult to know how they sat in the same room; but of course Darwin never let on about his secrets, while everyone knew Carlyle's. Here was the guest who considered Martineau's attempt to embroil others in 'the Black Controversy . . . a thing lamentable in the extreme'. Who cared whether 'Mungo' (the West Indian slave) eats 'his squash as a stupid Apprentice instead of stupid Slave!' Darwin found Carlyle's outbursts frankly 'revolting'.[46] The Negro-phobia and slavery-justification could only have confirmed him on his course.

Nature, like traditional society, was static in the 1830s, with the place of each creature ordained and held fixed by God. For anyone who argued otherwise there were severe social penalties. Darwin in his evolutionary notebooks, having made one creature the progenitor of another and released them from Creative bondage, was sailing against the prevailing wind and now became ill in the choppy waters. Aunt Sarah Wedgwood had seen him in Shrewsbury, her niece Emma heard, and pronounced him to be 'very unwell', with heart 'palpitations', though 'she says he does not appear at all invalidish.' In June 1838 Emma came to London herself, staying at Erasmus's house with Fanny, Hensleigh and the children. Darwin was still poorly: 'Some little Species theory, & lost very much time by being unwell', he recorded in his diary that month; it was a connection now well established. Carlyle's visits probably didn't help. Around the long dining table at 'The Darwin & Wedgwood Arms' (Erasmus's house), Emma just managed to make out his 'broadest Scotch' brogue, which was just as well when, as seems likely, the conversation turned to slave apprenticeships.[47]

Immediatists had predicted that the apprenticeship system following emancipation in 1834 would be abused. It was. Planters tried to extract the last ounce of enforced 'free' labour before the apprenticeship period expired. Privileges held under slavery were ended, harsh working hours set, brutal floggings condoned. Jamaica endured the worst of it. Despite the mediation of special magistrates, the former slaves lived like 'emancipated prisoners'.

A radical new organization, the Central Negro Emancipation Committee, headed by the Midlands Quaker merchant Joseph Sturge, campaigned for the instant abolition of apprenticeship. The anti-slavery public reawoke and for months petitions again rained on Westminster. Abolitionist motions were debated in both Houses. Brougham's failed in the Lords, but the Whig-controlled Commons voted on 22 May 1838 by a wafer-thin majority for immediate freedom, causing Wedgwood rejoicing.[48] Throughout the campaign, Maer supported Sturge's 'moral Radical party' with donations, subscriptions to his *Emancipator* paper and a petition from their anti-slavery society in time for the crucial vote.[49] Aunt Fanny Allen, with memories of the Haitian slave revolution led by Toussaint l'Ouverture, thought 'the heart of the West Indian Pharoahs must be hardened that they may be punished . . . [T]he slaves will never be free till they have fought and won'. Her radicalism spoke for itself. Sturge naturally tried to interest Aunt Sarah in bankrolling his West India Land Investment Company to endow freed apprentices with their own independent settlements.[50]

Carlyle couldn't care less. Martineau would write on Toussaint l'Ouverture shortly, making a 'beautiful "black Washington"' of what Carlyle considered 'a rough-handed, hardheaded, semi-articulate gabbling Negro' in his 'horriblest' phase of '*Black* Sansculottism'. She might lament that men don't fling themselves, with 'due impetuosity, into the Black Controversy', but Carlyle was oblivious to the state 'stupid' 'Mungo' was in.[51] Contentment, not improvement, was the black man's only hope, an indigestible thought at any Darwin–Wedgwood table.

If Carlyle's views on 'Blacks', 'cased in a beneficent wrappage of stupidity and insensibility', were not fit for discussion, all Darwin–Wedgwoods knew what was. By good fortune, *The Life of William Wilberforce* had just come out in five fat octavos, written by his reverend High Church sons Robert and Samuel. The family found themselves threaded through the book – Wilberforce plotting abolition at old Etruria with Emma's grandparents, father Jos, cousin Charles's mother, teenage Aunt Sarah and the other aunts and uncles; Wilberforce backing Uncle James Mackintosh in the Commons and receiving his praise for attacking 'all the forms of corruption and cruelty that scourge mankind'. There were 'expressions of great respect' for Aunt Jessie's husband, the Swiss historian Sismondi, who promoted anti-slavery in France, and with whom Emma had just spent three weeks in Paris.[52]

If ever a reminder was wanted about how obsessed Darwin's immediate culture was about slavery and the black human predicament, this circle provided it. The propagandizing Martineau, full of torture stories from the front-line States; Carlyle, moving in reaction into his 'nigger'-lambasting phase; Erasmus and the Wedgwoods next door, revelling in the family's role in the Wilberforce/emancipation story. It was the air Darwin breathed as he continued investigating the racial underpinnings of human evolution.

Just hearing an 'atrocious' rumour that the Wilberforces in the *Life* had smeared Thomas Clarkson – anti-slavery's first and greatest itinerant organizer – as an 'importunate paid agent' had the family fired up again. They all knew, and some remembered personally, how grandfather Jos had greeted Clarkson's flying visits to Etruria. Emma agreed that the 'dull sons' had been 'very spiteful about poor old Clarkson who is blind & 80 years old'. This alone should have made them 'careful not to hurt him', as 'their father never w[oul]d' have. Emma simply rejoiced that at least Clarkson's work was done: 'the apprentices will be freed as Jamaica has at last settled to free them on the 1st Aug'! It was a sad old Clarkson who sent his pamphlet against the *Life* to Maer. Full of 'unmingled pain' at having to defend himself rather than the defenceless, Clarkson's *petite histoire* marked the end of half a century of struggle. Jos replied humbly: 'I have ever venerated your character & your entire devotion of your entire self to the cause . . . which you now see triumphant'.[53]

On 1 August 1838, thousands of blacks attended thanksgiving services and paraded, not only through the streets of Kingston, Jamaica, but as far afield as Philadelphia and New York.[54] British emancipation was finally complete, just as Darwin's effort to emancipate all life from its Creative chains was beginning.

It was lucky that the 30-year-old Emma was as staunch on these matters. She had caught Darwin's eye during her stay next door. He now saw this cousin as the 'most interesting specimen in the whole series of vertebrate animals'. A postal courtship led to an engagement, of which Lyell was the first to hear outside the family. Erasmus took Martineau out in his cab house-hunting for the couple, and Carlyle redeemed himself a little by declaring Emma 'one of the nicest girls he had ever seen'. Emma for her part found Charles 'the most open, transparent man' she ever met (something of a problem for a man holding disquieting views), and 'perfectly sweet tempered', with some 'minor qualities that add particularly to one's happiness, such as . . . being humane to animals'. His 'great dislike' of theatre-going meant 'we shall have some domestic dissensions . . . unless I can get Martineau to take me sometimes.'[55] She had no idea yet what awaited her.

Darwin's father advised him carefully to conceal his doubts about religion lest Emma fret for his 'salvation'. (The Doctor understood devout Wedgwood women, having married one himself.) But sharing so much of an outlook, Darwin thought candour the better policy, and a week after the engagement he went ahead and told her of his notebook heresies. Such shocking beliefs were a negation of her deeply intuitive faith. He was erasing the line between body and soul. To him, morality and religious feelings were inherited from beasts rather than Breathed into the body. What need, then, for revelation of religious truth in the Bible? If Jesus's resurrection did not reveal the promise of immortality, how could she and Charles belong to each other for ever? Traditional Unitarianism, as espoused by Martineau, saw no necessary conflict here, and Darwin's views might have been squared with it. Not so Emma's Anglicanized Unitarianism, with its belief in an immortal soul. She sought reassurance and 'every word' he sent by return was a comfort. He said that he did not consider his 'opinion as formed' (too late was he heeding the Doctor's advice), which gave her hope.[56]

The emancipation issue resurfaced when the group went to a party hosted by Fanny and Hensleigh's old Clapham neighbour Sir Robert Inglis, 'a fat John Bull gent[lema]n; of pepticity, of energy, honesty and limited mind', to Carlyle's way of thinking. But then he sneered at the 'hundred mortals' present, 'almost all of the male sex; barrister and preacher-looking people; dull and commonplace'. Sir Robert congratulated Emma on her engagement. He 'shook me by the hand "till our hearts were like to break" ', and she wondered when this particular Tory would let go. She wasn't predisposed towards him (no Whig ever was). The MP had voted to set aside the May vote for immediate abolition, ostensibly in the interests of the apprentices.[57] Even now it felt like betrayal. Fanny Wedgwood disliked Sir Robert too, thinking he cared more for Church and Crown than morality. Darwin told Fanny that Inglis was a 'noble fellow' for his motives, at least. Unfortunately Emma was in earshot. 'It never can be noble . . . to do what you have no right to do though from great generosity'.[58] The slaves had deserved immediate freedom, not more suffering. This was one track, at least, she would keep him on.

Charles gave Emma the latest issue of the radical *Westminster Review*, with Martineau's stunning essay, 'The Martyr Age of the United States'. For all its 'little Harrietisms', the piece enthralled Emma. It was the first extended account of American abolitionism for British readers. Its climax came in the 'holy' martyrdom of the Revd Elijah Lovejoy, Presbyterian preacher and newspaper editor, whose press was wrecked after he damned the lynching of a free black man at St Louis. He had crossed the Mississippi River and taken over a paper in the 'free' state of Illinois, where again, twice, his press was destroyed by an anti-abolitionist mob. On notice to leave town, he stood on his right of free speech:

> I know I am but one, and you are many . . . You may hang me up as the mob hung up the individuals at Vicksburg; you may burn me at the stake as they did M'Intosh at St Louis; you may tar and feather me, or throw me into the Mississippi, as you have often threatened to do. I, and I alone, can disgrace myself; and the deepest of all disgrace would be, at a time like this, to deny my Master by forsaking his cause. – He died for me, and I were most unworthy to bear his name, should I refuse, if need be, to die for him.

Days later, Lovejoy was gunned down by a drunken mob attacking his press. Emma told Charles that this impassioned plea was 'the best' part of the essay, and she pressed it on her relatives at Maer.[59] It was clear that Southern slavery had immediately occupied the lacuna created by West Indies emancipation. From now on, thanks to the women, all eyes were focused across the Atlantic.

Darwin, about to get married, urged his sisters to pray for the freedom of 'our poor "nigger"' (himself) even as they prayed for the American slaves. He was content to submit to Emma, but the worry of his work was telling and she despaired of his 'looking so unwell & being so overtired'. Marriage as metaphorical slavery was a tolerable joke in a family whose abolitionism was absolute, and Emma settled into calling Charles her 'nigger', while he made himself her 'happy slave'. His sister Caroline (now married to Emma's brother Jos III) may have found the banter objectionable but it carried on for years. Emma even pictured Charles, peering out at the garden behind their new house on Upper Gower Street, as 'looking through your window on your estate & plantations'. Other times 'my own dear Nigger' would be the master and she his 'chattel'.[60] The air of anti-slavery was exhaled in pet names as much as in notebooks.

*

With marriage imminent, the question of sex hovered about his notebooks. What would their offspring be like? His and Emma's minds – instincts, memories, moral sense, emotions – would be passed on with their physical characteristics. The children would get a double dose of Wedgwood blood as the couple were first cousins. What would their breeding prove? How could it be understood in terms of 'my theory' (as Darwin called his notebook evolutionary project)? He observed his and Emma's mutual attraction. He knew the power of sex, but that power suddenly and unexpectedly became creative from 28 September 1838 as he started to read *An Essay on the Principle of Population* by Fanny and Hensleigh's old family friend the Revd Robert Malthus.

Malthus and the despised Malthusian workhouses had been ubiquitous in Darwin's Whig world. When Darwin was in his teens, Sir James Mackintosh had been a law professor in the East India Company's college at Haileybury, teaching young sahibs how to govern the subcontinent. Free-market economics was expounded here by Professor Malthus (though not so as to conflict with the company's trading monopoly). The industrial Wedgwoods held with Mackintosh and Haileybury that God's laws worked naturally to optimize wealth and reduce poverty, without interference. Legislation was needed to eradicate *moral* evils such as slavery, but *natural* evils arising from excess population and hunger were God's blessing in disguise; they made men work hard and control their lusts, thus keeping the population down. Poverty should be alleviated only for those unable to help themselves. Or so argued 'old Pop' – for population – Malthus. It was all in the family: his daughters were close to the Mackintosh children, and one daughter, Emily, had been a bridesmaid at Fanny and Hensleigh's wedding.[61]

The Malthusian 'principle' involved a feedback loop: overpopulation through couples having too many children led to competition for resources and rid the world of the unsuccessful. Martineau (who had met him) had been the Whigs' chief propagandist on the poorhouse consequences of ignoring Malthus. She wrote novels and penny pamphlets to spread the word. Darwin knew her propaganda tracts about couples who rushed into relationships and ended up with too many mouths to feed, needy, living in squalor, pushing up taxes and dragging society down. Ever-threatening hunger was God's way of teaching people about sexual restraint and improving their lives.

Darwin had himself mouthed this Whig mantra. But what struck him now was its implication for those *incapable* of restraining themselves. What misery for wild animals, forever multiplying and struggling for food. The death rate was unimaginable. Cut-throat competition from over-population caused by the animal sex drive worked by 'thrusting out weaker ones'.[62] This was the 'mysterious law' governing the extinction of species – the nub he would hereafter call 'my theory', *natural selection*.[63] In the struggle for life, 'the chosen' are those with some adaptive edge, some physical or mental advantage, which means they survive to leave offspring: they were being naturally selected.

But there was still a question hanging over racial evolution. Could *all* human characteristics be explained by this sort of selection, and by expunging every maladapted individual? This might leave 'superior' individuals, better fitted to survive; but was every aspect of hair, skin colour or facial shape physically adaptive?

Ultimately sexual success, via a struggle and selective survival of the offspring, had seen humans evolve from ape-like ancestors. Perhaps it became clearer because Darwin was about to contribute to the process, to add his and Emma's offspring to Britain's booming population. He plunged deeper into the question of racial and sexual differences. The niggling doubt that natural selection might not be able to explain the *origin* of all the quirky and diverse racial characters dogged him. Now, at the height of his notebook exuberance, he got to grips with it. If Brazilian Indians were so physically dissimilar to African Negroes, yet lived in supposedly the same tropical climes, it hinted that some other cause may be coming into play; something perhaps that made facial shape unique, hair type particular, skin colour different – features that conceivably had nothing to do with adaptation to the environment at all, or therefore natural selection. Was there another mechanism to produce the Brazilians' and Africans' 'colour & shape'?[64]

Darwin's thoughts oscillated between race and beauty. He took a masculine view: What makes a woman beautiful? By whose standard? And where does the *beau idéal* come from? Having once doubted whether animals have 'notions of beauty', he now believed that beauty was always in the eye of the beholder – an eye that had evolved. Ideas of beauty differ as much as bodily forms and, indeed, change and evolve with them. 'Our acquiring [. . .] our notion of beauty & negroes another' were parts of the same process by which the visible, aesthetic racial differences – skin colour, hair type, and physique – had emerged. This was the crux. Each race possesses its own *beau idéal*. Voyaging round the world, dreaming of 'white petticoats', he had found this to be true. A Negro born out of his native land would 'think [a] negress beautiful' anywhere, just as the homesick *Beagle* itinerant lost in the tropics had longed for 'an English lady . . . angelic & good'.[65]

Darwin was no idealist. Beauty lay not just in the beholder's eye; it was incarnate in each race, making the sexes mutually attractive. Fearing himself 'repellently plain' and wondering what Emma saw in him, he began to consider how external (or 'secondary') sexual characteristics evolved. As usual, he sought the answer by analogy, focusing first on vertebrate males like himself. Tone-deaf, he noted that 'cock birds attract females by song', which augured ill for him, but then – his telegraphic notes stuttered on – 'man [is] more hairy than woman' (he himself was hirsute), and he happily pondered the charms of plumage. Cocks being 'all warlike' reminded him that male animals everywhere were 'armed & pugnacious', whether buck deer, bull seals, or his *Beagle* shipmates who marched ashore at Montevideo carrying pistols. And the doe's 'passion [. . .] to the victorious stag', he jotted after deciding to propose to Emma, was 'analogous to the love of woman' for 'brave men'.[66]

Still, birds were becoming his main interest, being the most prized and bred for beauty. Here if anywhere lay crucial clues to the origin of sexual and racial characters. While some fanciers pitted their cock birds in a bloody arena, others paired males to see 'which will sing *longest*'. The singing went on to attract a hen's attention and 'drive away [the] rival'. Cocks of other bird species displayed beautiful plumage instead, trying to upstage one another and gain a mate. And what about the hens? Planning to propose, Darwin dwelt on the bloodless forms of competition, unsure of the female's role when males flutter around her rather than fight. She was generally smaller and 'weaker' and, like the young of both sexes, had less pronounced features and drabber tones. Yet she was not inert. Before his engagement he even toyed with the idea that, instead of being 'merely [. . .] attracted', females might 'fight for [the] male', with the 'most vigorous' of both sexes pairing, but this egalitarian notion – reminiscent of feisty Martineau – was only a passing fancy. He was satisfied that pea-hens 'admire [the] peacock's tail, as much as we do', though he still marvelled that Emma found him handsome. Why did she? Indeed, how 'does [the] Hen determine which [is the] most beautiful cock' bird?[67]

Before long he had the missing link between natural selection and the *beau idéal* mechanism. As befitted a son of the Shropshire gentry, with its lifelong passion for gundogs, show cattle and fancy breeds, Darwin was fascinated by domestication – in humans and animals. He ploughed through cattle husbandry manuals, and probably got the idea of 'picking' desired traits from them. He used tame breeds to understand inheritance and wild behaviour. He studied the way dog fanciers singled out the pups they wanted to mate in order to enhance some ability or anatomical aspect. This *artificial* selection in the kennel was the counterpart of *natural* selection in the wild. Breeders select as nature sifts – each acquires the desired offspring to shape the next generation. 'As in races of Dogs, so in species, & in Man'.[68] In 'Man' Darwin had only to extend the breeders' analogy to see *all* animals as self-breeders, picking out their own mates, making aesthetic choices, creating fancy varieties of themselves, for *sexual* selection proper to be born. All the disassembled elements were finally in place in his notebooks.

The landlubber's two years ashore had been an extraordinary whirligig. From a bronzed old salt fired up by sights of slavery, genocide and brutality, he had changed into a gentle recluse hiding his scientific response. His private notebooks had plotted one zigzagging path from the brotherhood of man and the unity of flesh denied by slavers to its ultimate 'common descent' upshot – in fact millions upon millions of brotherly 'common descents' going back through history, a whole evolutionary panoply made up of siblings and common stocks, with everything connecting by descent with everything else: mice and men, amoebas and mushrooms.

Common ancestry had been his innovation: a chartable pedigree for the whole of life, and not just for the human aristocrats. How this 'tree' of life had grown was in large part cracked by the time he snapped shut his last evolution notebook. He had 'natural selection' in place, but there were doubts whether it could account for human aesthetic traits. Now 'beauty' was entering the picture: and the need for a further cause to explain those alluring characteristics – from plumes to physiognomy, songs to sexy shapes – that made the males and females of each race mutually attractive. These were the thoughts of a bachelor on the verge of getting married.

6
Hybridizing Humans

Married in 1839, with babies beginning to arrive and his own literary first-born out (the *Journal of Researches*, about the *Beagle* voyage), Darwin began to retreat from society. The couple brooded together, their house on Upper Gower Street a protective womb. Invitations were parried: 'a civil note from Miss Martineau inviting us to a party ... which we have declined'; 'cards from Sir R. & Lady Inglis'. Thank God, the FitzRoys had moved out of town. Darwin was 'anxious to avoid' the captain after FitzRoy acquired a painfully pious wife. 'Friends & relatives' could be a trial. Fanny, Erasmus and Hensleigh would come to dine, but if Emma's morning sickness wasn't the cause, it was Charles's churning stomach that kept others away. Carlyle made him worse – 'quite nauseated with his mysticism, his intentional obscurity & affectation' – and the scoffing Scot was dropped. Even the arrival from Switzerland of Emma's genial Allen aunts, Fanny and Jessie, escorted by Jessie's husband Sismondi, was too much for Darwin, and Emma would be left to entertain.[1] The air of a recluse was already beginning to hang around the heretic.

Old Edinburgh friends were drifting apart too. If procuring Indian children for educational purposes was silently shunned by Morton, he and Hodgkin continued on a less fraught level to exchange boxes of minerals. But their tardy letter-writing showed the hairline cracks opening. Then came the complaints, with Hodgkin continually asking Morton in Philadelphia 'why my letter ... was not applied to', although he imagined 'that sufficient reason might be found in any of the various stations which [thou] holds, of husband, father, friend, Physician, lecturer, perpetual secretary of the Academy, Natural Historian, Poet, drafts man &c &c'.[2] It was a recognition of Morton's rise in Philadelphia, and Hodgkin would shortly have to add 'craniologist', for that would be how the future would remember Morton, and it was Morton's work which would ignite the growing scientific racialist movement in America. The drifting would accentuate as Friend Morton, moving up the Philadelphia echelons, emerged at the top a more conservatively acceptable Episcopalian.

Rather less acceptable to Friend Hodgkin, one fears. Hodgkin pushed on with his Quaker philanthropy. He had niggled Morton on the subject of the extinction of human tribes since their student days. Jameson's other old pupil Darwin was one of the few who had actually seen the extermination of native peoples at first hand: the atrocious genocide under General Rosas, Tasmanians all but wiped out, aborigines succumbing to European diseases. The British Government was well aware of the dire predicament of many aboriginal peoples, so much so that a House of Commons Select Committee on Aborigines in British Settlements had examined the problem from 1835 to 1837, chaired by Thomas Fowell Buxton, MP, the abolitionist stalwart. They had been pushed into it, partly by the Sixth Kaffir War that followed colonist incursions from the Cape – all of which, again, Darwin knew about at first hand.

From Canada to Calcutta, witnesses had come to detail the grim costs of colonization. Their testimony had yielded rich anthropological evidence, particularly from southern Africa. The Revd John Philip had been summoned repeatedly to answer over 200 questions, and the committee endorsed his 'propagation of Christianity' at the Cape as the 'one effectual means' of protecting the 'civil rights of the natives'. Lord Glenelg, the Claphamite colonial secretary who had censured Philip's nemesis, Goveror D'Urban, was vindicated for his principled intervention on behalf of the Xhosa. While some witnesses thought the behaviour of Christians was more cause than cure of their suffering, the committee's final report in 1837 must have warmed both FitzRoy's and Darwin's hearts by defending the civilizing work of missions. Indeed, the MPs called for new measures to resolve frontier conflicts and thus relieve missionaries of their 'political' work as 'consular agents';[3] such was also the argument implicit in the FitzRoy–Darwin article.

Hodgkin appeared before the committee. Unusually for an abolitionist, he was also an advocate of African colonization by freed blacks – educated, Christianized and liberated from an ingrained prejudicial system, particularly that in America – while Buxton wanted whites themselves to evangelize the continent.

Hodgkin and Buxton had moved quickly after the Select Committee reported to institutionalize these concerns in the British and Foreign Aborigines' Protection Society. Its motto '*Ab Uno Sanguine*' ('Of One Blood') spoke collectively of their faith in the Adamic unity of the earth's peoples. In the flush of success after the emancipation victory they naively hoped to reorientate British colonial policy. A society which saw civilization go hand in hand with Christianization, and which set out to expose colonial atrocities and promote an enlightened legislative policy to enshrine 'native' rights, obviously drew Wedgwood financial support.[4]

Those who shared Darwin's anti-slavery concerns were noticeably more outraged that so many indigenous races stood imperilled. And they were more prepared to act – the Aborigines' Protection Society publicized the gory details of colonial barbarity. This made the science of race from the start controversial. As a result the natural history of aboriginal races – let alone their fate – wasn't a topic welcomed by the elite gentlemen of the infant British Association for the Advancement of Science. It struck them as too divisive, its proponents too critical (not least of Eurocentric attitudes) and – judging by Hodgkin's Society – too inflammatory in denouncing the evil perpetrated by British settlers. Nonetheless both Hodgkin and *Researches into the Physical History of Mankind* author Dr James Cowles Prichard managed to inveigle their concerns into the 1839 meeting.

The BAAS convened in a different city each year. It was the public face of science, with a carefully manicured image of political and religious moderation (so necessary in troubled times). All of this was to take a battering in 1839 when the crowds converged on Birmingham, and not only because a 'fringe' ethnology sneaked on to the agenda. The turbulence was peaking during an economic depression; so was radical politics. In the first Birmingham city council elections the Tories had been 'mangled and minced' and the radical Birmingham Political Union held sway. The BAAS was in two minds whether to cancel its Birmingham event, but it went ahead with thinned attendance. Cosmopolitan savants descended to link up with local dignitaries for their festival of science; unfortunately, socialists and working-class activists demanding the vote also arrived in town for their own conventions. With riots having broken out only a month before, peace was ensured on the week of Darwin's visit by 'police staves and cavalry sabres'.[5]

The last thing the BAAS managers wanted were disquieting harangues aimed at overseas policy. Humanitarian concern for 'the feeble races of Mankind' was one thing. But Hodgkin's habit of mooting 'wrongs to be atoned for' was quite another. And his suggesting that less effort went into saving aborigines than restocking vanished capercaillies on Scottish shooting estates was a step too far for the gentry running the show.

Hodgkin and Prichard presented aboriginal extinctions as an 'irretrievable loss' to science, and their emotive language did make a public impact. The real worry was what the world would later call 'genocide', the extinguishing of 'whole races' since the time of the Conquistadores. These were peoples lost for ever, when so much had yet to be learned of their culture and language. 'Wherever Europeans have settled, their arrival has been the harbinger of extermination to the native tribes', Prichard lamented. The increase in British colonization only threatened to accelerate this destruction. During the depression of the late 1830s and early 1840s there was a massive exodus from Britain. Workshops shut and the surplus paupers left: 400,000 yearly were shipped out to Australia, the Cape or America. No indigenous tribes were safe when 'the whole of the unpeopled regions of the earth' were said 'to be British ground' – for 'unpeopled' meant unpeopled by Europeans. Human flotsam the emigrants were considered, but where they washed up the effects were devastating. Prichard claimed that within a century the complete annihilation of all 'aboriginal nations of most parts of the world' would be complete.[6]

Prichard was ringing alarm bells. Some were not so much alarmed as resigned. Darwin's student friend, now a cotton king, William Greg, thought many aboriginal peoples had been 'destined by Providence for early extinction', although God alone knew why they had been 'created merely as temporary occupants to fill the void, till pushed out of existence in the fulness of time by other races of more commanding energies'.

Meaning the Caucasians: 'commanding energies' was Rugby headmaster Thomas Arnold's image. Caucasian brutality ensured that 'savages' who could not be civilized would follow the fate of the mammoth. This showed how little real Christianity resided in Europe, Greg said in the *Westminster Review* in 1843. He took up Arnold's idea of moving centres of civilization through history, each passing on its heritage, but he twisted it surprisingly. Greg had visited the cotton pickers in the West

Indies. Here Africans – childlike, 'docile, gentle, humble, grateful, and commonly forgiving' – bore the true Christian hallmarks. As imitative souls their acculturation was assured; no extinction for them. Blessed by the 'virtues of peace, charity, and humility', black people might actually find it 'natural and easy' to take the European Christian mantle into the future.[7] Not one of nature's humanitarians, Greg had nevertheless produced a scenario worthy of the philanthropists, although it probably wasn't what Arnold had in mind.

Nor Darwin. It was now his turn to stand apart. While he shared Greg's resignation at the exterminations, he substituted an evolutionary explanation for Greg's plan of Providence. Whatever Darwin's views on the Christian redemption of Tahiti, he now took a very different tack on the fate of indigenous black peoples. King Cotton might produce docile plantation images; Darwin's were drawn from the Kaffir Wars in South Africa. All along he had applied his ideas to the human races, but the consequences of reading Malthus in late 1838 would be profound.

Before that moment, Darwin had envisaged the 'future destinies of mankind' more benignly. Some 'varieties are becoming extinct' (the Tasmanian situation was perilous, he knew), although he was sanguine about others, and 'the negro of Africa is not loosing ground'. He talked of 'the tribes of the interior [. . .] pushing into each other from slave trade, & colonization of S. Africa', but the result was 'the tribes become blended', preventing that total 'separation which otherwise would have taken place', and which 'in 10,000 years' would have left the 'Negro probably a distinct species'.[8] His eye was looking only to the encroachments in the interior causing a blending of the tribes being forced together.

There was evidence that whole-population merges – Dutch with Hottentots, or Irish Celts with English Saxons – might actually result in a stronger stock with greater potential. Human racial blending might be *good* in the short term. Darwin's marginal scores on a human breeder's handbook, Alexander Walker's *Intermarriage* (1838), show him following up the point.[9] (There was little he wouldn't read, and this was particularly to his taste – a how-to guide on mate selection for the Victorian mannered classes, which he fitted in among the sheep and hamster literature.) On a slip at the back of his copy of *Intermarriage*, Darwin jotted: 'on advantages of crossed races of Man'. This referred to Walker's quotation of the Scottish Fusilier Lieutenant John Moodie's statement[10] that the 'offspring of the Dutch by the Hottentot women' were superior to both parent stocks in both body and intellect. Such voices carried weight. Moodie was a brother of the Colonial Secretary in Natal. And his observation coincided with the Revd John Philip's testimony before the 1835 Select Committee on Aborigines, which paid glowing tribute to these people of mixed Hottentot/Dutch descent, living in Griqua Town on the Colony's northern border. (He was their protector and spokesman: their central town was actually called 'Philippolis'.) They were acceptable in the colonists' eyes for their God-fearing, anti-slavery sobriety, for appreciating British values of property and pay, and for their actions in keeping down the local 'banditti'. They were in effect acting as the Colony's northern eyes and spies.[11] Darwin, even if he hadn't heard of the Griqua while the *Beagle* was at the Cape, was now making contact with a paradigm people often cited in the unity-of-human-species literature.[12]

However, *after* reading Malthus, Darwin's imagery became much bleaker. Malthus's depiction of human competition for scarce resources highlighted how wars and famines act as a 'great check amongst men'. It galvanized Darwin into rationalizing the darker side of tribal contacts. The traveller had witnessed wars or their upshot on the Patagonian plains, in New Zealand and the eastern Cape. By 1838–9 he was primed to see them in a new, Malthusian light. Competition was all in Whig society and across the world; population pressure sifted and sorted the most successful, ensuring progress. So the ideologue who was concerned with ending slavery ironically began naturalizing the competition of white minds with dark bodies.

Darwin's scenario was becoming a battlefield: 'When two races of men meet, they act precisely like two species of animals.— they fight, eat each other, bring diseases to each other &c'. But while animals compete in bodily strength, human warfare is 'more deadly', ensuring that a race with the 'best fitted organization, or [. . .] intellect' will survive. Intellect gave whites the edge in Australia – he assumed – dooming the Aborigines, whereas the Negro's resistance to malaria may benefit him in Africa and the West Indies.[13] The good to come out of this for Malthus was that this conflict contained population numbers. For Darwin it also allowed improvements to the breed: as the weak went to the wall, the survivors – those with an adaptational edge, or, among humans, superior intellects – passed on those advantages, to be built on by later generations.

He didn't see the incongruity as his science took on a Malthusian life of its own, shaped by the race-judging attitudes of his culture: the civilizational goal, the superior intellects, expansion as a means of progress. His science was becoming emotionally confused and ideologically messy. Malthus's 'grand crush of population' resulted in conflict and conquest, and Darwin began to naturalize the genocide in these terms. He was assuming an inevitability that had to be explained, not a socially sanctioned expansion that had to be questioned.

Darwin was turning the contingencies of colonial history into a law of natural history. An implicit ranking – with the white man accorded the 'best' intellect – ensured the colonist won when cultures clashed. Already Darwin was accepting it as an evolutionary norm. Wedded so early to his evolutionary matrix, this supremacist image would itself be brought to justify later ethnic-cleansing policies, however abhorrent to Darwin's own humanitarian ideals. Darwin's gentlemanly class equated a greater intellect in whites with their cultural achievements. From Argentine barracks to Australian sheep stations, it seemed self-evident that work-shy, untrustworthy 'savages' were good-for-nothings according to civilizational norms. In a 'Taxonomy of primitivism' – where 'ugly' physiognomies (to European eyes) disguised still uglier moralities, and cranial deviations revealed degraded intellects – they came off very badly. So seemingly obvious was all this, that it gave 'the colonial gaze the character of scientific truth'.[14] And Darwin was fortifying that truth, even while he was making the races kin and refusing to place humans above other animals.

White ideals of government, property, money, technology and industry (in every sense) were severe standards. It was a peculiarity of the British people that while one half were vehemently defending blacks from being traded and kept as slaves, their expatriate kin in Australia and the Cape were cleansing hinterlands of nomadic nuisances in the name of economic progress. One has only to look at Darwin to see how uneasily the two ideologies nestled together. Here is the plains-roaming naturalist, at once horrified, yet musing on the upshot of General Rosas's Pampas atrocities – an on-the-spot note that never made it into his published *Journal*:

> If this warfare is successful, that is if all the Indians are butchered, a grand extent of country will be gained for the production of cattle: & the vallies of the [Argentine rivers] will be most productive in corn.[15]

So there were moments when Darwin, too, imagined the 'civilizational' gains from Indian eradication. But he presumably shocked himself on rereading this passage, and marked it for expunging in print.

By biologizing colonial eradication, Darwin was making 'racial' extinction an inevitable evolutionary consequence. Disappearing natives were put on a par with the fossils underfoot: Argentine dynasties had turned to dust before, the megafauna with its giant capybara *Toxodon* and ground sloth *Megatherium*, whose fossils he had found. Races and species perishing was the norm of prehistory. The uncivilized races were following suite, except that Darwin's mechanism here was modern-day massacre. On the edge were the Tasmanians, as Darwin recounted from his stay in Australia. The race was reduced by shootings and countrywide dragnetting, as used by 'the great hunting-matches in India', and even then the blackfellows seemed oblivious to 'our overwhelming power'.[16] The two hundred remaining had been corralled and were kept on Flinders Island from the mid-1830s. Here the local missionary George Robinson ensured the last nomadic stragglers died as penny capitalists, familiar with wages, work and savings.

Other peoples had already gone. Strange ones, too. High up in the Andes, near Lake Titicaca, were traces of a civilization whose inhabitants had had peculiar swept-back skulls. Low-brows they were literally, but not in reality, for their massive Cyclopean sepulchres were of 'very remarkable architectural beauty'. The secretary to the British Consul-General in Lima, Joseph Pentland, taking altitude readings in the newly independent Bolivia, had reported back to the Foreign Office and the BAAS on these tomb relics of an 'extraordinary race'. With their long heads, they were nothing like the Aymara people currently at those altitudes. Indeed here was 'a race of men very different from any of those now inhabiting our globe'.[17]

Seemingly unique, they were piquing interest, not least in Darwin, for he jotted in an evolution notebook:

> The peculiar skulls of the men on the plains of Bolivia—strictly fossil . . . they have been exterminated on *principles*. strictly applicable to the universe.—[18]

The Titicacans had left only inaccessible monuments and the inevitable skulls racked in Britain's unofficial mausoleum, the College of Surgeons in London. (Darwin might have seen these crania, extraordinary for their 'great depression of the forehead', on his visits to the Hunterian

Professor Richard Owen, then describing the *Beagle* haul of Pampas fossils.) Races were being annihilated, but it was according to explicable '*principles*'. Pentland envisaged the current Aymara Indians sweeping in from Asia and destroying these old Titicacans. This process was comprehensible on Malthus's analysis; now it was becoming integral to Darwin's evolutionary mechanism, in which the preponderance of animals were cut down by nature's selective scythe.[19] Extinct human races were a sign as much as extinct mammals.

All of this made Darwin receptive to Prichard's warnings at Birmingham. Humanitarians such as Prichard might have feared the destruction wreaked by settler migrations, but Darwin dispassionately wove these bleak images into his evolutionary scenario. They became evidence for his bloody biological mechanism. On Prichard's speech he wrote:

> A profound consideration of method by which races of men have been exterminated (see Pritchards paper) [. . .] very important. it seems owing to immigration of other races [. . .] now it is this very immigration which tends to make the destroyers vary.[20]

He was turning it round to make immigration, invasion and struggle *essential*. They were the honing process, the crucible: the victorious 'destroyers' survived to breed, while they further adapted to their new-won terrain. Imperialist expansion was becoming the very motor of human progress. It is interesting, given the family's emotional anti-slavery views, that Darwin's biologizing of genocide should appear to be so dispassionate. True, close up, his heart strings were tugged, and one feels the churning emotions in his lament:

> It was melancholy at New Zealand to hear the fine energetic natives saying, they knew the land was doomed to pass from their children.

Or at his hinting to the blood-drenched Rosas that the bayoneting of Indian women seemed 'rather inhuman' – or his horror that such 'atrocities' could take place 'in a Christian, civilized country' in the first place.

He noted that, to the perpetrators, it appeared the 'justest war' because the 'Barbarians' were being cleared for higher ends.[21] Ends kept coming into it – economic, civilizational. Religious even, as in Greg's 'plan of Providence'. Even those in the pulpit could share this view: one, witnessing the last of the Tasmanians, saw it as 'a universal law in the

Divine government' that 'savage tribes [should] disappear before the progress of civilised races'. Shards of such rationalization pierced Darwin's thought, but for him the 'progress' was writ on a grander canvas – the benefit was to the entire species. Extermination was an axiom of nature – 'strictly applicable to the universe', as he said. Nature herself moved forward, crushing skulls underfoot. 'The varieties of man seem to act on each other; in the same way as different species of animals – the stronger always extirpating the weaker', he wrote in his *Journal*. Natural selection was now predicated on the weaker being extinguished. Individuals, races even, had to perish for progress to occur. Thus it was, that 'Wherever the European has trod, death seems to pursue the aboriginal'.[22] Europeans were the agents of Evolution.

Prichard's warning about aboriginal slaughter was intended to alert the nation, but Darwin was already naturalizing the cause and rationalizing the outcome. Nevertheless he did take advantage of events. Having tried to rouse passions at the Birmingham BAAS meeting in 1839, Hodgkin and Prichard proposed a committee to collect facts on disappearing aborigines through a questionnaire for travellers. The Paris Ethnographical Society had a questionnaire in place: what a 'stain' on Britannia's character if she, with all her dominions, was 'left behind by other nations'.[23] France was the foil to shame the BAAS organizers into action.

Given such a chance to insinuate his notebook desiderata into the committee brief, Darwin could not resist joining. To them, he was a seasoned gentleman traveller familiar with Fuegians whose expertise was welcome. Perhaps he joined in the long discussion at Birmingham arising from Prichard's inauguratory paper. No doubt his agenda was somewhat different, so although many questions centred on skull and body shape (the doctors' sphere), or language (the philologists'), as well as superstition, ceremonies, art and laws, others did track his interests: on inter-tribal marriages, which parent the children of such marriages took after, whether such marriages affected health or longevity, the numbers of births and deaths, disease, parasites and so on. And since Darwin had himself asked prominently in another notebook about this time, 'Do the Savages select their dogs', one imagines his fingerprints all over the queries referring to the modifications of domestic breeds by aboriginals.[24]

He realized that as tribes diverged, partly through self-selection of mates, they also selectively modified their tame animals, changing these too. But since these were modified to meet local conditions or to comply with divergent tastes, as tribes spread through different regions they changed their domestic animals quite divergently from one another. There was no overall parallel between the paths of human racial evolution and those of their domestic breeds. But one thing he never doubted: just as the human races were geologically young and had not formed species yet, so it must be with their farm breeds. Thus races of humans could interbreed, as could their races of cows. He hoped to obtain statistics through the Hodgkin questionnaire, but the results were slow to come in.

The lack of interest among the BAAS leaders in the questionnaire is testified by their miserly funding for it – £5![25] And ethnology's propensity for generating controversy confirmed them in their scepticism.

At the BAAS meeting in 1841 at Plymouth, which saw Hodgkin discuss distributing their questionnaire, the bombastic, slave-owning Dr Charles Caldwell from Kentucky – a phrenologist and champion of separate origins for the human races, the man praised for having written 'one of the most triumphant refutations of Prichard's "Physical History" ever published' – touted a rival approach. It was a sign of hardening attitudes intruding from America. Anatomical study, said Caldwell, proved that the African bore 'a nearer resemblance to the higher Quadrumana [apes] than to the highest varieties of his own species'.[26] Caldwell, now almost seventy, had published Erasmus Darwin's *Zoonomia* in America; indeed he had had grandfather Erasmus's imprimatur for his own medical papers. Caldwell might equally have appealed to old Enlightenment rationalists when he criticized the Testaments and trumpeted the '"Elder Revelation," (the Word of God, spoken through ... nature)'. Soon others would share his belief that 'whether all men sprang from the same primitive root' or not was a question of science – to be discussed by naturalists in the same way that they treated the origin of any genus of horse or bull.[27] Moses had nothing to do with it.

But there the similarity with any Darwin ended. Caldwell foreshadowed the belligerence of the new doctrinal diehards of the South. His 'powerful little work', *Thoughts on the Original Unity of the Human Race* (1830) attacked Prichard's unity-of-human-species bible, *Researches into the Physical History of Mankind*. It portrayed Prichard and his ilk as dangerous to religion for trammelling minds: 'If it be not wrong in the Deity to frame some species of men inferior to others, it cannot be wrong . . . to declare and endeavour to prove it.' Nor would such recognition of inferiority 'pervert' the *noblesse oblige* of enlightened gentlemen; for just as 'Man protects' feeble woman, argued Caldwell, so the inferior slave is looked on more pitiably and kindly.

Caldwell set up many of the premises that later ran through Southern racial thought. His Enlightenment conundrum was this: the creation of animal pairs was logistically absurd. It ran roughshod over predator–prey considerations and ecological balances. The first pairs would simply have eaten each other or starved to death. Rather, whole populations must first have appeared where the terrain and climate suited. So it was for humans. The different species were rooted *in situ* from the start, adapted to distinct niches. That 'the Caucasian, the Mongolian, the African, and the American Indian' *are* species was attested by the fact that they differ from one another more than many monkey and horse species do. In fact, it was easier to tell a 'Hottentot' and white man apart than 'any two species of the cat kind'. Caldwell's reductio ad absurdum ran thus: if the four species of *Homo* – looking so different – are 'descended from an original pair', so similar species of monkeys, cats and horses must also go back to common monkey, cat and horse ancestral pairs. Caldwell was trying to make Prichard look ludicrous by thrusting him into the transformists' camp. If a Caucasian can produce a Mongolian by migrating, then the 'Ourang-outang' could produce a Malay. It would be a common accusation levelled at Prichard: that he was imputing an illicit power to nature. Illicit because there were no known physical causes which could achieve this effect.[28] Prichard denied that *species* transmute, as Darwin knew. But Darwin was taking this next step – seeking causes that *could* indefinitely transform species.

So Caldwell was arguing that the various human species were unrelated and had no common origin. Now, since species don't generally interbreed, except to produce mules occasionally, any mixed human offspring must be *hybrids* of limited viability. Caldwell believed that the children of 'the mulatto [Spanish 'mule' human], the hybridous production of the Caucasian and the African', cease ultimately 'to be productive, and the breed becomes extinct'. Prichard, and following him Darwin, knew that the limits of interbreeding and viability of offspring might be experimentally tested using backyard breeds of fancy, fowl and farmstock. But for Caldwell there was no comparison between tame animals and humans. Whereas hogs and poultry have clearly changed under domestication, men have not. To prove it he pointed to non-outbreeding Jews and Gypsies as easily recognizable across continents and climates. These peoples were 'fatal' to Prichard's unitary theory. Then Caldwell looked to ancient records which showed fully formed dark Ethiopians 3,445 years ago, when the Caucasian Noah was supposed to have walked down the gangplank to repopulate the earth only 700 years earlier. The various human species had *always* been distinct.[29]

Attacking Prichard's biblical justification for racial unity was bad enough to those on the BAAS podium. Worse was Caldwell's defence of slavery. Darwin might have had a thousand Brazilian slave-owners in mind, or he might have been thinking of the likes of Caldwell, when he scrawled in an evolution notebook 'Do not slave holders wish to make the black man other kind?' A rich complex of emotions lay behind this onslaught on Caldwellian-style denigration. Plantation apologias were loathed by Darwin, and his common descent of black and white, and ultimately *all* life, stood out in response.

Nor did Caldwell's provocation end there. Hodgkin pleaded – with Caldwell present at the BAAS in Plymouth – for Indian chiefs' sons to be packed off to Britain from mission schools. Caldwell was scathing on this. Only whites had the capacity for real culture. Aboriginals were 'Motionless', like other animals, unimproved since time immemorial. None had produced a 'Cicero, a Bacon, or a Shakspeare'. It hardly pleased Hodgkin to be told that Red Men were an untameable species locked into a niche, and that 'civilization is destined to exterminate them' along with the buffalo. Since every effort to 'educate the Indians, has but deteriorated them', said Caldwell, the extinguishing of the tribes was a mercy killing of a tortured feral people unfitted to the creeping civilization. The booming Caldwell portrayed this as the real natural history of man, which bleeding-heart philanthropy was unnaturally distorting. Humanitarian intervention was 'misspent', dragging out the misery of a people adapted to woods and prairies. Losing this habitat, 'their extinction will be a dispensation of kindness, not severity.'[30] To the Quaker Hodgkin, the slave-owning, extermination-happy Caldwell, an overbearing six-foot-two Southern orator, a man equally self-righteous and inflexible, must have been loathsome.

The BAAS leaders' fears were realized: ethnology was becoming explosive. Nevertheless, Darwin had his questions in place on printed ethnographic leaflets, whose real purport no one could have guessed. And with Hodgkin's distribution expertise (honed over two decades of anti-slavery pamphleteering) shifting numbers via the learned societies, sending them to émigrés, and supplying expeditions starting for the Red Sea, the Niger, South Africa, the Canadian far west and Russia,[31] the whole operation seemed God-sent.

The Darwins were finely tuned to the fates of these expeditions. Ironically the worst disaster among them had the biggest repercussions, not only for Darwin's science, but for the whole philanthropic movement.

The Niger expedition began optimistically enough with the foundation of the African Civilization Society. After the ending of slave apprenticeship, British moralists sought new conquests. Eyes scanned the globe for dusky peoples on dark continents needing the light of liberty and the blessings of free trade. Joseph Sturge founded the British and Foreign Anti-Slavery Society in 1839 to outlaw the demand for slaves worldwide (eventually other abolitionist bodies folded or merged into his). But Buxton adopted a more business-like strategy: he established the African Civilization Society to halt the supply of slaves at source. Commerce again was the key: he wanted trade in goods to replace trade in lives. By mobilizing Africa's natural resources and native ingenuity, the British would show that legitimate business was not only more profitable than slaving; it also reaped eternal rewards. The society's Niger mission seemed straightforward: explorers would push inland to make treaties with local chiefs, set up trading posts and a model farm, and collect scientific information.[32]

Darwin's disillusioned Edinburgh friend Greg, sick of his 'dog's life' as a mill-owner, and watching the family firm break up in 1841, saw through the plan. Greg feared that few of the civilizers 'will ever return'. But his real objection was economic, pushing him further from Darwin's humanitarians. A Custom House axiom had it that no trade could be put down that turned a 30 per cent profit; the textile magnate did the sums and wrote an anti-Buxton book on the answer: the slavers out of Africa were turning a 200 per cent profit. Making a cold monetary calculation (the sort that would have given even a free-trader like Darwin shivers), he argued that only when American or Brazilian planters saw West Indies commodities produced at less cost than their own slave-harvested products would they free their own slaves on self-serving grounds.[33]

Still Buxton's scheme attracted the best people. The first public meeting of the African Civilization Society in June 1840 was thronged with robed and ermined worthies. Prince Albert presided with a score of prelates and a dozen peers beside him. Lords Glenelg and Palmerston, the flamboyant Foreign Secretary, sat with Sir Robert Inglis, near the plain Thomas Clarkson, Esq. Buxton announced the committee membership, which included Captain Robert FitzRoy. Since, as Emma said, 'some deaths' were expected, which would be 'no great comfort to those who die or to their friends', no Darwin or Wedgwood joined.[34] Days later, at the World Anti-Slavery Convention in London – with the 80-year-old Clarkson presiding at the opening session (his last public appearance) and Martineau invited as an honorary American delegate – Buxton sought funds. The Niger Expedition would open a 'highway into the interior of Africa', removing obstacles to the 'diffusion of Christianity and science' and rendering the continent a 'salubrious residence for European constitutions'.[35] Money rolled in and three ships were built.

On this 'Bible and plough' expedition up the Niger, all three commanders carried the Hodgkin–Darwin questionnaire (one, Captain Bird Allen, was known to Darwin, having supplied him with information on coral reefs). So too did the draughtsmen and naturalists – the ships carried two botanists, a geologist, mineralogist, seedsman, and Louis Fraser, curator's clerk at the Zoological Society whom Darwin's friends recalled as 'a rough kind of lad'. Accompanying them were two Church of England missionaries and a party of black West Indians, with African labourers and crew: 302 souls in all. They reached the mouth of the Niger in their £64,000, state-of-the-art iron paddle-steamers, two named in honour as *Albert* and *Wilberforce*.

Six weeks later, a relief vessel found the *Albert*, with a black American freeman from Liberia at the helm, struggling back down river, its mission aborted, and moving 'like a plague-ship, filled with its dead and dying'.[36] 'River fever' ultimately killed thirty-nine Europeans, including thirteen officers. The tragedy had enormous repercussions: the African Civilization Society was wound up, Buxton died discredited and the redemption of Africa by what Dickens derided as 'railroad Christianisation' (through commerce) received a serious setback. Thenceforth the emphasis would be on bringing Africans to Britain for missionary and medical training.[37] Worse, racist literature would exploit the calamity by noting that the black crew walked though this white 'grave-yard' unscathed. Why had they survived? – because they were a separate species primordially adapted to the Niger miasmas.[38]

One survivor, Louis Fraser, went on to become museum curator at the Zoological Society and Darwin's occasional correspondent. Another, the *Albert*'s geologist William Stanger, would meet Emma in 1842 at Lady Inglis's and recount how he had mastered the engines from a manual and had run them night and day for ten days to get the steamer back into open seas, although oddly 'all the painful sufferings had gone out of his head like a dream'. A third survivor, intrepid master-at-arms John Duncan, came back emaciated, despite which he volunteered to return to penetrate the Niger and Lagos rivers inland. Such a venture still seemed a moral and commercial imperative, and Darwin subscribed a guinea to the cost even 'though it be murder on a small scale'.[39]

The replies to Darwin's queries would obviously come at some cost. But as a result of the disaster the questionnaire acquired a new importance for the Aborigines' Protection Society. With the public disillusionment, it shifted its policy towards scientific study of aboriginals, at least as a prelude to protection. The disaster moved Hodgkin's philanthropists in a scientific direction – in Darwin's direction.[40] While Darwin was now convinced that there was some aspect of black people's constitution which rendered their race different.

This 1839 BAAS meeting impinged on Darwin's work in another way: it spurred him to mine the latest (3rd) edition of Prichard's *Researches into the Physical History of Mankind*. On leaving Birmingham, Darwin went to Maer and Shrewsbury, 'languid & uncomfortable'. Here he read Prichard. He had meant to for a year. A memo in 1838 reminded him to make an abstract of Prichard on those areas – colour, cranial shape, hair, intellect, female genital organs (there were no taboo subjects) – in which the races differed.[41] Now he devoured the two volumes. No book had underpinned the edifice of human racial unity with such extensive zoological, medical, ethnographic and linguistic piles before. Mostly it was the evidence drawn from zoology that attracted Darwin, as well as the travellers' tales that were pushed into proving that a single human stock had diffused across the world to result in today's races. Prichard was a kindred spirit: he might have shown his age in defending Adamic unity and biblical chronology, but he was a moral ally. His global collation of animal facts showed Darwin that he worked in a surprisingly similar way.

But while Darwin's evolutionary studies also underwrote abolitionist values, his needs were different. Prichard stockpiled evidence to prove the races one; Darwin went beyond that to plot hypothetical mechanisms that would splinter them, push them apart – in short, point out how they could have *originated naturally*, as Prichard believed they did. The son of a doctor, a failed medical student, showing the first signs of genuine malady that would quickly turn into convenient invalidity, Darwin naturally investigated relative susceptibilities to disease in the races. He probed the causes of diseases, how they became hereditary, why they manifested in certain offspring or jumped generations. Nothing too odd was overlooked: the old revolutionary-era surgeon Henry Cline might have published only one book, on animal form, but Darwin spotted his observation that farm animals that were bred for bigger size were 'more liable to disease'. Races were like families: far into the future, some will have many descendants, others few; and 'Who can analyze causes [of this, whether they be], dislike to marriage, heredetary disease, effects of contagions & accidents'.[42] He started hunting up medical statistics for the 'same races' but in 'diff. countries', to see if their proneness to disease varied. If so, it might help them evolve into different races in the future.

It was in the back of his mind as he read *Researches*. Prichard was intent on proving that different species had different diseases; since humans all suffered the same diseases they were of one species. But even if no disease 'is peculiar to one race', there may be different predispositions to disease in different races, as there are in families and individuals. Darwin latched on to 'predispositions'. Was this the key to racial separation? Because the human races stand closer together perhaps they only show incipient predispositions to different diseases. Prichard's was 'surely the philosophical way of looking at this': the nearer 'races of men merely show predisposition different', he wrote.[43] But that was as far as he got.

Ultimately Darwin never could turn up much evidence for the 'disease' scenario, and as it faded the ground was finally cleared for his major theory – ultimately to be called 'sexual selection' – to rise unchallenged.

Others were reading Prichard too, and not only Southern gentlemen were refuting him. One of Darwin's informants at the Zoo was himself unhappy with Prichard's idea of primordial migration of black (or even white) men to the far continents where their anatomies transformed. This was the expert who described a number of Darwin's mammals from the *Beagle* voyage, William Charles Linnaeus Martin (no disguising the fact that he had a naturalist for a father).

Martin, newly appointed assistant curator in the Zoological Society's museum, was an expert after Darwin's heart – on monkeys and human races, and about to start publishing on domestic breeds. But it wasn't his account of Darwin's short-limbed fox from Chiloé island or irregularly plated armadillo that had caught the Darwin family's eye. Rather his word on Darwin's cats, and his belief that one, small-headed and long-bodied, shot by an old Portuguese padre near Rio for Darwin, may have been a new species. In that event, he said, he proposed to christen it *Felis darwinii*.[44] The *Morning Herald* report had tickled the family pink, 'especially Papa', wrote Catherine Darwin. Alongside the descriptions of Darwin's Galapagos finches by the artist and Superintendent of the Zoo's Ornithological Department, John Gould, the 'Felis Darwinnia [*sic*] is mentioned'. 'Papa wants to know what gratitude the Zoological have shewn you; they aught at least to make you an honorary Member'.[45] (Unfortunately it turned out not to be a new species, but the jaguarundi, or 'otter-cat', so called because it hunts in streams.)

Darwin sat in on meetings. As he watched Gould disentangle his South American birds, so he would take in Martin on monkeys[46] (Martin's forte: he was monographing the Indian species at that moment). Martin was a former apothecary or surgeon not only skilled with the scalpel: he knew as much about orang-utan expression as loris dissection. Darwin chatted with him and recorded his anecdotes about the Zoo animals in his notebooks.[47] Martin was trusted enough, in fact, for Darwin to cite him extensively in his later books.

Thus when Martin spoke it was as a serious naturalist, and he spoke now on Prichard. After the retrenching Zoological Society made him redundant in 1838, Martin turned publisher's hack and penned his major work, with the cumbersome title *A General Introduction to the Natural History of Mammiferous Animals, with A Particular View of the Physical History of Man, and the More Closely Allied Genera of the Order Quadrumana, or Monkeys* (1841).

The Negro, he observed, seemed to bring out the extremes in ethnologists. The venerable Prichard was given due deference, but Martin could see no evidence that natural causes were 'capable of producing distinct races'. There was no likelihood that heat had blackened the Negroes' skin, 'thickened their lips, crisped their hair, and elongated the jaws and the heel', nor that Europeans in the tropics were turning black. Just as the 'Race-horse, breeding with the Race-horse, produces a Race-horse', uninfluenced by the climate of their stables worldwide, so human émigrés breed true wherever they live. Indeed, whether humans constituted races, or species, was a moot point (although tellingly Martin classified them as *Homo ethiopicus*, *Homo hottentottus*, and so on). Only by piercing 'the darkness of antiquity', the veil of 'by-gone time, could we work out a history of our species, commencing with Man's first existence on the globe', and solve the question of origins.

But Martin's was a complex image of humanity. While he might have doubted Prichard's central tenet, he was quite prepared to accept Prichard's belief that the various African races could lose their 'supposed badges of inferiority,' given an education, and rise 'in the scale'. Their minds were adaptable enough in a cultural sense, however fixed their anatomies. Part of Martin's proof came from Darwin. That even the most degraded races could rise (with British help) was shown by his discussion of the Tahitians:

> Mr. Darwin speaks in very favourable terms respecting the natives of Tahiti, and regards them as far superior to the New Zealanders, both mentally and physically; a superiority to be attributed to the humanizing influence of British exertions, in reclaiming them from the barbarous habits and customs which, till recently, placed them amongst the most degraded of the human race . . .[48]

So not only were Martin's publications a resource for the growing school advocating the plurality of unrelated races,[49] but he was equally drawn on by those advancing a single origin. And Martin had enough humanitarian blood in his veins to sense that accusations of 'intellectual inferiority' left the Negro vulnerable to the slave-masters in their 'acts of cruelty and oppression'. He endorsed studies which showed the Negro brain was as large as the white man's, and thought that counterclaims were based on samples of 'the miserable remains of an enslaved people, bodily and spiritually lowered and degraded by ill treatment'. The savage ways of the 'lowest' races were, concluded Martin paraphrasing Darwin, capable of being 'ameliorated by civilization'.[50]

*

But this equality of brain size was fiercely contested. Cranial capacities were Morton's growing concern in Philadelphia. Hodgkin's old Edinburgh friend was in 1839 the new Professor of Anatomy at Pennsylvania Medical College and a pre-eminent palaeontologist (who knew his fossil oysters). He was riding high when he introduced his epochal book, *Crania Americana* (1839), seemingly out of the blue.

This illustrated his other talent, for collecting human skulls (having been fired, no doubt, by his Edinburgh experiences). Some $10–15,000 had gone into building his modern Golgotha, and students from the earliest days recalled seeing huge numbers of skulls in his office – 867 were counted by 1840. The effusive, irreligious United States Consul at Cairo, George R. Gliddon, had enthusiastically plundered tombs down the Nile as far as Nubia, shipping him 137 skulls, including ninety mummy heads. But Morton's speciality was Native American tribes. *Crania Americana* described the skulls of more than forty Indian nations. Morton's mausoleum was praised at home as 'the scientific glory of the United States'.[51] On the back of *Crania*'s plaudits, Morton was elected one of the Vice-Presidents of the Academy of Natural Sciences in 1840.

Although Hodgkin had Morton made a Corresponding Member of the Aborigines' Protection Society before *Crania* appeared, one wonders about his feeling afterwards. For while Hodgkin might have hated the bombastic Caldwell, Morton was impressed, and in Kentucky Caldwell *loved* the *Crania Americana*. Morton was moving away from Prichard, Hodgkin, Darwin and the English humanitarians. He was not merely rejecting the notion of mankind's origin from a single pair, but its corollary: the 'unnecessarily inferred' belief that human races must have been brought into existence by the 'vicissitudes of climate, locality, habits of life' and so on. The Noah's Ark imagery in Prichard was also under attack: Morton doubted that an 'all-wise Providence' would send a universal deluge, forcing a subsequent dispersal of people across the whole globe against desperate odds. Morton argued for the creation of aborigines in their homelands, as befitting divine wisdom: his twenty-two great families of man consisted of nations that were initially unique and created on the spot, even if subsequent mixing had occurred. His commentators at the Academy talked of 'that imperishable unit, the Genus Man',[52] as if divergent humans were so many species. And although Morton was careful never to say as much in *Crania*, this line was hardly discouraged.

Crania Americana was a test of whether 'American aborigines, of all epochs, have belonged to one Race, or to a plurality of Races'. Morton's conclusion was that (barring Eskimos) all the Indians from cold Canada to arid Paraguay were of a single sort. This meant that one cranial type occurred across many different terrains from sub-Arctic to torrid zone. Climate could not have formed these people, because they were uniform across such a vast range. It was an attack on Prichard's thesis, and reinforced in a paper to the Boston Society of Natural History in 1842, published as *An Inquiry Into the Distinctive Characteristics of the Aboriginal Race of America*. Nearly 400 skulls of North and South American Indians – from Peruvian cemeteries to the Western plains – proved they were of 'one race'. Even the long-headed Titicacans were not unique, he now believed, but peoples who bound their infants' heads, as tribes such as the Flatheads on the Columbia River were known to do. Obviously, the problem with characterizing them all as uncivilizable 'savages' was the great, vanished 'Toltecan' dynasty – the one that had left buildings and sculptures in Peru and Bogota, the pyramids of Mexico and the fortifications of the Mississippi Valley. But Morton downplayed the importance of these Toltecan monument-builders. He supposed that this high point in 'refinement' was only possessed by a 'handful of people'; most remained savages, who eroded this ancient civilization and left their descendants on the modern plains as a sinking race. All in all, 'extinction appears to be the unhappy, but fast approaching doom of them all'.[53]

Crania Americana added to the evidence to support the eviction of Native Americans from their lands. It was a huge resource for racial analysis and Dr Caldwell responded in kind in a Kentucky journal. Fleshing out the animality revealed by these savage skulls, he emphasized how different tribes 'wreak their personal vengeance, or conduct themselves in war'. A phrenological analysis revealed them as fierce, degraded, indolent and resistant to civil restraint, while the Caribs were even revealed to 'covet extermination, which is nearly accomplished'. Caldwell went further than Morton. The redskins were destined to lose 'in war with a race of men *superior to themselves*'. Like the millions of Hindus kept in check by a few thousand East India Company men, the Native Americans had found 'their conquerors and masters are Anglo-Saxons – that variety which stands decidedly at the head of the Caucasians, and is their highest caste'.[54]

Humanitarians might challenge Caldwell and Morton. They might push for an Asiatic origin of America's aborigines, and religiously sanctioned unity for all of God's children. They might counter Morton's image of warring indigenes with lists of civilized atrocities; or notice Choctaw and Cherokee progress as proof of potential; or mention that the locals who greeted the Spanish in Mexico and Guatemala were more scientifically literate and architecturally gifted than the woad-covered Britons who greeted Caesar. But these humanitarians were preaching to the converted, and too many were leaving the church. Those left in the pews – including the Wedgwoods and Darwins – had been baptized into such comparative views. Hadn't the 1835 Select Committee, questioning Hodgkin, pointed out that the Britons resisting Rome's legions were in as great a state of 'degradation' as today's Maoris? Such relativism shaped Darwin's evolutionary vision, helping to explain his taunt later in face of Lyell's hauteur: 'to me', said Darwin, 'it would be an infinite satisfaction to believe that mankind will progress to such a pitch, that we sh^{d}. be looked back at as mere Barbarians'.[55]

The humanitarian aisles were emptying in the 1840s, particularly in America. The new ethnology was empowering for medical men in a nation battling 'savages' on its frontiers, while a slavery system suddenly found a new scientific sanction.

Morton's statements certainly caught the pro-slavery eye. Gliddon's Egyptian skulls and hieroglyphs showed that 'the Caucasian and Negro races were as perfectly distinct in that country upwards of three thousand years ago as they are now'. And when Morton published plates of Gliddon's mummies in *Crania Aegyptiaca* (1844), it didn't take much to suggest that the Negroes' 'social position, in ancient times, was the same that it is now; that of servants or slaves', making servitude look like a Providential system. Morton's hard-edged ethnology was equally appealing to the Boston intelligentsia gathered around the *North American Review*. (This was a conservative rag at the time, which 'no Southerner ever has accused of Abolitionism', even if the *Review* complained about the moral repression of the slaves.) To the New England elite Morton's precision measurements guaranteed 'fairness and accuracy'. The Caucasian Egyptians and Negroes *were* distinct; the feudal Toltecans *were* overlords of the savage Indians, and resisted 'amalgamation' as strongly as the classes in 'aristocratic Europe' ever did. The emphasis remained on stratification: civilization was sustained by the ennobled few, perilously so in Toltec culture, which offered a metaphoric warning about the overrunning of Southern society, should the blacks ever be freed. *Crania Americana* was simply the 'most valuable addition' to the race question for a generation.[56]

But what really met with approval was Morton's recalculation of brain sizes. He flatly contradicted the Heidelberg anatomist Friedrich Tiedemann and claimed that there was a large spread in brain sizes, from 87 cubic inches in white men down to 78 in Ethiopians. This brain ranking drew more comment than the rest of the book put together, and it continues to the present day.[57] But the point was it met prejudiced expectations. A year before Morton's publication, a leading Edinburgh phrenologist announced that 'all modern-day anatomists are correct' in affirming the Negro 'inferior'.[58] With Morton the stereotype found its hard data. Tables of skull-plate measurements gave his cranial capacities their authority. Most commentators were now convinced that not only were the races separate from time immemorial, but they could be impartially ranked. In 1840–41 the *American Phrenological Journal* praised Morton's 'magnificent work' and publicized his hierarchy under the title: 'Superiority of the Caucasian Race'. Morton had shown the 'blindness' of Tiedemann's humanitarian lackeys.[59] The *Journal* used Morton's figures to make black men a separate species and move redskins closer to reservations. Racial Anglo-Saxonism was in full swing – with phrenological explanations for historical subjugation, and anatomical predictions of future extinction.

This was the Edinburgh phrenologist George Combe's territory, and it was Combe's review that helped paint Darwin's picture of *Crania Americana*. The dour Combe was coincidentally touring the United States in 1838–40, lecturing and reading American heads. His books had heralded his arrival, with local prefaces to the US editions given a hard racist edge, far beyond anything in Combe's original.[60] Growing numbers did consider blacks irreclaimable, and Combe's *System of Phrenology* (in its fourth edition by 1836) had done little to dissuade them. Even if Combe abhorred slavery, his *System* had portrayed blacks with small organs of 'Conscientiousness' or 'Justice', which showed why judges in India and America were right to bar native testimony in colonial courts. It also suggested that Africa's kidnapped 'sons and daughters', with their diminished sense of fairness, were oblivious to the injustice of slavery.[61] Combe's deterministic system could easily be used to bolster the South's 'domestic institution' (and was).

But meeting slavery in the raw, even in Washington, DC, seems to have shaken Combe. In Maryland it might take its 'mildest form', or so said the Kentucky Senator Henry Clay (in a speech against abolition which many saw as his opening shot in the 1840 presidential election campaign). But like 'piracy, murder, and fire-raising', said Combe, it was still an abuse and, unchecked, would end 'in blood and devastation'. The 'high-toned' rhetoric in Congress about 'universal freedom' sat ill with local appeals in newspapers for slaves for 'the Louisiana and Mississippi market'. Worse were the Senatorial discussions on slaves as property. Such commodification seems to have convinced Combe of the institution's baseness. One wonders, too, about Combe's reaction to the virulent strain of pro-slavery phrenology that he encountered in some cities. Clay's rabble-rousing certainly upset him, particularly the claim that abolition would lead to 'a war of extermination . . . between the races'.

Washington lay on Combe's road to Damascus. He had a change of heart: formerly he too had thought that war would follow liberation; now he back-pedalled. The counter-evidence was piled on: blacks, he found, had bigger brains in the Free States (where they could exercise their intellectual faculties); they were attentive at his lectures in Quaker halls (having been permitted to stand discreetly by the door); their sermons were delivered in impeccable English. Then, having a black woman's brain dissected, Combe found that the faculties of Benevolence, Conscientiousness and Reflection were actually bigger than those of Native Americans. This explained black people's patience, trustworthiness and, in a complete reversal, sense of justice. So 'the very qualities which render the Negro in slavery a safe companion to the White, will make him harmless when free'.[62] Phrenology, reconsidered, was declaring Clay's scaremongering 'unfounded'.

So it was a cautious Combe, sensitized to slavery, who anonymously penned the review of Morton read by Darwin. Combe's about-face didn't stop him taking a dim view of the 'ferocious' Indian, however. Like so many, he marvelled at Morton's skull collection and had even lent Morton Blackfoot Indian skulls to figure in *Crania Americana*. Morton, for his part, had the 'distinguished' Combe add a phrenological appendix to the book. Here Combe confirmed what was, in effect, Darwin's problem: naturalists had to explain shades of difference in the races under *similar* environments. Not that Combe investigated Native Americans *in situ*. His analysis was based on Morton's skull illustrations. Nothing more, he thought, was needed to provide 'an authentic record in which the philosopher may read the native aptitudes'. These proved to him that national attainment could only be determined by mental endowment. It could not be shaped by 'soil and climate', because Europeans progressed in the New World where local tribes stagnated.[63]

Even having contributed to *Crania Americana*, Combe didn't scruple to review it. His panegyric appeared anonymously in the *American Journal of Science*, run by Yale's Benjamin Silliman, who refused to treat phrenology as 'ridiculous and absurd'. (Like bona fide phrenologists, Silliman noted its practical value in identifying potential criminals – which probably explains Combe's hearty welcome in New Haven by Connecticut's Governor and the Chief Justice of the State.)[64] This became possibly the most syndicated review of them all. It introduced many to Morton, in Darwin's case through a de-phrenologized and bowdlerized reprint in the *Edinburgh New Philosophical Journal*.

Darwin would dip into Morton periodically from now on, but he was initiated into it by this review, which fixed the emotional and moral attributes of a race from the dawn of its appearance. Darwin knew the way the wind was blowing – very much in Prichard's face. This review confirmed it again, with its derogatory racial ranking based on innate abilities. And as if to confirm the ultimate disparity between key races, the review gave black and 'red' people unique pigmented layers in the skin. No epidermal layer could be *produced* by sun or heat or cold. It backed Morton's belief that each race was so adapted 'by an all-wise Providence' from the start. Neuroanatomy explained why the Indian nations were incorrigible, incapable of abstract reasoning, unchanging since the 'primitive epoch' and resistant to 'missionary labour and private benefaction'.[65] Nothing was outside phrenology's grasp as a parlour explanation of racial encounters.

It must have shown Darwin yet again that scientific society in an expansionist age was diverging dramatically from his path. He no longer saw climate as paramount, either. But a hardening pluralist line, so useful to Southern slavery and Northern craniologists, was cutting the ground from under his sexually selective proto-mechanism for dividing up the races from an ancient unified people.

*

The Darwins stood firm even as events looked bleak. They were outraged by the British government's attempt in 1841 to cut the sugar duties that had favoured the emancipated West Indies (the tariff had effectively blocked cheaper slave-produced sugar from elsewhere). Wasn't it preferable even for poor Britons to pay more for non-slave sugar than to let in slave sugar and thus push up the demand for slaves again? Darwin's old friend Greg took a less ethical line: a temporary tariff was necessary to allow the West Indies planters time to optimize production using black or imported Irish labour and thus undersell slave cane or cotton. The Darwins, who could afford sugar at any price, stood firm. 'We Antislaveryites are all taken aback by the Government's plans', Emma protested.[66]

Matters grew worse as a diplomatic row threatened Anglo-American relations. Although the transatlantic slave trade had been outlawed by Congress in 1807, an internal slave trade flourished along the south-eastern and southern coasts of the United States, with major ports at Charleston, Mobile and New Orleans. Human cargoes were the life-blood of the plantation system, but vulnerable at sea. In 1841, slaves en route from Virginia to Louisiana hijacked the American brig *Creole* and forced the crew to land at Nassau in the Bahamas, an emancipated British colony. All 135 blacks (except those accused of murder) were given asylum and set free. Southern slave-holders screamed for compensation. The US Secretary of State, Daniel Webster, demanded the slaves' return on the ground that, under American law, they were legal property. In London, the Foreign Office noted calmly that the mutiny took place in international waters where American law did not apply. As tempers frayed the Darwins grew 'alarmed about America going to war with us'. At stake also, as Aunt Jessie heard, was the internationally agreed 'right of search', aimed at halting slave traffic on the high seas. But whether over 'the right of search or not giving up the slaves in the "Creole" ', Emma assured her, if war came it 'will be in a good cause'.[67]

Whether denouncing American slavery or boycotting its products, Darwin was all the while busy undermining its scientific props, its denial of common blood and common heritage. Yet these props were being strengthened by the year, and one man did more than all the others to reinforce them.

In Mobile, Alabama, Dr Josiah Nott was a true Southern gentleman. He had a kindly hand for the sick and callous thoughts about the place of 'niggers', of whom he owned nine. A congressman's son with impeccable credentials, Nott was inspired by Morton's *Crania Americana.* From South Carolina, he too had gone to study at the University of Pennsylvania with Morton's old mentor, Dr Physick, before migrating to Mobile in 1836. Here his taint of irreligion was camouflaged by dedicated work on yellow fever. He grew wealthy amid the surging population, treating cotton traders (and their valuable property), and started airing his views on '*niggerology*', which suddenly put him in the driving seat of the new science.[68] Here – in the slave town that would run out preachers for anti-slavery sermons – his became the scientific voice of race separation.

Mobile presented many faces. Its bustle of cotton traders, English merchants and visitors each saw a side that flattered their attitude to abolition. Parts of the town were elegant, with gorgeous gardenias scenting the air, particularly along the shore. Under them Southern belles promenaded, explaining to foreigners how 'they cannot do without slaves', and that 'public opinion' ensured their good treatment. Yet mingled with the fragrance was the stench of burning flesh, and manacled ghosts stalked the old harbour, once Alabama's prime port for shipments from Africa. Just before Martineau arrived in Mobile in April 1835 two slaves had been flayed alive 'in a slow fire, in the open air, in the presence of the gentlemen of the city'. Apparently it was for the rape and murder of a white girl, although the slaves were arrested after an unrelated incident. The details were obscure, but Martineau's hostess put it down to white patrician 'licentiousness' among the black girls, forcing retaliation, for which the two unfortunates paid with their lives. The story, as so often, had a sexual theme, but then abolitionist propaganda often gained its force from near-voyeuristic trawls in 'the deepest sinks of a licentiousness' known only in these torrid Slave States.[69]

Miscegenation was the subject that obsessed Nott. Whether blacks and whites could produce viable offspring over generations of intermarriage was becoming a contested question. Many on the anti-slavery side never doubted that they could. Hearing of Henry Clay's belief that Negroes and whites, whom God had separated, must 'be kept asunder', Combe answered:

When Providence intends to prevent races from mingling, he renders the product of their union unprolific, as in the case of the mule. The slave-holders have impressed on the slave population striking evidence that no such prohibition exists between the African and European races.

Overseers cohabiting with slave girls and selling their fair children; white masters seducing slaves (and white mistresses stringing the women up for 'permitting' it); the owner's elegant illegitimate daughters being sold into prostitution on his death: these were the stock-in-trade tales of British travellers to prove that 'licentiousness and tyranny have met together'. Because such stories were lurid and legion, and because 'mulattos' were everywhere, there was a belief in the total viability of the intermarried generations. 'The white tint distinguishable in thousands' of slaves proved it, said Combe. He then aggravated his hardline backers more (so that he had trouble getting his *Notes on the United States* published unexpurgated) by describing the 'quadroons' (one-quarter blacks) as 'handsome and talented'.[70]

Nott challenged these assumptions in an explosive article, sparking a debate on hybridity that ran deep into Darwin's domestic animal literature. Defence of slavery was the cornerstone of the South's increasingly coherent philosophy. With the abolitionist onslaught from the 1830s, Dixie had renounced any Revolutionary heritage based on equality. The reaction had quickened with Nat Turner's slave uprising in Virginia (1831). After Turner's execution, vigilantes routed out abolitionists and set up plantation patrols. In Nott's Alabama, crackdowns ensued. Even before the insurrection, when Nott was growing up, South Carolina had jailed free black sailors on their shore leave to prevent any contagion. Pro-slavery justifications were settling into a pattern as Nott started to write, thanks partly to a growing Southern publishing industry. (With its rise went the bans on slaves being taught to read.) Christian support for slavery in the South may have been overridingly biblical – Israel had practised it, St Paul had sanctioned it – but through Nott's statistics, Southerners would now know how 'natural law' and an increasingly prestigious science fortified their conceptions of hierarchical order and social responsibility. Talk had turned from 'rights' to 'duty'. Care and Christian stewardship were slavery's new catchwords. The 'inferior' blacks in the South were portrayed as better off than the 'wage-slaves' sacrificed to Mammon in the northern factories. As the Founding

Fathers' ideals were rejected, an organic view of the peculiar institution was coalescing in the South, a view – backed by Morton's, Nott's and Gliddon's science – which saw slavery sustained through prehistory and ordained by God.

How to prove the real slaves were better off than the North's free 'wage-slaves'? Juggling questionable mortality statistics and the 1840 Census, Nott asserted that the mortality of free blacks in New York was higher than for slaves in Baltimore. But these ledgers of death also suggested that mulattos were shorter-lived than either 'pure Africans' or 'Anglo-Saxons'. On the back of this 'finding', Nott ran his article in 1843 (in an impeccable medical journal) under the title 'The Mulatto a Hybrid'. The piece itself was a bit bald, merely five sides, and mostly cannibalized. But the conclusion, indeed the subtitle – 'probable extermination of the two races if the Whites and Blacks are allowed to intermarry' – threw down the gauntlet. The weakened, disease-prone mulatto was a decreasingly fertile time bomb. The doctor had seen it in New Orleans, where old mulatto families had 'run out so completely' as to leave estates devoid of heirs. The women were 'bad breeders' and worse mothers. Indeed, presented in sexual-aesthetic terms, the question in a white journal for white readers was loaded:

> Look first upon the Caucasian female with her rose and lily skin, silky hair, Venus form, and well chiseled features – and then upon the African wench, with her black and odorous skin, woolly head and animal features – next compare their intellectual and moral qualities, and their whole anatomical structure, and say whether they do not differ as much as the swan and the goose, the horse and the ass . . .

Planter lore had resurfaced with a medical vengeance. With infertility increasing through the generations, this amalgamation of 'Anglo-Saxon' and 'Negro' types would wipe out all purity and any future for civilized humanity.

There was a certain family irony to this. Nott's equally scoffing brother Henry had had a far more relaxed attitude to philandering among the black 'wenches'. He once acknowledged a friend's 'true philosophy' in taking 'the goods the gods have provided' – the friend having 'like Desdemona "battened on a moor" in absence of better pasturage'. But to the doctor this was vile; miscegenation of the 'Anglo-Saxon and Negro races' intolerable. Not really *races*, because a 'hybrid', unlike a 'mongrel', as Darwin cautiously reminded himself (terminology was a trigger word in these explosive areas), was a mix of *species*. Nott's mulatto was the 'degenerate, unnatural' progeny of 'two distinct species – as the mule from the horse and the ass'. This mule of a man was 'doomed by nature'. That such mule-men existed at all – being a 'violation of nature's laws' – had to be explained away by Nott.[71] His solution was to suggest that species hybrids were commoner than naturalists credited. Some were sterile, others, among geese, goldfinches and canaries, goats and ewes, were not, and many, like the mulatto, lay in between.

Nott's papers, and his pushiness, helped to reinvigorate the disputes on domestic animal-crossing which were to mark the ensuing decade. Indeed the flaring controversies over human hybridity drove the debates. Darwin was already engaged, and he was far from immune to developments in Nott's Deep South. He had his emissary to the Slave States: his urbane mentor, the gentleman through whose scientific eyes Darwin was trained to see, Sir Charles Lyell.

7
This Odious Deadly Subject

At the pinnacle of his profession, Charles Lyell had never crossed an ocean, met a 'wild man' or lived among slaves. His life was richly upholstered. Even when geologizing on the Continent, he travelled in the best circles. The terrain might be coarse, the volcanic craters fiery, but his view was always from above. So also it was in the 1840s when twice he sailed, hammer and wife in hand, to visit Americans on their native soil. For every rough railway trip or steamboat passage, after all the rock-climbing and clambering over hills, a dinner party awaited them, attended by liveried servants.

Bidding farewell to his pupil Darwin, that 'congenial soul' with financial independence and interests 'precisely the same' as his own, Lyell first sailed for Boston in July 1841 to deliver the Lowell Lectures on geology. An aspiring politician, Charles Sumner, met the couple and drove them about, as he had done Harriet Martineau years before. Sumner defended her 'scorching' *Society in America* and was drawn to her abolitionism. He would help form the Free Soil Party to halt the spread of slavery and devote his career as a Massachusetts senator to the cause of racial justice.[1] Among the Lyells' new friends, he would be the exception.

Moving effortlessly into Boston society, the Lyells found their hosts religiously liberal and socially conservative. The Unitarians and Episcopalians (American Anglicans) cared more for the sanctity of property than the sacred rights of man. Anti-slavery was fashionably respectable, but abolitionism – demanding fellow Americans forfeit their 'property' – was still anathema, the more so with the escaped slave Frederick Douglass lecturing around the state, stirring up trouble and recruiting support. Boston's cynosure, George Ticknor, a retired Harvard professor, welcomed the Lyells into his Park Street home. Here they met his great friend Daniel Webster, the US Secretary of State, whose defence of property in slaves endeared him to Southern planters. John Amory Lowell, trustee of the eponymous lectures, took them to see his spinning mills in the family's eponymous town, where slave cotton was made into textiles for sale back to the South. In suburban Cambridge, the Lyells met Ticknor's brother-in-law, the Revd Andrews Norton, 'the Unitarian Pope', and in Harvard's chapel they heard a dry Unitarian discourse on the text 'Thou shalt love thy neighbour as thyself'. In an Episcopalian service, Mrs Lyell felt a pang at seeing 'seven coloured women' receive the Communion cup separately, but she found the Unitarians segregated too. (During the Lyells' visit four years later, the textile magnate Abbott Lawrence took them to the Brattle Square church, where an elderly ex-slave, Darby Vassall, sat in the 'negro-loft above the organ'.) Blacks could vote, hold public office and serve on juries in Massachusetts.[2] But they were not loved as their white neighbours loved themselves.

Lectures finished, Charles and Mary Lyell headed south for the winter. Across the Potomac River into Virginia and down the eastern seaboard their eyes opened. But not in horror, as Darwin's had in Brazil; rather these were wide-eyed visitors to a far country who unexpectedly encountered gentlefolk like themselves.

A sign, 'Men and Women bought and sold here', marked entry to the Slave States. Immediately the complexion of society changed. The number of blacks had doubled or trebled in a generation. Invariably polite, black people were held on a tight leash, with 'passports' to control their movements after dusk lest they fall prey to 'abolitionist missionaries'. Having heard in Boston about these busybodies, Lyell listened to mild-mannered planters damn them as 'incendiaries, or beasts of prey', fit only to 'shoot or hang'. As the weeks passed, he began to share their resentment. With slaves 'as excitable as they are ignorant', and soon to outnumber whites because of the planters' indulgent paternalism, 'the danger of any popular movement' was, he admitted, 'truly appalling'. 'Severe statutes' had to be enacted, 'making it penal to teach slaves to read and write', strict laws passed against 'importing books relating to emancipation', all because of the abolitionists' 'fanatical exertions'. While Lyell thought their 'influence and numbers' overrated, he agreed with the planters. The arrogance of 'anti-slavery speakers and writers on both sides of the Atlantic' was proof, if it were needed, that 'next to the positively wicked, the class who are usually called "well-meaning persons" are the most mischievous in society'.[3]

This was what Boston liked to hear and, writing to Ticknor, Lyell played up the abolitionist evils endured by the planters. Were he in the masters' shoes, such 'meddling' would only irritate him too. Not an 'anti-abolitionist'[4] when he arrived, he now blamed the abolitionist zealots for causing slavery to be shored up. His conversion was Martineau's in reverse.

The Lyells steamed into Charleston, South Carolina, formerly the continent's chief slave port and the last to end the transatlantic trade. Here blacks were the majority and a white conscript militia kept guard. Local dignitaries queued to greet the couple at their hotel, including Dr Samuel Dickson, a Yale-educated medical professor, and his colleague, Dr Edmund Ravenel, who had been Morton's fellow student in Philadelphia. Ravenel took them out to see his 3,000-acre cotton plantation, The Grove, where they found a hundred slaves gathered at the mansion kitchen for a wedding. The happy couple were of 'unmixed African race' and the bride and bridesmaids 'all dressed in white'; an Episcopalian clergyman officiated. Later the Lyells learnt that the ceremony was a sham. It wasn't legal. Legal marriage would interfere with the 'right of sale' – paramount when 'slaves multiplied fast' and masters had to 'separate parent and child, husband and wife'. At Dean Hall plantation, where Ravenel took them in his steam launch for fossil-hunting, the point of human husbandry became clear. Each day, when the call went out, forty 'piccaninnies' came running to the kitchen door to get their dinner. The future toilers in Colonel William Carson's rice fields required a proper feed. 'It is really like taking charge of so many animals on a farm', Mary Lyell mused, even while worrying about the children's 'moral & religious state'.[5]

'Remarkably cheerful and light-hearted . . . talkative and chatty as children, usually boasting of their master's wealth, and their own peculiar merits' – so slaves in general appeared to Lyell. One of them, asked whether she was owned by a certain family, 'replied, merrily, "Yes, I belong to them, and they belong to me."' Was this not proof of her master's kindness? For all the horrors laid bare by abolitionists, Lyell found it 'impossible to feel a painful degree of commiseration for persons so exceedingly well satisfied with themselves'.

The more he was impressed by their cheery ways, the easier the comparison with servile English clods; the slaves were certainly 'better fed than a large part of the labouring class of Europe'. His sympathy wasn't reserved for the slaves, but the planters who bore the responsibility of keeping them: sensitive souls who, 'from motives of kindly feeling', were often 'very unwilling to sell' their multiplying stock. 'We must never forget that the slaves have at present a monopoly of the labour market; the planters being bound to feed and clothe them, and being unable to turn them off and take white labourers in their place.' Thus some owners were 'constantly tempted' to live beyond their means and face bankruptcy.

The South baffled Lyell. Here were 'two races so distinct in their physical peculiarities as to cause many naturalists, who have no desire to disparage the negro, to doubt whether both are of the same species, and started originally from the same stock'. How could they be reconciled and live in harmony? Perhaps 'negro reservations' could be set up where 'free blacks . . . might form independent states' and improve themselves. The districts would be chosen for their torrid climate, 'insalubrious to Europeans, but where the blacks are perfectly healthy'. The planters hooted the idea down: the 'niggers' would only sink back into 'savage life', regroup and mount raids on vulnerable whites. They were not Lyell's 'amiable, gentle, and inoffensive race' who over successive generations might aspire to 'moral and intellectual' equality.[6] Blacks were fit only for servitude.

Here finally Lyell drew a line. He knew Negroes could be improved. He disliked slavery – it affronted his liberalism as much as Darwin's – nevertheless six weeks in the South had transformed his 'feelings towards the planters'. The 'aristocratic' culture of the 'genteel' South – a country within a Union, populated by 'true gentlemen', whose standards were based on character and property – had secured one more supporter.[7] His prejudices were 'entirely eradicated'.

There for the geology, admittedly, Lyell had little time and perhaps less inclination to reflect deeply on the tragedy of the South. That science was implicated in it he knew from naturalists who doubted whether black and white were 'from the same stock'. He knew that the races were not created separately; those mixed-race people he saw showed blacks and whites were kin. But perhaps the time *would* come when the whites' intellect had so far outstripped the blacks' that a naturalist would call them 'two species'.[8]

One naturalist who had doubts on this score was an old acquaintance of Darwin's. Perhaps Darwin had advised Lyell to look him up. Originally a New Yorker, the Revd John Bachman had been called to St John's Lutheran Church in Charleston. There was no better climate than the South for his consumption. Darwin had met the pastor in the summer of 1838 when, his health 'shattered', Bachman had cruised to London. Here was a man after Darwin's heart, who raised ducks, and was a pigeon fancier and a keen horticulturalist – he had already published on vultures and bird migrations, and was monographing his way through American hares, shrews and squirrels. He had talked Darwin's talk: domestic breeds and hybrid ducks (having successfully crossed a guineafowl with a peahen himself, although the offspring was sterile). In summer 1838 Bachman had been in and out of the Zoological Society museum. He had also presented it with thirty-six native animal skins and sorted out their American shrew moles, dwarf mice and squirrels, causing grateful metropolitan experts to name a new 'Bachman's hare', *Lepus bachmani*, after him.[9] At meetings and in the crowded museum, Darwin and Bachman had clearly got on. Darwin's evolution notebooks left a trail of recorded conversations about how hares and birds vary across the Rockies, how flycatchers were extending their ranges across America, how Bachman's squirrels turned white in the north. It had all been grist to Darwin's theoretical mill.

Lyell now too found an encyclopaedic cataloguer of new American species. Bachman described in a thousand words what Audubon pictured in each of his accompanying lush plates. (The co-workers were in-laws; two of Audubon's sons had married two of Bachman's daughters.) Bachman swept his hands across a map, showing Lyell the boundaries of the mammalian ranges east to west, zones which existed even though there were 'no great natural barriers' to migration. Climate alone had limited the ranges. Humans, however, had broken through in their relentless westward march. Bachman may have been a slave-owner and anti-abolitionist, but he believed in human unity as taught in Genesis. All humans were of one stock descended from Adam and had ranged freely over the earth since the Ark landed on Mount Ararat. The Negroes were descended from Noah's son Ham, whose offspring God had destined to be 'the servants of servants'. As such, God-fearing whites had a special responsibility of care as the masters.[10] Minus Genesis, this was not far from Lyell's view. Blacks and whites belonged to one caring hierarchical family, however and whenever its members arose.

After lectures in Philadelphia and New York (where the *People's Press* favoured him with a phrenological profile), the Lyells resumed their geological junket, travelling as far west as Cincinnati on the Ohio River, separating the Slave and Free States. Criss-crossing the mighty waterway was a web of clandestine routes with safe houses where abolitionist 'conductors' helped fugitive blacks escape to freedom. Thousands got away on this 'underground railroad' each year, although many fell prey to planters' posses. The sight of 'four runaway slaves . . . chained together', under armed guard and jeered at as they were dragged back to certain punishment, made Mary 'feel quite sick', though her husband assured her it wasn't 'much worse than deserters taken back to a regiment'. While the mulattos were thought to be feckless, Lyell noted that in Ohio many mules proper were 'commended for their longevity'. It surprised him that the hybrid of horse and donkey enjoyed 'a portion of the ass's length of days', and he sent word to Darwin, laughing, 'What a perfectly intermediate creature!'

They travelled to Canada along the fugitive slave route, then back to Boston to bid farewell to the Ticknors. It was here four months earlier that Lyell had first met Morton, chairing a meeting of the American Association of Geologists. Lyell pronounced his inventory of American Cretaceous fossils 'for the most part very correct'.[11] He little realized how much more important Morton's cranial Golgotha would be to the future.

In 1840 Darwin closed his evolution notebooks and let his theory mature. Going public was not an option; there was too much 'prejudice'. He might desegregate the races through a common ancestor, might underwrite abolitionism by uniting all peoples, but putting apes in their family tree – making men soulless beasts – that would be tantamount to treason in a Anglican society sustained by creationist props. The fabric that 'totters & falls' would not be the creationist cosmos but his reputation. He ended his private jottings on a defiant note by comparing the instincts of child and chimp.

His health was getting worse, with frightening fluctuations. In 1840 he went to Shrewsbury and Maer for a family break. While cogitating about species, he became violently ill and stayed there for months recuperating.

Dr Darwin's gloomy prognosis resigned him to chronic invalidity. The invalid and his wife-cum-nurse Emma went house-hunting in the country. Fresh air, space for more children, room to grow plants, keep livestock and experiment – if a long-term palliative could be found, this was it. In the summer of 1842, as Britain's cities boiled over, hungry and angry during the economic depression, they left for an old parsonage at Down, a Kentish village, fifteen miles from the metropolis and far from the nearest railway. Moving was like closing the clasps on his clandestine notebooks. Privacy was assured.[12] Living at Down House like a country vicar, he would keep society at bay and work quietly on his theory.

Shortly before the move he dashed off a rough pencil sketch of natural selection in 35 pages. Over the next two years, while writing up the geology of the *Beagle* voyage, he 'slowly enlarged & improved' the sketch, and by summer 1844 he had a 189-page essay ready to be fair copied and sent to the press. But it was only to be published over his dead body, he told Emma. He preferred her to pay 'some competent person' to flesh out the essay, correct and enlarge it from his notes – spend a few months on the job. The 'best' editor would be Lyell. He would 'find the work pleasant & . . . learn some facts'.[13]

In these earliest private drafts of his theory, the audacity of the notebooks was reined in. Darwin's materialism can be glimpsed: mental changes produced by habit are said to be inherited 'through their intimate connexion with the brain'. But his examples came from horses, dogs and pigeons. He looked to the origin of domestic animals, which, in the case of dogs, was from more than one wild species. And he glossed the complex problem of the sterility of hybrids, and what the odd exceptions proved. Humans inhabit the earth like any animal and have similar instincts, as 'savages' show clearly; but the humans appear here without their ancestors. These drafts were safe, to an extent.

If Darwin scarcely touched on humans, he did concentrate on something unique to them – they breed animals. Each tribe selects and enhances the domestic stock it finds most suited to its circumstance (favouring shorter-legged sheep in mountains or thicker-pelted in the cold). Each race, 'slowly, though unconsciously', makes its own breeds. These are peculiar to itself. What humans thus achieve artificially, Darwin argued, the struggle for existence does naturally. Animals with an adaptive benefit, suiting them better than their competitors to local conditions, tend to leave more offspring. He also saw a 'sexual struggle' involved, with males competing for females. The 'most vigorous' bucks, best equipped for fighting, sire most fawns; the gaudiest courtship display or most striking singer lures the hen. Darwin now called this a 'second agency',[14] and it would merge into what, much later, he would dub 'sexual selection'. But he had yet to analyse the process or use it to explain the origin of human racial differences.

Nothing mattered to him more than 'man', the animal he had seen in its remotest habitats: wild men were still a shock. His notebook speculations might never see the light of day, but he continued to deepen his insights, trying to understand how this ape-descendant came to dominate the earth so variously. Travelogues – tales of encounter – like his own, were a favourite source. But his appetite was voracious. He got through scores of books every year: history, biography, theology, philosophy and novels; and natural history from animal husbandry to zoology. In 1841 he added Martineau's *The Hour and the Man*, celebrating the black revolutionary Toussaint l'Ouverture. Shrewsbury and Maer had large, well-stocked libraries catering for the family's catholic tastes. On visits, Darwin would spend days mining the leather. Stanhope Smith's 1788 *Essay on the Causes of the Variety of Complexion and Figure in the Human Species* defended Calvinist theology and American democracy with a case for Adamic unity. It might have seemed unpromising, but Smith, the late president of what would become Princeton University, argued for racial divergence by natural causes, and it was 'well read', even if it wasn't referred to again.[15] Less useful was *An Account of the Regular Gradation in Man* (1799) by a Manchester man-midwife, Dr Charles White. His 'hypothesis' (to Prichard frankly 'absurd') had species on a static staircase ascending from plants to people, with Europeans at the top and Africans just above the apes. Lice gauged the difference: those on 'negroes are blacker, and generally larger'.

'Poor trash Lyell' was Darwin's cryptic comment on this book. He meant Lyell had called White's book 'poor trash'.[16] Lyell was 'the one man in Europe, whose opinion of the general truth of a longish argument' Darwin was 'always most anxious to hear'. But the one 'longish argument' Darwin never discussed was for natural selection: Lyell knew nothing of Darwin's theory. Yet Darwin bought and mined each new edition of Lyell's *Principles of Geology* (now in its sixth edition), looking to Lyell's statements about humans as never before.

It now seemed so obvious: Lyell stood too high to see the lowly – he denied primeval man had ever existed. Humans to him were a recent creation, perhaps the most recent. They had arrived as if by a 'sudden passage of an irrational to a rational animal' – Darwin poked an exclamation mark in the margin against the statement of difference in kind. Moreover, according to Lyell, they had probably arrived in a civilized state, already 'intellectual and moral' and with a 'far higher dignity' than any beast. Another exclamation mark. Clearly Lyell had never set eyes on primitive humans: Darwin had *met* them in Tierra del Fuego, and in Australia where he found other archaic survivors. He scribbled his cryptic verdict: 'The introduct of man, only greater change than any species ornithorhyncus' – which meant that man's arrival was no more marvellous than a platypus's.[17] An argument might be raging over the platypus's egg-laying, but no one denied its uniqueness, and it too had an unknown history. To Darwin, mankind's appearance was no more and no less extraordinary.

In 1845 Darwin had finished writing the last of the *Beagle* geology works when Lyell's *Travels in North America* came out. At the time Darwin was preparing a new edition of his *Journal of Researches into the Natural History and Geology of the Countries visited during the Voyage of H.M.S. 'Beagle'* and adding a paragraph about how human skeletons in South American caves proved that 'the Indian race' had lived there for 'a vast lapse of time'. He scoured the *Travels* for more evidence, noting the limestone-embedded human skeleton from Brazil that Lyell had seen in Philadelphia. He was suddenly pulled up sharp by Lyell's plantation talk. Darwin's shock and dismay were palpable: 'your slave discussion disturbed me much . . . it gave me some sleepless most uncomfortable hours', he wrote Lyell, 'but as you would care no more for my opinion on this head, than for the ashes of this letter, I will say nothing'. But he could not let go, and added a snooty factual correction: the Fuegians 'never used a hollowed tree, but bark sowed together' for their canoes. Lyell's mistake had come from Morton's *Inquiry Into the Distinctive Characteristics of the Aboriginal Race of America*, a work which Lyell considered 'luminous and philosophical'. Morton had never seen a Fuegian; Darwin had, and it showed in his imperious response: 'D^{r}. Morton is so far wrong' about Fuegian canoes.

Lyell's reply has not survived. Perhaps he explained that tact was necessary: that slave-owners had welcomed him and he was about to make a second tour through the South where he would depend on their hospitality. Confronting them like Martineau would have been both ungracious and self-defeating.[18] Anyway, wasn't *Travels* dedicated to George Ticknor, Boston's anti-slavery eminence?

Darwin still tried to hold back in his rejoinder: 'I will not write on this subject; I sh^{d}. perhaps annoy you & most certainly myself'. But to no avail, it all flooded out: 'I have exhaled myself with a paragraph or two in my Journal on the sin of Brazilian slavery: you perhaps will think that it is in answer to you; but such is not the case'. Darwin was being coldly polite. He hated the 'sin' and despised Lyell for not hating it as much. Darwin scored Lyell's statement in the *Travels* which mentioned uncritically how a slave-owner 'defended the custom of bringing up the children of the same estate in common, as it was far more humane not to cherish domestic ties among slaves' (because they could be sold off), and he scored again Lyell's sentiment a few lines further on that 'the effect of the institution on the progress of the whites is most injurious'. This was too much:

> How could you relate so placidly that atrocious sentiment about separating children from their parents; & in the next page, speak of being distressed at the Whites not having prospered; I assure you the contrast made me exclaim out.

What did Lyell really know of family life, thirteen years married and childless? The Darwins, already with five children by the time this was written in August 1845 (the last having been born only six weeks previously), thought it unforgivable to write so dispassionately about children being torn from their parents. The new edition of Darwin's *Journal* would denounce Lyell and everyone else complicit in such abomination. Who among them had seen its wickedness, the stinking cargoes, the branding irons, the terror on men's faces, the cringe of helpless children? 'I have remarked on nothing' in the amended *Journal* that 'I did not hear on the coast of S. America', he told Lyell. Darwin's 'explosion of feeling' left them both probably shaken. 'I have broken my intention & so no more on this odious deadly subject'.

He had the publisher send Lyell the revised *Journal* for reading en route to Boston and to carry through the South like a talisman on his second trip. Opening the book, Lyell found a fulsome dedication to himself, but the sting was in the *Journal*'s tail, five hundred pages on.[19] Here Darwin gave full vent to his anger:

I thank God, I shall never again visit a slave-country. To this day, if I hear a distant scream, it recalls with painful vividness my feelings, when passing a house near Pernambuco, I heard the most pitiable moans, and could not but suspect that some poor slave was being tortured, yet knew that I was as powerless as a child even to remonstrate . . . Near Rio de Janeiro I lived opposite to an old lady, who kept screws to crush the fingers of her female slaves. I have staid in a house where a young household mulatto, daily and hourly, was reviled, beaten, and persecuted enough to break the spirit of the lowest animal. I have seen a little boy, six or seven years old, struck thrice with a horse-whip (before I could interfere) on his naked head, for having handed me a glass of water not quite clean; I saw his father tremble at a mere glance from his master's eye. These latter cruelties were witnessed by me in a Spanish colony, in which it has always been said, that slaves are better treated than by the Portuguese, English, or other European nations. I have seen at Rio Janeiro a powerful negro afraid to ward off a blow directed, as he thought, at his face. I was present when a kind-hearted man was on the point of separating for ever the men, women, and little children of a large number of families who had long lived together. I will not even allude to the many heart-sickening atrocities which I authentically heard of;—nor would I have mentioned the above revolting details, had I not met with several people, so blinded by the constitutional gaiety of the negro, as to speak of slavery as a tolerable evil. Such people have generally visited at the houses of the upper classes, where the domestic slaves are usually well treated; and they have not, like myself, lived amongst the lower classes.

Lyell could see himself here, though Darwin, recalling the row at Bahia, was surely rounding on FitzRoy as well: 'Such enquirers will ask slaves about their condition; they forget that the slave must indeed be dull, who does not calculate on the chance of his answer reaching his master's ears'.

Old scores were being settled. FitzRoy, now Governor in New Zealand, his recall just announced, would receive a homecoming gift of Darwin's *Geological Observations on South America* rather than the *Journal*. Afterwards their letters petered out. FitzRoy had always been a lost cause. But Lyell might still be brought to his senses, and Darwin turned his angry slashes on Lyell's text into the *Journal*'s parting shot:

Those who look tenderly at the slave-owner, and with a cold heart at the slave, never seem to put themselves into the position of the latter; – what a cheerless prospect, with not even a hope of change! picture to yourself the chance, ever hanging over you, of your wife and your little children – those objects which nature urges even the slave to call his own – being torn from you and sold like beasts to the first bidder! And these deeds are done and palliated by men, who profess to love their neighbours as themselves, who believe in God, and pray that his Will be done on earth! It makes one's blood boil, yet heart tremble, to think that we Englishmen and our American descendants, with their boastful cry of liberty, have been and are so guilty . . .

Never before had Darwin expressed himself so strongly. Three generations of Darwin–Wedgwoods rose up inside him to condemn the mentor whose judgement in science he so respected. Yet as every believer knew, redemption was possible. Just as Britain had 'made a greater sacrifice, than ever made by any nation, to expiate our sin',[20] said Darwin, so Americans could do God's 'Will' and emancipate their slaves.

Before Liverpool's ascendancy as a cotton port, Bristol dominated the nation's trade in slave commodities. Bristol was built on sugar. Goods shipped to Africa had been exchanged for slaves. Sent to the West Indies, they had reaped the harvest that was shipped back in barrels to line Bristol's pockets and rot Britain's teeth. At the time of abolition, at least three-fifths of the port's commerce had depended on slave sugar. Everyone was touched or tainted by it as the cash trickled down. Bankers and merchants grew fat, their churches and chapels praised God for the glut. The sugar aristocrats were mostly Anglicans, while many influential Nonconformists, their consciences pricked, dissented from the slave trade as well.

As the abolition movement had gained its first momentum, the Unitarians erected a spacious chapel in the heart of Bristol where anti-slavery would be preached. Lewin's Mead Meeting House stood beside a sugar refinery, a stone's throw from the docks where the hogsheads rolled ashore. Here the young James Cowles Prichard had first seen men of colour and had met his wife, the minister's daughter. Prichard had studied medicine at Edinburgh with her brother, John Bishop Estlin, who had returned to Bristol as an eye surgeon and anti-slavery campaigner. Estlin Senior's successor at Lewin's Mead was Lant Carpenter, who taught anti-slavery to Darwin's Edinburgh friend W. R. Greg, got up Parliamentary petitions and, despite his caution on emancipation, propelled one teenage worshipper, Harriet Martineau, towards immediatism. Carpenter's eldest son, William Benjamin Carpenter, was apprenticed to Dr Estlin. After travelling with him in the West Indies, Carpenter followed him to Edinburgh into medicine. The path from devout Nonconformity via Edinburgh to anti-slavery was well worn.

W. B. Carpenter was not so radical as to demand emancipation overnight. Like Lyell, he had grown up on the 'lukewarm' wing of anti-slavery. With Dr Estlin on St Vincent, he had seen the 'benevolent' side of plantation life. Not witnessing any cruelty, he had supposed that only religious instruction was needed to counter the immorality into which the blacks seemed to be sinking. His experiences reinforced the family's belief in a 'gradual and "safe" emancipation' via the apprenticeship system.[21] But William's position would change as American slavery came to the fore.

Eighteen forty-four was Carpenter's year. He too had trained under Robert Grant, learning how life was connected into a continuous stream of forms. A rising star, he finally moved from Bristol to London in 1844, gained a coveted fellowship of the Royal Society and followed Grant in the Fullerian chair of physiology at the Royal Institution. He was a formidable medical and social critic (hack work was obligatory for men of science without Darwin's deep pocket). He churned out articles for the liberal *British and Foreign Medical Review*, pieces that mirrored a starchy Unitarian rationalism by promoting a cosmos of 'law and order' rather than Creative whim. He was Darwin's man in more ways than one. That year Carpenter even bucked the reviewers by praising the 'beautiful' *Vestiges of the Natural History of Creation* for its dignifying image of the development of life through time. The book, an anonymous potboiler, argued that continual miraculous tinkerings with species were unnecessary.[22] The process had started at Creation and unfolded 'naturally', in accord with a divine plan.

Sneering Carlyle dismissed Carpenter as a 'sort of apostolate of the microscope', a philosopher with a fashionable skill in an age enthralled by microcosmic revelations. That skill made careers, however. As Darwin was publishing descriptions of his *Beagle* remnants – his translucent spiny *Sagitta* from Chilean waters (later shown to be an arrow-worm) – in the workaday *Annals and Magazine of Natural History*, Carpenter was continuing his *Annals* series on the microscopic structure of shell. And so it was as a microscopist that Carpenter first contacted Darwin (again, in 1844), offering to section his Chilean rocks to search for fossil shell fragments.[23] Thus began the lifelong acquaintance of two naturalists whose Unitarian anti-slavery beliefs would only grow fiercer.

'Fiercer' for Carpenter meant a rearguard action to protect his hero Prichard from Southern slurs. Carpenter had gone over to radical abolitionism, and his was the first public fightback against transatlantic slave science. More resolved in 1844, he politely but publicly put down the New York Unitarian pastor Orville Dewey. Dewey was a conservative anti-abolitionist, preaching in his self-built Church of the Messiah on Broadway. He was a gradualist who, while damning the 'stupendous immorality of the slave-system', thought that black emancipation would lead to something worse. Abolitionists were pushing the country in that direction and threatening the Union, which to Dewey was sacrosanct. To save it, he once said, he would sacrifice his own brother or child into bondage, which brought the inevitable riposte.[24] When the fugitive slave William Craft, shipped to Britain for his safety by abolitionists, was received in Carpenter's circle, they suggested Dewey take his place back on the Georgia plantation.

Fresh from a trip to England, Dewey was enraged by a British public who derided America's rhetoric of freedom when it held blacks enslaved. However much he loathed slavery, he hated British hauteur more. He made it seem like a British insinuation of national depravity. But what really galled Carpenter was Dewey's planter-apology: his insisting that 'manumission . . . did the colored man no good; that he was . . . worse off for his freedom'. Meaning, the 'depressed' and isolated pockets of freed slaves would be 'separated from us by impassable physical, if not mental barriers; refused intermarriage, refused intercourse as equals, be it ever so unjustly; how are they ever to rise?' They'd be better off in Haiti or the West Indies, where as a majority they 'could rise to their proper place as men'.[25]

If Dewey was inclined to 'take John Bull by the horns', many were inclined to ascribe the horns to Dewey. In the *Christian Examiner*, Carpenter dismissed Dewey's separationist statement as 'neither scientifically nor historically true'. How could there be impassable barriers between races 'if their origin is the same'? Carpenter entertained no 'shade of doubt' that Prichard had proved the point, undercutting any notion that blacks and whites were 'distinct species'. Prichard's *Researches into the Physical History of Mankind* traced the races 'to a common stock' and would have us 'look to various external circumstances (such as those which have produced the various breeds of our domesticated races) as the cause of the diversities'.

The tie-up between human racial variation and domestic breeds was firming. Darwin had never been alone in grasping this, but his was a more complex pattern than Carpenter's. In his unpublished 1844 essay, Darwin was actually prepared to accept 'most of our domestic animals having descended from more than one wild stock'. This was because he envisioned primitive human groups on different continents each domesticating the local wild dogs, pigs and cattle. So wild wolves, which had evolved into separate species in Siberia and America, might each be the source of local domestic dogs. On the other hand, even in the essay, he was already wary of those who excessively multiplied the number of aboriginal stocks that formed domestic breeds.

In 1844 Carpenter's view (shared of course by Darwin) was that once naturalists cracked how fancy and farmyard stock was formed, they could explain human diversification. Animals with pliable constitutions, said Carpenter, which include 'all our domesticated races', and which spread over the globe, often exhibit 'great variations': among these

> man unquestionably stands first, the dog probably next, and then our horses, sheep, and cattle. Will any one affirm that there is more difference between a Negro and a Caucasian, than between a greyhound and a mastiff; or that the education which, continued through a succession of generations, developes certain faculties and habits in the dog, shall be less effectual in man?

In his new two-volume textbook *Zoology* Carpenter ascribed the inferior-blacks-as-separate-species view to those 'who wished to excuse the horrors of slavery or the extirpation of savage tribes' – making the stance as 'immoral' as it was unscientific. Races, he said, could hardly be 'species' if 'we have every reason' to think that domestic breeds were traceable 'to a common origin' (which was more than Darwin was prepared to publish at this time).[26] If domestic races were, the human races too descended from common stock.

Over the next decade, Carptenter increasingly studied the variability of certain test species, using his microscope to expose the variable shapes of the lentil-like, limestone-entombed nummulite shells and their living relatives. The morally loaded issue of natural human variability remained part of the subtext for this filigree work. But that was no surprise, for he repeatedly admitted that his attention had first been drawn to 'the variability of species' by Prichard's human studies.

While Carpenter shared one premise with the American doctors of divinity and medicine – that civilization distinguished blacks from whites – he denied another: that blacks had no aptitude for it. Given educational opportunities, they would come up to par. The potential was there, even if some centuries would be required to raise 'the New Holland Savage, or the African Bushman, to the level of the European'.[27] As proof he pointed to post-emancipation Jamaica under Governor Sir Charles Metcalfe's proactive policies. Here within ten years a prosperous black population had 'risen' enough for its leaders to 'be admitted to the Governor's balls and parties, as well as to receive Government patronage'.

Dewey's views were symptomatic as a reaction set in. The notion that blacks could not be educated to the rank of gentleman was gaining ground, even while the immortals who had fought for black equality and black freedom were proving themselves not so immortal. The last of them, the tall, dour but totally committed Thomas Clarkson, had become more and more sympathetic to America's most fervent and immediatist abolitionist, William Lloyd Garrison, whose fiery paper, the *Liberator*, called for the nation's destruction, if necessary, to end slavery. (At home Garrison was himself called America's 'Clarkson'.) Clarkson was now so idolized that a craze for making keepsakes of his hair had led to his accelerating baldness. His last years had been spent writing against American slavery, and in recognition Garrison turned up at Clarkson's door in August 1846, accompanied by the eloquent ex-slave Frederick Douglass. The frail 86-year-old Clarkson died a few weeks later, on 26 September 1846. Alone of the anti-slavery leaders, he was not buried in Westminster Abbey (it was thought his Quaker friends might object). Yet to Darwin he was the greatest of them all.

Fewer now stood out to fight, but among them was the young Carpenter. He took strength in the traditional 'coloured' college contingent at Edinburgh, who 'have never been excluded from social intercourse on account of their hue'. He took strength in the 'black student in the Temple [one of London's elite Inns of Court, training barristers], who is keeping his law-terms, eating his dinners, and associating in the usual manner with his fellow-students; and I have not heard that he has manifested any of the fancied disqualifications which are erected as barriers between the two races in America'.[28] But for all of Carpenter's protestations, it was not hard to see the barriers going up. Slavery apologists were acquiring a voice. Britain was being urged to save the black souls rather than fashion black gentlemen. Carpenter's was a rearguard action, but such voices would become rarer with the passing years. The unity-of-race thesis, secretly extended by Darwin to a unity-of-life thesis, was wading increasingly against the tide of new scientific thought.

Stemming the tide had its Canute-like aspect. Not only in the Slave States was the new scientific mood evident. The flamboyant Robert Knox had no truck with Carolina slavery, but every truck with its hard-boiled racial science. The most influential and widely attended anatomy teacher from Darwin's Edinburgh years was now an itinerant hack. He started a crowd-pulling countrywide tour in 1844, lecturing on 'The Races of Men'. From the industrial crucible Newcastle in the north-east, south via Sheffield, Warrington, Manchester, Liverpool and Birmingham to Colchester he went. He set up in Philosophical Institutes, agitating the race question. Audiences were told unequivocally that 'no race will amalgamate with any other', and the results of interracial unions, the 'mules, or mulattoes, as they are called, nature will not support'.

Whether in Chelmsford or Charleston, the message was the same: the '*amalgamation of races*' was unnatural. 'When mulattoes intermarry, they seem to die out in two or three generations', said Knox. Even locally, Saxons and Celts have been in 'eternal disunion'. And Knox – with his one remaining eye focused on topical politics – saw the North German Saxons soon breaking away from the Hapsburg 'sclavonian' domination. Race was everything. Knox dismissed the 'prosing twaddler of detail' (meaning Prichard), whose vast compendia 'on the natural history of man' were as castles in the air. Knox had by now lost all faith in the climate's ability to alter 'the physical or psychological characters of the races'. Red Indians were not Caucasians burned copper colour by the sun. Such 'sickening, silly follies' had to go. So while civilization does express 'the literature, science, and art of the race', it cannot *alter* the race. Time offered no more hope. The 'Jews, the Copts, and the gipsy' – peoples who had not married out – proved the races unchanged within an historic time scale.[29] And Egyptian tomb chambers showed that the Negro race was unaltered 3,000 years on.

The social stereotypes imported into phrenology and ethnology in the Edinburgh of Darwin's years were now reinforced by Knox with a vengeance. Where the Saxon was 'the very pattern and pink of perfection, of cleanliness, and of method, of economy and regularity' – in truth, the 'perfect democrat', which made America 'the destiny of the Saxon race' – the barbarous Catholic Celt was 'reckless, a waster and destroyer, incurably indolent . . . personally filthy . . . a despiser of the law'. Just as history depicted a 'great struggle between the dominant races for supremacy', so a great killing field lay before the Saxons as they invaded the tropics. Extinction awaited its occupants:

> Already . . . we have cleared Van Diemen's Land of every human aboriginal; Australia, of course, follows, and New Zealand next; there is no denying the fact, that the Saxon . . . has a perfect horror for his darker brethren. Hence the folly of the war carried on by the philanthropists of Britain against [human racial] nature.[30]

Yet Knox conceded that the whites never will actually 'colonize a tropical country': the Niger fiasco taught that. Aboriginals, with 'a constitution adapted to labour under a tropical sun', will hold these outposts if they succcssfully retreat from white guns. Here was the last hope, in the Torrid Zone of Amazonia or the Congo, where the indigenous peoples 'continually recover their pristine vigour and numbers, rolling back the white invasion'. Here the blacks will hold out, without great buildings or great works.[31] In the dense forests lay their salvation.

The tour wound up in London, where Knox's exhibitionism reached its peak in the philanthropists' favourite venue, Exeter Hall, site of great anti-slavery rallies. The page-one *Times* advert in 1847 announced that he would 'introduce to the notice of the physiologist and man of science Five Bosjiemans or Bush people – two males, two females, and infant, the only specimens of this singular race of human beings that ever visited Europe'. The notice was ostensibly aimed at medical professionals, but voyeurism was expected to cover the costs. Topicality helped sell tickets. The shilling spectacle was pitched at those absorbed by 'the exciting events now going on in . . . the [latest] Kaffir War'. These 'Pigmies' – portrayed in the press as a filthy, brute-like, property-less people, and dressed for the part accordingly in a 'festering bundle of hides' (as Dickens described them) – were savages being deliberately de-nobled. The image of an animal-like race was being visually crafted. Knox capitalized on the public's taste for freak shows, following P. T. Barnum's Tom Thumb tour the previous year. His raree show had one advantage. It gave gawpers perhaps the last chance to see 'Pigmies' before the 'probable extinction of the Aboriginal races'.[32]

The gold-waistcoated Knox predicted future race wars, and after the 1848 revolutions in Germany, France and Italy he would point to them in vindication. They were certainly the occasion for the *Medical Times* to publish his lectures and announce that 'history, during the last six months, may be considered as but one continuous advertisement of our course'. Wherever the revolutions, 'the great element of . . . European reorganization is race'. No longer so extreme, such ideas were gaining ground. The *Medical Times* echoed Knox: history was no chapter of accidents, nor guided solely by 'principles or interests'. The new hegemony-grabbing medical ethnologist, bumping up his social worth, was claiming that he possessed the real key to its underlying trend. Knox's racial explanations were as valuable 'in a political' as a 'scientific point of view'.[33] They held policy implications in the Empire for the Irish, Jews, half-castes and aborigines.

The sands were shifting under the Prichardians. Their bloodline for all races back to Adam was whiffing of quaint biblical attitudes. Their special pleading smacked of protectionist philanthropy, not practical anatomy.

For Carpenter, mixed-race humans *were* fertile. Black and white parents shared an ancestor (however distant; he agreed that Egyptian hieroglyphs testified to a long separation). That made them *races*, not species. But in his 1844 essay Darwin was more equivocal on what distinguished the two. In fact, he flatly stated that 'we must give up sterility . . . as an unfailing mark by which *species* can be distinguished from *races, i.e.* from those forms which have descended from a common stock'.

The traditional view, that infertile offspring proved the parents separate species in *all* cases, was wrong. Darwin knew that occasionally cross-species unions *were* fertile: there was a gradation of fertility according to closeness of the species and 'constitution' of the parents' bodies. Cross-species hybridity did occur, but it was disruptive and anomalous, and he ruled it out as part of his transmutatory mechanism. It was also very rare, although 'the experiment [in species-crossing] is very seldom fairly tried', so no one really knew how rare. Darwin's horticultural source on these matters, the Revd William Herbert, Dean of Manchester, had even shown in his hothouse experiments on water lilies and amaryllis that, in very odd cases, hybrids could be 'decidedly more fertile than either of their pure parents'. So this was probably less of a definitive test now for humankind in Darwin's eyes. The fertility of offspring said little about whether parents were separate species. But he would have agreed with Carpenter that mixed-*race* offspring were sometimes even 'superior' to the parents, as in the thriving melting pots of 'South America and Hindostan'. Darwin in fact went further. He thought in 1844 that race crossings provided a 'copious' source of new races, and that in subsequent generations these mongrels would 'vary exceedingly', providing the raw material for selection to adapt them differentially.

What *exactly* happened to crossed human races had been intensely interesting to Darwin since 1837. Authorities thought 'that offspring of Negro & white will return to native stock', he jotted, which colour depending on whether the 'mulattos' back-bred with white or black people. Andrew Smith was sure this was so at the Cape, and others used Herbert's lily experiments as parallel examples. This left Darwin jotting notes to himself to cross plant races as well as those of domestic 'pidgeons. fowls, rabbits', to see how good an analogy with mulattos they really were.[34]

Carpenter's own mixed-race allusions were a taunt to his American antagonists. Bristolians were habituated to black sailors and their 'often very handsome' white wives. These couples spoke volumes against 'impassable' barriers. He acknowledged the 'instinctive repugnance' of many Americans to such interracial contact, but was 'at a loss' to explain it, except as a prejudice induced by a society habituated to black slavery. Part of the repugnance he put down to social station. To him a black–white liaison was no more unnatural or 'impassable' than that 'between the daughter of a peer, and the son of a ploughman'. Carpenter was deliberately treading on contemporary sensibilities by conflating social distinction with racial pride. He lit a fuse by predicting that, not long hence, 'the daughter of an American merchant might find the descendant of the despised Negro not unworthy of her attachment'. This was outrageous to many Americans. What made it worse was the insinuation that the mother country had outgrown such things, when

Americans thought *themselves* the ones who had broken Old World restraints.

For his pains Carpenter was given the lie direct. A riposte by Dr Samuel Dickson in the same *Christian Examiner* reaffirmed every point of Dewey's. Dickson's hospitality to Lyell in Charleston had been gracious, but he showed none towards Prichard, who was damned as an 'advocate', distorting truth for traditionalist ends. Hard facts lay with Morton now, 'of whom American science is justly proud'. In depicting the races set in Egyptian stone, Morton had exposed Prichard's 'innumerable errors'. 'When we first meet with the Negro, – in the world's infancy, – he is just what we find him now; conquered, subject, in servitude'. Subservience was written into his constitution.

This was white South Carolina speaking, with its 300,000 black slaves. Here Nott's voice – declaring that the 'Mulatto is incapable of "keeping up his number," but must decay and disappear' – was echoed by Dickson. (Nott returned the compliment, declaring Dickson one of the South's 'leading medical men' who had helped to prove that extinction awaited the illicit Saxon/Negro mutant.) The mulatto was the real problem: a hybrid 'doubly despised; partly because of his ancestry, and partly because he is of course a bastard'. Needless to say, Dickson's harangue was sparked by Carpenter's solution of 'unrestrained amalgamation by intermarriage'.[35] The suggestion that intercourse, social and sexual, would save the day, if not the race, was 'insulting'.

When Lyell returned to America in September 1845, he tackled this hybridity question with a new urgency. In Boston to give his second series of Lowell Lectures, he and Mary rejoined the Ticknor circle. Then winter chased them south again via Washington and the White House, where they were received by President Polk's wife.

Lyell carried a letter from Darwin. He had received it in Boston and it continued their dialogue. Darwin had briefed Lyell before he left with 'questions about negro-crosses'. Now he had another arising from Edward Long's old pro-slavery *History of Jamaica* (1774). Long saw Africans as subhuman, but some white men showed no natural repugnance to mating with them and spawned 'spurious offsprings of different complexions', which were, like mules, sterile. Darwin, in total disbelief, put it to Lyell: 'I see Long . . . says he has never known two mulattos have offspring!!!!! Can you obtain any comparative information on the crosses between Indian & Europaeans & Negros & Europaeans?' Darwin wanted proofs of fertile crosses.

Lyell was being tested, maybe punished for being soft on the 'sin' of slavery, and Darwin's next task for him was still more 'disgusting a subject'. Darwin had long heard that the lice on Negroes were supposed to be larger and blacker than those on whites. Different races of lice living on different races of men made adaptive sense, but what if the lice were of different *species*? For comparison, Darwin needed lice from 'negros born in N. America', and he asked Lyell to get them 'through some medical man . . . without disgusting yourself much'. Darwin knew how Lyell hated getting his hands dirty. Still taunting perhaps, he told of visiting the botanical Dean of Manchester, that 'great maker of Hybrids', and how afterwards he had dined with the South American traveller Charles Waterton and his half-Arawak adoptive daughters, 'two Mulattresses!' from Guyana, whose mother was an Indian.[36] Waterton himself had married a third daughter. (The girls were daughters of the planter Charles Edmondstone, whose ex-slave John had taught Darwin bird-stuffing.) Mongrelization was commoner than Lyell thought, even among planters, and Darwin made sure he knew it.

After taking travel advice from Morton in Philadelphia, Lyell was primed to identify the mixed-race South that he missed on his previous visit. Mulattos, quadroons and rarer mixtures were everywhere. In Charleston at Christmas, the Lyells took communion at St Philip's Episcopal Church before the cup passed to coloured parishioners. Some were 'quite black', Mary observed. But she was now also looking to shades. One or two were as light as her own sister, '& with hardly a trace of the African features. It does seem very hard & cruel that this badge of serfdom should never be effaced'.

Worse than serfdom, it was a badge of bastardy in states that outlawed mixed marriages. 'Licentious intercourse with slaves' left 'no possibility of concealment'. Fathers stood guilty as accused and the children suffered. 'The female slave . . . thinks it an honour to have a mulatto child, hoping that it will be better provided for than a black child', but the mulatto bears his colour, Lyell noted, like an 'indelible stain' and passes it on to generations 'born in lawful wedlock'. These, he heard, might not be very numerous, and he recorded the fact for Darwin. 'The mulattoes alone represent nearly all the illicit intercourse between the white man and negro of the living generation', yet after 150 years of slavery here they made up no more than two and a half percent of the population.[37]

Before leaving Charleston, Lyell chatted again with Bachman about his great work on American mammals. Bachman had just finished the first volume, describing over 120 new species, brilliantly enlivened by Audubon's plates. This time Bachman pointed out a distribution pattern that might be explained by God's creating in 'specific centres' on opposite sides of the Rocky Mountains, although Lyell thought there could be more to it. From Darwin's new-edition *Journal*, which he carried, Lyell knew that the distribution of finch species in the Galapagos archipelago seemed to raise questions about 'that great fact – that mystery of mysteries – the first appearance of new beings on this earth'. (This was Darwin's first published statement hinting at his deeper thoughts.) Lyell looked to the patterns across the Rockies and around the world, 'the limitation of peculiar generic types to certain geographical areas', and followed Darwin as far as admitting that a 'higher law governing the creation of species' might help explain those patterns, even if such a law 'may, perhaps, remain a mystery forever'.[38] Whether he told Bachman so is unknown.

Certainly word went back to Darwin: at Savannah, Georgia, Lyell told him about finding fossil ground sloths. Then the Lyells enjoyed a fortnight's hospitality on a plantation with five hundred slaves before pushing down through Alabama. Sundays were for sermon sampling. Lyell attended a Baptist service with 'about 600 negroes of various shades' in the congregation, 'most of them very dark'. He was the only white man. In other churches, Negroes sat by permission, with master and slave worshipping together but in separate pews. To Lyell this represented an immense step in the slaves' 'progress towards civilisation'. Hearing the master's message 'that the white and coloured man are equal before God' raised the slave 'in his own opinion, and in that of the dominant race'. Actual equality, though a distant aspiration, was slowly being realized, he thought. The 'impassable gulf' was being bridged over, log by log, as 'the humble negro of the coast of Guinea' showed himself to be one of the most 'improvable of human beings'.

In fact, most of the slaves Lyell encountered seemed happier, healthier, more secure, and some of them more intelligent, than the residents of an 'average English parish'. Were it not illegal, an English spinster here would do better marrying 'a well-conducted black artizan' than a 'drunken and illiterate' Irishman. Or her chosen might be of mixed race – Lyell saw both 'Africans and mulattos' learning skilled trades.[39] There was no doubt that many of the blacks were doing well. The question, as Lyell saw it, was whether the survival of slavery and the Southern way of life was a price worth paying for continuing even this piecemeal and partial elevation of blacks. Lyell thought so; Darwin, he knew, did not.

The Lyells witnessed slaves auctioned in Montgomery, Alabama, where horses were sold the next day. In February 1846 they arrived in Mobile on the Gulf of New Mexico. Here Dr Josiah Nott was possibly the supplier of the lice from his black patients that Darwin required. Were that so, then it was a fine irony: the white-supremacist doctor supplying black lice for an abolitionist studying racial origins.

Nott had just returned from visiting Morton, whose hoard of human crania – now upwards of a thousand – emboldened his own '*niggerology*' and developing views on human hybrids. Nott also found Lyell 'very full of the subject' (not knowing that Darwin had briefed him) and well-read.[40] Lyell knew the argument that intermarriage produced inferior people, dragging down both white and black. He had studied Nott's piece in the latest *Southern Quarterly Review*, which claimed the Negro race lacked 'pliancy':

> Lyell and others tell us, that very few generations are sufficient to effect [in Negroes] all which can be effected in [domestic] animals, and we are informed, also, that these changes are effected with great certainty and uniformity; but history, from the time of Herodotus to the present, affords no positive evidence of these changes in man.

Egyptian walls showed the Negro race unchanged in 3,000 years, Morton's skulls that black brains were deficient. Left to themselves, Negroes degenerated; emancipation would be 'their ruin'.[41]

Lyell, scribbling notes, conceded that blacks could 'only be civilized through slavery' and thought that the system, if humanely managed, would eventually bring Negroes 'up to the Caucasian standard'. Nott disagreed; such naivety was born of a foreigner 'travelling so fast' across America that he had missed the truth. Ten years of medical practice in Mobile had taught him otherwise:

> The races of men, like animals, in a wild, uncultivated state, may, if docile, be tamed, educated, and vastly improved, but there are limits set to each by nature, beyond which no advance can be made. Although there may be an occasional example where a negro will show a degree of intelligence and capacity for improvement beyond the mass, yet no negro has ever left behind him any intellectual effort worthy of being preserved.[42]

For that reason, 'amalgamation' was even worse than emancipation. Mulattos had a 'hybrid intellect'; they made 'more upish' servants – 'not content, "too smart"', supposing themselves 'superior'. In vain Lyell objected that improvements would also be inherited, that mixture with pure Africans had diluted the civilizing process, that prostitution reduced the mulatto's fecundity, that without reliable statistics, their longevity was still open to question. Lyell's notes ended ominously: 'Dr. Nott wishes there were no negros, no necessity for them'.[43]

Later that day the irony of Lyell's position emerged when he called on Nott's religious nemesis, the Revd William Hamilton, minister of Mobile's prestigious Government Street Presbyterian Church. In popular lectures, Nott had spelt out for Mobilians the implications of what Lyell had just heard. The science of Genesis was wrong; Prichard, the 'great orthodox defender' of human unity, played fast and loose with sacred texts. The best modern science, appealing only to 'analogies, facts, induction and to the universal and undeviating laws of Nature', showed the 'Genus, Man' to consist of at least two species, the 'Caucasian' and the 'Negro', which combined at their peril. Hamilton would hear none of this. He considered the mulatto 'more intelligent than . . . the full negro' and wanted slave marriages legalized to keep families from being split up. Familiar with the *Principles of Geology*, he applauded Lyell's belief in human unity, citing, as Lyell had, 'the learned work of Dr Prichard'. Ten years' ministry in Mobile had convinced *him* that blacks were *not* innately inferior. 'From time immemorial the negro has been an oppressed race, secluded from all elevating influences. Remove the pressure, and who shall say of what even the negro intellect shall not prove capable?'[44] Surely this was a man after Lyell's heart?

Not quite. Everything Hamilton believed was based on the Bible. He interpreted it literally and made Noah's family of eight the ancestors of all humanity. After the Flood, at the Tower of Babel, God brought about 'the present diversity of race and of language' in time for black faces to appear on Egyptian monuments. Prichard had required much too long for diversification to occur; by his method the races could not have been formed by the time of ancient Egypt. So Hamilton made God 'directly interfere',[45] miraculously, to create the races at Babel. He believed his was a unique solution. To Nott it was laughable, and he said so.

The Bible also allowed Hamilton to tolerate slavery. The Israelites kept slaves, so did the early Church. In an 1844 sermon, *The Duties of Masters and Slaves*, he admitted that the Golden Rule ('Love thy neighbour as thyself') referred only to what it would be reasonable to expect at a given time. Thus Jesus's message to planters was to love your slaves as you would feel entitled to be loved if you were a slave yourself. For such teachings, Hamilton was vilified, as Lyell noted. Abolitionists smeared him as 'a hoary ruffian – a pious hypocrite – a bloodthirsty villain'. If you prove that the Bible endorses slavery, they roared, 'you prove God to have outsataned Satan'.

The abolitionists were wicked men; slavery was not sinful, said the Southern faithful. The barbs grew ever louder as Lyell left Mobile for New Orleans. In 1844 the white Churches had erupted over slavery: Methodists and Baptists split north and south, and Hamilton's Presbyterians were poised to follow. The Unitarians seemed to stand for sanity as Lyell triangulated his own beliefs, still seething at 'the fanatical party of the North', whose 'interference and insults' so hurt his friends among the Southern gentry.[46] Why didn't good men such as Hamilton and Bachman believe the right things for the *right* reasons? Was there no better basis for human unity and 'Negro' progress than the fictions of Genesis? Did science point inevitably to Nott's racial pessimism? Was there not hope for blacks in the fact that whites had advanced? If all were members of one progressive species, surely there *was* hope.

In the great cotton port of New Orleans, the Lyells first saw the racial rainbow that so astonished Darwin in South America. Carnival was ending; the streets teemed with Creoles. These people of French and Spanish descent shrank 'as much as a New Englander from intermarriage with one tainted . . . with African blood'. In segregated theatres quadroons, the offspring of whites and mulattos, sat in an upper tier of boxes as a 'select and exclusive set', and in the marketplace every sort of human cross mingled amid a bewildering babble of tongues. The social strata fascinated Lyell as much as the geological, and for a moment he thought he glimpsed a new world in the making:

> Amidst this motley group, sprung from so many races, we encountered a young man and woman, arm-in-arm, of fair complexion, evidently Anglo-Saxon, and who looked as if they had recently come from the North. The Indians, Spaniards, and French standing round them, seemed as if placed there to remind us of the successive races whose power in Louisiana had passed away, while this fair-couple were the representatives of a people whose dominion carries the imagination far into the future. However much the moralist may satirise the spirit of conquest, or the foreigner laugh at some vain-glorious boasting about 'our destiny', none can doubt that from this stock is to spring the people who will supersede every other in the northern, if not also in the southern continent of America.[47]

No longer seeing in black and white, Lyell saw the South changing on a scale from black *to* white as the fairer race prevailed. It was a vision that 'Anglo-Saxons' increasingly shared, Darwin included.

Lyell's attitude to slavery had not changed. 'If emancipated', the blacks 'will suffer very much more than they gain', as the rootless northern Negro seemed to show. If the Free States 'really desired to accelerate emancipation, they would begin by setting an example to the Southern States, and treating the black race with more respect and more on a footing of equality'. Extremism was manifesting everywhere now, even inside science. In Philadelphia, Lyell met Morton twice – the man who, more than any other, would catalyse the new segregationist anthropology. Before they left America, the Lyells, with the Ticknors, heard an anti-war sermon by Dewey.[48] It was timely, with the Slave States now outnumbering the free and fears for national unity growing.

8

Domestic Animals and Domestic Institutions

By now the pluralists – those who believed in primevally divided human races – were driving the study of domestic animals. That study accelerated rapidly from the mid-1840s. Morton took up the question after cordially contacting Nott in 1844. Nott himself was busy debunking Moses in his popular lectures in Mobile. He insisted that the 'plurality of species in the human race does no more violence to the Bible, than do the admitted facts of Astronomy and Geology'.

In the Southern environment slaves became not only figurative 'others', but literally another species. Thus Nott argued for a 'Genus, Man', made of so many fixed species. And he argued it through a caustic mix of biblical disputation, Egyptian chronology and Morton's craniology – with their corollary that Negroes were and will always be servants and slaves. Since no species, plant, animal or human, 'can be propagated out of the climate to which they are adapted',[1] he believed that the blacks ripped from African soil would fare ill in America, especially in the colder northern states, unless well cared for (as the slaves in the South were said to be). To prove it, he delved into the mortality statistics from insurance companies covering free and enslaved blacks in New York, Philadelphia and Charleston. In his hands they showed a decrease in deaths with each move southwards, and a further decrease when slaves alone were counted. It lead to the consoling conclusion for plantation apologists that 'even Cholera and Slavery combined [in the South] . . . are far less destructive to the negro, than liberty and climate in Boston'.[2]

This sat well in *DeBow's Review*. Where better than the premier business journal of the South, when slaves were commercial investments? More, the journal vehemently upheld a slave-holding agrarian economy, even if it did advocate diversifying away from a pure cotton/slave base into industry. By the time Nott entered its ranks, the defence of Southern institutions was becoming paramount, and his contributions were appreciated. James DeBow himself, Professor of Political Economy at the University of Louisiana, even offered Nott his professorial platform to continue the lectures. The result, claimed Nott, was that 'all the title articles I have written on *niggerology* have been eagerly sought for at the South'.[3]

Tucked away inside these agitprop pieces on '*niggerology*' was a key component. It concerned dogs (and sheep and chickens). Nott tackled head-on Prichard's analogy between the supposed Negro-to-Caucasian transformation and the changes in domestic animals. He granted that the different dogs, from Newfoundlands to poodles, were a striking illustration of their plasticity in human hands, but pointed out that the humans who had fashioned these dogs themselves remained unchanged from Egyptian antiquity until today.[4] *Human* races were simply less 'mutable'.

This was untidy and unsatisfactory: 'What applies to one animal' (as Darwin believed, and most zoologists concurred) 'will apply throughout all time to all animals'. Or, as a more youthful Darwin jotted, a 'Man acts on. & is acted on by the organic and inorganic agents of this earth. like every other animal.'[5] Humans weren't exempted or different or less 'mutable': animal, plant and human races should be treated alike. Morton held the same view. Hence he had a problem with Nott's preferential treatment.

There was another reason why Morton had to tidy up Nott's position. If mulattos were hybrid species, then nature had to be looked at afresh, because traditionally species were defined as infertile when crossed, barring the odd exception. Nott was making a counter point: that species *could* cross – that goldfinches and canaries cross-bred successfully, as did goats and ewes – and now Morton, agreeing more and more, would extend the list of fertile hybrids.

He was entering increasingly acrimonious waters. Nott's pugnacity had raised hackles and he had suffered ridicule as a naive naturalist. His provocative lectures proved a flashpoint. The question of human plurality exploded into the open in 1845, as the first of the 'Unity' versus 'Plurality' articles appeared in Charleston's *Southern Quarterly Review*, the prelude to swamping numbers over the next decade across the American reviewing spectrum. Three further ripostes and rejoinders to this first article alone ran in the *Southern Quarterly* within a year. (Being another religious, conservative, agrarian, pro-slavery organ which spoke for Southern values, the *Southern Quarterly* might have liked Nott's Negro-inferiority arguments, but since the review 'spat venom' at Enlightenment rationalists it would have execrated his secular stance.)

Nott was initially panned as flippant and superficial, and taken to task equally on biblical chronology and biological hybridity. The doctor might know yellow fever close up, but not nature. Sloppy was the word – he didn't even know what a 'genus' was, judging by the fact that he lumped the 'Ourang Outang, Apes, Baboons, &c' in one genus![6] The debate from the start was fairly dirty, with this first reviewer, indeed one of the few, taking a Prichardian line on mankind's 'power and pliancy of constitution'. (A reviewer close to Darwin's zoo acquaintance, the Revd John Bachman. Indeed, Bachman's observations on London Zoo's hybrid East Indian ducks, which were sterile after one generation, are mentioned in the review.) The old clerical naturalists were speaking out against the secular Nott. The reviewer, close to the hen-coop, Bible in hand, would no more countenance hybrids being common in nature than that the nature of the mulatto was a common hybrid. Ducks were picked over, Nott's errors picked up, and Bachman's hands-on hybrid experiments on guineafowl and doves were reported to prove sterility the rule. And since the mulatto was no exception to nature, 'the presumption from prolificacy is, that he is not a hybrid'.[7]

But in this Southern atmosphere proclaiming the breeder a superior authority did not guarantee a win, as Nott and Morton would prove. All sides now – in the tumult of 'Unity' versus 'Plurality' reviews – would refocus the debate on to the analogy and pliancy of domestic stocks and hybrids.

Morton joined an already heated dispute. The further he shifted Nott's way in 'favour of the doctrine of *primeval diversities* among men', the more necessary it became to follow up Nott's study of hybridity. After all, the mulatto, though a hybrid, *was* clearly fertile (even if only, as many Southerners believed, to a limited extent). So it became crucial to prove that animals too could casually hybridize. Morton had to discredit that old resilient definition of a species, which included its inability to produce fertile hybrids.

Not that naturalists knew this as a hard-and-fast rule. Darwin was quite prepared to accept hybrids. In fact his notebooks show him seeking for some rule to explain the *degree* of hybrid fertility, whether closeness of the parent varieties or species – that is, recentness of their evolutionary separation – or some factor to do with specific constitution or situation. So that when he came to make his closet cosmic claim for evolution, Darwin scribbled in his telegraphic way:

> My theory would give zest to recent & Fossil Comparative Anatomy, & it would lead to study of instincts, heredetary. & mind heredetary, whole metaphysics.— it would lead to closest examination of hybridity to what circumstances favour crossing & what prevents it—[.]

Only evolution, in short, could begin to tackle this vexed question.[8]

But that was Darwin in private. In public others were making hybrids an ideological battleground. If humans were so many species and their hybrids were not exceptional in nature, as Morton now agreed, pluralists would have to prove that fertile animal hybrids were commoner than supposed. That lions and tigers really did produce fertile ligers and tions. That Darwin's otter-cat would unite with domestic mousers. Even, said Morton, chancing his arm, that sheep could cross with goats, hogs with deer, or cats produce litters with pine martens. If he could cite instances of fertile unions despite the 'remoteness of the genera', so much the better. It would make the human-cross sit comfortably among much stranger miscegenations.

The only first-hand account Morton gave was of a chicken and guineafowl cross, on a local farm. But his 1847 paper was full of wilder claims, in which the few unexceptional cases gave credence to the many controversial ones. Most of the latter were trawled from the anecdotes and travelogues of old authors. And some not so old: Prichard's *Researches* was mined for mentions of animal hybrids, 'although to my view,' said Morton, 'they conflict strongly with his main position'. Morton's list of the exotic crosses in the *American Journal of Science* (run from Yale) took the debate mainstream. It acquired a new importance, coming from the Vice-President of the Academy of Natural Sciences at Philadelphia – one of America's foremost men of science, shortly to be President of the Academy. Nor did its title hide the intent: 'Hybridity in Animals, considered in Reference to the Question of the Unity of the Human Species'. He was throwing down the gauntlet: if diverse animals could hybridize, then fertile mulattos were no proof against their parents being separate species.

The piece was reprinted in Britain, making it impossible to miss. Not that Darwin could miss it anyway: Lyell, increasingly the *agent provocateur*, passed the *Journal* on for his opinion. Provocation or no, Darwin politely gave it. This was 'a merely tabulated compilation' from older, untrustworthy compilations. Morton should have known better. He was 'too credulous'. And here was the cardinal sin for a naturalist: he accepted second-hand accounts and 'has not gone to his original source'.[9]

To quash Lyell's lingering admiration, Darwin pointed out some crasser mistakes. Take the tropical American curassows, chicken-like fowls with helmeted males in some species, domesticated for their flesh. According to Morton these all interbreed 'readily', 'giving rise to a progeny that is reproductive without end'. The compiler he cribbed from seemed to think that permutations of these hybrids would produce new varieties '*ad infinitum*'.[10] This garbling was unforgivable. The original source – as Darwin pointedly informed Lyell, displaying his fluency in chapter and verse – was the premier Dutch zoologist Coenraad Temminck. No one knew more about the world's fowls. Temminck was the unimpeachable authority, with a three-volume monograph, not to mention more recent publications on the East Indies fauna (as a result, zoos housed numerous *temminckii* species, from junglefowl to monkeys, named in his honour). The British Museum had bought his stuffed-bird collection, said to be the greatest in the world, and their avian exhibits were named according to his classification, the 'best and most generally received'.[11] *This* was the expert one quoted, as Darwin knew, not some hack compiler. Temminck had been sent curassows, and his *actual* assessment of their hybrids, Darwin told Lyell, was that 'a great number are quite sterile, others have bred once, & a smaller number have produced a good many young,' but among the latter there was no telling whether the hybrids had been self-crossed or back-crossed with the parents, so the observation was of little value. (Darwin could have gone further: Temminck actually had trouble distinguishing their species, because these table birds were so variable in colour, making it difficult to know if the chicks really were hybrids.)[12] Morton was shown to be shoddy and imprecise.

And even Temminck *could* be wrong. He might be venerated, but during vicious disputes even the unimpeachable are impeached. His fowl monograph was creaking, having been published back in 1813–15.

A lot had happened since then, but only an experienced naturalist with his finger on the pulse would know. As a result Darwin couldn't stop rubbing Lyell's nose in it, adding a PS to his reply about supposed wagtail crosses, 'in case you sh^{d} believe this from the high authority of Temminck': these were now 'doubted by good authority'. Each error, pressed into such appalling service, angered Darwin. Morton separated the common and carrion crow, believing them different species, thus making a hybrid of birds of the same species! and 'so I c^{d} go on, but will not waste your & my own time'.

Darwin's exasperation was visible. A decade studying such crosses had left him assured, and it showed as he destroyed Morton's credibility before Lyell's eyes. Between Morton's fowl, finches and crows was a real hotchpotch of the interesting, the valid and the fabulous, regurgitated without much critical thought or first-hand knowledge. Darwin, warier these days of Lyell's gullibility, simply ended: 'In conclusion, therefore, I do not think D^{r}. Morton a safe man to quote from . . .'[13] Lyell was left in no doubt that Darwin was now *the* authority.

Morton raised a further critical point: that each *breed* of domestic animal was descended from a separate wild species, or was an amalgam of these primitively distinct stocks. As such he was trailing another growing trend. This time his authority was the retired English military engineer and naturalist Lt. Col. Charles Hamilton Smith, known for his work on antelope (of which he named numerous new genera). Hamilton Smith was a brilliant watercolourist with an eye for detail, who had used his military postings well. His books were well-thumbed by a riding-and-hunting nation, and no less so by Darwin. He had scrawled all over Hamilton Smith's natural histories of *Dogs* and *Horses*, leaving a mass of annotations. In these books Hamilton Smith made today's domestic horse the hybrid of five distinct stocks, and domestic dogs had as many discrete ancestors.[14] The separate ancestral species were primeval. Unlike Nott, trapped in ex-Cairo minister George Gliddon's historical chronology of Giza tombs and Pharoanic dynasties, Hamilton Smith was *au fait* with fossil dogs and horses. He was dragging the debate backwards into *geological* time – something that would be important when it came to understanding deep human prehistory.

Here was another influential naturalist. The rough-edged traveller and prolific author William Swainson rated him one of the four 'greatest authorities on the classification of the Mammalia', which put Hamilton Smith up with the immortals Linnaeus and Cuvier. Indeed, Georges Cuvier's divinity as a comparative anatomist rested largely on his crowning *Animal Kingdom*, and when this was compiled in English in sixteen volumes (to become the compendium that Morton most relied on), Hamilton Smith brought it up to speed. He was known to all the savants. Thus when the rising star, Richard Owen, the 'English Cuvier' himself, went to the Plymouth meeting of the BAAS to announce the 'Dinosaurs' in 1841 (or rather, when he didn't, for it is now thought he introduced dinosaurs into his subsequently printed speech in 1842), he stayed with Hamilton Smith, who lived in the town. Owen's wife actually knew the Lieutenant Colonel from Cuvier's *soirées* in Paris.[15] So the old gent was well connected and well versed in field studies: in other words, another formidable authority for Darwin to contend with.

Morton drew out Hamilton Smith's conclusions. There had been no common ancestor for domestic dogs. Had there been – had mastiffs, greyhounds, Newfoundlands and spaniels all developed from one primeval wolf – we would have no trouble accepting 'the progressive transmutation of species' as a result of climate or habit. After all, if humans can bend these dogs into such different shapes, nature could certainly have done so. But it hadn't: like most naturalists Morton considered the prospect of a human-bestializing transmutation ludicrous. One dog type cannot turn into another, in nature or in the kennel. That is why Hamilton Smith was so appealing when he suggested that our dog 'breeds' had probably come from many wild species, including the wolf, jackal and dingo. The deep-chested *dhole* or Indian wild dog was probably domesticated as the greyhound, and some extinct hyena-relative became the strong-jawed mastiff. Darwin's old captain, FitzRoy, sent notes to Hamilton Smith which suggested that the Fuegian dog, like other South American canines, had a black palate, making it the possible ancestor of terriers. All these dogs had then hybridized to produce the interminable shapes, sizes and, sometimes, absurdities of domestic dogs today. No mutations to split some *single* wild stock were required – only separate pedigrees and subsequent hybridization.

This was Hamilton Smith's thrust in *Dogs*. And his evidence was interesting. It ran to things like nipples (in a passage marked by Darwin). Each wild species has a fixed number, which might differ from one species to another, but they fluctuate from three to five pairs in domestic bitches, which sometimes have different numbers on each side. It suggested that our breeds were amalgams of different wild species. A sceptical Darwin scored these passages. 'I doubt any *hybrid* having unequal mammae', he scribbled. But the nature magazines that he read singled out Hamilton Smith's multi-species ancestry for man's best friend without blanching. It was a talking point, but nothing bad was being said. Reviewers reported uncritically Hamilton Smith's belief that kennel breeds came 'from genuine wild dogs of more than one homogeneous species,' just as the nag on every London street was a blend of many aboriginal horses.[16] Fewer and fewer were prepared to query it.

Dogs, gamely trailing their masters, became a litmus test for race relations. Everyone from the aborigine-protector Hodgkin to Nott now looked to the other end of the leash. The normally stuffy Hodgkin in 1844 uncharacteristically wowed the BAAS with his buffoonery, as he unsuccessfully donned wigs to emulate shaggy breeds (the press had a field day, reporting his final triumph as an old English hound, which his countenance naturally resembled). It was all in an effort to show how canine modifications could highlight the climates to which their masters had been exposed during their historical migrations. In short he was pedigree-chasing to uncover the tribal and racial affinities of their owners.[17] (He presumed a sort of parallel change as dog and master were equally exposed to new climes, a notion Darwin had jettisoned. Darwin believed that active human *selection* would create dogs that were different in each tribe, whatever the similarity of climate or habitat.)

Anyway, from Friend Hodgkin, theatrically using Arctic-circling Spitz dogs to relate their fur-clad owners, across the board to Nott, who was studying dog species to split humans, there was *one* consensus emerging: that fathoming the origins and changes of domestic breeds was a key to discovering the 'natural' relations of the human races.[18] And behind this was a more fundamental consensus. All agreed that donkeys in fields and dogs in kennels and chickens in coops had a *history* – as Darwin had long assumed.

Darwin had known Hamilton Smith from early days, having breakfasted with him in Plymouth while waiting for the *Beagle* to sail. The youngster even then found him 'a very clever old Gentleman', and Darwin had later tapped the soldier's well-travelled brains on sinking coral islands. When *Dogs* came out Darwin left a trail of squiggles on Hamilton Smith's trendsetting book. But his summary on the cover said it all: 'The analogy of sheep & Cattle makes me doubt Col. Smith hybrid view of dogs'. Hamilton Smith accepted that coarse-fleeced sheep contained goat's blood, which discredited him in Darwin's eyes – or at least his wilder claims about hyenas as the ancestor of one domestic dog type.[19] Darwin doubted that all the world's domestic dog breeds living and extinct could have been hybridized from such vastly different ancestors. Variety came not from 'hybridizing', but from crafting and conscious or unconscious selection over centuries, and Darwin assumed a more pliant bone and gristle in the original and closely related stocks. But he was among a diminishing number of doubters.

However old Colonel Smith was, he was no fuddy-duddy. Like Darwin, he was happy to switch between domestic and human races. And having served with the 60th Regiment in Jamaica for a decade, as well as in Canada and Europe, he had first-hand knowledge of differently hued peoples. His *Natural History of the Human Species* (1848) was expressly written to follow up *Dogs* and *Horses* and carry his ideas to the crown of creation. It was something of a break-out book, smashing the Noachic confines that restricted so much of the ethnological literature. Hamilton Smith was adventurous when it came to human antiquity – far ahead of the timid Lyell, putting the old soldier more on a par with Darwin on the question of a human ancestry extending into geological time. The book bolstered pluralism's hard-edged, 'naturalistic' image. The expansive geological time-span for man made the rival biblical scenario of descent from ancestral pairs look crimped and old-fashioned.

No original 'pairs' for Hamilton Smith – he wouldn't have domestic curs derived from the Ark's inmates, any more than humans from Noah's sons, much less Adam and Eve. Nor were the original nations created from one another by changing climate or food. For him, all animal and plant species 'must have been co-existing' from their first appearance in perfectly balanced populations. The 'dogma' of human races diverging from a common ancestor was dismissed. Each had its separate origin, although, in contrast to cats, dogs and pigs, these unrelated human races interbred with less success (like Nott, he doubted the mulattos' ability to last five generations). But it was the timing of the first appearances of the various human species that was sensational.

In 1848 Hamilton Smith was among the first to admit that ancient humans had lived alongside extinct mammoths, woolly rhinos and cave bears. Others dismissed the human remains from caverns and fissure-fillings as recent. Not him. He had no qualms about Dr Lund's human bones excavated in Brazilian caves; Lund's word that they were in the same condition as the forty-four associated species of extinct mammals was good enough. Nor about Dr Schmerling's skeletons in Liège caverns, or so many others across the world, all mixed with lost species of rhinos, bears and giant elk. This West Country gentleman routinely accepted – long before most major players – that the human remains and flint knives from the local Torquay, Brixham and Plymouth caves in Devon were coeval with their caches of bear and hyena bones. The embedding and subsequent stalagmite formations in places suggested that these were not recent skeletons washed in. A few of Brixham's human bones, he observed, had actually been gnawed, presumably by hyenas.

To cap it, the localities showed evidence of smoke, indicating that the grottoes had sheltered fire-making 'troglodyte savages'. He was not afraid to sketch a mental picture of the 'youthful' planet, all jungly forests, lush palms and marshes:

> Imagination might behold remaining Pachyderms on the borders of lakes; huge ruminants swarming on the plains . . . Hyaenas by the borders of the wood, or glaring from opening caverns; and, perhaps, a distant solitary column of white smoke ascending from the forest, the certain indication of Man's presence, as yet humble, and in awe of the brute monarchs around him; possessing no weapons beyond a club, nor a tool beyond a flint knife . . .[20]

So pluralists in the 1840s could offer a 'naturalistic' diorama of the dawning human world. And this at a time when the many unitarists (with their single origin for the races) admired by Darwin were still evoking images of the Garden of Eden.

More fascinating yet was Hamilton Smith's timing of the emergence of the various human races. As an old acolyte at Cuvier's court, he was *au fait* with French 'recapitulationist' embryology – a study of the growing human foetus, developing through sequential fish–reptile–mammal stages, to culminate as a human baby at birth. Close observation of the final days' changes showed something more to the leading French embryologist of the day, Etienne Serres: a brain passing through 'the woolly haired' phase, then one typical of an 'intermediate Malay and American' person, after that a Mongolian, and finally maturing as a Caucasian type.[21] This sequence mirrored the order in which these races appeared in time. Thus just as fish preceded reptiles and mammals in the rock strata, so Negroes preceded Malays and Red Indians, followed by Mongolians and finally white men in the more recent geological era.

Not only were some human nations less 'developed' anatomically, they were actually more ancient. It was all too much for most unity upholders. While Hamilton Smith's *Natural History of the Human Species* generated a huge commentary, at least one unitarist with his Adamic limitations showing saw Hamilton Smith's belief 'in the existence of fossil men' totally discredit his argument.[22] Nothing more damaged his credibility in an age when so few could countenance a geological history for mankind.

That our dog and horse breeds arose from different wild species was becoming a commonplace by 1850. Partly catalysed by the human-race debate, the subject of domestic origins fed back into it. The upshot was that the surge of publications on 'The Unity of the Human Race' broached dogs, horses and chickens with a knowing air.

There was more in Bachman's Morton-defying *Doctrine of the Unity of the Human Race* (1850) about hybrids than biblical begats, and this was typical. Half the book was given over to farm and wild breeds. Bachman was well-travelled and well-respected, a field and museum naturalist of renown. (In London in 1838, remember, he had briefed Darwin and liaised with the metropolitan elite. These gave him his credentials in the Morton affair, and he traded heavily on his work with 'the best naturalists of Europe and the World'.) As a result, his suffocatingly persistent refutation of Morton made the *Unity of the Human Race* one of the most authoritative tomes on tame and wild animal varieties, and hybridity, before Darwin's *Variation of Animals and Plants under Domestication* (1868). It shows the ubiquitous intertwining of the subjects which made Darwin's own books – the *Variation*, which was intended to have a 'man' chapter, and the *Descent of Man*, in which the human chapters were swamped by discussions of animal varieties – look unexceptional then, but rather odd now, and hard to explain out of this human-race context.

Bachman even roped in Darwin's reference in his *Journal* to the 'niata' cattle bred by the Indians of the Rio Plata. These oxen had such short heads and upturned lips that they acquired a pugnacious, 'ludicrous self-confident air', like bulldogs of the cow world. It was proof of how the most striking changes in the skull could be produced; and if in cattle, why not in humans?[23] Bachman even covered key patches of Darwin's ground. The plurality threat sent the Lutheran minister equally to human parasites. Not only to the intestinal worms shared by all the races, and proof against plural origins; Bachman was also on to head lice – the sort which Lyell had brought back from America at Darwin's request, and which Darwin himself saw as an indicator of host relationships. Bachman admitted that these lice did differ among blacks and whites, but only in colour, and that colour, he thought, the louse got from chemicals in the scalp.

Fancy pigeons were prominent in Bachman's book. All animal breeds fluctuate in 'form, internal structure, instincts, and habits'. Looking at skulls alone was not enough: every variation, from physiology to instincts, had to be taken into account. And of all the tame breeds, it was the top-knotted, bustled and pouting pigeons exhibited at county fairs which displayed the most divergent variations in shape and behaviour.

Bachman accepted that all the fancy pigeons came from one ancestral type, the wild rock dove (*Columba livia*). Of course he was begging the question. This derivation was the point to be proved. He denied Morton's, Hamilton Smith's and Temminck's claim that each tame pigeon or fowl variety sprang from a separate wild species. Bachman, a pigeon breeder himself, directed 'the advocate of a plurality in the human race, to an examination of the varieties that are found in the domesticated pigeons'. Human and pigeon breeds were of a piece: prove that these extraordinary pigeons had all come from one ancestor, and the case would be so much easier for humans. Here was 'the runt, nearly of the size of a common hen, and . . . the tumbler, but a little larger than the [American] robin', yet both derived from Europe's rock dove. It was also the origin

> of the rough-legged pigeon, the crested pigeon, the laced-winged pigeon that cannot fly, the hairy pigeon, whose feathers appear converted into hairs, the fan-tailed, the cropper, the powter, the carrier, and an immense number of others, the enumeration of whose varieties would fill a page . . . we would ask [naturalists] to point out those distinctive characters, by which the varieties of men are divided into many species, and the varieties of the pigeon are all included under one species . . .[24]

Holding to unity so religiously, Bachman ridiculed those who called these show favourites separate species. There never was 'so gross an absurdity'. But fewer were now prepared to agree out of hand. This separate identity of *all* varieties is precisely what the growing numbers interested in plural human origins sought. The point being contested was opening up. But who to trust? and why? The questions raised by pouters and runts would require a fresh approach before a comforting closure would again settle on the loft.

Bachman gave the lie direct to Morton's 'fertile' hybrids. He countered the examples one by one. Curassows: he knew from first-hand experience the difficulties in unravelling these Guyanan birds, or getting them to breed. (The English had had no better luck: a prize offered for their successful clutch-rearing has lain unclaimed.) He had been surprised, like the English naturalists, at the 'infinitely' variable colour of these pot fowls in Lord Derby's collection. Back home, he had eventually procured some himself, but again no two birds were actually alike. These fowls changed so easily under domestication that Temminck and Morton had been duped into multiplying 'species' and making 'hybrids' from simple variants. They never were 'hybrids'. By his own reckoning, Bachman owned more proper hybrids of different species 'than any other individual' in America, and all were sterile, except one – the China/common goose cross – and it was only temporarily fertile. His conclusion went further than Darwin's: hybridity *was* a good test of parentage. If the offspring are fertile, the parents are of the same species. 'Consequently the fact that all the races of mankind produce with each other a fertile progeny . . . constitutes one of the most powerful and undeniable arguments in favour of the unity of the races.'

Such practical knowledge and chapter-and-verse recitation knocked the wind out of the pluralist sails for a while. Even Nott wrote to Morton that 'old Bachman is a pretty hard customer and has hit you some hard digs under the ribs.' While another Nott-apologist, himself son of an old Charleston family, Robert Gibbes, wrote similarly of the 'stir' the *Doctrine* was making: if nothing else, Bachman 'deserves great credit for his researches'.

Though Bachman's wife owned house slaves, the debate was still framed by slave-holders' concerns. Bachman never doubted that the Africans were 'inferior', judged by their historical 'incapacity for self-government'. The Scriptures formed his guide to the 'duties both of masters and servants'. The slaves were as children, whom it was our Christian duty to support, not soulless carthorses to be worked to death. Hence he slyly lashed the pluralists for giving 'the enemies of our domestic institutions' the impression of the slave-holders' 'prejudice and selfishness, in desiring to degrade their servants below the level of those creatures ... for whose salvation a Saviour died, as an excuse for retaining them in servitude'. Of course Bachman knew that Morton was in Nott's bad company, where slavery was lashed to species separation, and where the Genesis-bigots were considered 'prostitutors of science'.

Never daunted, the Lutheran out-rationalized the rationalists. God had created only once, Bachman reasoned, but had stamped on to his creatures – humans included – the power to vary a little, anytime, anywhere. So variants could appear whenever needed quite naturally. Thus the 'varieties of men would be formed without a miracle', like the varieties of chickens that farmers fashioned for the coop. Bachman was painting himself as more naturalistic to compete with the pluralists' aggressive anti-clericism. He envisioned new islands, volcanic perhaps, like the Galapagos, populated without divine intervention. No miracles were required, just colonization from some nearby mainland, whence new variants could 'gradually adapt themselves to the climate'.[25] In Bachman's view, the pluralists required a new miracle for every race and variety, human, animal, plant, domesticated or wild, a million years ago or at the moment. The unitarist didn't. For a moment, the pastor of St John's Church looked like a naturalist after Darwin's own heart.

It remained true that one of the best places to pick up contemporary chicken lore was a human-race book like Bachman's. One begins to see that Darwin's farmyard obsession was not so out-of-kilter after all. But his own position was clearly a narrowing consensus (closing around biblical die-hards) and his evolutionary solution to race relations a unique way to cut the Gordian knot. What makes Darwin's position interesting was this company he kept. Pluralism offered a viable naturalistic image of life, coloured by Caldwell, Nott, Gliddon and Hamilton Smith's anti-clerical triumphalism. This, with their cultivated data-crunching, and struggling-for-unpalatable-truths ethos, made them seem modern to the new professionals clambering up the echelons of London's scientific society. Superficially, Darwin's rival moral approach, grounded in the sanctity of anti-slavery and akin to Prichardism, itself founded on biblical bedrock, could look positively antique. Or it would, if ever he exposed it.

Certainly in America, victory was already being celebrated by the plurality camp. Prichard's *Researches* was portrayed 'like the Parthenon', as a great ruin, full of antique grandeur. It toppled because it stood on the quagmire of racial unity and common origin – 'so revolting to universal taste'. And with the edifice in rubble, what of Prichard's social prescription? – 'that negroes, being of the same species, are capable of the same civilization as white men, and should not be enslaved by them'. Modern science, said the *Southern Quarterly Review*, had established that 'the highest grade of civilization [the Negro] ever reaches is in a state of slavery'. Prichard had laboured for 'forty years' to move a 'mountain'. And he had failed. The *United States Magazine and Democratic Review*, influential and onside (and promoting Carlyle's tirades as a 'bold' argument 'in favor of re-enslaving' the West Indian peasant 'Quashee'), also awarded its laurels to the pluralists. The journal was looking to reshape policy towards Indians and blacks by eschewing the sickly sentimentality which it saw shrouding the unitarist cause, and given this aim it declared the battle over. The winners: that unholy caucus of secular Notts (with whose unholiness many were unhappy) and Frederick Van Amringe (who invoked a more acceptable biblical justification for human species separation). The grounds: 'few or none now seriously adhere to the theory of the unity of the races'.[26] So much for Bachman.

If America was losing the war, Britain wasn't far behind. While ethnologists increasingly opted for separate human species, substantiating views about the multiple origin of farm stock were shooting through the breeding community.[27] The breeders' craft was Darwin's bread-and-butter. It was the source of stud lore about size, shape, parentage and mongrelism – or hybridity, as it seemed in peril of becoming.

Darwin had an inside track on these debates among the breeding lobby. Take pigs. All the protagonists on the human question bandied around the results of one 'T. C. Eyton, Esq.' They promoted or contested his finding – that Chinese, African and English pigs all interbred. Yet Eyton showed that the number of dorsal, lumbar and caudal vertebrae varied between these pigs (often dramatically: the total number of vertebrae in the spine could range from fifty to fifty-nine). To the pluralists this made these pigs anatomically distinct species, and if different species of hogs could interbreed, so could different species of humans. Otherwise it would allow incredible 'plastic powers' to nature in the sty, as farmers gave some races more backbone. Not so, argued Bachman, a breeder himself. He had no problem with domesticity bending a single species this far, producing varieties with additional vertebrae. And, anyway, he reckoned that Eyton had only mated different varieties of the same wild boar that was spread widely across Europe, Asia and in Africa, where it is known as the 'Guinea hog'.[28] (By contrast, the Ethiopian hog and warthog were truly different genera. But they were hardly known in London – as Bachman was aware, having arrived in town shortly after Eyton's pig paper was read to the Zoological Society.)

To the protagonists this 'Eyton' was an unknown entity, a name, and useful only in so far as he could be dragooned into the paper war. Not so for Darwin – for this was Tom Eyton, one of his dearest Cambridge friends.

'Hail fellow, well met' hardly describes their camaraderie. The same age, the same background, the same defects (neither young man could carry a tune, or even recognize one, which caused hilarity among the students),[29] these boys had rollicked together at Cambridge, in an insecty sort of way. They had entomologized in Wales, climbed Snowdon, and Darwin had ridden to Eyton's hounds. But it was a smaller world for the gentry than even this snapshot implies. 'Eyton of Eyton', as Darwin often called him before third parties, showing that status mattered as a testimonial of trustworthiness, was the son of the high sheriff of Shropshire, twenty-third heir of the Eytons of Eyton. Even as Shrewsbury School friends they had hunted and fished together. And when Darwin landed in Brazil he encouraged Eyton to follow him out. With no luck, for scalpel and callipers were to be Tom's calling. He reported back to South America that he was working 'hard at the birds both English and foreign' and already had 'one of the finest collections in this country'. He was boiling down 'fish birds & animals' and had 'near a hundred skeletons'. Independently wealthy, like Darwin, he too was making a career of his avocations. Such skeletonizing – rendering birds into bone and measuring them to determine differences – was a skill he would teach Darwin. Slightly horrified, sister Susan Darwin had found 'T. Eyton's room . . . one mass of skeletons'. And Catherine had reported Tom's marriage in a way that showed growing priorities on both parts: 'I think you will find them a very nice couple to stay with, when you come back; Tom Eyton, I am sure, is very fond of you; he has written to you already *three times* at Valparaiso, and is going to write to you again. – He has got some Chinese Geese . . .'.[30]

Darwin, fresh off the *Beagle*, had not given all of his birds to John Gould to describe. He had skimmed off numerous birds in spirits and skins for Tom. As a published ornithologist (interested, like so many of the gentry, in farmstock and game, hence his definitive monograph on ducks), Eyton had added an appendix to Gould's birds volume of Darwin's *Zoology of the Voyage of H. M. S. Beagle*. There is even a hint that Darwin had borrowed one of Tom's own finches to augment Gould's description of the *Beagle* Galapagos birds.[31] The fellow Salopian was clearly enmeshed in critical Darwinian events.

Eyton's letter, listing the vertebral count of various swine, had been read at the Zoological Society on 28 February 1837 – the meeting at which Gould described Darwin's Galapagos mockingbirds. Pigs were Eyton's forte. Darwin asked his advice and sought his help: hence Eyton's name peppers Darwin's evolution notebooks from the word go. In these private musings, Darwin marked the pig paper 'VERY GOOD', and what he meant was VERY GOOD for him. It instanced a staggering variation in the skeleton of interbreeding stocks. That's what Darwin needed for natural selection to get a purchase on. Darwin didn't know why the vertebrae differed: perhaps some had lengthened and subsequently divided.[32] But the fact that they did was grist to his mill. Eyton's tables of bone measurements pointed up variation, which was what Darwin needed to split new species from a common ancestor. It was of secondary concern whether these pigs were varieties or species (merely a continuum for him). If natural barriers could separate these spinally differentiated boars in Asia, China and Europe, the pigs could follow divergent paths. They were the potential ancestors of new genera or families.

Eyton was a dab hand at crossing species, one of the few whose results Darwin really trusted. Those outside Darwin's provincial family circle were much more sceptical. Eyton's data were thrust into the heated debates over human racial distinctions – with their slaving overtones – and even Eyton had trouble steering clear.

He actually announced his initial hybridizing results casually on the periphery of the human debate. When the savants met at Bristol for the British Association gathering in 1836, Carpenter discussed fellow townsman Prichard's views on humans and his basing of the definition of 'species' partly on the infertility of their offspring (humans were races, therefore, because they were inter-fertile). At this Eyton stood to announce that he had crossed a farmyard gander and Chinese goose – two totally different species – and that the goslings *were* fertile. The same with a Chinese boar and an English farm pig. Eyebrows were raised all round because it suggested that Prichard's justification of human unity was not safe: separate species could interbreed. Zoologists questioned Eyton's 'want of proper caution in performing the experimenting'.[33] This spurred Eyton to write a pointed goose paper for the *Magazine of Natural History*, designed to place his results 'beyond all suspicion'. His conclusion was this: either the birds were the same species – not impossible, because their differences were 'not greater than those existing between the races of man' in England and China (the standard of how much change could be achieved by 'local circumstances')[34] – or attempts to define 'species' in terms of hybridity had to go. He thought the latter, because his interbreeding pigs had differing numbers of vertebrae, and there was absolutely nothing to match *that* in human racial differences.

Tom's paper sparked more evolutionary jottings by Darwin, starting with acceptance: 'Hybrids propagating freely', was his first comment. Eyton's Chinese/English goslings were fascinating to Darwin, not least for the questions they raised: would continued breeding return them to one or other parental type? (no, they were always midway); what sorts of species would cross? Only those from *distant* countries, according to Eyton – and Darwin scored the passage – that is, species unfamiliar with one another, so there was less repugnance. Darwin rather doubted this, and he taxed the best hybridizer of lilies in the country, the Revd William Herbert, on whether it was easier to hybridize distant plants. Herbert too doubted that distance makes the heart grow fonder.[35]

Unlike Bachman and the Prichardians, Darwin lapped it all up, without for a moment thinking that Tom's goslings supported human pluralism. He would always cite Eyton's geese as one of the few 'well authenticated cases of perfectly fertile hybrid animals' known to him. To the end of his life, he was still describing Tom's success as the 'most remarkable fact as yet recorded' on the subject.[36] It was a rarity, but such crosses had to be accommodated, although *not* in relation to human plurality. These goose species, like goose varieties, were still descended from the same ancestral stock in Darwin's eyes: he was – it is now a platitude – a unique 'common-origin' evolutionist, quite on his own. Darwin wanted to know how the diverging occurred, and what caused the back-breeding normally to stop and hybrid sterility to start. Tom was unwittingly supplying more questions: why hadn't sterility cut in here to keep his particular geese species on their separate paths?

Darwin accommodated the hybrids within his Prichardian migrate-and-adapt framework. As the seasoned traveller of the pair, he did still engage with Tom's views on humans. Where the Shropshire-anchored Eyton divided the world into 'Zoological Provinces', with one human race characterizing each, Darwin the circumnavigator knew this to be simplistic. The *Beagle* voyage had shown him a messy nature that couldn't be neatly divided along preconceived lines. Australia had similar Aborigines on east and west coasts, yet very different plants; New Zealand and New Caledonia the reverse: 'two races of men' but similar plants.[37] Darwin's sophisticated alternative would be built on many more variables, in time and space, and take account of sheer chanciness, as newcomers made the best of each situation in their migrations. He had a lively sense of historical contingency lacking in his friends.

It is amazing what connotations these experiments carried, but even pig data fed into the earnest debates over mankind's responsibility before God. In a dozen books on the 'Unity versus Plurality' of the human races published in the late 1840s and 1850s, domestic animals and the 'domestic institution' were a hair's breadth apart. But Darwin himself in his notebooks had portrayed animals as slaves and damned those who viewed slaves as animals. He too shared the view that creation was all-of-a-piece, stretching from human institutions to animal instincts, and that there would be political lessons to be learnt from nature. He knew how the beasts of burden were piled high with ideological baggage.

Even pottering vicars were turning traitor. The parson in question this time was another old Cambridge contemporary, the Revd Edmund Saul Dixon. He had settled into a parish in Norfolk and was 'a first-rate man', according to Darwin. Dixon was one more of those lore-givers who comprised Darwin's esoteric sources. Darwin shared Dixon's obsession with poultry, and his despair that so many 'respectable' naturalists had a sniffy attitude to farm and fancy.[38] Rummaging around among such backyard sources made Darwin different, and so successful.

Darwin's appropriately festive read on 25 December 1848 was Dixon's *Ornamental and Domestic Poultry*. It was typical of the coop and loft books he loved; 'very good & amusing', he called it. It was about 'the "origin" of our domestic races', and any dog-collar who put the living coop above dried exotic skins was going to get Darwin's sympathy. But there the similarity ended. Dixon had apparently 'started with a great idea of the powerful transmuting influence of time, changed climate, and increased food' in producing varieties, but had ended convinced that *every* breed on the farm, in the coop and in the loft was an aboriginal creation. Clearly, Darwin (in lost letters) had tried reasoning on the variety and hybrid question, but, wrote his reverence,

> Mr. Darwin's discovery, the result of his great industry and experience, that 'the reproductive system seems far more sensitive to any changes in external conditions, than any other part of the living œconomy,' confirms my suspicion of the extreme improbability of the origination of any permanent, intermediate, reproductive breed by hybridising.[39]

Darwin believed that changed climate influenced the reproductive system to produce slight variants, but for Dixon the results were ill-conceived monsters destined to die out. Darwin did goad him into trying to breed hybrids to test their fertility, which evidently started Dixon's later geese-crossing experiments, although the results simply pushed him further away.

Nonetheless, Darwin was one of the few to have descended from Olympus, and the breeders appreciated it. Indeed, knowing that Darwin was mired in marine shelled-creatures (he was waylaid for eight years, 1846–54, working on barnacles), 'and has been well nigh driven to despair by the slipperiness of their character', Dixon implicitly advised his friend to forget the nasty shells and get back to birds. Little did he know how much he might regret the suggestion. Like all but a handful of his closest confidants, Dixon had no inkling of Darwin's bent. As with so many of Darwin's contemporaries, Dixon abominated evolution because it threatened to turn humans into soulless brutes with amoral leanings, led more by earthly gain than heavenly reward. Darwin and Dixon might have shared a poultry obsession, but it was driven from diametrically opposite directions. Dixon thought that elite naturalists with their Prichardian racial descents had been led astray because of their disdain for the grubby sty and poultry shed. He thought the sty and shed contained evidence *against* a single ancestor for all fancy pigeons, or a common descent for all the diverse ducks, or, by extension, a joint starting point for all the wild races. To Dixon, 'the domestic races of birds and animals are not developments, but creations', a statement Darwin marked, implying disagreement.[40]

In the *Quarterly Review*, Dixon slated the evolutionary *Vestiges of the Natural History of Creation*, with its hurtful account of monkey-generated 'thinking man'. He also castigated a future Darwin sympathizer, Edward Blyth (Curator of the Asiatic Society's museum in Calcutta) for his own 'thirst after "origins" ', that is, *common* origins in domestic fowl. Then Dixon reached the pay-off. Hamilton Smith was right. The main domestic races are each derived from a distinct species, in dogs, in pigeons, and in *people*. Blyth out in Calcutta couldn't trace all of his various Indian turkeys and chickens to one wild ancestor because there wasn't one: these were aboriginal creations. Domestication wasn't 'a sort of harlequin's sword: touch a creature . . . and you convert a clown into a columbine'. Neither artifice nor nature can fork lineages. Neither environment nor domesticator's wand can mutate breeds. 'Then follows the fact that the Indo-Portuguese population do not turn into negroes, nor even derive "their exceedingly dark complexion from the permanent influence of climate . . .".' It was obvious to Dixon that proper farmyard science would quickly thwart those 'who pursue the studies of Dr. Pritchard'. Darwin disagreed and marked his copy of *Ornamental and Domestic Poultry* so as to leave no doubt. Making every farm breed 'aboriginal' would scotch Darwin's whole enterprise: 'Look at the oxen of every different country of Europe – look at dogs of d[itt]o – look at men – if their variations are denied – my work might be closed'.[41]

'Mr Dixon of Poultry notoriety' was how he was known to Darwin's circle – the man 'who argued stoutly for every variety being an aboriginal creation'. Today, looking back, it is impossible to comprehend why pigs, pigeons and poultry could have evoked such passions until one realizes the connection to race. Domestic chickens lay at the far end of an attenuated analogical thread stretching all the way up to MAN. What went for poultry went for God's chosen being. In an age rejecting any civilizational power inherent in black brains, poultry studies seemed naturally to underscore the separateness of all races.[42] The moral was clear, and backed by experiment – in this instance encouraged by

Darwin himself. Dixon began mixing geese, or reporting crosses among geese and pheasants in Lord Derby's menagerie, to see – he was now wobbling in his doubt – if fertile hybrids *could* be fashioned between aboriginals, which might explain the gradations of intermediates found on the farm today.

Dixon was among the growing number who believed that tumblers remained tumblers and black men had always been black men and white men white. All biological commonality was denied. It was a burgeoning leitmotif of the age. Domestication among enslaved stocks down on the farm lay behind the rash of books for or against human unity.

One of the least abrasive was Charles Pickering's *Races of Man*, typically huge and once portrayed as 'amorphous as a fog, unstratified as a dumpling and heterogeneous as a low-priced sausage'. Sausagy consistency never worried Darwin (he could write that way himself). He devoured it come New Year 1851. In fact the book was better received than this description might suggest, as much for its technical matter as its ambivalence on the one-versus-many-origins issue. Darwin read it through a second time. Pickering was a fellow circumnavigator. He had also been one of Morton's contemporaries at the Academy of Natural Sciences, but his voyage with the Wilkes United States Exploring Expedition, which took him from the South Seas to the Andes, gave him much greater authority. This was boosted by subsequent trips to Egypt, Africa, Arabia and India to study their peoples and faunal and floral distribution. (He obligingly posted lists of plants he found on coral islands to Darwin, who was one of the world's atoll experts.) Pickering raised the number of human races to eleven and the complexity he found was evident in his mass of facts, just the sort Darwin liked. But Darwin scoured *Races* single-mindedly for snippets on the history of domestication and the geography of introduced species. The battleground for him was the common origin of tame varieties. Prove that and the pluralist case collapsed, making descent by selection much more credible.

Evidently Pickering had toned down his predilection for plural origins in the final draft. (As a report of the United States Exploring Expedition, *Races of Man* was subject to Library of Congress approval, and one anti-slavery senator on the committee had already berated Pickering's 'strange notions' in the first drafts.) So muted was it that New England journals and an old England reprint could purloin *Races of Man* for the angels and pit it against the 'ridiculous [pluralist] theories'. Pit it against those who denigrate our 'African brethren', those evil men who 'have "laughed and wondered if a Negro's soul could feel"' and who have likened him to a brute 'for the purpose of degrading him' in the 'lands over which the black plague-cloud of slavery even yet is permitted to remain'.[43]

Tugging and distortion were typical because the debate engendered such moral outrage, with the American South as its new focus. The latest 'nigger' outbursts by Darwin's old acquaintance Carlyle were being snapped up by Southern-sympathizing American reviews. His harangue on the 'twaddle' of philanthropists, coming as it did from England, was rebroadcast all the more loudly. *DeBow*'s saw 'the case of Quashee . . . disposed of with a master hand . . . When British writers can so speak, it is time for Northern fanaticism to pause and reflect'. The *United States Magazine and Democratic Review* saw it as a sign that 'a powerful re-action has taken place in England in regard to the policy to be pursued in relation to the blacks'. Given the tensions, the din could only stiffen the slave-holders' resolve. Carlyle could call slavery a 'precious thing', although in this instance he said of Britain's freed West Indians, without much conviction,

> You are not 'slaves' now; nor do I wish, if it can be avoided, to see you slaves again: but decidedly you will have to be servants to those that are born *wiser* than you, that are born lords of you . . .[44]

Others saw the outburst as 'unspeakably wicked'. John Stuart Mill recoiled at this arrogant 'work of the Devil'. Intriguingly, while some now withdrew from Carlyle's company, complaining of the damage being done, Darwin's brother Erasmus didn't. Still in and out of Carlyle's house, the bachelor-about-town merely heaved a sigh about these 'occasional discourses beneficent whips or whatever the title is. I feel rather alarmed, as one can't stand a great deal of such stuff.'[45] A weary shrug would have been the farthest from his younger brother's mind. Carlyle's condescending sneer would likely have made him quake with rage.

Harder racist attitudes were spreading through the classes down to the gutter. Bruisers such as Knox in England and Nott in America were defiantly secularist. Now urban 'infidel' activists tapped this freethinking vein to exploit pluralism at street level. Here it became even more transparently ideological.

Notions that the existing races of men 'must have originally sprung from *perfectly separate stocks*' (as a propagandist piece of street literature, the *Infidel's Text-Book*, had it) were fashioned into proletarian weapons aimed at the Adam and Eve credulity shoring up the Church's hated authority. It could turn into an agenda for action, notoriously so in the case of the agitator Charles Southwell. He had started a pointedly atheistic set of evolutionary articles in the illegal street-corner rag, the *Oracle of Reason*. (The first number in 1841 had featured an illustration of a hairy fossil man – itself subsequently seen as a 'racist fantasy' – long before any gentleman naturalist, bar Darwin, believed in such a beast.) After a career in Church-bating, initiated by his martyrdom in prison for blasphemously damning that 'revoltingly odious Jew production, called BIBLE', he emigrated to New Zealand. The Church Missionary Society had long held sway here. (FitzRoy, the old governor, had himself consulted it on Maori affairs, never having moved from his pro-mission stance on the *Beagle*.) Southwell's racism exploded on seeing bishops and missionaries back Maori land rights over settler claims. He boosted the *Auckland Examiner*'s sales by deriding the mission's 'amalgamation of races' doctrine. Then he sent fellow radicals at home tirades to be fired at the Exeter Hall philanthropists, carrying the implicit message that, since the 'savages' cannot be civilized, they had best be exterminated.[46]

Having sailed home with FitzRoy proclaiming the missionaries' good work, Darwin found himself again thrust on to the evangelical side. That was the side of common humanity through shared ancestry, which many now linked to an exploded Prichardian ethnology and pulpit naivety. The cutting edge of science was becoming less and less hospitable to Darwin's descent from a single stock.

As abolitionists regrouped and racism hardened, naturalists looked to the wild to understand slavery.

Marching, invading, slave-driving Amazon ants, rearing the young of captured species as menials: decades before, their astonished discoverer thought such behaviour an 'almost incredible deviation of Nature from her usual laws'. But some were now doubting its unnaturalness. Articles proliferated in the 1840s on these red ants and their slaves. All anticipated the readers' incredulity and all shared the same strapline – 'to heighten the wonder, most of these slave-dealers are ruddy or reddish, while those which are captured to become their servants are black!'[47] The words were William Swainson's, a pernickety Customs-collector's son who made his way with a pen, but they could have come from any number of writers. Slavery in another species startled a generation flushed by its own abolitionist successes, no less than its harder-headed sons. Endlessly one reads of their 'amazement', and always how the 'kidnappers' are pale 'and the slaves, like the ill-treated natives of Africa, are of a jet black'. Evangelicals thought it a horrid lesson, and that evil resided everywhere; others that it was 'unnatural'; while an older generation noted that Providence ensured that the enslaved 'Negro' ants, being taken only as larvae or pupae (unlike the African captives), were at least spared the pain of being torn from their loved ones.[48]

Much of Swainson's account was rehashed from others, including its battlefield scenes. These verbatim jungle notes carried the authenticity of a breathless *Times* report, with its evocation of carcass-strewn carnage as the pupae were dragged into bondage. It was presented as direct reportage, and Swainson *was* widely travelled, having collected in Brazil many years before Darwin. But that was a long time before. Now he was a disgruntled old hack, pumping out voluminous nature books full of these second-hand accounts, and about to emigrate himself. He was derided by plummy Oxford-educated naturalists as an uncultured specimen-haggler. And all agreed that, by churning out the books to make ends meet, he could be 'abominably careless'. But what equally galled Darwin was his 'nonsensical' theoretical explanations, about how natural groups of all animals and plants could be arranged nesting in circles of five. For Swainson there were five races of man, discretely arranged in five 'true zoological divisions of the earth'. Yet they were to be imagined connected in a circle, so that each human race shaded into the next at the boundary. The supposed five great stocks of birds were placed likewise, with one characterizing each zone. Swainson probably expected ants to fit the five-fold pattern as well. To Darwin it was classificatory madness, arbitrary to a degree. It left him with a 'disinclination to believe any statement of Swainson's'.[49]

Darwin preferred to get his reports of ant behaviour first-hand from the 'highest authority', and that meant Frederick Smith – a real field naturalist and voracious describer of bees, wasps and ants. Smith was the Curator to the Entomological Society during the 1840s and assistant in the Zoological Department at the British Museum in the 1850s, where Darwin visited him for discussions. Ultimately, Smith's reputation was capped by his astonishing discovery of slave ants in England (in 1854),[50] a find that set Darwin searching for himself. Smith would become Darwin's prime source when the time came to write up natural selection. Although Darwin had ignored them in his evolution notebooks, with the ants' new paradigmatic status, he would have to explain the origin of their slaving behaviour, and in a way that made it seem neither unnatural, nor Providentially mitigated, *nor* a parallel to human atrocities. This was probably the more necessary now, because of events in the United States.

Swainson's regionalizing of the human races and his ridiculing the idea of white men migrating to Africa and acquiring black skins or thick lips, found him admirers across the Atlantic.[51] His words were picked up by pluralists who would also put the slave ants to good use. Among the first was William Frederick Van Amringe, a New York lawyer who explicitly encouraged the red ants to nip Prichard. Slave ants were 'one of the most exciting topics of discussion of the present day', not least because they showed the double-edged sword Prichard was wielding. These ants 'systematically make predatory wars' on target species 'for the sole purpose of procuring slaves to perform the servile drudgery of their habitations': they proved the danger of Prichard's use of animal analogies to establish human kinships.

That was Van Amringe's warning in his 736-page *Investigation of the Theories of the Natural History of Man* (1848). Make Prichard's approach valid and

> we have an example, by a most unerring law, derived directly from the Creator, manifested in the instinct of these insects, that slavery is permitted, if not ordained. It is remarkable, too, that the resemblance . . . to human institutions of slavery is perfect, not only in regard to the genus, but to the color of the beings enslaved: and not only to the color, but to the comparative social, and . . . mental conditions of the masters and slaves; for the domestic economy of the [red] . . . ants exhibit an advance in comfort and security, beyond the condition of the negro ants, the fair representative of the comparative advance in the civilization of Europe over Africa.[52]

He actually believed that no invertebrate grubbing around could help us understand humans. Van Amringe even went so far as to deny that zoology itself had anything to do with the origin of the races. The climate could not have formed them. They were a biblical given – the white Shemites, yellow Japhethites, red Ishmaelites and black Canaanites –

to be ranked from white top to black bottom. Each had its physical distinctions and temperament, but only a lawyer could have characterized them so off-handedly, with the 'Shemitic species' given 'the strenuous temperament; the Japhethic the passive; the Ishmaelitic the callous; and the Canaanitic the sluggish'. The species were inviolate, God-given, psychologically separated by a chasm from animals. There was no transition, no vile transmutation of black man into white – the very concept was 'humiliating'.

For Van Amringe, human-specific differences made the 'temporarily fertile' mulatto as unnatural as an orang-utan–chimpanzee cross in a zoo. With patriotic pomposity, the lawyer pleaded 'in the name of the American people, at least, if not of the whole human family' against Prichard's search for human racial analogies among the animals.[53]

Reviews caught Van Amringe's drift and unpacked the consequences of a unitary origin for all humans:

> If a European can gradually be changed into a Bushman, or vice versa, (according to the theory of Dr. Prichard, who contended that all men were originally black, and that the white race is only a congenital variety,) from naked black Hottentot savages into Bacons or Miltons – what necessity was there for creating Adam and Eve, since a gradual progress of the animal creation, as fossil remains show them to have existed, would gradually have produced human beings, when the earth became prepared for their habitation?[54]

Darwin had happily slid down this slippery slope a decade earlier. But Van Amringe would go this far with Darwin: that the great creative law which kept the races apart was a constitutional difference in the appreciation of *beauty*, and he devoted a whole chapter to proving it. In fact, many of the pluralists took this line – obviously playing up its opposite, the 'repugnance' of white man for black.[55] Darwin wasn't alone in putting aesthetics into his science.

Van Amringe wasn't happy with it, but he knew that the big human question, Unity or Plurality, was being made to hang on fancy and farm breeds. And given that 'the science of breeding domestic animals had arrived at such perfection in that wonderful island' (Britain), he wasn't surprised that Prichard and his hold-outs lived there.[56] Religious attempts to steer the debate away from domestic analogies were unsuccessful. The ability to change, interbreed and trace the ancestry of domestic stock would be one of the leading characteristics of the human race debate for a decade. The momentum was unstoppable: all sides now had a vested interest in the problem.

The Darwin clan itself had no doubt whatever of the big issue. The rather vain and very eminent Queen's Physician Dr Henry Holland spelt it out. Holland was a generation older, and shared his own descent with Darwin from common maternal great-grandparents. He was also Darwin's doctor during these critical years. At times of acute stress – while writing his heretical evolution notebooks, after his father's death – Darwin lay on Holland's couch. Thus at the end of his creative evolutionary phase (about 1840), Darwin was discussing his deteriorating health with Holland, while tapping his brains: marking passages on 'Hereditary Disease' from Holland's *Medical Notes and Reflections*, and so on. Darwin finally sent Holland a string of questions on inherited diseases among individuals and nations, asking also about the physical appearance of racially 'cross-bred' humans.[57] Though he found the royal physician 'dreadfully conceited' and rather dismissive 'of the labours of others', their views of one another clearly fluctuated, and Holland would eventually board the adulatory bandwagon following his second cousin and revel in their 'long and intimate friendship'.

Sickness and depression following his father's death forced Darwin back to Holland's consulting rooms.[58] It was in January 1849. As chance would have it, the doctor at that moment was himself paying tribute to Prichard. The greatest defender of human unity, James Cowles Prichard, had just passed away, in December 1848, five weeks after Dr Darwin. The two men, potent symbols, had been intertwined at intellectual levels. Prichard's books had engaged with the Doctor's own father, Erasmus Darwin. Or rather, Prichard had attempted to distance his views of limited racial change from the libertine's evolution of all life (a rational heresy from a revolutionary age, described in *Zoonomia*). He achieved arm's-length separation by demarcating his mechanism for diverging the human stocks from Erasmus Darwin's slow climatic cause: Prichard's eventual fast-track process relied instead on predetermined, inheritable changes to transform ugliness into beauty, black bodies into white. But death had done with all of that. It had struck both worlds, Prichard's and the Enlightenment evolutionist's, and a new horizon was looming.

Even Holland gave signs of it. He thought Prichard too hasty in dismissing all talk of transmutation. It might be wrong (he thought it was), but it posed interesting questions about the definition of species,

> and the changes more or less permanent of which they are susceptible, either from natural causes, from education, or from forced union with each other in the production of hybrids. The topic is one of deep interest, carrying us by divers paths into the midst of the most profound questions which can legitimately exercise our reason. It is associated closely with many of the natural sciences, as especially with all that relates to the physical history of man.

Holland was another, like Carpenter, brought up among Bristol's Unitarians. It showed in their rational cause-and-effect outlook and materialist explanations. (Since God worked through natural causes, these were nothing to fear but something to celebrate – even, perhaps, those leading to life's development through geological time.) And like Carpenter he was an unflagging supporter of that other luminary of the slave port, Prichard. In fact, Holland had been educated in Bristol by Prichard's father-in-law, the staunchly anti-slavery Unitarian minister John Prior Estlin. Now, on Prichard's death, the physician grandee paid homage. He wrote a 40-page *Quarterly* review of Prichard's exhaustive *Researches into the Physical History of Mankind*, the monumental five-volume set only completed in 1847.

Ostensibly, the review was also to embrace Hamilton Smith's *Natural History of the Human Species* for the sake of balance, but balance was one thing it didn't get. A single, fleeting mention saw Smith off. Darwin's cousin acknowledged that the big question of the day, the one 'which may be said to govern the whole subject', was whether the human races were 'the offspring of a single stock, or have descended respectively from several original families'. The whole review centred on it and the whole review was an instinctive knee-jerk support for Prichard's answer. The reply to the question, whether 'the perfect Negro and the perfect European, seeing the strong contrasts and diversities they exhibit, can be rightly deemed of the same species', was a foregone conclusion. Anatomy, physiology and interracial fertility answered 'yes'. But the sheer 'exuberance of the subject' of domestic analogies showed that these were set to hammer the answer home. Here, among the dogs and fowls and pigeons – with their astonishing array of shapes and bustles and bone sizes – lay the convincing proof of human unity. This final word 'will be understood by every one'.[59] Darwin knew it.

9
Oh for Shame Agassiz!

One man was ultimately responsible for drawing Darwin out on the human race issue: the future doyen of American science, Harvard professor Louis Agassiz.

Agassiz had been the elitist of the elite Continental naturalists. He was himself an heir to Cuvier's throne, a fossil fish expert when such things were interesting to the gentry. But when it came to humans – and Agassiz came to them when he stepped off the boat in America in October 1846 – he would put a definitive stamp on racial segregation. Through Agassiz the strength of the Morton–Nott fist landed directly on Darwin's jaw. Agassiz helped in large part to trigger Darwin's annus mirabilis in 1854–5, when the quiet Down House recluse worked out his response on the human racial question.

Darwin's relationship with Agassiz had begun in September 1846, when the British Association for the Advancement of Science had rolled into Southampton for its annual meeting. Crowds had poured into the port. The odd visitor, like the smart young assistant surgeon Thomas Henry Huxley, billeted in the hulks nearby in Portsmouth harbour awaiting his own round-the-world voyage, was there to learn the tricks of the dredging trade. Lyell turned up to talk about the Mississippi Delta, his first outing after nine months in America. Agassiz, his old friend, came from Switzerland to talk fossil fish. And Darwin emerged from Down to catch up with Lyell and test the choppy waters for his theory.

There was one tiny desideratum before Darwin could think about publishing on natural selection. Everything from the *Beagle* was in print or accounted for except a minuscule burrowing barnacle from Chile. Anatomizing it wouldn't take long. Or so he thought before Dr Joseph Hooker, the young botanist describing his Galapagos plants, challenged him to describe 'minutely' the differences among *all* the barnacle species. Hooker was becoming one of Darwin's closest scientific confidants. He too had been a voyager, not admittedly as a dining companion to a captain, but as an ill-paid assistant surgeon on HMS *Erebus* (1839–43), sent to explore the Antarctic Ocean and its islands (he had carried a copy of Darwin's *Journal* and marvelled at what was needed to follow in Darwin's footsteps). A camaraderie had developed between these travellers. Hooker was now writing up his *Flora Antarctica* at the Royal Botanic Gardens at Kew, where his father was Director. Becoming an expert in one entire group of animals, he said, would win Darwin the right to talk on an abstruse issue like the origination of species. Hooker's caution carried weight because he was one of the select few who knew of Darwin's belief in transmutation. (Even confiding in Hooker had been painful to Darwin, tantamount to 'confessing a murder'.)[1] In a culture which considered transmutation abhorrent and a self-empowered nature blasphemous, only a fool would publish without sound credentials. Anyway, Hooker had pointed to a problem. Natural selection needed a constant supply of variants in order to be able to pluck out and polish the better-adapted ones. But how much did species vary? No one really knew. Darwin needed to, and that was another reason to extend his study.

Agassiz was being lionized at Southampton. For years the great naturalist had been causing Darwin consternation. He had postulated a former 'ice age' in Europe that had glaciers gouging through the Scottish Highlands, spoiling Darwin's prize solution to their famous landscape puzzle: in Darwin's first major scientific paper (clinching his fellowship of the Royal Society), he had argued that the strange parallel 'roads' or terraces on the slopes of Glen Roy were ancient beaches, left as the land was lifted in stages above sea level. Agassiz considered these 'roads' the shores of successive glacial lakes created as ice dammed the Glen. Sketchy and based on cursory fieldwork, Agassiz's theory was still making converts.[2]

It was such a setback that it had probably caused Darwin's confidence in his other, evolutionary, theory to ebb. But it had flowed again in 1842, as he had dashed off the first sketch of natural selection, flushed with the conviction that he *was* right about Glen Roy, whatever role glaciers played.[3] At Southampton he still thought that 'there was never a more futile theory' than Agassiz's glacial lakes. But the Swiss savant's ice sheets were sweeping all before them, and Darwin remained wary of

Agassiz, even after giving in to glaciers many years later. He hated to be trumped, it made him 'horribly sick'. He was hurt more by the prospect of being 'proved wrong' on Glen Roy than on 'almost any other subject'[4] – and 'any other' meant evolution.

At Southampton Agassiz redeemed himself. He spoke up for barnacles at the Ray Society, which published specialist natural history tomes. He told the society – Darwin was a member and memorized his words – that 'a monograph on the Cirripedia [barnacles] was a pressing desideratum in Zoology'. What could be more useful to a sea-faring nation? In Southampton Water they encrusted Britain's wooden shps, and the problem had to be tackled scientifically. Darwin decided to examine as many species as possible. Within months the Ray Society project had taken on a life of its own. Everyone sent him specimens, including Agassiz.[5] Eight years later, Darwin had described the entire world's barnacles, living and extinct. When he received the Royal Medal of the Royal Society in 1853 for the feat – a sort of scientific knighthood and his licence to speak on bigger things – he had Hooker to thank for goading him and Agassiz for starting him off.

After Southampton, Lyell took Agassiz to Liverpool by train to see him off to America. The trip was of Lyell's engineering. Agassiz was at a crossroads: his wife had left him, he was in debt. Lyell had persuaded the cotton-industrialist and science benefactor John Amory Lowell to have his friend follow him as Lowell lecturer.[6] Agassiz never looked back. Over the next three decades, the man who covered Europe in glaciers would become America's most influential naturalist and Darwin's greatest New World rival.

There was more to Agassiz than ice. He was warm, and to Lyell's friend George Ticknor his bonhomie seemed inexhaustible. His love of God's creation was so great that even high and dry Lyell found it contagious. Erudite and ebullient, he charmed his audiences. Boston society stood in awe of him, America agog. 'Fashion has set her seal upon Agassiz's lectures', a Bostonian observed. Harvard accordingly netted this big fish who knew more about fishes than anyone. In 1847 the textile tycoon Abbott Lawrence saw Agassiz installed as the first professor of geology and zoology in the Lawrence Scientific School that he was sponsoring. Lowell, on Harvard's governing body, negotiated the appointment. After Agassiz remarried in 1850 into a leading mill family, he stood on the commanding heights of American science. His oeuvre was backed by the wealthiest and most conservative elements of the New England slave-cotton industry.[7] What Agassiz desired, his patrons delivered. And he epitomized what they wanted in a Yankee man of science: independence, objectivity and spirituality, with a democratic flair.

His spirituality was not what it might have seemed to most Americans. Son of a sixth-generation Protestant pastor, Agassiz learnt traditional theology before the romantic *Naturphilosophie* blew him from his biblical moorings. Nature became his bible, revealing the Creator's intent. Agassiz divined it in the patterns of life – the progress of fossil species through time which seemed to parallel the development of the embryo. And so it was that Professor Agassiz, in his guarded way, set about explaining the mind of God to America: how life's history revealed the execution of God's Plan to place spiritual man on earth. This history stretched over vast geological ages, punctuated by cataclysms (and glaciations were perfect for the job), each wiping out all life before a new wave of miraculous repopulations. And with each, man's appearance came a little nearer. So there could be no continuum of species back through history: Agassiz's ice sheets saw to that.

British geologists were familiar with Agassiz's strange ways. At Southampton a party (including Lyell, and possibly Darwin) had gone dredging in the sound. The jaunty, versifying dredger par excellence, Edward Forbes, a specialist on seabed molluscs, diagnosed the haul as common shells which were *varieties* of those found as fossils in some of the red sandy rocks along Britain's coast. But Agassiz allowed no passage of life back into pre-glacial time, represented by these rocks: he pronounced the living shells in the dredge a new species, with their similar-looking fossil counterparts a separate, extinct species. If Agassiz's catastrophic geology was all 'moonshine' to Darwin's old Cambridge teachers, the implications were still more so to Darwin. Watching events, he already had Agassiz pegged as sustaining 'the strongest arguments in favour of [the] immutability' of life.[8] For Agassiz there could be no obnoxious *mutations* of 'lower' into 'higher' forms through time: while there had been a succession of ever 'higher' types, from fishes to humans, these were explained romantically as the unfurling of God's thoughts – there was no material or evolutionary connection between one fossil and another, they were only related via the Divine Mind, which brought each new species miraculously into existence.

Personally, Agassiz was a religious freethinker like Darwin. He made peace with his new wife's Unitarianism and fitted perfectly into her patrician culture. When pressed, he would say that Genesis was wrong and the Flood story was just watered down ice-age science. None of that was news to Unitarians, who interpreted the Bible liberally, but few would follow Agassiz far into idealist metaphysics, and certainly not Lyell. Lyell's *Principles of Geology* had been one long argument *against* the existence of global catastrophes. Nothing gave Lyell 'more pleasure' a bit later than to hear of 'Forbes' beautiful essay on . . . the identity of the crag [fossil] shells with living ones', simply because it negated Agassiz's claims.[9] On the other hand, unlike Agassiz, Lyell didn't think that life had progressed, for a good reason. Strip the icy cataclysms from Agassiz's doomsday scenarios, and what is left? The march of increasingly complex life-forms, culminating in orang-utans, savages and ultimately civilized Anglo-Saxons. This risked playing into evolutionist hands, as Lyell surely warned Agassiz.[10] Someone would sooner or later explain such an ascent by evolution, degrading man and denying God. But Lyell and Agassiz both believed in spiritual man and they led the world in earth science. Agassiz's Plan might be risky, but he and Lyell stood shoulder-to-shoulder against transmutation and remained close friends.

Before delivering the Lowell Lectures, Agassiz had met Morton in Philadelphia. In fact, he met him near the time Morton was reading the Academy his papers pitting hybridity against human unity, so they might naturally have discussed the subject. It had been an unforgettable first visit to the city, and not only because of Morton's cranial 'Golgotha'. The shock for Agassiz came after, at the hotel. Here he had his first close-up experience of blacks, and he was nearly sick. There stood an army of African waiters, 'their black faces with their big lips and their grimacing teeth, [and] the wool on their heads' proving them a 'degraded and degenerate race'. Agassiz had never met a black person, never mind at dinner; now 'elongated hands' with 'large curved fingernails' stretched towards his plate. The result was visceral, instantaneous and shocking (and expunged from his *Life and Correspondence*). His revulsion was hard to disguise. Transfixed,

I could not tear my eyes away from their appearance in order to tell them to keep their distance. And when they put their hideous hand on my plate in order to serve me, I wished I were able to distance myself in order to eat my morsel of bread elsewhere . . .

Such was his gut reaction. He hardly dare tell his mother in Switzerland of the 'terrible impression' they made. It was impossible 'to quell the feeling that they are not of the same blood as us'.

Suddenly he understood how the black problem threatened the future of the United States. He damned the philanthropists and slave-holders alike – the former wanted to make black people citizens, yet they refused their own daughters in marriage. The son of Africa could only thrive in the tropics; that's where he should be, not in slavery. 'What misfortune for the white race to have tied its existence so tightly to that of the negroes' in the American states. 'God save us [in Switzerland] from such contact!', he wrote to his mother.[11] The 'hideous' sight of black people had pushed America's premier zoologist into Morton's arms.

Carrying a copy of *Crania Americana*, Agassiz went back to Boston to perform a public about-turn. In Europe he had written that the human species consisted of different races, each inhabiting its geographic zone. Now he agreed that each race was *created* in its own zone – blacks and whites had separate origins. Unable to parry the waiters, he would make sure their ancestors 'keep their distance'. Only unfeigned 'references to the Creator' and a pointed refutation of transmutation saved him from censure for this theological gaffe. Or so thought Asa Gray, Harvard's professor of natural history, then trying to 'keep clear of slavery' by confining his new *Manual* of American botany to the northern flora. Gray had theological reasons for rejecting multiple racial origins – moderate Calvinism taught him the unity of mankind in Adam – but he admired Boston's great guest nonetheless. Like so many, he thought it was possible to reject Agassiz's conclusions without faulting 'his spirit'. In an age when Americans held many conflicting interpretations of Genesis, Agassiz's reputation did not suffer greatly. Gray, however, had 'heard the fire bell in the night'.[12]

Darwin and Agassiz were now pulling in diametrically opposite directions. Darwin had pestered his publisher to obtain an American copy of his second-edition *Journal* of the *Beagle* voyage, fearing the editors might have toned down his fit over slavery. The edition had appeared in 1845, and Darwin didn't have copyright. Lyell was writing his *Second*

Visit to the United States with its gentle slavery compromise, which might have increased Darwin's concern that his own lambast had been censored. (It hadn't.) While Darwin feared the power of the pro-slavery lobby, the pro-slavers he feared were successfully lobbying Agassiz. The South welcomed Agassiz with open arms. His first trip to a Slave State late in 1847 saw naturalists, Nott-sympathizers and plantation-apologists lining up to honour Harvard's man. At Charleston – a city buzzing with talk of Nott's pluralist views[13] – Agassiz gratified them by bending with the Southern breeze. In his lectures he made blacks and whites totally distinct species. He did so, moreover, at Bachman's own Literary Club. No wonder the pastor's *Unity* book oozed such contempt.

Lyell kept up with Agassiz's activity, and passed pandering messages via Ticknor in Boston: 'As to Agassiz saying the negro's brain is like a child's fourteen or sixteen years old, if I am not greatly in error, Owen says the same of the adult male stolid and uneducated agricultural labourer. Tell Agassiz this, and see if it is new to him.' The urbane Lyell was drawn like a moth to a flame on these matters, unable to avoid this sort of haughty banter. It showed again as his two-volume *Second Visit* was published in 1849. Where Darwin had worried about the plantation lobby expurgating his volume, Lyell feared the reverse: New England abolitionists denouncing his judicious portrayal of the plantation aristocracy. He expected them to pan 'that part of my work'.[14]

Some Southern journals were rather pleased at Lyell's judiciousness on slavery's unavoidable difficulties, especially because English authors stepping inquisitively over the Mason–Dixon line could be so censorious. They were prepared for the worst after Dickens' notorious visit, and his *American Notes* with their outrageous quotes from local newspapers:

> Ran away, a negro woman and two children. A few days before she went off, I burnt her with a hot iron, on the left side of her face. I tried to make the letter M.

> Ran away, my man Fountain. Has holes in his ears, a scar on the right side of his forehead, has been shot in the hind parts of his legs, and is marked on the back with the whip.

Dickens shared Darwin's wrath: no one doubted that there were 'many kind masters', but

Slavery is not a whit the more endurable because some hearts are to be found which can partially resist its hardening influences; nor can the indignant tide of honest wrath stand still, because in its onward course it overwhelms a few who are comparatively innocent, among a host of guilty.[15]

After this the South was pleased to have 'the reflections of a *gentleman*' rather than the 'slanders of a vulgar cockney'. Lyell had sympathetically studied 'the strata of our social economy' and as a result 'bears willing testimony to the happy condition of the slaves'.[16]

It was well that Darwin responded only to the Northern section of Lyell's book, and that 'makes me long to be a Yankey'.

Lyell tactically positioned himself between Charleston, Boston and Down. The Lyells continued paying visits to Darwin and plying him with American books. One was by the Unitarian minister Theodore Parker, whose stirring sermons Lyell had heard in Boston. Parker, like English Unitarian militants, saw true religion expressed in action to alleviate the suffering in society. Nor did he mince his words: he damned an apathetic clergy who were 'willing to send their mother into slavery, pressing the Bible into the ranks of American sin'.[17] He was agitating and embarrassing Lyell's polite middling classes towards abolitionism. There was no doubting the leaning of these books, which would have been to Darwin's taste.

The years around 1850 seemed dark. It was a moment when the French Count Gobineau, from a culture prone to regular revolutions, could see racial causes of past political catastrophes, just as the disillusioned Knox in Britain saw them explain future ones. Gobineau's four-volume work was imported into America by Nott and translated in Southern fashion by his 21-year-old Mobile-based friend Henry Hotze as *The Moral and Intellectual Diversity of Races*. Hotze, like Agassiz, was another mannered Swiss racialist who would serve the slave-owners well. 'Southern fashion' meant purging the unitarist sentiment, removing criticisms of slavery and making the single-volume condensation compatible with pluralism (a point Nott reinforced in an appendix).[18] It was then fit to be addressed to 'The Statesmen of America'. Nott's appendix, needless to say, gave equal time to hogs and dogs, horses and hybrids, in his usual Bachman-baiting fashion. But that was de rigueur by now.

Like Nott, the cynical Robert Knox in Britain was denounced 'as an atheist, Deist, infidel' for his caustic comments on the clergy. The cynicism was distilled into Knox's *Races of Men* in 1850. It was extreme, sarcastic and enough to make the 'Saxon boors' who read it tremble. 'Unlike the work of Colonel Smith', reckoned a commentator, the inflammatory book, 'if taken up, will probably, by many, be laid down again either with aversion or dread.' It made race the sole determinant of human history and Prichard the naïf who had duped the world. Knox not only rewrote British history as a racial sparring ground, but also his own history, to make him a racial determinist from his earliest Edinburgh days. His book's biting strapline said it all: 'Race is everything'. It governed political relations; it was inflexible, unaffected by religions or climates, and unchanging, for 'the races of men . . . are not convertible into each other by any contrivance whatever'.[19]

Knox's radical views were perplexing to many. He actually hated slavery but the slavers loved him. He had the races segregated and stereotyped (and 'drawn in caricature', one pundit conceded), each in its own geographical domain. There was, however, one guarantee against his predicted genocide, claimed Louisa McCord, a flinty, conservative, South Carolina 'plantation mistress', and spokeswoman for the South (in fact, the only woman essayist speaking publicly in the South). Knox had cried apocalyptically,

> The hottest actual war ever carried on – the bloodiest of Napoleon's campaigns – is not equal to that now waging between our descendants in America and the dark races; it is a war of extermination – inscribed on each banner is a death's head and no surrender; one or other must fall.

One institution safeguarded against this 'bloody climax', McCord claimed: the 'providentially established institution of slavery, blessed, thrice blessed and beautiful in its harmony with creation!'.

True, as Knox predicted, with white expansion, 'The yellow race, the feebler, will naturally yield first; then the Kaffir – he also must yield to the Saxon Boor, on whose side is right, that is, might; for, humanly speaking, might is the sole right'. But slavery *shielded* the weak against this, said McCord. The benighted black man was 'incapable of civilization, but good-tempered, and with considerable intelligence and activity'; rather than extinction, crushed by 'the white races', what better 'destiny could God, in his merciful wisdom, have marked out for him than the one which he occupies under our institution of slavery?' Freed, he goes the way of the American Indians. Enslaved, he is saved. In her post-Knoxian scenario doomsday is averted. Faced by 'extermination or slavery', the latter, God's beneficent gift, was the black man's salvation.[20]

No one in America expected an early end to slavery. The abolitionists were themselves deeply divided, victims as much of their own strategies as of political repression and mob violence. William Lloyd Garrison and the immediatist radicals had polarized rather than mobilized the nation, moved hearts not senates. They still preached moral revolt, but two decades since the *Liberator*'s launch, and, despite support from English converts such as Erasmus's erstwhile belle Harriet Martineau and Darwin's future supporter W. B. Carpenter, immediate abolition looked impossible. Moderates had accomplished little more through political engagement. With no significant electoral base they were powerless to bind up the Union or even to neutralize the radicals who would see it ripped apart.[21]

The United States suffered from being less united with more states. Seventeen of the thirty-one states in 1850 were 'free soil', giving them an edge in Congress. One in seven of the whole population was black, in the South one in two; and here slaves outnumbered their owners by ten to one.[22] The figures changed continually because the country was on the move: blacks fleeing north; planters pushing into Texas with their slaves; incoming refugees from Irish famine, European revolution and plain hardship. All trudged across the plains to Oregon or to the new Free State of California in search of gold. In this broken land, people and power were shifting west at the expense of the native tribes, shattering their lives as well. North and South agreed on one electoral fact: if slavery could not expand, it would die.

Darwin never visited America, but he half-dreamed on occasions of emigrating. He avidly read travelogues: after de Tocqueville on American democracy came *Four Months Among the Goldfinders*, a Utah history of *The Mormons*, the *Forest Scenes* in a Canadian wilderness diary, and, with the situation deteriorating, he finally ploughed through the six volumes of *Society in America* and *Retrospect of Western Travel*, by Harriet Martineau. Her narratives were laced with heart-rending reports of slavery, first-person accounts so damning that she was vilified in the States as a 'tool of a nest of poisonous radicals'. Perhaps Darwin glimpsed himself in this predicament. He gave the *Memoirs* of the slave emancipator Thomas Fowell Buxton the rare accolade 'very good'.[23]

There were seven Darwin children under the age of eleven in 1850, and they too were supplied with romantic reading. A favourite story-time destination was the United States. With the republic bitterly divided and drifting deeper into crisis, even churches split along sectional lines as abolitionists – notably Quakers, Methodists and Unitarians – flouted federal law by smuggling fugitive slaves to the North and later Canada on the 'underground railroad'. One main line ran through the central state of Ohio, where the English Quaker relatives of Mary Howitt, a popular children's author, had settled. She told their story in *Our Cousins in Ohio* (1849), a cosy portrait of an English family living as neighbours to liberated blacks, their children playing in the lush countryside only miles from the Slave States' border. Darwin bought the children a copy, which he inscribed 'Charles Darwin Down'. Worried about their future – in 1850 an eighth was on the way – Darwin in his emigration daydream plumped for the 'middle States' as 'what I fancy most'[24] – New York, Pennsylvania, maybe even Ohio: free soil situated between New England's snobbery and Lyell's beloved South.

The daydream faded as America veered towards open conflict. In 1850 the casus belli came into focus. Was slavery to be contained in the South or extended to the West? With Southerners threatening secession to protect their 'way of life', Congress hammered out a 'compromise' package, giving western territories the vote on whether to be free and – the South's most cherished victory – enabling the legal recapture and return of slaves escaped to free soil. The legislation, enacted in September, decided nothing except the grounds of future strife. When the Kansas and Nebraska territories were created in 1854, armed pro- and anti-slavery settlers rushed in to swing the vote. Fraud led to murder and armed conflict, troops intervened and by 1856 'bleeding Kansas' had become a byword – it would take a local civil war to keep Kansas free.

The draconian Fugitive Slave Bill incited fierce resistance. (Ticknor's friend and Southern sympathizer, Daniel Webster, had shamed himself for ever by endorsing it.) Fugitives and freemen alike were suddenly turned back into prey – 20,000 escaped to Canada – as federal marshals stalked them. Some bounty hunters opened their own slave trade, dragging free Negroes south for sale. The rendition was rigged: no warrant was needed; the sworn testimony by a claimant sufficed to secure the return of his property. Citizens in the North were now responsible for enforcing slavery, for upholding the right of white Southerners to own black Africans.[25] Nothing could have been better calculated to unite abolitionists. Even moderates had to choose between defying an unjust law and breaking with conscience. The Unitarian abolitionist Parker himself protected fugitives in his home and served on Boston's Committee of Vigilance, which helped escapees. It was he who armed the fleeing William Craft with a revolver and then smuggled him and his wife out of the country, to be picked up in Liverpool by Carpenter's relatives. Parker was even a member of the committee that abetted John Brown's plans for armed insurrection. Thus began a decade of defiance, via legal challenges and armed resistance. What immediatists had failed to achieve in decades was becoming a violent possibility.

At such a moment came Agassiz's apotheosis as a race theorist. The American Association for the Advancement of Science met twice a year, and in March 1850 Charleston hosted the event. Agassiz arrived with a sop for the South, an intellectual compromise between Northern science and Southern sensibilities. His address showed how his un-Genesis-like view of life's history could nevertheless come to the aid of white Christians by backing up their beliefs about the inferiority of blacks.

In that month's *Christian Examiner*, the Unitarian thunderer (with members of Harvard's Board of Overseers assuming editorial roles), Agassiz had laid the groundwork by challenging the concept that all life spreads from 'one common centre of origin'. Practically every pluralist had made a badge of such a denial, but it was the stature of the savant now denying it that was significant. Agassiz had also excavated further into the rocks. The fauna in each zoogeographical region of the earth, he argued, had been placed there from the beginning of the current period. And so had the fauna in every preceding period. True, the *oldest* known fossils from widely separated regions did seem to approximate one another. Go far enough back and the ancient life of the Cape, South Australia and Europe would seem 'almost identical'. But a common origin? 'By no means.' (Darwin, reading Agassiz's article, stabbed an exclamation mark against this statement.) There were still slight differences. So the animals and plants in each region had never been connected to those of any other. They had simply been wiped out periodically and new, more modern-looking successors created on the spot.[26] Darwin scored such passages in contempt.

The article touched on humans. At one point Agassiz provocatively noted:

And what is not a little remarkable is the fact, that the black orang occurs upon that continent which is inhabited by the black human race, whilst the brown orang inhabits those parts of Asia over which the chocolate-colored Malays have been developed.

So the creative progression had run from chimps to black men in Africa, and orang-utans to Malays in Asia. The lines of successive creations were independent on the different continents. This foretaste of human–ape geographical tie-ups wasn't taken any further here, but it was a portentous sentence.

Within days of the *Examiner* piece appearing, the Southern slave port of Charleston saw Agassiz grasp the nettle at the AAAS. Nott stepped off the podium on 15 March 1850. He had talked about Jews and their unadulterated history and unaltered physiognomy since the great Pharaoh's day, as proof of their own separate origin and identity. Next came the man everyone wanted to hear (as local palaeontologist and plantation-apologist Robert Gibbes reported to Morton) – the one whose human pronouncements were expected to be *ex cathedra*. Agassiz followed Nott faultlessly. He confirmed that human races were locked to their zoological provinces. They were aboriginal creations, rooted to the spot, not some adapting, colour-changing migrants from elsewhere. The emerging 'American School of Anthropology' (as it would be called) had found its émigré echo. The races could be ranked 'scientifically', said Agassiz. As usual blacks languished at the bottom because they were unable to form a civilization in their African province. Even if Agassiz thought there was an equality 'before God', or a higher spirituality linking the races, he ceremonially crowned the pluralist view, to Bachman's dismay and Nott's delight. 'With Agassiz in the war the battle is ours', Nott crowed to Morton. He expected Agassiz to write a book on the subject to 'blow out old Bachmans brains'.

Radical Unitarians must have wondered at the propriety of publishing Agassiz's speech in – of all places – the July issue of the *Examiner* ('superb', Nott raved to Morton). It didn't matter about the snub to Genesis – radicals had long jettisoned that. But even if Agassiz dismissed charges that his views 'tend to the support of slavery', it still must have rankled for Unitarian abolitionists, however tepid, to see their house organ giving succour to racial segregation and humiliation, knowing how this played down South.[27]

After the meeting Nott's supporter Dr Gibbes escorted Agassiz round the great estates near Columbia, north of Charleston. Nott and Gibbes went back a long way, having jointly founded a preparatory school of medicine in Columbia in 1833. Gibbes still practised here, hand in glove with the plantation aristocrats. He tended the slaves on these estates: Benjamin Taylor's Edgehill plantation, and Colonel Wade Hampton's 18,000-acre Sand Hills. Gibbes's work on typhoid, like Nott's on yellow fever, had vast financial implications for these cotton barons: Hampton, one of the richest men in the South, acknowledged it. With 3,000 slaves, he had saved tens of thousands of dollars' worth of slave lives because of Dr Gibbes' studies. Physician and plantation owner were often tightly tied. Given, also, Gibbes' interest in fossils – and in particular South Carolina's ancient sharks – one can understand why Agassiz, the world's fossil fish expert, should gravitate to him. And Morton: one hears far less of him as a palaeontologist. Yet as Vice-President of the Academy of Natural Sciences, he read his own paper on fossil sea urchins at the same meeting as he read Gibbes' on Eocene fish. One colossal shark – at least its four-inch tooth, all that was known of it – Gibbes respectfully christened *Charcharodon mortoni*.[28] So there was a meeting of minds. But it was as a fellow slavery supporter that Gibbes chimed most. He toed the Morton–Nott line on the separate human species, their initial appearance on the earth in discrete provinces and viable populations, and the absurdity of Genesis pairs.

Agassiz spent eight days with Gibbes. There was an air of fear-inducing brutality on some plantations, where slaves outnumbered whites twenty to one. It could only have bolstered Agassiz's condescending view. He toured the Taylor and Hampton estates selecting representative men and women of different tribes, stipulating pure African descent. He wanted them daguerreotyped – scientifically recorded, stamped and ranked. These 'objective' images were to emphasize black physical differences, just as Nott's statistics were designed to draw out physiological ones, and Morton's cranial mausoleum their osteological debasement. It was another quantification; an exercise to firm up social categories on scientific principles. Gibbes had the slaves brought to the daguerreotypist, stripped, photographed, and their names, origins and ownership recorded. 'I wish you could see them,' he told Morton.[29] The results were haunting, tragic pictures: blank, resigned faces stared out of them. Fifteen silver daguerreotype plates still survive, the oldest known images of named slaves. This record for taxonomic and ultimately political purposes was no less a record of oppression. Here they were, forcibly stripped to the waist (men and women), which completed the image of degradation and subjugation. (Imagine a Southern belle being stripped by a black photographer: he would most certainly have been lynched.) The political control of science lay equally naked in these compelling pictures.

From the slave port of Charleston, Agassiz also collected barnacles for Darwin. Along with them went his latest book, *Lake Superior* ('is not that an immense Honour!', Darwin asked Lyell with a hint of sarcasm). But the real message from his donor was far more disquieting. While the barnacles in nauseous vaporizing spirits were making Darwin literally sick, Agassiz's 'races of man' diatribe simply increased the stomach-churning. 'Agassiz's Lectures in the U.S.' uphold 'the doctrine of several species, – much, I daresay, to the comfort of the slave-holding Southerns', Darwin told Lyell.[30]

If Agassiz's pluralism among animals stopped any talk of bestial transmutation, among humans it stopped what he seems to have feared even more: black men sharing a white ancestry, with its prospects of social equality and revolting miscegenation. Such detestation was endemic in the literature: amalgamation would 'degrade the whole human family'; it would 'revolt the feelings of every member of the superior race'. Even if 'disgust bears the imprint of desire' – there was a strange sexual undercurrent to much of this – it was visceral more than erotic, a shock to the system, like Agassiz's first encounter. Racial bastardization was part of a 'maudlin philosophy' of 'negro-ology' 'oft times amounting to treason' (said the 500-page segregationist *Negro-mania*, published in 1851). No equality, and nature shouted against amalgamation.

> Will the white race ever agree that blacks shall stand beside us on election day, upon the rostrum, in the ranks of the army, in our places of amusement, in places of public worship, ride in the same coaches, railway cars, or steamships? Never! never! nor is it natural or just that this kind of equality should exist. God never intended it.[31]

Darwin had ceased looking into the mind of God, but his own moral horror of injustice was just as vehement. By the time he wound down his barnacle work in 1853, the call for domestic-breed evidence to tilt the race-origin and slavery debate was crystal clear. The Morton–Bachman row had rumbled on into 1851. Both sides had become entrenched. Morton the pluralist – in one of his last pronouncements – clung to Cuvier's view that fancy pigeons were an amalgam of rock doves and wild species, and Bachman the unitarist as steadfastly denied the fact. For him there was a single common ancestor, as there was for humans.[32] Such views were echoed by all the protagonists across both camps. Morton defended his ancient authors but quickly joined them, dying himself in 1851.

It is ill-recognized but a truism: concepts like 'unity of descent' and 'common descent', so familiar to us today through Darwinian saturation, emerged in the first half of the nineteenth century in debates over race relations and slavery. The concepts were at the centre of countless contested books and reviews. 'Unity' and 'common' implied consanguinity and closeness, precisely what the 'American School' of anthropology denied. Indeed, between British-colonial slave emancipation in the 1830s and the American Civil War such emotive racial signifiers were largely the preserve of these two hostile camps. The terms were familiar through the multiple editions Prichard's works. In these, languages were traced to common origins which 'can bear no other explanation than that of an original unity of descent'. Such was denied by Agassiz and his followers, for whatever the spiritual unity in the eyes of God, it 'may exist without a common origin, without a common descent, without that relationship which is often denoted by the expression "ties of blood"'.

Inside London's Ethnological Society it was the topic of concern. The Ethnological was Prichard's and Hodgkin's successor to the Aborigines' Protection Society, and initially solidly unitarist. But cracks had started appearing after Prichard's death in 1848. Not only financial cracks, although these were bad enough as interest waned and members dropped out (in mid-decade there were thirty-two, by 1858 only six turned up to the anniversary meeting). Worse was the developing ideological schism. One remaining member in 1853 dared suggest that 'the real unity of mankind does not lie in the consanguinity of a common descent, but has its basis in the participation of every race in the same moral nature, and in the community of moral rights, which have become the privilege of all'. So the need for proof of a real, *material* 'unity of descent' was imperative. And since this 'unity of human descent' argument had carried over into domestic-breed studies, the solution was obvious in the fifties. Bachman summed up what the cognoscenti knew: the surest way to tackle shameful ideas of 'plurality in the human race' was to undermine it by a study of 'the varieties that are found in the domesticated pigeons'.[33]

There was egging-on from the race sidelines for anyone who could get a new handle on the domestic-breed issue. The reviews of the Morton–Bachman fracas had split predicably, with antagonists shouting 'case not proved' at Bachman. Detractors still bridled at Bachman's 'dictatorial assertions' when they wanted 'irrefutable facts'. Nott's bald statement in *DeBow's* in 1851 was, 'It is still a dispute amongst naturalists, all over the world, whether all the varieties of each of our domestic animals are of one or more species.' Back in Norfolk, Dixon insisted that the answers to the profoundest questions of the origin and diversity of life lay in poultry and pigeons: in fact, racial origins 'can scarcely be studied so well in any other field'. The occasion cried for new, original work on the origin of breeds, on hybrid fertility and on dispersal. Even Agassiz cried for it. He could not deny the 'striking' changes animals undergo when acclimatized and domesticated, only insist 'there is no subject which more requires a renewed and careful investigation'.[34] By the 1850s, the loft, the farm and the kitchen garden had become *the* battle-ground for good naturalists and bad propagandists at loggerheads over human origins.

Where extreme pluralists had each species fixed, and each variant aboriginally created, Darwin had proved that *every* species, at least among the barnacles he studied for so long, was continuously and 'eminently variable', indeed 'every part in some slight degree of every species' was.[35] It actually made sorting his specimens difficult: where was the cut-off point between an extreme variant and a new species? The pluralists were wrong on this point.

Prichard too had taught that variants were commoner than thought, and that domestication increased their likelihood in animals and plants, as civilization did in humans. Prichard's devotee Carpenter, by the mid-1850s, was a powerful bureaucrat in science: in 1856 the Registrar to London University, where he would help usher in the new Science Faculty and BSc degrees. Variability of species, he admitted, was a doctrine 'very early' impressed on him 'by Dr. Prichard'. So Prichard's human-race books had softened up Carpenter. He studied microscopic, multi-chambered, shelly planktonic creatures called *Orbitolites*, and in 1855 Darwin refereed his Royal Society monograph on them.[36] Darwin was gratified to find that they were variable in every aspect. This made a mockery of the pluralists' more extreme demand: that every variant was itself an aboriginal creation, if not a separate species.

Responding to the draconian Fugitive Slave Law came, of all things, a novel – more than a novel, 'a verbal earthquake, an ink-and-paper tidal wave', the shattering anti-slavery blockbuster *Uncle Tom's Cabin* (1852). Harriet Beecher Stowe may have caricatured Negroes and played on white prejudice, but she wrote passionately, full of pathos, with a belief in black capacity acquired while living in Ohio among former slaves. No political tract save the *Communist Manifesto* ever had such a reception. Translated into at least twenty-three languages, *Uncle Tom's Cabin* became the century's best-seller; in Britain alone it sold over a million copies and revived the anti-slavery movement. In 1853, Stowe's tumultuous welcome in England sparked a women's petition reminiscent of the old days, its 576,000 signatures urging an end to America's 'peculiar institution'. Emma Darwin's conscience was equally pricked, for she advised her eldest, William, at Rugby School, 'to get Uncle Toms Cabin from the library'.[37] It meant that Darwin, upon finishing the barnacles, plunged into the bigger racial-origin and distribution problems in a climate of renewed abolitionist determination.

Agassiz was becoming Darwin's *bête noire*, his views summing up all that was rotten with the zoological pluralism underlying slavery. Lyell was ever the go-between. He visited New England for a third trip in late 1852, there to geologize with Agassiz around Boston. (He also learned that Agassiz's work in Florida supported Darwin's sinking mountain-top theory of coral reef formation, so 'I shall have much to tell him [Darwin]', Lyell reported.) Then there was a fourth visit in summer 1853, when he was again with Agassiz. Power was accruing to the professor by the year. He was recruiting eager protégés, and when it came to subscription lists for his zoological tomes he could command a staggering 'twenty-one hundred names before the appearance of the first pages of a work costing one hundred and twenty dollars!' By March 1854 a mild irritation had settled over Darwin: 'I seldom see a Zoological paper from N. America, without observing the impress of Agassiz's doctrine's, – another proof, by the way, of how great a man he is' (meaning, only a great man could get such 'bosh' accepted – a classic Darwin sneer).[38] The zoologist whom Darwin saw comforting the 'slave-holding Southerns' was going from strength to strength.

Suddenly one understands why Darwin jumped straight from barnacles to, of all things, seeds. There was nothing anomalous in the leap. It was the first part of his experimental attack on the entire origin and dispersal problem. Racial spread was anathema to Agassiz: he had flatly denied the explosive diaspora image, the hegiras of dark humans, migrations of animals, or dispersal of plants, as they diffused from some alpha point and adapted to their hinterland homes. He was categoric about it:

> It is inconsistent with the structure, habits, and natural instincts of most animals, even to suppose that they could have migrated over any great distances. It is in complete contradiction with the laws of nature, and all we know of the changes our globe has undergone, to imagine that the animals have actually adapted themselves to their various circumstances during their migration, as this would be ascribing to physical influences as much power as to the Creator himself.

Darwin, so proper in public, was scathing in private of the high and mighty laying down the 'universal law', as if it was a diktat about what animals or plants can or cannot do. 'How very singular it is', he blurted out to Hooker, 'that so *eminently* clever a man' should write such 'stuff & bosh as he does.'[39] In his quirky way, Darwin started toppling Agassiz's monolith by subverting its common-or-garden props. Darwin's mundanity was always his ingenuity. He had a sort of intelligent head gardener's approach to rarified existential conundrums. Now was no exception. He performed an unexpected flanking manoeuvre on Agassiz.

In November 1854 Darwin continued questioning Hooker on world plant distribution. Then he worked out distances across botanical zones and between islands and continents; calculated oceanic currents – then took the adage that seeds are killed by salt water and tested it. No one seems to have doubted that brine was fatal to seeds, not even his closest confidant Hooker (and he should have, as the incoming Assistant-Director of the huge imperial repository and seed bank, Kew Gardens). Precisely this had lain behind Hooker's objection to Darwin's manuscript on natural selection a decade earlier: how are islands colonized, given that wind-and-wave disseminations are an unknown factor? Doesn't it beggar belief to imagine that sea crossings could explain the same plants in Tierra del Fuego and Tasmania? Even before this date, an ability to withstand brine was something Darwin, so well-salted himself, had wanted to test. How had plants reached the Galapagos, never mind mid-ocean destinations? The *Beagle*'s plants from the Keeling Islands (now Cocos), south of Sumatra, had intrigued him: could their castaway seeds float? So his own views were themselves always of long germination. His evolution notebooks had been full of speculation on owls carrying mice to islands, thrushes bringing seeds in their stomachs; then there were the obligatory memos: 'Experimentise on land shells in salt water', and 'Soak all kinds of seeds for week in Salt. artificial water'.[40] Ideas had been thrown out left and right, half forgotten, to await a new day. That day was now and there was a new urgency.

The questioning became systematic from November 1854. There was nothing he would not question, and nobody. He even had a sailor tracked down who had been shipwrecked on Desolation Island (aptly named, as lying equidistant from Australia and South Africa) to find out if he had seen driftwood thrown up by the tide. Since Hooker had found Tierra del Fuego plants here, Darwin needed to know whether sea currents could account for their tremendous voyage. Others had evidence that birds-of-prey pellets contained seeds from their victims' last meals. Or that seeds were transported on ice floes. In March 1855 Darwin started 'salting' seeds himself, popping local produce seeds – cabbage, cress, carrot, radish, lettuce and celery – into saline. Some bottles went in the garden, others to the cellar (packed in snow, to see if temperature affected the result). After a week samples were transferred to dishes on the mantelpiece: all sprouted. After two weeks they still did. Poor Hooker was tortured with weekly reports, and was good-naturedly pressed into guessing survival times. For his pains Darwin started receiving Kew's exotic seeds in return. Chimneypieces and ledges were lined with fifty-odd bottles, all becoming smelly if the water wasn't changed. Then there were uncountable dishes with sprouts, and Darwin was so busy testing samples that he was cancelling appointments. The house became a nursery. And not only his house: friendly vicars were roped in, bags of seeds were sent to sit in Channel water, and the salting extended to other kitchen gardens and to other kingdoms, or at least to snail and lizard eggs – what else he wouldn't say, for his other saltings were 'so *absurd* even in *my* opinion that I dare not tell you', as he laughed to a vanquished Hooker.[41]

The results on the other hand were deadly serious. One implication was spelt out in Darwin's opening statement to the *Gardeners' Chronicle* in May 1855:

> As such experiments might naturally appear childish to many, I may be permitted to premise that they have a direct bearing on a very interesting problem, which has lately, especially in America, attracted much attention, namely, whether the same organic being has been created at one point or on several on the face of our globe.[42]

By this time the cress, carrots, celery, lettuce and radishes were germinating after forty-two days, while a haphazard list of tender, exotic, cultivated and wild seeds were doing well after a month. Only one species, a clover, was killed outright from the initial twenty-three species tried. And for rhubarb and celery, the germination time was actually accelerated by the soaking.

To calculate distances travelled Darwin used the standard *Johnston's Physical Atlas of Natural Phenomena*, which graphically depicted plant regions according to Humboldt's statistics. He found it confusing (although he learned a 'good deal' from its other maps, which included an 'Ethnographic Map' of the five 'leading races', adorned with representative skulls to illustrate Caucasian, Mongol and Negro disparities – showing at what a deep cartographic level the new anthropology had penetrated).[43] He computed from the *Atlas* the average Atlantic current at 33 nautical miles a day: in 42 days it would carry his seeds 1,300–1,400 miles. Never mind that they tended to sink in his tanks: it was apparent that more buoyant pods, capsules and whole plants might drift all the way to the mid-Atlantic Azores.

Agassiz's fellow Harvard professor, the botanist Asa Gray, was astounded and immediately saw the pertinence. 'Why has nobody thought of trying the experiment before! Instead of taking it for granted that salt water kills seeds'. No one had because it needed Agassiz's threat to push someone like Darwin, who knew the issues at stake. Each of Agassiz's species – plants, animals and humans – was aboriginally created over its entire range, and anti-slavery naturalists like Gray and

Darwin knew how this science was sustaining the evil institution. Gray wasn't slow off the mark after receiving Darwin's paper. 'I shall have it nearly all reprinted in Silliman's Journal [the *American Journal of Science*], as a *nut* for Agassiz to crack'.

By this time (mid-1855) Darwin had fingered Gray as a potential Harvard ally to be cultivated, his anti-Agassiz man in place: a unitarist, a dispersionist and more. Gray, like Darwin, considered the appearance and disappearance of varieties as normal events. Varieties were natural transients not aboriginal creations. Their 'inevitable mingling' and the back-breeding of wild stock prevented them becoming a permanent variety. However, he said, 'Interfere with nature by domestication & segregation, and they spring up as fast as you please'. This was cheering to Darwin, and the letter (which had been sent to Hooker and passed to Darwin) also revealed that Gray was himself taking on Agassiz, by 'trying to show him that his own data' did not lead to 'the conclusions he sometimes draws'.[44]

They were all wary of Agassiz now. Hooker had written his *Flora Novae-Zelandiae* (1853) with one eye on the devil. In choosing between aboriginal or dispersing species, admitted Hooker, 'Agassiz had a prominent seat in judgment before me'. Hooker screwed up his courage to 'combat' Agassiz's aboriginal creation theory – that way lay confusion, with 'every slight difference' chaotically signifying a new species. And 'Oh dear, oh dear', Hooker admitted to Gray, 'my mind is not fully, faithfully, implicitly given to species as created entities'. By then Darwin was beginning to soften Hooker up and to show how variability, if not mutability, could broaden the scope of botany. The upshot was to strengthen Hooker's own belief that 'all the individuals of a species . . . have proceeded from one parent'. They vary more than was thought and their distribution has been by natural causes. Darwin's continual prodding had paid off. Indeed, when Hooker won the Royal Medal in 1854, it was in part for his botanical attack on the science's knottiest problem. As the Royal Society's president said, it was 'one of the most difficult questions of natural science, which is now acquiring the prominence to which it is so well entitled, – I mean the question of the [place of] origin and distribution of species'.[45] In this instance too, Gray goaded Agassiz by sending a huge abstract of the *Flora* to the *American Journal of Science*.

But the botanists were still chary, with Agassiz so 'extraordinarily clever'. 'I have long been aware of Agassiz' heresies', Hooker told Gray. 'His opinions are too extreme for respect and hence are mere prejudices. They are further contradicted by facts. Lyell and I have talked him over by the hour. Lyell and Agassiz are great personal friends.' Gray was 'more and more' glad that his own 'general notions, formed in the closet', were confirmed by Hooker: 'we maintain the *orthodox faith*, in the midst of "right-hand defections and left hand falling away" ', Gray replied. Faith in the common descent of human races, that is. One had only to look at tailless Manx cats or five-toed Dorking fowl to prove 'that you can't tell beforehand what the possible limits of variation in a species are; or to be . . . against those who, like Agassiz, say that the differences among *men* (the largest domesticated of all animals) are too great to allow of their origin from a common parentage.' Gray did not overly worry about *how* new races (plants or humans) broke the continuum of '*genetic resemblance*' to start their own bloodline.[46] It was enough that it had happened. Hearing this, Darwin knew his views wouldn't scare Gray off.

Agassiz was 'a good genial fellow' and Gray feared 'it often does more harm than good' to issue public rebukes; nonetheless he would, 'at the proper place & time put on record' his own rival views, never wanting it to be 'afterwards supposed I had fallen in' with him. This was partly what Darwin wanted to hear. But there was a niggle, and it involved compromising with the devil. He told Hooker, 'I cannot quite understand why you & he think so strongly that it "does more harm than good to combat such views".'[47]

For Darwin, the combat had begun. The seeds, for all their smelly frustrations, had proved a point. His counter-experiments ran on, with a pepper taking the prize for sprouting after 137 days in brine. And the targeting was ever more precise – and global. Her Majesty's consuls were tapped. One sent seeds washed up in Norway; they turned out to be Caribbean (and to Hooker's 'unutterable mortification' they grew). Another provided a list of Azores plants so Darwin could salt their British equivalents. Half the time Hooker was the one in Darwin's sights. Having travelled in the HMS *Erebus* and *Terror* surveying expedition to the Southern Ocean, he was an expert on the 'fragmentary' flora of the region – what he saw as scattered remnants pointing to the existence of some former supercontinent. How else to explain anomalies like the

Edwardsia tree occuring in Chile and New Zealand and nowhere in between? In *Flora Novae-Zelandiae*, just before the salting started, he had asked again, whence came *Edwardsia*? 'The idea of transportation by aerial or oceanic currents cannot be entertained, as the seeds of neither could stand exposure to salt water, and they are too heavy to be borne in the air.' As Darwin was getting underneath Agassiz's prejudices, so he was undermining Hooker's conventionality. Hooker was coming round, but slowly: 'I believe you are afraid to send me a ripe Edwardsia pod', Darwin goaded him, 'for fear I shd float it from N. Zealand to Chile!!!'[48]

Nothing was sacrosanct. He picked through the mud on ducks' feet (people posted feet to him!) to find seeds and snail eggs – showing that birds could fly in new plant and invertebrate colonists. He watched a dead bird, its crop stuffed, float for a month, and poked through bird-droppings looking for seeds to germinate. Retrospectively it seemed so obvious. But it had taken a determined attack and an utter lack of faith in land bridges between continents and multiple centres of creation. There was no need for aboriginal species or aboriginal humans to be created over their entire range, for dispersal *was* viable, even across oceans. Lyell had long accepted a single creative explosion for each species, followed by radiation outwards. Darwin's work now had Lyell prophesying that 'The multiple creation of Agassiz will one day rank with spontaneous generation' as heresy. He added: 'I long to see your . . . species-making modification system' – that is, Darwin's theory of natural selection. Agassiz's aboriginal creation theory and Darwin's 'selection' stood mutually threatening. Hooker realized it, knowing that 'multiple centres' were 'worse for your theory than any thing else'. Darwin agreed: 'You say most truly about multiple creations & my notions; if any one case could be proved, I shd. be smashed'.[49]

If the seeds were 'smelly', that was nothing compared to the experiments he was concurrently running – the *big* programme, and the second part of his experimental attack on the racial question. This was literally noxious, with corpses rotting in evil brews of caustic potash and silver oxide. The stench of putrid flesh was making Darwin 'retch so violently' that he had to stop and ask Eyton's expert help. 'It really is most dreadful work'. He was reducing pigeons to skeletons; not just the odd one, but wholesale. Even if the human-race clamour hadn't demanded domestic pigeon studies, Darwin might have turned his attention this way eventually. Pigeon crossing was glossed once or twice in his evolution notebooks, but now he was engaged in an urgent zoo-scale operation,[50] committed to proving a point using fancy pigeons, not to weighing up the evidence.

Backstreet breeders were largely ignored by gentlemen naturalists, but they had a fine eye and an experienced touch. In a thousand grimy terraces they selected and mated choice pigeons with desired traits – to make money, to show off, for the sheer delight of it. A parallel process characterized the Darwinian wilds: the struggle among competing variants pushed the weak to the wall, leaving only those with advantageous traits to multiply. In the fancy-pigeon world, the 'losers' suffered predation just as much – Darwin's unwanted fledglings were taken by the cook (thinly-sliced bacon, pigeon squabs, sweetbreads and cocks' combs went into a 'Pupton of Pigeons' vol-au-vent). Breeders only kept those squabs with some accentuated tail or ruff feathers, or whatever was required. That's how the strange 'jacobins', 'pouters' and other fancy peculiarities became ruffled and puffed and coiffed. It was an analogy that made evolution itself something of a domestic science.

Late in 1854 Darwin had begun stocking his country house with dogs, rabbits, ducks and geese, dead or alive. Perhaps it wasn't so odd that he was listed as a 'Farmer' in Bagshaw's *History, Gazetteer and Directory of Kent*. But he was no conventional one, despite his fifteen acres of hayfield behind the house or the 300 acres in Lincolnshire he had bought as an investment.[51] His concern was not so much with improving his pedigree stock as understanding it. Of all the tame animals, it was pigeons which could be shown most convincingly to have come from a single source – he would measure them, show how the breeds diverged from one another and how the fanciers had managed the trick of getting them to look so different. He attacked the task with gusto and, apart from skeletonizing, even more delight. It was a case of 'in for a penny, in for a pound' – or a pound each, which is what his birds cost him. The first loft went up in late March 1855. Every European breed of pigeon was his goal (many procured from one of the judges of the big fanciers' shows).[52] Thirteen or fourteen pairs were settled in by November, when he sought Eyton's help because of the smell of rendered flesh, offering him a dead dog's head for his pains (no ordinary dog, of course, but Chinese, for Darwin was also following up the dog breeds that Nott made so much of).

By December domestic pigeons – or their skins – from every corner of the world was his goal. Letters went off to Asia, Africa and South America – to diplomats, collectors, botanic gardens directors and East India men. They were a veritable who's who of the colonies: consuls and collectors from China to Persia, doctors in Antigua, Natal and Gambia, imperial wallahs well remembered such as Rajah Brooke, and not so, such as our man at San Domingo, Sir Robert Schomburgk (an expert on British Guyana and Carib culture, who had explored Waterton's Essequibo River to its source, in the process discovering the giant water lily). Most were known personally or through a contact. Schomburgk shows us the sort of man Darwin favoured. His works on Guyana's wildlife, reefs and geology were featured in Darwin's books. Schomburgk himself acknowledged Darwin's help on coral islands in his own *History of Barbados* (1848) – the first post-emancipation history, which celebrated a 'former slave population converted into a happy peasantry'. Schomburgk's brother Richard had collaborated with Darwin on the BAAS committee drawing up the ethnographic queries in 1839. Sir Robert himself contributed to the human-race literature. He supplied Darwin with information about 'savages' and their dogs (and yes, the Arawak Indians *did* cross their domestic dogs with wild breeds to improve their hunting ability).[53] And Schomburgk was well-enough known to have been invited to Down.

Consular staff interrupted busy schedules to chase local pigeons. Parliamentary papers show that Edmund Gabriel, Her Majesty's Arbitrator in the British and Portuguese Mixed Commission Court at Luanda, Angola, was actively adjudicating property confiscated in connection with suppressing the slave trade while 'making an extensive collection of fowls &c for you',[54] Darwin was told.

Darwin's racial round-up of pigeons extended to the four corners, all to show that – like the people who kept them – pigeons descended from a single stock. The nation's long reach had become Darwin's also. His station and attainments allowed him to call on Victoria's far-flung diplomats and émigrés, even with a request he thought most would find laughable – for feathers. It might 'appear at first foolish', he told the minister to the court of Persia, 'but I trust that you will think the final object worthy'. Laughable or not, he got his skin-and-bone because they did think it worthy. As he explained in his letters, he was working on 'the variation & origin of species', or 'the origin of varieties & species';

and given the contemporary brouhaha over the zoning of life and aboriginal place of creation of each species, a catch-all term like 'origin of species' could easily be interpreted to mean the *place* of each origin (especially as he was writing to such distant outposts). He wasn't giving anything away – he certainly wasn't declaring himself an evolutionist.[55] To study the 'origin' of species in 1855 was permissible in implying the place and time of local creation. Agassiz had made it fashionable. Imperial diplomats were happily dragooned into such a contemporary enterprise.

'I have thought that I sh^{d}. do my work best by carefully attending to a few small groups of varieties', Darwin added to the Persian minister, '& I have taken especial pains with the domestic Pigeon, Poultry, Ducks, & Rabbits'. But the first is what he really wanted. He scoured the globe for breeds lost in Europe but flourishing in remote colonies such as the Cape, or for varieties uniquely fashioned in Ceylon or India. He continued:

> Pigeons if really descended from one wild stock, as after much consideration I believe to be the case, have varied most wonderfully in almost their whole organization. – Persia is reputed to be the birth place of several races, as Carriers, Tumblers Pouters, Bussorah Runts &c; & the comparison of, for instance, a Persian with an English Carrier would be of extreme interest to me.

To this end he requested all the long-standing breeds, or choice bits. 'The *whole* bones of legs & wings, & as much as possible of the skull, should be left in the skin. – Each specimen should be ticketed with native name, habitat, & any procurable information on habits.'[56] It wasn't often that Her Majesty's consuls got such requests.

Nevertheless they sent crates of carcasses to his 'chamber of horrors'. In the more salubrious air of the garden he now had two large lofts full of living specimens. He counted them up in June 1856 and was astonished at the tally: eighty-nine, with more arriving monthly. They included all manner of rarities – a miniature racial menagerie to parallel that rainbow spread of men he had met on his voyage.

Unlike men, pigeons could be experimented on, to prove his point. As fast as the birds bred he plucked them for the pot and measured their bleached bones. He wanted to know whether the young looked less extravagant and more like the ancestral dove. (He hated this grisly part of the business. His daughters too doted on the birds, seraphic beauties plumed like the latest fashions. But it was the stewpot for them, and a pricked conscience: 'I have done the black deed,' he would confess, '& murdered an angelic little Fan-tail & Pouter at 10 days old'.) What with comparing carthorse and racehorse colts and barrels of 'puppies of Bull-dogs & Greyhound in salt', he had to admit 'I am getting out of my depth'.[57] Cadavers floating in witches' brews, and crushed boxes arriving in the post with ducks' intestines poking through: it was becoming a regular charnel house.

In fact it was a major research programme, run by one man. And it served multiple functions. Thus the nestlings were slaughtered so he could discover at what age certain selectable traits appeared. But what he was intent on proving was that fancy pigeons the world over were descended from the humble rock dove. (A species known to most of the northern world – its range extends from Britain to Japan.) This was the controverted point running deep through the burgeoning human-race literature. Many new anthropologists denied it. Few pigeon fanciers actually believed it (most assumed each of the eleven major 'races', in Darwin's classification, began as a separate creation), nor did elite breeders like that 'excommunicated wretch' the Revd Edmund Dixon. The unity of descent argument, said Darwin, 'falls without the least impression on some people, as on M^{r}. Dixon of Poultry notoriety, who argued stoutly for every variety being an aboriginal creation'. That's why, Darwin told his cousin Fox, 'I am anxious to get as many precise facts as I can about crossing, both [to prove the unity], & generally for the comparison of mongrels & Hybrids'. Pigeons pointed up the evolution problem in microcosm. Darwin had settled on them precisely because the evidence of their descent from one stock was 'far clearer than with any other anciently domesticated animal'.[58]

He proved it, first, by doing something no self-respecting fancier should. He cross-bred his pedigree pigeons. He wanted to be able to state from extensive first-hand knowledge that the offspring were 'perfectly fertile', which made belief in their 'aboriginal' separation 'rash in the extreme'. He went so far as uniting in one bird no less than five pigeon races. And still there was 'unimpaired fertility', although it was known that 'complex crosses between several species are excessively sterile'.

So these were *not* species. Of course, crossing championship birds didn't help his standing among fanciers, where purity was prized for show purposes. He knew it:

I sat one evening in a gin-palace in the Borough amongst a set of Pigeon-fanciers, – when it was hinted that Mr Bult had crossed his Powters with [giant] Runts to gain size; & if you had seen the solemn, the mysterious & awful shakes of the head which all the fanciers gave at this scandalous proceeding, you would have recognised how little crossing has had to do with improving breeds [by fanciers] . . .[59]

But he had a second – and far more obvious – way of proving that pigeons descended from one stock. It was prompted, perhaps, by another collector Darwin pestered for pigeon skins. He was 'a very clever, odd, wild fellow' out in Calcutta, Edward Blyth. They had met in England, and, admitted Darwin, 'I liked all I saw of him'. This first impression was vindicated by the score of letters which he received from 1855 to 1858, full of details on Eastern breeds and wild ranges. No naturalist knew more of the pigeons and fowls of India. Perhaps Blyth was even more attractive because Dixon had denounced him for stating that all the fowl breeds 'come from one wild bird'. (Blyth stood with Darwin on this.) As a result Darwin's letters were probably a little more explicit than might otherwise have been the case. None of Darwin's outbound letters has been found (they may be in Calcutta still), but we can gauge his reference point from Blyth's reply, dated 21 April 1855. Blyth responded: 'The subject of the races of domestic animals has never yet been fairly taken up with reference to *Ethnology*, upon which it is not unlikely to throw some important lights.'[60] Blyth, fourteen years himself 'cooped up' like one of his pigeons in India's 'huge overgrown capital', was gratified to learn that the fancy-pigeon analogy of human racial descent 'has been undertaken by one so competent'.

Blyth mentioned the fowls pictured in 'Etruscan, Greek, & Roman' tombs. They all seemed to be '*the same type*' and very like the wild bird, the '*Persian bird*', as Aristophanes called it, which suggested its embarkation point. There was nothing on varieties in the Old Testament or Homer, so the 'Persian bird' must have been turned into so many varieties during the Middle Ages. The *real* history of the world's fancy pigeons, as Darwin cleverly realized, might be reconstructed directly from breeders' manuals going back centuries. It turned into a typically Darwinian manoeuvre, so obvious in hindsight. He ran down dusty books of pigeon lore. The musty texts were notoriously patchy, and he could barely make out much before 1600, by which time the major breeds were in existence. But thereafter he could 'trace the gradual changes in the breeds', their modifications and extinctions.[61] He could show – contrary to Nott and Gliddon's claim from their tomb pickings – that there *had* been historic change. After 1600 each variety had gone on evolving with human tastes. Pouters became thicker legged, and jacobins better hooded. English carriers (which arrived from Persia) had their bills lengthened. Short-faced tumblers – appearing in the early eighteenth century – had seen the beak shorten still more. Darwin now had direct proof that pigeons had changed.

How they were changed he learned from the breeders. With practised eye and skilled hand, they selected from minuscule variations in beak or feather. To get an insiders' view of this esoteric craft, Darwin joined the pigeon clubs, high and low. High meant an exclusive gentleman's club in the West End, just to his social taste. As to low: 'I am hand & glove with all sorts of Fanciers, Spital-field weavers & all sorts of odd specimens of the Human species, who fancy Pigeons'. These 'little' men puffing clay pipes – and insisting he puff with them – taught him their skills.

They also told him how difficult it was to cross varieties and establish a permanent half-breed: these mixes kept reverting to one or other parental type. While the fertility of the crosses proved unity, the near impossibility of establishing half-and-half *intermediate* breeds showed the error of those who proposed that all the domestic dogs or poultry or pigeons could have resulted from 'the crossing of a few aboriginal species'. In the same way it stripped the ground from under the anthropologists who saw the human racial continuum similarly created by crossing a few aboriginal humans.[62] Permanent changes were not brought about by crossing varieties, but by constant selection of individuals over untold numbers of generations, in nature, as in the farmyard.

Later, however, he would accept that intermediates might stabilize after many generations, six at least for pigeons, more than eight for dogs, but it required enormous perseverance on the breeder's part. Not that this stopped him teasing dog breeders. He took delight in telling 'purists' that some pedigree dogs were themselves, dare he say it, 'mongrels'. The Scotch deerhound for one, but, he laughed to Fox, 'Dont tell any Scotch so, or I shall be murdered'. The Irish deerhound he thought was a 'Scotch & a Mastiff'. To test it out, he found people who had back-bred the hound with a pedigree Scotch to lose all trace of the mastiff, so successfully that one expert was ready to put down big money for a championship puppy.

There was a point to all this. Darwin had proved that you couldn't trust fanciers on their purity claims, because of their blinkered single-generation vision and vested interests. And it was a 'very valuable case for me', because, while the pluralists made every domestic race a separate creation, no one could now 'imagine that the real Scotch Deer Hound was a pure & distinct aboriginal race'.[63] It had been made by a slight mixing, much selecting, then breeding in and in. Dogs, again, were a microcosm of nature, where fixity and purity were Creative chimeras. Neither a 'Deer Hound' nor an 'Anglo-Saxon' was anything other than a little crossed and long selected type. Each just happens to be separated at the moment by temporary geographical (or in the hound's case, human) barriers from its cousins.

Darwin had long believed that if breed-mixing were stopped in a wild or tame species, then continued selection pressure would ultimately allow the race to breed true.[64] In humans, he thought, it was largely stopped because of aesthetic reasons (whites prefer mates with white features, blacks with black). Nevertheless, his dog researches in the mid-fifties suggested that each human population could well have met and absorbed members of others in their long evolutionary journey – as his unitarist and dispersionist views allowed – and we would never know it.

The pigeon family tree became a Darwinian paradigm and a perfect metaphor for mankind's family tree. Darwin could start putting the pigeons back on their genealogical branches. When the time came, he would have his lead-in to the *Origin of Species*. He would hook the public with a fashionable topic bearing a moral tale. After setting up his target,

> The doctrine of the origin of our several domestic races from several aboriginal stocks, has been carried to an absurd extreme by some authors . . .[65]

he would let pigeons lead the fight-back. The social context of Darwin's offensive has slipped away and been subsequently lost, but recovering it makes sense of the project in Darwin's moral world. It scientifically undermined the ethnographic drive for segregation and 'aboriginal' homelands so comforting to the 'slave-holding Southerns'. Darwin, good at out-flanking, was reclaiming the high ground. His study might have taken on a life of its own, but in tracing all the varieties back to a single dove, he would essentially do for pigeons what Prichard had attempted for mankind – unite the races.

Some people needed telling. Darwin invited Lyell to Down in April 1856, offering to 'show you my pigeons! which is the greatest treat, in my opinion, which can be offered to [a] human being'. Knowing Lyell's sympathies, we might imagine an ulterior motive. That weekend Darwin revealed his theory of natural selection and confessed his heresy on human evolution. Having viewed cultured man as 'Time's Noblest Offspring', Lyell now faced a pedigree sullied by the prospect of ancestral black blood and a more revolting ape parentage. His private notes testified to his agony as he tried to swallow his racial pride. Still he clung to counter-evidence. If domestic dogs came from more than one wild species could we not envisage 'many races like Man being equal to species'? It was a good point. 'You remember,' he asked Darwin, 'a passage in Agassiz about negros being black where the Chimpanzee is black, & yellow in Borneo? where the orang is yellow'. But Darwin stuck with his pigeons and unity. Of all the theories, Agassiz's, 'that there are several species of man', did not help 'us in the least'. Even if domestic dogs came from multiple species (as Darwin conceded), his pigeon varieties were traceable to a single origin. Lyell left from his tour of the lofts staggered: had Darwin's fancy pigeons been wild, an ornithologist would have recognized 'three good genera and about fifteen good species'.[66] So much spread in one species – it tipped the scales in favour of a common human descent. If the strategy convinced Lyell, Darwin knew it would work in the *Origin of Species*.

Darwin's hotline to the breeders was no less buzzing than his hotline to the diplomats who dealt with the day-to-day aspects of emancipation. Another who looked around the pigeon lofts was Governor-General Robert Mackintosh, newly retired from his post in the Leeward Islands, and 'looking like a Man fresh out of a Madhouse'.[67]

Sympathizers could understand why. The Governor was an in-law (in fact, Emma's sister-in-law Fanny Wedgwood's brother). The only son of Sir James Mackintosh, he had married into a Boston textile baron's family, the Appletons – another of those clans, like much of the Unitarian upper-crust, with father defending the 'lords of the loom' and the children damning the 'lords of the lash'. Robert's ill-mannered wife Molly had given Emma's Aunt Fanny Allen such 'a horror of the whole [American] race' that she wanted no contact with anyone 'nearer than Newfoundland'. But Molly and Robert had been endowed with cash and Darwin–Wedgwood values, and they had joined the family circle even before Robert landed his governorship of the Leeward Islands, in the West Indies. Subsequently he had been promoted to viceroy in Antigua's capital.[68]

Tidings of a backward sugar colony held huge interest for the in-laws. The Governor considered absentee planters to be Antigua's 'master evil'. They ruled through local agents amid swirling undercurrents of racial tension. Many free blacks chose to till their own soil. With high child mortality and cheap American slave sugar depressing wages, this led to a chronic labour shortage on the plantations. Cane-workers shipped in from Africa had to be assigned island by island and the sick treated to stem epidemics.[69] The problems were endless. And always there was the human traffic – the lad kidnapped while fishing, sold into slavery and traced to Havana, whose family begged for his release; the 'poor woman of colour', abducted as she bathed, enslaved and abused in Puerto Rico, with three children and two grandchildren. Every issue had required the Governor's attention. No wonder he looked like a lunatic. After visiting Erasmus in London, Robert, Molly and the children spent ten days at Down House.[70] Here they saw Darwin's own exotic racial menagerie, being bred and cross-bred, all to prove they were siblings under the skin as black and white men were brothers.

The Governor's ways and Darwin's means were now being openly lampooned. Thomas Carlyle laughed 'immoderately' at Molly's Yankee manners, but then noisy impertinence was expected from him. Old admirers had been irritated by the racket even before his *Occasional Discourse on the Nigger Question* had been published. This howling jeremiad about West Indian emancipation ridiculed the sort of society that Mackintosh governed. Bondage had been beneficial for 'Quashee', keeping him at work; smallholdings were bad for lazy Negroes. Immigrant labour would bring more African indolence, more misery, creating a 'black Ireland' in the Caribbean, unless whites regained the whip-hand.[71] Carlyle's venom increased his alienation from humanitarians even as the conflict in America was renewing their 'rosepink sentimentalism'.

By contrast, the Darwin–Wedgwoods were if anything more committed to ending American slavery. Certainly Fanny Wedgwood's passion for radical causes deepened over the years, unusually in a woman of her rank. Underneath 'that refreshing quiet', she had a 'lava of living fire', ready to erupt on behalf of Italian republicanism, higher education for women, anti-vivisectionism, female suffrage and, her first great cause, black emancipation. A lifelong friendship with Marianne Thornton from Clapham, who would recall the Saints gathering in her father's library, helped keep the evangelical spirit alive even while old abolitionists were passing. Emma's rich Aunt Sarah Wedgwood, who had left Maer to live near Emma and Charles, was now a wizened spinster living in austerity. She still tended her charities and sent the Anti-Slavery Society £40 or £50 a year, evangelical missions more.[72] Until her death in 1856, family parties at Down House brought back memories of Maer's anti-slavery enclave.

Blyth's letters continued, as wild as the man himself – reams of facts and streams of domestic consciousness: on the blending of dog ranges and overlapping varieties, on the derivation of domestic dogs from multiple wolf species, on his Indian hybrid fowls rarely hatching, on bull's hummocks, the zoological place of human monogamy, on different species of orang-utan and what that said about the 'unity or distinctness of certain human races', on '*Man* (zoologically considered)' being 'of the Old World' monkey type, as opposed to a New World monkey type, which Darwin knew negated Agassiz's notion of some 'aboriginally American race or species of *Homo*'. Agassiz's home-grown American man might be impossible, but even Blyth was inclined to believe in more than one Old World human cradle: in both Africa and South-East Asia he imagined 'naked & untutored man & woman' emerging at a time 'very much more ancient than is generally supposed'.[73] It was another setback for Darwin, but he still relied heavily on Blyth's packed factual letters. Calcutta, where Blyth was curator of the Asiatic Society's overcrowded museum, stood at the crossroads for the British in Asia and he was ideally placed to collect information. Darwin marked the letters and took notes, and these details on the subcontinent's fauna would resurface in his own books. He might even have bought some of his pigeons from Blyth, who was trading heavily in exotic stock to supplement his income.

Blyth himself distinguished natural '*Races*' from '*breeds* (artificially & intentionally produced)', 'why I know not', said Darwin.[74] It was because Blyth saw two fundamentally distinct operations at work, one natural, forming wild races, the other an artificial, deforming, domestication process, and he was differentiating the results. Darwin had moved away from other naturalists on this question. He had artificial and natural selection pegged as analogous. Both involved pruning and picking and he saw no reason to separate their products.

Besides normal breeds, like European cattle and sheep, which had been changed slowly and resembled their wild ancestors – and the regular human races fell into this category – Blyth was also fascinated by the 'monstrosities'. Examples included the short-legged 'otter sheep' in Massachusetts, kept and bred because the flocks could not jump fences, or 'penguin ducks' which waddled upright like penguins. These were 'sports', deformities straight from the womb, and made congenital, as Blyth admitted, by selective breeding. There were human examples too, including the 'Porcupine family' – a famous family, shown on stage, with flanks and abdomen of bristly skin like a shorn hedgehog – or people born with six fingers or toes. Blyth even wondered if selection from '*the choicest specimens*' of the 'Bojesmans' or albinos might make new, strange human races. He quickly made plain his source for all this: 'Have you seen Knox's curious volume on the races of mankind?' he asked Darwin. Darwin had been too engrossed in barnacles to read it. So, on to his list of 'Books to be Read' it went: 'Knox Races of Mankind a curious Book. (Blyth). in London Lib^{y} – (read)'.[75]

Like Darwin, Blyth was fascinated by odd snippets. He mentioned that the Australian outback explorer Paul Strezlecki's 'curious' belief that, once having a child with a colonist, an Aborigine woman could not conceive with her own race, had been 'satisfactorily' refuted. Again Darwin noted the reference. And Blyth's repeated reference to Prichard, particularly his view that domestication made dogs more intelligent (as civilization did humans), led to another Darwin reminder: 'Pritchard (I must read)',[76] meaning reread and reacquaint himself. Darwin owned Prichard's previous edition, now he picked up the popular condensation, *The Natural History of Man*.[77] Early 1856 accordingly saw Darwin's quickening interest in racial affairs.

What reinforced this and catapulted Darwin into the heart of the race-and-slavery literature was Lyell's next move. Sir Charles, that conduit to Charleston, passed two of Bachman's pamphlets to Darwin.[78] Which two wasn't stated, but since between September 1854 and July 1855 Bachman had published a huge, four-part review in the *Charleston Medical Journal* of Nott and Gliddon's clarion-call book *Types of Mankind* (1854), and had the printers strike off three parts as pamphlets, two of which rounded on Agassiz's contribution, we know what Darwin was reading. Bachman had sent his anti-*Types* broadside to Lyell, who had passed it on.

Types of Mankind was the culmination of the separate-species and separate-homelands literature. With its coloured maps, fold-out charts and 362 woodcuts showing groups of human and animal species all primordially distinct, what could be more impressive than these 700 closely printed pages? Not only physically huge, the book was hugely expensive, about $150 in today's money, yet an enormous seller. It started off with over 700 pre-paying subscribers, and had roared through seven editions before Darwin even glimpsed Bachman's pamphlets. Agassiz had not gone on to write his own book, as Nott had hoped, so Nott and Gliddon had put together *Types of Mankind* in homage to the dead Morton, with an opening contribution by Agassiz as imprimatur. The Harvard professor wasn't an outsider roped in, or an innocent duped; he was hand-in-glove with the Mobile and Morton men. For Nott's 'private character' he had the 'highest regard'.[79] Nott was a Southern gentleman of integrity. Even the vulgar Egyptologist Gliddon Agassiz declared a more congenial companion than the 'unity' bigots with their closed minds.

There was taunting aplenty to anger Bachman, and everybody else. Calling Hottentots orang-utan-like was considered libellous in Europe, such 'shameless exaggerations' being 'advanced in the interest of the slave-holder and slave-dealer, and accepted only in America'. Gliddon's cat-calling was worst, especially when he ended up calling the Bible a 'fetiche'. No new versions would be needed, Gliddon prophesied, because the educated were making 'new religions for themselves'. This was guaranteed to leave a minister like Bachman shaking, but then he was one of Gliddon's 'biblical dunces'. *Types of Mankind* was rank blasphemy to Bachman: irresponsible science unconstrained by a better Book. Twice he turned to the claim of a 'fossil man dug out of the delta of the Mississippi, "57,000 years old"'. This outrage, and Morton's prediction 'that "man will be found in the fossil state, and that he walked the earth with the Megalonyx [giant ground sloth] and Paleotherium [dog-sized horse ancestor],"' made *Types* absurd.[80] Fossil men were an impossibility in Bachman's view. On the subject of human antiquity the minister was out of step.

Bachman's first pamphlet attacked Agassiz's view in the opening essay that the chimpanzee and gorilla differ no more than 'the mandingo [West African] and Guinea negro'. But he wasted shot criticizing Agassiz's use of the newly found gorilla, when no one was sure how distinct it was (on its debut in the 1840s it had been christened *Troglodytes gorilla*, making it a big chimpanzee). Better, said Bachman, to compare 'domesticated men' to domestic pigeons, the perfect analogy for the 'varieties in the human race'. But he failed to follow through and we can see why Darwin was disappointed. Bachman was constantly sidetracked: here on bird song (catalysed by Agassiz's taunt that languages had no common origin, any more than the sounds made by related birds), there by the jibe that unitarists were closet transmutationists. His second pamphlet was one giant digression on Agassiz's 'Tableau' – his folding pictorial plate showing the human and mammal types in the eight 'natural provinces', rendered in an atrocious freehand by Gliddon. The human faces were farcical: the caricature of a 'Creole Negro' so resembled 'an orang-outang' that it clearly betrayed Gliddon's intent.[81] Bachman was unable to contain himself, making his reply look hasty and ad hoc. The mess he made was quite apparent to Darwin.

As unitarists Bachman and Darwin were countryfolk who had once met and were now separated by much more than the Atlantic. They should have been talking the same language, and it is obvious why Darwin would have been interested in this attack on the slavery-sustaining pluralists. But Bachman wasn't using his hogs and pigeons to advantage. 'I am surprised', Darwin added to Lyell, 'from what I remember of Bachman, that he did not do the thing better.'[82] Bachman was blinkered by religion, blinded by rage and making a hash of his case. Darwin would bypass the lot of them.

He bought his own copy of *Types of Mankind* and perhaps suffered a pang of remorse for deriding his old acquaintance. For four days Darwin ploughed through the tome, just before Christmas 1855. He indignantly showered Agassiz's essay on aboriginal men in their zoological provinces with angry jabs: 'How false . . . what forced reasoning!'[83] And, as for Agassiz deducing 'primordial' human races from Morton's hybrid studies, that 'begs the question'. Agassiz ended by announcing that 'there is no evidence whatever' for 'common origins' – common descent, said Agassiz, was 'contrary to all the modern results of science' and led to a damning, godless evolution. Flip the coin, and he was saying that modern ethnology was morally purifying: it showed that the 'human races, down to their specialization as nations, are distinct primordial forms' created by God in their homelands. (Against this Darwin jotted the refutation: 'Look at same race in United States & S. America' – meaning two of Agassiz's provinces actually had the same peoples thriving.) To Agassiz the homelands were 'determined by the will of the Creator', not by the dispersal and adaptation of some common human stock.

At the bottom of the page Darwin jabbed 'oh proh pudor Agassiz!'[84] Oh for shame Agassiz!

The notion that humans were so many species was gaining ground. For Darwin the cross-breeding and blending proved that humans *are* races, but even to call them 'species' was not fatal from his 'common descent' point of view, because they still all evolved from the same ancestor. This was the gist of another note of Darwin's at the end of *Types of Mankind*: they would still be 'descended from common stock', so it will 'come back' to the same thing.[85] The blood ties remained in a descent-and-dispersion scenario, whatever naturalists and demagogues called local human types.

Darwin was quietly confident, but this battering was pushing him deeper into hostile territory. He was forced to follow through all the citations in Nott and Gliddon, and his reading lists swelled with pluralist books.[86]

Blyth's sumptuous stream of facts was the confirmatory backdrop to the 'domestic' question. The torrent of exotic scraps coming out of Calcutta meant that Darwin was becoming dependent. Blyth's description of the similar cooing in all pigeons was another insight to be played up. Pigeons would become Darwin's star presentation. By themselves they were safe, but to insiders they were the proof needed to clinch the long-running 'Unity versus Plurality' debate. They were the model of the way races, however different they appeared, could be traced to a common stock. For pigeons or people, this model held. As the young Darwin had said, man and beast were 'netted together'.

Lyell, having finally been told about the mechanics of natural selection, began chivvying Darwin to write it up, fearful lest he be scooped. Do 'publish some small fragment of your data[,] pigeons if you please & so out with the theory & let it take date – & be cited – & understood', Lyell pleaded following his visit to Downe (the village name was now spelt with an 'e').[87] After almost twenty years of worry and vacillation it was difficult for Darwin to take the plunge. But he complied. On 14 May 1856, he began drafting a sketch, which swiftly turned into an opus, suitably entitled 'Natural Selection'.

10

The Contamination of Negro Blood

America began haemorrhaging in spring 1856. As Darwin started writing up his evolutionary theory, Kansas lay torn between rival free-soil and pro-slavery legislatures, with Southern congressmen and the President himself conspiring to keep the latter in power. The bloodletting started on successive days in May. An old acquaintance was the first to suffer.

Charles Sumner had lived in London at the end of the 1830s, staying at Darwin's club, the Athenaeum, and making many friends in Darwin's circle. After hosting the Lyells in Boston, Sumner became a Massachusetts senator, and now on 19 May 1856 he delivered a ferocious attack in the US Senate on his Southern colleagues for abetting an electoral 'crime against Kansas'. He taunted one refined gentleman with taking 'a mistress ... who, though ugly to others, is always lovely to him; though polluted in the sight of the world, is chaste in his sight; – I mean the harlot Slavery'. Such verbal violence, worthy of an ultra-abolitionist, was not unprecedented in the chamber. Sumner's violent punishment was. Sitting quietly afterwards, huge at six feet four inches, he was accosted by a senator wielding a hard rubber cane. The South Carolina planter thrashed him about the shoulders like a slave, the heavy gold handgrip twice splitting his head open. Blinded by his own blood, Sumner flailed and roared until he fell unconscious on the Senate floor, the cane shattering in its final blows. His degradation was deliberate. It showed insolent abolitionists that they deserved no better than uppity blacks, and the South roared its approval. Darwin's paper, *The Times*, gasped.[1]

Shortly after, a federal marshal leading a 700-strong pro-slavery posse thundered into the free-soil haven of Lawrence, Kansas, sacking the town, breaking presses and killing an anti-slavery supporter. Days later, a revolutionary abolitionist, John Brown, and his guerrilla 'jayhawkers' took revenge: five pro-slavery settlers were shot, stabbed and hacked to death. *The Times* now talked of a 'civil war'. In June, the battle-cry became political – 1856 was an election year – when the anti-slavery Republican Party, with Sumner as its martyr, put up the heroic explorer John C. Frémont for president at its first national nominating convention. Frémont's report of his famous Western expeditions sat on Darwin's shelves, and he drew on its account of migrating buffalo when exploring 'instincts' in his 'Natural Selection' manuscript.[2] *The Times* was happy with Frémont's candidacy and at first condoned the use of force to stem the pro-slavery attacks. Its correspondent reported 'thousands of men marching in solid column through' New York, 'singing a rallying song to the Marsellaise, and raising . . . the midnight chorus, "Free soil, free speech, free press, free men, Fremont, and victory"'. In July, federal troops broke up the free Kansas legislature; in August, a frontier war erupted, leaving hundreds dead and property worth millions of dollars wrecked; in September the troops were in again. The Northern states looked weak, hostages to Southern demands. When the Democrats – a 'slavery extension party' in *The Times*'s view – swept to power in November, the governor of South Carolina, under pressure to secure black labour for plantations in the West, called for the unthinkable, the reopening of the African slave trade.[3]

At this moment, *Types of Mankind* was thriving, with an eighth edition imminent. A white supremacist tract it might be, but the tome – given Agassiz's imprimatur – had to be reckoned with. Most reviewers, impressed by its fashionable Egyptology and newfangled ethnology, thought *Types* contained some truth about the races. In the South a biblical case for slavery took precedence, and Gliddon's attacks on Genesis were deemed superfluous and flippant; but some still thought a little religious adjustment could be accommodated. One pro-Republican magazine endorsed a mediating view: if the science of *Types* was accepted, Christian teaching could still be saved. Morton, Nott and Gliddon might have dispensed with Adam and Eve, but they had not touched the core truth taught by Agassiz, that God created the races 'of one blood . . . not genealogically, but *spiritually*'.[4] Even abolitionists, who generally saw slavery as a moral rather than scientific issue, might live with that.

Twice in recent years Lyell had been back to visit Agassiz at Harvard.

They had talked coral reefs and God's creative Plan, but disagreed on how the Plan was made incarnate. Life's 'progressive development' was anathema to Lyell. He knew that Darwin was working on natural selection, but little enthusiasm went into Lyell's own dossier on evolution as he grappled with its implications. It lacked the 'let's-suppose-it's-true' thrill that shot through Darwin's old pocket notebooks. As Lyell opened a second and a third notebook, he stared bleakly into the face of a gibbering ape. Where Darwin saw a common ancestry humbling the mighty and subverting slavery, Lyell saw degradation, a loss of nobility. Darwin's moral advance would, for Lyell, be mankind's ruin. As the fabric of his lofty world-view tottered, he dreaded the fall. He filled page after page with worries, wondering 'what if it were true?' If mankind's history were the earth's history, then whither the soul? If Shakespeare or Newton could grow from a mewling brat, then couldn't a white man develop from a black savage, or a chattering chimpanzee? Nobility lay in the result, not the journey. Or did it? Round and round he went, with Darwin nipping at his heels. There seemed to be no escape from painful 'night thoughts' about chance and design, law and providence, body and soul, death and eternity, men and apes. But Lyell equally shunned Agassiz's multiple centres of creation, though agreeing that the crowning event in time was the recent appearance of man. By what means he appeared Lyell didn't know, but it was no miracle, *pace* Agassiz. 'The mystery of the genesis is as great in every theory yet suggested',[5] Lyell announced in a Boston lecture, knowing that Darwin was working on a theory.

Lyell had come primed for his fourth visit. On the voyage from Liverpool, he paced the deck with Harriet Beecher Stowe's husband, who was returning early from the English publicity tour for *Uncle Tom's Cabin*. The upright Lyell still harboured aristocratic Southern sympathies. '"Uncle Tom's Cabin" will, I hope, do more good than harm on the whole [meaning it might hurt a violent abolitionism more]. It is a gross caricature, because the very great number of kind masters, and of families where the same negroes remain for generations, is carefully kept out of view.' But he did admit, probably to Calvin Stowe's relief, that 'nearly all' of the evils described by his wife 'do now and then occur in a population'. Seeing *Uncle Tom* on stage in Boston with George Ticknor perhaps softened his heart, and then Harvard stiffened his spine. Agassiz had just returned from lecturing in the South; at

Mobile Nott had persuaded him to write for *Types of Mankind*, and they had compared notes on racial differences.[6] Whatever Lyell heard from Agassiz about the conversation, it could only have reminded him of his own encounter with Nott.

Agassiz defended Nott, who had wished that 'there were no negros', for he could see 'no necessity for them'. Yet all the while Agassiz was teaching that Nott's redundant race had been created spiritually equal to whites. Such contradictory nonsense grated on Lyell more and more: for all his Southern sympathies, he was being driven into Darwin's arms. Black and white belonged to the same family and, given the chance, Negroes could be brought up to the Caucasian standard – this belief was as much a part of Lyell's Christian heritage as Darwin's. But the only viable defence of human unity – the only alternative to Agassiz's multiple creations – was beginning to look like Darwin's evolutionary 'common descent'. Having refuted old French versions of *transformisme* through nine editions of his *Principles of Geology*, it was with trepidation that Lyell confessed his longing to see Darwin's theory in print.[7]

Always Agassiz's spectre hovered about Lyell. Believing in a progressive, simple-to-complex creation, 'Agassiz & the more orthodox' with their '6-days theories' had painted a picture of history too easily appropriated by the evolutionist. And by putting 'Man into the same system', as its end product, he too became vulnerable, the outcome of a mechanism that 'developed the Orang out of an oyster'. It was believers in Genesis, with their six-step staircase up to man, who 'led towards C. Darwin'. Set the staircase moving, let species change, and 'all the hypotheses for the separate & independent Creation [of species] can be turned in favour of the Transmutation theory',[8] Lyell moaned, even the story of Adam.

He tried to picture how man's origin looked to Darwin. His mind's eye watched the races pass in reverse-time, 'like the last scene in Macbeth', the 'negro type' receding 'indefinitely into the past'. What then? Did the first human appear overnight like Adam? Or 'sh^d^. we trace back Man . . . to several stocks or starting points?' If so, 'where w^d^. Man commence & in the form of what race, Red, White or Black? savage or civilized, superior in stature & form or inferior or of average beauty'? If Agassiz's theory of many aboriginal races was difficult, Darwin's idea of making them all one family posed its own problems. Racial divergence from a single stock would 'require time so vast' that the history of civilization would seem but a moment. And what of the stock the races descended from, was it of 'moderate or average ability'? Or low? Indeed, would God have allowed humans to live for ages as 'the African negros & Hottentots' now exist?[9]

If mankind 'started from one pair', Lyell jotted, and we consider creation through Darwin's eyes, then the diversity achieved by humans, judging by 'the bushman, the negro[,] the Caucasian', was as great as in other species. Humans have splayed out 'like any other mammal'; in that sense 'the races of Man' were 'equivalent to the varieties in the Dog', indeed everything living was an extended race. The whole 'variety of Creation' was simply 'the formation of races miscalled Species'. And, jumping from past to future, who was to say that 'the genus Man' won't 'in time give rise to many species' of beings 'higher in power of mind'?[10] Lyell was titillated and terrified.

He was 'coming round at a Railway pace on the mutability of species', Darwin supposed. But it was a rough ride for Lyell, and he warned Darwin time and again 'to be cautious about man'. The moral bond Darwin had gained by having the races united through common descent came at great cost for Lyell. For the patrician it meant a loss of privileged status. Lyell desperately sought compensation. Pondering a pedigree littered with dark-skinned inferiors and soulless beasts, he looked for the coming of redemptive 'higher' beings in an evolutionary 'afterlife'. But this was paltry comfort and, anyway, he hardly understood how such beings might come into existence. Although Darwin had told Lyell only recently about 'natural selection', while showing him round the lofts in 1856 (he was only the second naturalist, after Hooker, to be let in on the twenty-year-old secret; that was how tight-lipped Darwin remained), it still hadn't completely sunk in. Lyell kept musing in his notes on 'the work of time to enable any one race to give rise to others by slow & gradual adaptation'.[11] He was still thinking in old terms of some sort of automatic climatic adaptation – rather than Darwin's selection. Clearly Lyell faced a long road ahead.

The irony was, of course, that it was Lyell who advised Darwin to rush a 'short sketch' into print. This chafed the most. Darwin simply could not squeeze his findings into a small work. He opted for a huge one, 'nearly as complete as my present materials allow', one that would withstand criticism and demolish Agassiz's primordial centres of racial creation. He knew that it had to deal a fatal blow through overpowering detail.

The daily post showed what opposition he faced. Take the pointed letter in June 1856 from an asthmatic, obsessive, shell-specialist, Samuel Woodward. He had been a staff member, first at the Geological Society (as sub-curator), then at the British Museum, and until now Darwin had only had administrative contacts with him. Woodward had been the fetch-and-carry minion. He would bring rock samples to Mr Darwin's visitor's desk, or post fossil barnacles to Downe. He was slow, but in proud clerical fashion scrupulously knowledgeable. Such drudges with facts at their fingertips Darwin tapped eagerly. In an age when gentlemen owned conchological cabinets, this Gradgrind's three-part *Manual of the Mollusca*, with its nuggets on fossil and recent shell distribution, raised his profile. The last part fresh out, Darwin pored over this 'capital book on shells', annotating it.[12] Then he made contact to tax Woodward on whether his molluscs varied more or less through the geological epochs. Their correspondence blossomed for a brief moment in mid-'56, as Darwin begged facts on shell distribution.

Woodward had claimed in the *Manual* that oceanic islands were remnants of old sunken supercontinents, their populations the relics of empires long drowned. Darwin did not believe this: his theory rested on the assumption of islands as volcanic, rising lands (like the cinder-scorched Galapagos), ready for colonization. In fact he was 'fairly *rabid*' on this topic, so his questioning was intense. Woodward steadfastly held his line on the disappearing Atlantises. He took the problem to his boss, the Keeper of Geology, George Waterhouse – the *one* man long privy to one of Darwin's innermost thoughts. Waterhouse had monographed Darwin's *Beagle* mammals (recognizing nineteen new mice in the haul, which helped convince Darwin that mice, like his Galapagos mocking-birds and finches, were representative of each island – a prerequisite to developing his theory of evolving island races). Uncharacteristically, Darwin had then confided to Waterhouse that 'classification consists in grouping beings according to their actual *relationship*, ie their consanguinity, or descent from common stocks'. It sounds innocuous today, but it was an inflammatory admission then, and ignored as aberrant (and probably offensive) by Waterhouse. Darwin's first rule was, assemble groups which have 'inherited' their traits 'from a common stock'.[13] The secretive Darwin had given glimpses of his views to very few colleagues. A lapse like this could cost him dear.

Woodward knew that Darwin was working on the variability of life and awaited 'the publication of your *specific* researches!' Perhaps this implied he knew more, for the chapel-goer who believed that a naturalist must look with a 'spiritual eye' to fathom God's care for his 'feeblest creatures' now recommended that Darwin read a religious book on God's sea creatures, with its warning to the 'panter after fame' who would arrogate to himself 'specious theories'.[14] Woodward talked of the support he had garnered for his island theory, and how 'Dr Pickering must hold something like the same notion – from his chapter on the Probable scene of the Creation of Man'. This talk of human creation led Woodward to the nub of the matter. Pickering in *Races of Man* had placed the human cradles in the jungle lands of the orang and chimpanzee (in the East Indies and West Africa). To Woodward this suggested a wretched ancestry: 'He hesitates between the *Area* of the Orangs, & that of the *Chimpanzees* & seems inclined to make the first man black!'[15] Africa and/or East Asia, it mattered not: both were full of sable skins. Woodward accepted separate realms of white and black men, much like Agassiz's. Whether or not he was playing on Darwin's 'consanguinity' confession, he now wrote plainly that

> It is fortunate for those of us who respect our ancestors & repudiate even the contamination of Negro blood – that Agassiz remains, to do battle with the transmutationists.

This was intolerable to Darwin. He replied by return, saying he would not be begging 'any further favours'. Then he gave a rare, if understated, glimpse of his bullish state of mind:

> I am growing as bad as the worst about species & hardly have a vestige of belief in the permanence of species left in me, & this confession will make you think very lightly of me; but I cannot help it, such has become my honest conviction . . .[16]

Darwin opted for 'Negro blood' and notoriety rather than creation and slavery. And that admission effectively terminated their correspondence for the decade.

Agassiz's long arm unsettled Darwin. With Woodward touting Agassiz's name in London, nowhere was safe. Darwin needed allies in high places, particularly in America, in the anti-Agassiz camp. Darwin's 'big Book' on 'variation & species' would be ready in a few years – he said so discreetly now, testing the waters.[17] And since finishing the barnacles, he had been quietly gaining confidence that, come the day, a few able men would vouch for his integrity. No one was helping more than Blyth, whose huge letters, wittingly or not, showed how animal and human races had evolved together by selective breeding. Darwin had Hooker's ear and Lyell on his toes.

But there was another naturalist whom Darwin had his eye on. Thomas Henry Huxley was an old salt himself – his *Rattlesnake* voyage to the Great Barrier Reef and New Guinea had left him an expert on jellyfish and just about every other lowly invertebrate. Now ensconced as a teacher at the School of Mines in Piccadilly, he was a perfect target. Huxley was a fearful talent and a scathing wit, a brilliant propagandist, ready to poke a finger in every pie. But Darwin worried about his views, particularly his dismissal of all progress in the fossil record. On the human question, Huxley too looked suspect.

Just thirty years old, Huxley was from a younger, harder-headed generation at ease with an ethnology that characterized blacks as irreclaimable. As an anonymous *Westminster Review* columnist in 1854, he actually called *Types of Mankind* 'remarkable' and declared Morton among the 'first rank', his death a 'severe loss'. Huxley liked the book's data crunching, was untroubled by its Bible-bashing, and opposed slavery less because of any sympathy for blacks, but more because it dehumanized the slave-holder. Although he wasn't won over, he still considered Nott and Gliddon's 'Diversitarian' answer to the question of origins 'by far the most elaborate and efficient brief' ever devised, and as powerful as their opponents' case. The argument from Egyptology was 'a good one', for it illustrated the permanence of human varieties – more than human, it showed the stability of 'turnspits, greyhounds, and hounds' (also pictured on the tomb walls) over 4,000 years, all very telling against the 'Unitarian' believer in a common stock.[18] The idea of peoples settled globally before the Genesis time-frame appealed to an upstart who had jettisoned the Bible and now traded on the fact.

Types of Mankind came across as advanced, rationalist thought. For asserting the independence of science, it was winning young adherents. Rising men like Huxley stood at a turning point in the newly optimistic 1850s. More and more colleagues were drawing pay for teaching science and they sought a legitimating literature, cut free from religious moorings.

Not that any words from *Types* could legitimate slave-holding, in

Huxley's eyes. The 'incessant though concealed reference to the slavery question' through the book was 'bad in taste and worse in logic'. America's 'domestic institution' was an abomination which no science ever supported. With Britain still in the grip of 'Uncle-Tom-mania', Huxley excoriated these ethnographic apologists of the murderous slave-owner Simon Legree in *Uncle Tom's Cabin*. Were the Negro 'a distinct species, or even a metamorphosed orang', it would not justify enslavement. But Huxley was no philanthropist and he lived in a new age increasingly hostile to black aspirations. He shared none of Darwin's 'Man and a Brother' sympathy: that was a relic of a bygone age. 'We do not prosecute the drover or the cabman because we believe the poor maltreated ox or horse to be our brother', Huxley noted, coldly. His was a very different rationale for getting 'the Legrees of the south' to end 'their atrocities'. Slavery was evil, not because of the slave-owner's 'hypothetical cousin-hood with his victim which may or may not exist, and which, at any rate, we shall never be able to prove', but because this brutality 'degrades the man who practises it'.[19]

Darwin, of course, was straining to prove this cousin-hood. Although his name would become indissolubly linked with Huxley's over the next five years, he had a profoundly different understanding of science and slavery. Darwin worried about Huxley and invited him to visit Downe in late April 1856. Huxley spent the weekend being ritually grilled about his views. He saw which way the wind was blowing, and he and Hooker (who was there too) grew 'more & more unorthodox' about the stability of species, presumably in response to Darwin's destabilizing questions.[20] Their conversion would have been capped by the obligatory pigeon tour.

Huxley probably hadn't known Darwin long enough to be let into the secret of natural selection, but he got the evolutionary drift. He had been wrong-footed and promptly made amends. His next lecture at the School of Mines, in May 1856, asked the big questions. Where do species originate – from 'created' pairs? – but 'we have not the slightest scientific evidence of such unconditional creative acts', nor could we. For him, an indefinite modifiability of pre-existing species was the only alternative, and, whatever the difficulties, 'by far the most satisfactory solution'. Hybridism 'is almost valueless' as a guide. For that guide we need look elsewhere. Huxley had come away swayed, as Lyell had, by the pigeons. Look at the 'great changes' that could be effected in fancy stock, he told his students. Such extraordinary sculpting – said the teacher fresh from Darwin's lofts – that the ensuing breeds 'may differ as widely from one another as do many species'.[21]

Darwin was pulling in flanking acolytes. They would be needed in the stormy years ahead. But where to recruit in the New World? One good advocate might stand up to Agassiz and win him a hearing. Washington and the West had nobody of stature; the South was politically under siege, and Charleston naturalists were embroiled in their own fight, defending scriptural slavery against Bible-defying abolitionists and racial unity against the Bible-denying *Types*. New England alone remained. In American academe, the ivy grew no higher than at Harvard and Yale. Darwin turned to Agassiz's friend James Dwight Dana at Yale College, who took a tougher line against slavery.

He was a scientific kindred spirit. Not only was Dana equally well salted – he had sailed with Pickering on the United States Exploring Expedition – but he had written up the corals and crustacea collections, dogging Darwin's every step. (It seems crustacea and corals attract nervy sorts with an infinite capacity for detail; in this instance it showed – Dana's huge expedition monographs dwarfed even Darwin's: 3,117 pages in all, not including the plates. No wonder Huxley thought Dana's the most comprehensive mind in America.) Here was a fellow barnacle man, and such people weren't common. It was conducive to a long-distance camaraderie and two-way flow of specimens and monographs – especially after Dana agreed with Darwin's pet sinking-mountain-top theory of coral atoll origins.[22] A sensitive Darwin took to people who flattered his science.

Dana was from an old merchant family and impeccably evangelical. He had editorial control of the nation's most influential scientific periodical, the *American Journal of Science*, which had been founded by his father-in-law, whom he had just succeeded as a Yale professor. He wrote lucidly and could be 'very persuasive' – all in all, a 'very clever fellow', thought Darwin, and Lyell agreed. They backed Dana for foreign membership of the Royal Society. More importantly, Dana was close to Agassiz – maybe no naturalist in America knew him better.[23] This could be a boon.

Through 1856, Dana was engaged in a paper war with a learned exegete, Tayler Lewis, who argued from the Bible for a six-day creation and slavery. Lewis, a crusty classicist, was still dilating on 'the darkness which the foul spirit of abolitionism . . . has shed over some of the most beautiful ideas of Christianity' when Dana tackled him on Genesis.[24] Without referring to his own belief in abolition and Adam, Dana used his geological authority to undermine Lewis's Bible-only stance by giving science free rein over questions of early history, including those posed by *Types of Mankind*. Nevertheless, Dana thought that *Types* itself was largely 'unscientific', save for Agassiz's contribution, which Dana declared to be 'altogether out of place' in Nott and Gliddon's work. Agassiz responded tartly that Nott was 'a man after my heart . . . a man of truth and faith'. He had dealt with 'bigots' in 'about the same manner as you have with Professor Lewis', by invoking science. Gliddon, he admitted, was a loud-mouth, 'but I would rather meet a man like him . . . than any of those who shut their eyes against evidence.'[25]

Dana looked like the man for Darwin. He wrote one of his canny favour-begging letters, asking Dana if their friend Agassiz had noticed whether Kentucky's blind cave fish were peculiar to North America. He was hoping Agassiz would let slip some fact that would help undermine his aboriginal creation theory. Darwin closed on a personal note, unprecedented in his letters to Dana: 'I sh^{d}. *very* much like to hear a little news of yourself, & whether all things go well with you. Are you at work at any particular great subject? I should expect so, though no one whatever in the world has a better right to rest on his oars than you have. I never cease being fairly astounded at the amount of labour which you have performed'. After a paragraph about himself, the family and life at Downe, he closed.

Dana took the bait. He consulted Agassiz in person about the cave fish and, hearing they were 'strikingly American', relayed the verdict to Darwin, along with pages of distributional data. He also mentioned researching 'Embryogeny' for an article on life's 'plan of development' in North America. This was Agassiz's Divine Plan, as epitomized in the growing embryo, the foetal recapping of life's rise through geological time. Darwin knew that Dana could be 'dreadfully hypothetical', but his concluding 'warm wishes for your welfare' seemed promising, as did the personal news about his family.[26] Having taken the bait, Dana might be hooked, so in September 1856 Darwin gave the line a tug.

This letter was a calculated revelation. There had been none like it since his 'confessing a murder' letter to his friend Hooker a dozen years before.[27] He thanked Dana for the 'wonderfully interesting' facts about the blind cave life and then shifted the focus to Agassiz.

I shall be very curious to see your *Embryogeny of N. America*! . . . I wish with all my heart that Agassiz would publish in detail on his theory of parallelism of geological & embryological development; I *wish* to believe, but have not seen nearly enough as yet to make me a disciple.

I am working very hard at my subject of the variation & origin of species, & am getting M. S. ready for press, but when I shall publish, Heaven only knows, not I fear for a couple of years but whenever I do the first copy shall be sent to you – I have now been for 19 years with this subject before me; but it is too great for me, especially as my memory is not good. I have of late been chiefly at work on domestic animals . . . I am surprised how little this subject has been attended to: I find very grave differences in the skeletons for instance of domestic rabbits, which I think have all certainly descended from one parent wild stock. But Pigeons offer the most wonderful case of variation, & as it seems to me conclusive evidence can be offered that they are all descended from C[olumba]. livia . . .

I know that you are not a believer in the doctrine of single points of creation, in which doctrine I am strongly inclined to believe, from *general* arguments; but when one goes into detail there are certainly frightful difficulties. No facts seem to me so difficult as those connected with the dispersal of Land Mollusca. If you ever think of, or hear of, any odd means of dispersal of any organisms I sh[d]. be *infinitely* obliged for any information; as no one subject gives me such trouble as to account for the presence of the same species of terrestrial productions on oceanic islands . . .

You will be rather indignant at hearing that I am becoming, indeed I sh[d]. say have become, sceptical on the permanent immutability of species: I groan when I make such a confession, for I shall have little sympathy from those, whose sympathy I alone value. – But anyhow I feel sure that you will give me credit for not having come to so heterodox a conclusion, without much deliberation. How (I think) species become changed I shall explain in my Book . . . But what my work will turn out, I know not; but I do know that I have worked hard & honestly at my subject.

Agassiz, if he ever honours me by reading my work, will throw a boulder at me, & many others will pelt me; but magna est veritas &c, & those who write against the truth often, I think, do as much service as those who have divined the truth; so that if I am wrong I must comfort myself with this reflection. It may sound presumptious, but I think I have to a certain extent staggered even Lyell . . . Your sincere & heteredox friend Ch. Darwin[28]

Self-effacing, disarming yet tantalizing – typical Darwin. Prefaced by a feigned '*wish* to believe', this was his answer to Agassiz in a sugar-coated pill – domestic rabbits from one stock, fancy pigeons from the rock dove, dispersal rather than multiple creation explaining distribution – everything that might pique the interest of a believer in racial unity.

But as much as he needed an ally, Darwin wanted Agassiz discredited. He couldn't resist throwing the first stone. In a PS he attacked Agassiz's resort to the lack of fossils to support his discrete, unconnected steps in life's ascent, where the animals and plants of each epoch were destroyed before a new set was created. On this Darwin declared himself 'differing, as wide as the poles, from the great Agassiz'. In fact the 'geological evidence' could not be made to fit Agassiz's Procrustean Plan. Maybe Dana would tell Agassiz so, fight Darwin's fight.

Much more than evidence was at stake: 'We are all here now much interested in American politics – You will think us very impertinent, when I say how fervently we wish you in the North to be free'.[29] Free of what? This sudden jump in subject came at the end of the letter; it was perhaps brazen – a Briton opining in a US election year – but Darwin stuck his neck out anyway. Conviction overcame caution, as it always had when Darwin declared his heterodoxy on evolution. The letter was all of a piece. After the hints at Agassiz's racial pluralism, Darwin wanted Dana to know precisely what this implied. 'We . . . all here' – British abolitionists, the Darwin family – wanted the northern states 'free' from the stranglehold of Southern demands to protect slavery. This was the abolitionist call on the *Liberator*'s masthead, 'No Union with Slaveholders', the call for 'Disunion' that William Lloyd Garrison had dramatized by publicly burning the US Constitution on the Fourth of July 1854. At least Dana would know where Darwin stood. Slavery, race and evolution remained inseparable.

Darwin ploughed on with the 'big Book' through 1856, finishing a chapter every couple of months. Not only was he slowed by a houseful of children, a pregnant Emma (unexpectedly, in her 48th year) and a periodically erupting stomach, but by the sheer complexity of his subject. Despite the welter of information he moved inexorably forward, even if he did occasionally admit that the subject seemed 'beyond my powers'. He persevered, 'sometimes in triumph & sometimes in despair', fearing the book would be 'horridly imperfect . . . with many mistakes'.[30]

Portfolios of notes were plundered, broken-backed books dug out as he continued to ransack geology, plant geography, animal husbandry and a dozen other fields. Just juggling the material was a prodigious feat, never mind writing it up.

The strands of twenty years' research were coming together. They were like the yarn spun from slave cotton in Lancashire's mills, fibres wound tight enough in the new theory to withstand being torn apart by its critics. He only feared publishing a weak case, not that he might be wrong and Agassiz and the planters right. His belief in transmutation was as strong now as twenty years before, when he first saw common descent unify all races. But how to translate old insights into a robust theory of racial origins?

Darwin believed sex was the key: sex not only in the sense that it was the men of each race who, by choosing partners with certain features, were dictating what the race looked like, but sex in another sense. He wondered why there were sexes at all, how evolution had left us with males and females in the first place, and whether there was not some grand unifying theory to explain the origin of both sex and race differences.

He still had a 'theory of sexes' from an old, pre-wedding notebook. In this, life began with a bisexual hermaphrodite and ultimately evolved by splitting the male and female reproductive organs on to separate organisms, even though each kept traces of its bisexual past. Foetal growth, he thought, reflected this evolutionary route: humans start life as a hermaphrodite foetus, which develops into one sex or the other. 'Every man and woman' is an 'abortive hermaphrodite' because each sex retains some of the other's traits: 'both are present . . . but unequally developed', such as breasts and beards.[31] Darwin, imbued with the social prejudices of his age, concluded that males are generally more developed than females. The young of both sexes resemble each other until the age of reproduction, when the males spurt ahead and become bigger and feistier, leaving the childlike females lagging.

In the early drafts of his theory, Darwin worked out how this sexual differentiation occurred. He saw animals tending to manifest traits at the same age as their ancestors acquired them. Competition for scarce resources in nature led to the selection of the best traits to adapt animals to changing environments. That was natural selection, his primary agency of change. In addition, males competed for females, and this resulted in the 'second agency' of Darwin's 1844 draft. At sexual maturity, the 'most vigorous males' – in birds, those which had developed more appealing songs, looks or courtship displays – mated more successfully and left more offspring. Males continually honed and passed on their attributes – horns, antlers and so on – and courtship behaviours to their male offspring, who also manifested them at sexual maturity. These altered 'sexual characters', designed for fighting or displaying, were *added* to the adaptive traits males and females possessed in common.[32] So the females, as every Victorian paterfamilias knew, remained weaker, passive and suited to rearing the young.

In reacting to the racist books Darwin now gained a greater insight into the role that 'beauty' played in leading the races along their divergent paths. 'Beauty' was integral to the rival literature. *Types of Mankind* was a hymn to the 'manly beauty' of noble Caucasian faces, 'the perfection of the beauty of which is justly admired'. It praised the 'faultless' phrenological vault, whether in ancient Greece or modern Britain, whose Apollo features contrasted with the 'coarse and ugly' Negro physiognomy.[33]

Knox too had celebrated beauty in his anatomy books for artists (Darwin had read *Great Artists and Great Anatomists*). And in his *Races of Men* the disillusioned Edinburgh reject dismissed all but his own race on aesthetic criteria as much as any: Gypsies – 'a race without a redeeming quality . . . they have not the elements of beauty'; Jews – 'Why is it that the female Jewish face will not stand a long and searching glance? The simple answer is, that then the want of proportion becomes more apparent'; Slavs – 'remarkably deficient in elegance of form; external beauty does not belong to them'. These books posed an aesthetic teaser, which Darwin translated into his old relativist question. Why did white males find white females beautiful, and black males black females? Wherein lay 'beauty'? More to the evolutionary point, if the 'beauty' of form was non-adaptive – not aiding survival – how could it be selected?

Pondering these racial castigations helped Darwin fix firmly on his secondary evolutionary mechanism. Partners were choosing mates on aesthetic criteria, leading to a divergence in non-adaptive but desirable traits. A different ideal characterized each human race. Indeed, since all animals shared a sense of beauty, it ran right through nature – through all races. He scribbled a question on a note at the back of *Types of Mankind*:

What effect wd idea of beauty have on races in selection. it wd tend to add to each peculiarity. V. our aristocracy.[34]

Picking pleasant features was subjective in each race. Lips, ears, noses, forehead, they could all be changed by this self-selecting process, even though minuscule variations in the ear lobe or lip shape could probably not be seen as adaptations. Witness the British aristocracy, which had plucked the choice society belles each season for its own experiment in stately breeding. Darwin was a great reader of the society and court circular column of *The Times*. He taxed his correspondents worldwide, asking if the *beau idéal* of the New Zealand Maoris or African 'savages' matched that of Europeans. He meant 'whether we & they would pick out the same kind of beauty'.[35]

Darwin had the concept. Now he coined a term to encapsulate it. Its earliest use (that we have found) occurs – tellingly – in his scrap ruminations on Knox's *Races of Men*. In March 1856 Darwin conceded that it was difficult to detect any changes in the human races over historical time, using either Gliddon's tomb images or Andrew Smith's description of the oldest European families in the Cape. Change was obviously very, very slow, and – jotted Darwin in a note on Knox on 3 March – 'The slowness of any changes [must be] explained by [resilient] constitutions [and the chancy nature of natural] selection & sexual selection'.[36] *Sexual selection*: he had his catchphrase.

Reading *Types of Mankind* – which slashed Prichard's 'darling unitary fabric' to shreds and slated his *Researches into the Physical History of Mankind* as a ruinous distortion 'to suit the theological notions of the day' – reminded Darwin to brush up on Prichard. He jotted '1847' (the date of Prichard's latest edition) in the margin of *Types* and wrote 'Prichard Last Edition' on the back endpaper. This he crossed out when the great book arrived: five fat tomes with 2,500 pages, expensively priced at four pounds and two shillings.[37] It dwarfed *Types of Mankind* in every way, and Darwin set to, pencil poised.

From the time of his Edinburgh dissertation, written at the height of the anti-slave-trade movement, until his death in 1848, Prichard had argued for small-scale variation, within the human species, within historic time. Climate or 'external agencies' had produced all the races of mankind from one Adamic stock. He retreated, or sped up the process, or equivocated on some points, but over forty years he stood firm for human unity, particularly of black and white. As freethinkers and French pluralists (such as the transmutationist Bory de Saint-Vincent, whom Darwin had read on the *Beagle*) had argued for multiple origins, so *Researches* had grown – one volume in 1813, two in 1836, three more in 1847, then in 1851 the reprint that Darwin had bought. Critics called it armchair ethnology, but no one had seen its like. Prichard had ransacked whole realms of scholarship: anatomy, physiology, zoology, physical geography, history, archaeology, philology – nothing had escaped his notice. The last edition looked like an encyclopaedia. Darwin, churning out his own chapters, had not finished volume one when he suddenly realized – jotting at the back of *Researches* – 'How like my Book all this will be.'[38]

'My book' now had a title, 'Natural Selection', and like Prichard's it promised to run to several volumes. And at last it was becoming clear how 'man' would figure in the 'big Book'. Darwin read *Researches* at least twice during 1856, leaving a trail of marginalia and paper scraps. Beauty was his main concern: continual marginal snippets testify to the point. The 'Chinese admire Chinese beauty', likewise their Cochin relatives and the Siamese, though some Asian women might turn heads in Europe. Then again, one race's deformity might be another's ornament – for instance, the 'flattened head' in Morton's American Indians. Darwin had originally assigned an adaptive purpose to such traits. But it is clear that he was moving to place these and more – the complexion of 'Hindoos', the 'Bump in [the] Hottentot', the 'Shin Bone of Negros', and the 'skull of Australians' and 'shape of Pelvis' – in the realm of sexual selection. This is confirmed by another note left at the back of Prichard's first volume: a deformity might be 'an essential point in Beauty'.

As Prichard paraded mankind before Darwin in woodcuts and colour plates, the scope for choice expanded, and by the end of volume five Darwin realized what it could do. Distorted Malay skulls conformed to local beauty standards as much as the curious 'hair & colour' of other races. The only selection necessary to create these traits was artificial, relying on each race's sense of beauty. Now Darwin would push the concept much deeper into nature. Having grasped human racial aesthetics, he could start searching for a similar kind of non-adaptive selection throughout the animal kingdom. As he scrawled on a slip at the back of

Prichard's last volume, 'Man's Sexual characters [are] like [those of] tufted Ducks' – in each case, chosen. On the same slip, he penned, 'In my note on Man . . .' This was the first sign that Darwin planned to include humans in his book, if only briefly.[39]

He alternated reading and writing through summer 1856, leaving references to 'Ch. 6' in book margins, on scraps of paper and even on his old reread notebooks. Some scraps went into a special folder labelled 'Ch 6 Sexual Selection'. Evidently he planned to treat the subject alongside natural selection in the same chapter. He began scribbling thoughts on other books. On a French treatise on heredity, read in September, he made 'sexual selection' a rubric for the whole sexual differentiation process – covering his theory of separating sexes from hermaphrodite ancestors as well as human beauty and rutting males. Male and female flowers on the same or separate plants? See 'Ch. 6 Sexual selection'. Sexual differences 'throughout animal Kingdom (look at under Ch. 6)'. 'On animals, [such] as Bull & Stallion, having much more choice than [one] w[oul]d think. Ch 6'.[40]

More telling evidence turned up in his pirate edition of surgeon William Lawrence's *Lectures on Man* (1822). ('Pirate' because the book was declared blasphemous in court, for its materialistic attitude to human physical and mental development: books explaining humans naturally, as Darwin knew, were dangerous. Under English law, it lost its copyright and working-class atheists sold pirate copies, one of which Darwin owned.)[41] On the inside back cover, Darwin again jotted his neologism, 'Sexual selection', then listed 'Beard', 'Ears', 'Tattoo females', 'Lips', 'flatten nose' and 'steatopyg[i]a' (referring to the buttocks in 'Hottentot women'), then in ink 'Baboons steatopyg[i]a'. This cryptic jot pointed to Lawrence's ethnographic fancy that 'the Negroes and Hottentots approximate in some points to the structure of the monkey kind'. Where he mentioned 'the tremulous masses of fat' in the Hottentot buttock that, according to Cuvier, form 'a striking resemblance to those which appear in the female mandrills, baboons, &c', Darwin drew a knowing line beside the text. He was giving a new twist to an incautious and condescending old literature on the 'Hottentot Venus' (the Khoikhoi girl Sara Baartman who had been exhibited in raree shows in Cuvier's day, when 'freaks' and exotic foreigners were so many deviants). Darwin seemed to be thinking: 'Why shouldn't Africans who tattoo females and like flat noses also fancy buttocks like those of a female baboon?'[42]

It was all about *choice*, and to the Western gaze at the centre of civilization, strange choices led to stranger anatomies at the periphery. Standing even further from this centre, animals themselves must also select their otherwise inexplicable shimmering, gaudy and wattly features on aesthetic criteria. Back to birds, Darwin thought of their sexual characters, and glossed some of his old notebook remarks on magpies, jays and starlings. '*Sexual Selection*', he jotted, then wrote, 'If masculine character [is] added to species,. we can see why young & Female [are] alike': females remain undeveloped while males accumulate additional distinctive characters – gaudy plumage and a morning-chorus voice – by competing among themselves. In December 1856, when he cut out notebook pages and added them to the portfolios, he scribbled 'Good Ch 6' on this one, 'Keep'.

So what was Darwin projecting in his 'note on Man'? Working fast, he left few clues. In another notebook he pencilled 'Ch 6' beside his comment on how non-adaptive characters, such as the 'mammae of men', could be inherited. He found an 'excellent remark' in a book by Prichard's heir and successor at the Ethnological Society, the surviving unitarist flag-waver Robert Latham. (Or barely surviving: Latham, an Old Etonian, was now looking like a dissolute old clergyman: slightly sleazy – a shabby-genteel symbol, perhaps, of what Prichardism had come to. But he could churn out endless dry books ranking Her Majesty's ever-expanding number of subjects.) In *Man and His Migrations*, Latham discussed how the ranges of tiny 'Hottentots' and big 'Kaffirs' overlapped; the two tribes, so different in their features, had displaced and eradicated all intermediate types, to leave only these terminal twigs of the evolutionary human tree, as it were. Good example, '*quote* in Ch. 6?' Darwin scribbled in the back of his copy. It seems that 'Man' – the human races – were destined for chapter six of his treatise, which would also contain the explanation of their original divergence, 'sexual selection'.

By March 1857 he had five chapters done, some 90,000 words, and this sixth one under way. At the end of the month he stopped and drew up a table of contents for chapter six.[43] If there was to be a 'note on Man', this surely would show it. But it didn't. There was a blank space and a dangling page number where it should be.

Dana was a disappointment. For his heretical openness, Darwin received back a confession of faith in divine 'type-ideas' behind nature's façade.

Like Agassiz, Dana drew a parallel between foetal growth and palaeontological progress; even if not 'to the same extent as Agassiz', he still saw it as the creationist science of the future. Successive species through history were no more than so many motes in the Almighty all-seeing eye. Change was *not* brought about by a bestial transformation of one animal 'tribe' into another, but by repeated miracles. He offered to assist Darwin's search for the 'real truth', confident that the facts being compiled at Down House would lay 'the best of foundations' for Agassiz's 'laws'.

There would be no fifth column at Yale. 'Farewell – Floreat Scientia', Darwin wrote back with a note of finality. At the end of 1857, when Hooker confessed that he couldn't follow 'Dana's metaphysical ideas', Darwin offered his own verdict: 'Poor fellow, he believes in [the] 1st Ch[apte]r. of Genesis, so great allowances must be made for him.' The more immediate chapter for Darwin was his own, for he added a PS: 'I have just finished a tremendous job, my chapter on Hybridism.'[44]

The 30,000 words had taken him three months. Dana or no, the Americans would be trumped. It was the biggest chapter by far, and whatever Darwin's caveats – his admission that the 'whole subject is extremely complicated' – it pointed one way: species were invariably sterile when crossed, barring, as always, Tom Eyton's geese. Races – 'varieties'– were always fertile together, even if, very occasionally, to a slightly reduced degree. (He actually thought the results would depend on how long the varieties had been established: the longer, the more species-like their constitutions.) The sheer complexity made any 'universal' claim rash. Each case had its subtleties, to be appreciated only by 'carefully conducted experiments'. The chapter was much bigger than a mere refutation of the Morton–Nott camp, so overwhelmingly so that Morton was relegated to barbed footnotes. They showed him mixing up a Canada and a greylag goose when recounting reports about hybridity, and then garbling the results, claiming fertility where the original source talked of sterility. As to Morton's only first-hand claim, the cross of chicken and guineafowl, Darwin countered with the authority of the Zoological Society, whose own 'hybrid was quite sterile'.[45] Elsewhere Nott and Gliddon were simply accused of exaggeration in reporting things such as fertile goat–sheep unions.

Even if domestic dogs had descended from several wild species (as Darwin conceded), the original crosses would not have been as fertile as today's mongrels. But when it came to domestics, the exemplary 'case of perfect fertility between varieties, which has struck me most, is that of Pigeons'. Darwin's authority was now paramount: 'I have myself largely experimentised on the fertility both of simple & the most complicated crosses between the most distinct breeds'. All were fertile, and considering how different were the 'Pouter, Tumbler, Carrier, Fantail & Barb', it was telling that they bred while more similar-looking but different *species*, such as 'Gold, Silver & common Pheasant' either do not, or any young are 'utterly sterile!'

So strong was the pigeon evidence that he planned to start 'Natural Selection' with two chapters on tame animals. 'Variation under Domestication', they would be called, and they would emphasize the pliability of domestic life. Into these chapters would go his breeding know-how, to give 'Natural Selection' the confident authority absent from the Morton–Nott propaganda. He would show the ubiquity of infinitesimal variations, the age they appear in squabs, and how they are selected with the effect 'of adding up small changes'.[46] These chapters might run the gamut from cabbages to cattle, in astounding and compounding detail, but the pigeon loft was central. And its message was this: crossing fancy breeds proves that they are all of one species. Species can be stretched artificially so much that, were they wild, a naturalist might rank its races as different genera. But they weren't different genera: their ancestry (indeed that of cats and rabbits and ducks) could be tracked through ancient breeders' manuals. The breeds perched on the genealogical branches of their fancy tree, whose trunk was an ancestral rock dove.

This tree image explained how animals and plants should be grouped and classified: genealogically. Unity came from blood and ancestry. It was what Darwin had confessed to Waterhouse all those years ago. Everything could, in truth, be put in its place, and the need for Morton's '"primordial forms", adopted by Agassiz', or the independent creation of identical species on distant islands, vanishes.[47] The diaspora of life tells of the spread of ancestral voyagers, not beings 'primordially' created on the spot. It applied to birds, barnacles and bushmen alike.

By the time Darwin had finished the gigantic anti-Morton/Nott chapter (on 29 December 1857), Nott and Gliddon had published a new opus, *Indigenous Races*. Just as provocative, this one featured Gliddon's pull-out maps showing the 'Geographical Distribution of Monkeys, in their Relation to that of Some Inferior Types of Men'.

Ancient Indians and extinct prehensile-tailed monkeys lived where modern Indians and modern spider monkeys now do – they were fixtures in their American 'province of creation'. Gliddon was coming round to geology. He played up reports of fossil men (and rumours were rife), not least because it irritated the likes of Bachman. By Gliddon's theory, each inferior fossil man would be in his geographical place. That was the prediction – fossil white men in Europe, fossil Negroes in Africa. There had been a unique historical 'progression' of the entire fauna and flora within each zone. In the Caucasian province, simpler fossil monkeys might be expected alongside fossil men 'of lower intellectual grade', precursors of modern Anglo-Saxons; what were *not* anticipated here, of course, were ancestral 'Negroes'. They would have their own line in Africa. The palaeontology of Gliddon's pluralism was now interesting. He had fossil lines going back in *parallel* the globe over. There was no 'commonality' at root. Darwin's concepts of forking and racial divergence were absent.[48] There was no ancestral blood tie between Negro and white man.

Gliddon's expectations of fossil humans ran high after the discovery of the jaw and arm bones of the short-faced 'oak-ape' *Dryopithecus*, which had once lived in the Pyrenaean oak forests. News of it reached him in 1856. Even Lyell became excited about this human-sized ape. He thought it so 'near the negro' (and perhaps the Negro's ancestor) that it would fool a student at the College of Surgeons. Richard Owen – ex-College of Surgeons himself – wasn't fooled. He loathed transmutation to his very soul and dismissed the fossil as an extinct gibbon. In Gliddon's view this 'gratifying' find meant that 'fossil man, of some inferior grade, is now the only thing wanting to complete the palaeontological series in Europe.'[49]

Indigenous Races brashly captured the momentum of the age. It also captured the language. In a rebranding exercise, Gliddon now renamed his pluralists and their unitarist antagonists. The latter he called 'Monogenists', while his up-to-the-minute scientific group were 'Polygenists'. As keywords go, their meanings were loaded by the very terms of the debate. These weren't neutral epithets. Gliddon introduced the terms in his swamping 200-page contribution, 'The Monogenists and the Polygenists' and did a brilliant marketing job. Even Darwin was eventually forced to fall in with the terminology. Nor did later generations realize that the anti-black, pro-slavery, separate-homeland pluralists had stamped their authority on the language. The words historians now routinely reapply to earlier periods were coined in 1857 for a purpose. 'Monogenesis' came tainted with the 'religious dogma of mankind's *Unity*'. It was a morally worthless myth, to which a 'trembling orthodoxy clutches like sinking mariners at their last plank'. Monogenists, in this caricature, saw Adam and Eve delivered to earth by a 'bevy of black angels'. Modern science supported 'Polygenesis', a dispassionate and fearless exegesis of the rocks and tombs which pointed to a separate black and white ancestry.[50] Although Gliddon had to admit, even polygenism had ultimately to rest upon deep history, and that was truly terra incognita.

Thus stood the grounds of the bigger debate as Darwin rushed towards his debut, with the stomach-churning fear of exposing mankind's real origin from the beasts. It was the sort of fear that had kept him quiet for two decades: the sort that would put him in a sanatorium as the eve of exposure dawned.

Facts: he always needed more facts. He never stopped collecting them. Now he wanted specialist insights into the human physique. Army surgeons across the world were an untapped mine of information, like the breeders. He began asking whether troopers with 'light or dark complexions' were more resistant to tropical diseases. Sexual selection could explain so many odd traits, but he had always supposed that skin colour was naturally selected, a dark complexion being more useful than alluring in hot climates. That was roughly Prichard's view and Darwin had made it his own. Surprisingly, a staff-surgeon in Sierra Leone claimed the opposite: light-skinned 'sanguineous' soldiers withstood the climate far better than darker-skinned 'melancholic' men. In the Niger delta, the King of Warri – whose Itsekiri tribe had long been middlemen in the slave trade – had equally long experience with visiting Portuguese slavers and confirmed the impression: 'Ah! . . . it is the right age to bring white men to Africa, the younger the better, and he is a true child of the sun, his fire (light) hair will save him from many bad diseases; he and others like him, *will* live!!'[51] After the Niger Expedition fiasco, Darwin doubted it.

He pushed sexual selection further and further, pressing fellow naturalists to tell him whether they had ever observed male insects or crustaceans 'fight for females'. By this point, he had begun to suspect that the *males* were not the only active ones doing the choosing. Potent males might have the pick of the field, but the females did not always cooperate. If choice was really involved, perhaps females were pickier and less passive than he had thought.[52] As always with Darwin, it was a slow-motion leap; even to start it, he had to break through a Victorian prejudice, female passivity.

At some point, probably in mid-1857, he wrote up the evidence in hand for sexual selection, 2,500 words on ten pages. And he pencilled on to the contents page of chapter six, opposite the dangling page number, 'Theory applied to Races of Man'. Here at last was to be his portentous 'note', running from page 63.[53] The summer slipped away, with Darwin finishing chapters seven and eight. That was another 60,000 words by the end of September.

Even as he was rethinking sexual selection, so he was re-strategizing what it was safe and expedient to publish. He simply would not expose himself without overwhelming evidence. Having written to the affable collector out in the Far East, the socialist, Dyak-sympathizing Alfred Russel Wallace, requesting bird skins, he duly received them. Agreeable letters, too. One wanted to know whether Darwin's rumoured book (yes, even on the other side of the world the rumours were heard) would 'discuss "man"'. Darwin replied on 22 December 1857, 'I think I shall avoid [the] whole subject', as it was too 'surrounded with prejudices', even if humans do pose 'the highest & most interesting problem for the naturalist'.[54] Suddenly, mankind, the ultimate problem threatening fearful repercussions, had disappeared from the projected magnum opus. The *raison d'être* of much of Darwin's work was to be concealed. He would take his science out of the house and into the garden, and restrict his public talk to pigeons and plants, which were safer.

In public, *Hamlet* was to be performed without the Prince. Why? All the angry abolitionist energy Darwin had ploughed into saving the races, evolutionarily, ensured that humans were to be part and parcel of his 'big Book' right up until mid-1857. Of course his other chapters would contain elliptical references to habitual actions becoming instinctive among humans, from old ladies knitting to infants mimicking parental mannerisms, but these would be lost in the encyclopaedic detail. The fact is, the 'note on Man' was gone. Had Darwin heeded Lyell's caution, not to stir up the hornet's nest by mentioning the human races? The two were close; Lyell continued to agonize,[55] and he was astute in his softly-softly approach.

Just how astute we can gauge. The imperious Anglican Richard Owen, the 'English Cuvier' himself, spoke in February and April 1857. Owen execrated transmutation and had made a reputation of crushing such abominable science. Well patronized for his zeal and piety, the comparative anatomist was at his peak, the newly appointed Superintendent of the British Museum's natural history collections. Most of the showy resurrected monsters so tantalizing to the period – from dinosaurs to giant ground sloths and moas – were his doing. While many, including Huxley, loathed the man, thinking him arrogant, he was loved by Thomas Carlyle. The 'nigger'-baiting sage was attracted by Owen's reverent nature, and he thought this towering man talked enough sense 'to make me cry'.

In 1857 at the Linnean Society Owen classified humans as a distinct subclass, ostensibly because they had 'unique' brains, with features found in no ape's; but in truth he was insulating the one being with an immortal soul. If even Wallace in the Far East had heard what Darwin was doing, Owen at the British Museum surely had. By now, a small circle was privy to Darwin's thought, including the entomologist Thomas Wollaston – he too had been at Downe in 1856 when Huxley and Hooker started discussing evolution positively. Wollaston, who dismissed transmutation, was coincidentally arranging his beetle collection for Owen in the museum in 1857. And while Huxley was criticizing supernaturalism in his lectures, word had gone round the Royal Society about Darwin's project. So Owen's was probably a pre-emptive strike.[56] He created a top taxon to receive mankind, whose moral nature required a separate order of explanation. No wonder Lyell foresaw trouble and cautioned Darwin. Darwin probably saw it coming himself. He was aghast at Owen's classification, which would leave humans further from chimps than chimps were from a platypus. Ever the relativist, he wondered what 'a Chimpanzee w^{d}. say to this'. No one was surprised at what the brusque Huxley said; he called Owen's lordly classification 'a Corinthian portico in cow-dung'.[57]

But Darwin's withdrawal of mankind from the book had another possible rationale. His modus operandi had always been to amass an overpowering amount of evidence. He could not describe one barnacle;

he had to describe them all. In dealing with humans this was simply critical. Darwin's inflammatory racial divergence theory would touch on the deepest social taboos and, with Agassiz on the rise and huge polygenist books being praised, Darwin had to pile on crippling quantities of detail, as Prichard had in his magnum opus. His correspondence network worldwide had poured information into Downe, but time was needed to discern which racial features had survival value and which were sexual choices. He wanted to be absolutely sure that iridescent plumes and skin complexion did not adapt birds and humans to some niche; rather they were courtship accoutrements, possibly to charm choosy females. Off went more letters to the colonies on complexion and constitution.

The subject of mankind was 'surrounded with prejudices', as Darwin told Wallace. Any talk of human transmutation would invite a terrible backlash. Darwin was nervy at the best of times; in and out of spas, his stomach and swimming head growing worse, he dreaded broaching the subject. Now the perfect excuse to pull mankind out of the book presented itself. The line from Blyth, his major supplier of colonial intelligence, went dead.

Blyth's huge letters from Calcutta about animal and human breeds, their history, hybridism, colouring, competition and sexual traits, had become mainstays – ten in 1855, nine in 1856. Then just as Darwin, while reading Prichard, sought more information on the evolution of racial beauty, the letters dried up. India's First National War of Independence – what the British called the 'mutiny' by native sepoy troops in the East India Company – had broken out. The revolt began just outside Calcutta in January 1857. Blyth managed to get one letter through in April. A month later Delhi was besieged and the British women and children of Cawnpore were massacred in June. Blyth's next letter was written as a 'royal salute' was firing 'in honour of the arrival of the glorious garrison of Lucknow, *i.e.* the wounded officers, & the ladies and children'. The sight left him decrying the useless 'struggle of barbarism against a higher civilization ennobled by the application of all the sciences'.[58] During the bloodshed, army surgeons had their work literally cut out for them by sabre and cutlass, leaving no time to answer Darwin's letters about disease resistance. On top of this, another catastrophe struck Blyth personally. His wife of less than three years died suddenly from hepatitis in December 1857. It caused his own near-fatal collapse.

All letters stopped. Darwin, so full of empirical and emotional insecurities, made his decision. As if to confirm that the communications breakdown was the problem, he coyly told Lyell that he had 'got some rays of light' on 'the Races of Man', but the 'mutiny in India stopped some important queries'.[59]

One year after lobbying Dana, Darwin sought a new American ally, this time at Harvard – the botanist Asa Gray. They had met through Hooker years before and were now corresponding. No one abroad, bar Blyth, had helped more, and Gray's statistics on North American plants were going straight into the 'Natural Selection' manuscript. Hooker praised Gray for being 'pure and disinterested', meaning a stickler for facts and viscerally opposed to Agassiz's creationism. Darwin had read Gray's correspondence with Hooker, which had shown him as an anti-slavery ally, awaiting the right time to take on Agassiz.[60]

A farmer's son from upstate New York, Gray had cut his scientific teeth on Lawrence's 'blasphemous' *Lectures on Man* while training to be a doctor. Its arguments for the unity of the human species took moral root after Gray's conversion to a revived 'orthodox' Presbyterianism which abominated slavery. As a new Harvard professor in 1842, he had welcomed the heightening of 'abolition feeling' brought by the case of a fugitive slave from Virginia, George Latimer, imprisoned, tried and then manumitted after a ferocious public outcry.[61] Runaways had been causes célèbres even before the draconian new Fugitive Slave Act came into force.

In 1848 Gray married into a family soon touched by the hated statute. The Lorings were conservative Unitarians like the Ticknors. Daughter Jane was raised a 'country Calvinist' on her mother's side and was well-matched with Gray. Her father, the prominent Boston lawyer and member of Harvard's Corporation, Charles Greely Loring, initially had no sympathy for abolitionists. The family attended the fashionable West Church in Boston, however, whose minister grew less conservative with the years. The capture and rendition of slave Thomas Sims in 1851 rallied them all to the poor boy's cause, and Loring turned to the anti-slavery Republicans. With Sims threatening suicide and Theodore Parker thundering from his abolitionist pulpit, Loring's legal team, reported *The Times*, fought in vain to secure Sims' release. Hundreds of armed men marched the 17-year-old by torchlight, tears streaming, to the ship in Boston harbour waiting to return him to his owner in Georgia. At Savannah, the lad was publicly lashed to the legal limit, thirty-nine times.[62]

Quiet and reserved, devoted to a sickly Jane and the Harvard herbarium, Gray was no firebrand, nor even, he admitted, 'very much of an abolitionist'. He firmly believed that discoveries about God's world could not harm his religious convictions. And though he had 'other than scientific grounds' for resisting Agassiz, like Darwin he always stood on the 'facts'.

At Yale, Dana was caught in a cleft stick, committed professionally to Agassiz but theologically to Adamic unity. Gray tried to show his friend Dana the way out: Darwin's facts pointed to 'centres of radiation' for each new species, possibly even for groups of new species as in the Galapagos, rather than to Agassiz's blanket areas of wholesale creation. Every species in Darwin's world was born only once and in one place, mankind included. To Gray this was 'the orthodox faith', and he drove the point home to Dana. 'I am glad to hear that your idea of the unity of the human species is confirmed more and more', he encouraged Dana, grasping what was evangelically at stake, and then added, 'The evidence seems to me most strongly to favor it.' As Darwin drew closer to Gray and tipped his own hand, Gray in turn tipped off Dana about a coming 'resurrection' of the notorious 'development theory'. The new angle, Gray warned, would make belief in creation by evolution 'more respectable and more formidable' than ever before, had Dana the wisdom to grasp its significance.[63]

Agassiz's 'superficiality and wretched reasoning powers' disgusted Darwin. He laughed about how 'the great man' got round awkward facts, declaring 'Nature never lied' (meaning Nature never told Agassiz a lie). Gray laughed with him, knowing that there was nothing Agassiz 'cannot explain away'. It was the arrogance of someone who believed he understood 'Nature through and through'. Yet Agassiz's arrogance was not unlike Gray's more modest assumption that nature would square with his faith. And in his heart Darwin knew that he too had impervious 'theoretic notions' and managed his way round difficulties in 'the Agassi[zi]an sense'. Hadn't Darwin once himself admitted feeling 'an instinct for truth', which he saw parallel an 'instinct of virtue'? Still Agassiz was the butt: when Gray produced data on the ranges of large and small plant genera, Darwin rejoiced that 'the result was as it sh[d].

be, for as Agassiz says, nature never lies'! Gray *had* to be right; the data showed that migration and adaptive radiation, not blanket creation, explained the distribution of species across the globe. And what held for plants applied to humans, 'so long', Gray said, 'as we cling . . . to the idea of the single birth place of species'.[64]

A dozen naturalists, maybe more, had been told that Darwin was writing a book, but only two what was in it. A hefty opus was expected. Indeed, after 'steady progress', Darwin reckoned the manuscript would be ready for press – barring a recrudescence of his dreaded illness – in 1859. Huge at 375,000 words, in two tomes at least, *Natural Selection* would be a sitting target for a demagogue with a 'boulder' to throw. Darwin hoped Gray might intercept that boulder and even toss it back. So he baited Gray, telling him of the nineteen-year gestation and fact-assemblage from 'agriculturists & horticulturists', his study of 'domestic varieties' as the key to '*how* species change', his 'heterodox conclusion that there are no such things as independently created species' and about how all this 'will make you . . . despise me & my crotchets'. He reaffirmed that Gray's own distributional data showed species descending from one another rather than created in place. Agassiz as usual was the tacit target. Gray's reply, 'you begin . . . with good, tangible facts; and I am greatly interested to see what is to be made out of them', opened the floodgates.[65]

'My dear Gray', Darwin began intimately, dropping 'Dr' for the first time; then he uttered a sigh of relief. Gray's anti-slavery sentiments had left Darwin with 'the warmest feeling of respect', and no greater respect could he show than to take Gray into his confidence. Gray was entrusted with the 'secret', joining the two naturalists who had been told of the mechanics of natural selection, Hooker and Lyell. Asking Gray pointedly 'not to mention my doctrine', lest someone steal his thunder, Darwin posted a 1,200-word summary to Harvard.

Reading the manuscript, Gray thought it rather 'hypothetical', but no worse; he remained open and agreed to keep sending facts. But then what he read had already been toned down. Or, more precisely, it had been shorn of human races and sexual selection: such might be interpreted by an American, however liberal, as simply bestial – not only racial union, but orang-union. The degree to which Darwin had actually pushed 'common descent' – making humans and apes share an ancestor – could have killed Gray's acceptance of the whole theory.

Nonetheless, in any debate over transmutation the question of racial-and-then-ape-unity would be the first flashpoint for Americans. Agassiz had made sure of it. Agassiz's creationist biogeography remained in the works of pro-slavery, '*niggerology*' activists. (A letter from him was prominently placed in Nott and Gliddon's new *Indigenous Races*, which ensured his name on the title page. It offered proof that even the Malay and Negrillo peoples in the same South-East Asian islands must be separate species.) As such Agassiz's 'polygenism' remained integral to the agonized conflict over slavery as the nation girded itself for war.[66] His intervention ensured that 'polygenist'-creation versus 'monogenist'-transmutation would be an ideological issue as never before – it would, and could only, focus on mankind.

While scientific Yankees and advanced Southerners warmed to the new Agassiz-blessed ethnology, the slavery issue in and out of science became unavoidable. Gray realized this and was already taking Agassiz on. But uniquely placed though Gray was to counter his colleague's racial pluralism, he was devout, and when it came to a deeper human transmutation Darwin couldn't be sure of him. Darwin was incalculably better equipped to deal with this issue. But even he wanted more racial information. And so, despite his own revulsion at slavery, caution and events in India forced him to delay publishing on mankind.

11

The Secret Science Drifts from Its Sacred Cause

Not everyone shared Darwin's reticence. Another naturalist who had grasped the centrality of race to the transmutation debate was his Far East contact Alfred Russel Wallace.

Wallace had sailed abroad as a self-employed specimen collector in Prichard's day and had spent less than eighteen months in Britain since. His was a forties' mentality, all but innocent of Agassiz and *Types of Mankind*, and through the tropics he carried the evolutionary *Vestiges* in his mental kit. While bagging birds and netting butterflies, he had hoped to prove its 'ingenious hypothesis' that one species had naturally generated another. Wallace, helped by Lawrence's 'interesting & philosophical' *Lectures on Man*, came to the same conclusion reached by Darwin twenty years earlier: since human races were analogous to those of animals and naturally originated, all species must have been made in the same way.[1]

Humans had emerged from apes, and *Vestiges* even predicted where 'the cradle of the human family' was located. So Wallace had headed there: the Dutch East Indies. In the forests of Borneo and Sumatra he had expected to bridge the gap between orangs and humans. Within a year had come his short synoptic paper (1855), which argued that one species could only derive from another, 'closely allied species'. At the time Blyth had tipped Darwin off about this little article. 'What do you think' of it? Blyth had asked. 'Good! Upon the whole!' 'Has it at all unsettled your ideas regarding the persistence of species'? Darwin of course was way ahead and hadn't paid much attention: 'nothing very new', he scribbled, 'Uses my simile of tree'.[2] He was too engrossed to read the signs.

Ten thousand miles from home, Wallace rushed in where Darwin feared to write. The birds in the Malay archipelago were his livelihood, but the 'human inhabitants' were no less a find. Soon after disentangling 'two quite distinct zoological provinces', one typified by placental mammals, the other by marsupials (later famously demarcated by 'Wallace's Line'), he also encountered 'two of the most distinct and strongly marked [human] races'. The brown, straight-haired Malays had come from the Asiatic mainland, the black, frizzy-haired Papuans from a Pacific land mass long subsided. Or so Wallace insisted against those armchair ethnologists who took the two races for one (and even Agassiz lumped them into one 'province') or mistook their 'cross-breeds' for linking or transitional races. Wallace was embedded among his subjects, living with the Dyaks:

> I am convinced no man can be a good ethnologist who does not travel, & not travel merely but reside as I do months & years with each race, becoming well acquainted with the average physiognomy & moral character, so as to be able to detect cross-breeds, which totally mislead the hasty traveller, who thinks they are transitions!![3]

Even though the Malays and Papuans now showed 'no traceable affinity to each other', he was convinced that they shared an ancestry.

In letters home he was already locking horns over the subject with a Morton disciple, the old Arctic-whaler-surgeon turned provincial doctor, Barnard Davis. Davis had just co-authored his own adulatory *Crania Britannica*, complete with picture of Morton on the title page. *Types of Mankind* had trailed it; *Indigenous Races* had praised it. Darwin had read it, although probably unsympathetic to Davis' intent, which was to keep Anglo-Saxon history pure and free from Celtic taint. While Wallace was sending Davis common-descent 'posers', what came back each time was the pluralist line, that it was impossible to '*people the earth from one source*' – in short, unity of descent was defunct. 'I am sick of him', Wallace grumbled to a friend; for a counter-argument 'read "Pritchard through, & "Lawrence's Lectures on Man" carefully'.[4]

Slavery was endemic in the archipelago. For ages, the supposedly 'savage' Dyaks with whom Wallace lived happily had been cruelly exploited, their men slaughtered, their women and children captured and sold by Malay traders. While the natives enslaved one another, the Dutch Creoles enslaved the natives, although the Dutch government had, Wallace noted, 'long accorded them legal rights and protection'. On the island of Ternate, Wallace's English-educated Creole host owned 'above a hundred slaves', and in February 1858 it was a slave crew, 'mostly Papuans', who rowed Wallace to the adjacent island of Gilolo to collect birds and insects. He fell ill with malaria. In a hut near a village occupied by Malay traders, 'shut in by low hills' where natives of Papuan descent raised rice and sago, he lay suffering, pondering the fragility of life, the struggle between racial groups, indeed the struggle of all life for the means of subsistence. Between fits, he remembered reading a book by Malthus about the problem of feeding the excess number of people born. He thought of how hard it was for booming animal populations to feed themselves, or for the Dyaks and Papuans to obtain enough, living almost in a state of nature. 'Why do some die and some live?', he wondered, and the answer came to him: those best fitted to feed themselves survive. From them come new competing groups and ultimately new species.

When his fever passed, he returned to Ternate and wrote up the idea on the 'two succeeding evenings'.[5] Off went the manuscript to Darwin in the first mail steamer. For weeks it would be stowed away on the high seas.

At the Darwins', amid the 'constant come and go of relations', Emma's radical Aunt Fanny Allen, still razor-sharp in her seventies, followed events in America with a seasoned eye. In her heart she believed 'emancipation will work its own way', but she thought it might 'go quicker' if the abolitionists were 'more tranquil'.[6]

The American situation deteriorated sharply with the 1857 Supreme Court decision, in the case of the fugitive Dred Scott, that Negroes were not citizens but property and had 'no rights which the white man was bound to respect'. It came two days after the new Democratic President used his inaugural address to denounce all abolitionists. Months later, the same James Buchanan, former ambassador to London, threw himself behind a pro-slavery Kansas, paving the way for a Republican successor named Abraham Lincoln. As Americans hurtled towards catastrophe, *The Times* retailed their predicament for English readers, including Darwin. 'Another weapon in the hands of the Abolitionists', it called the Dred Scott decision, whose extremism pushed moderates into the abolitionist camp.[7]

For the other side Harriet Martineau was railing against the United States in the radical *Westminster Review*, prophesying dis-Union as its 'manifest destiny' for sanctioning slavery. Darwin's old dining companion, the cigar-smoking Martineau remained close to the Wedgwoods and retained her acute moral conscience. Her purest invective was saved for the South's Cuban slave imports and barefaced calls to reopen the African trade. The execrable Dred Scott decision, embedded in the legal 'foundations of a fair-seeming structure of liberty', would prove devastating. It recalled how her own rights had been threatened with brute violence in the South. She hit back mercilessly at 'the slave power', berating the 'intellectual barbarism' of a 'race of bullies' in their 'crumbling mansions' who sent poor white 'ruffians' into Kansas with 'arms and bibles in their hands' to 'propagate the institution which had ruined' them all.

This out-and-out abolitionist took no prisoners. Long after her baptism into American politics she still wielded a deadly pen. From the Lake District, beneath a portrait of her hero William Lloyd Garrison, she poured forth *Daily News* leaders on the moral bankruptcy of the South, the empowerment of the Negro and the greatness of abolitionists. The fiery response to her *Westminster* article convinced her that 'the whole course of American politics is determined' by the slavery question.[8] Small wonder that her ear-trumpet was mistaken for a megaphone.

To cap it all, she was now a self-proclaimed atheist, one of the first that England's genteel ladies had ever met. She was proselytizing on this too, hence Darwin's quip, 'There is no god & Harriet is his prophet.' Her lapse from Unitarianism was accepted, more or less. Aunt Fanny did wince, but Erasmus sympathized and Charles was unruffled. Like his grandfather, he saw Unitarianism as 'a feather-bed to catch a falling Christian' – she had just carried on falling.[9]

They all seemed to thrive on Harriet's moral resilience, God or no. Fanny Wedgwood was a devoted admirer and they corresponded copiously between visits. Martineau in 1858 confided that she was writing another dangerously topical essay. (Fanny knew the 'extreme importance' of concealing Martineau's identity from angry Americans.) That summer Martineau's 'old friend' Charles Sumner, who had shown her around Boston, was in Europe for '*surgical* treatment' after his vicious beating by a Southern senator. In England he had met many of the Darwins' friends, including their neighbour Lord Cranworth at his residence near Downe. About to call on Martineau, Sumner urged her to bury America's doom-mongers by laying bare the 'real results' of West

Indian emancipation. As a Garrisonian purist, Martineau distrusted Sumner's political anti-slavery, but 'the spirit of his public life' helped make up for it, and she thought 'if he recovers, he will be President'.[10]

Martineau's new article was her first for the *Edinburgh Review*, staple reading in Darwin–Wedgwood homes for half a century. Darwin himself apparently had been approached to write for the stately Whig quarterly and mouthpiece of political anti-slavery. His student friend W. R. Greg, a leading contributor, sent the offer through Fanny – almost £1 per page, more than the *Westminster* – which stroked Darwin's ego, although he was no hack.[11] He declined of course. His 'present book on species & varieties' came first, and just as well. Though the 'Natural Selection' manuscript undercut pluralism and slavery in its own devastating way, nothing from him in the *Edinburgh* could have bettered 'The Slave Trade in 1858'. It was Martineau's sharpest exposé since 'The Martyr Age of the United States'.

Fifty years since abolition, the slave trade was reviving thanks to public complacency. A new generation found anti-slavery 'old and tiresome'. 'Modern views and fresh interests' were sweeping in. Southern propaganda, with 'rapturous references to Mr. Carlyle' and appeals to the new science – that groundswell culminating in *Types of Mankind* and *Indigenous Races* – claimed that 'negro servitude to the white man is not human slavery, but the normal condition of the inferior race'. Martineau counter-attacked by appealing to the examples of Mackintosh and Wilberforce – the Saints, not science.[12] There was little modern science available to subvert the slavery cause.

That science was Darwin's night and day concern. Continual work and worry left him exhausted and repeatedly fleeing to a fashionable hydropathic rest home in Moor Park, near Farnham in Surrey. For a week or a month, he would give up 'Natural Selection' to take the cure, not caring, he told Emma, 'one penny how any of the beasts or birds had been formed'. She knew better. 'I loiter for hours in the Park & amuse myself by watching the Ants', he confessed. One break in 1858 turned into something of a breakthrough as he finally caught a glimpse on the sandy heaths of 'the rare Slave-making Ant & saw the little black niggers in their Master's nests'.[13]

The first notice of slave ants in England had only been published four years previously. That was by Darwin's trusted informant, Frederick

Smith at the British Museum, who had found them nearby in Hampshire. In fact the slaving *Formica sanguinea* only existed in select areas of southern England. This had set Darwin hunting – and now his ants went off for Smith's confirmation, or perhaps Darwin took them up to London. Morning or evening was the best time to see a marauding party, Smith advised; and during the summer, when the pupae were in the raided nest. So each summer the invalid Darwin would experiment: he found fourteen nests in all, and set about disturbing them, dropping pupae outside, then watching how the tiny black slaves (half the size of their red 'masters') reacted. Always the slaves defended the alien red colony, protecting the nest, carrying pupae to safety, as if to suggest that they 'feel quite at home'.

He speculated on the evolution of ant slavery in 'Natural Selection'. The ancestral nest-raiders were carrying off pupae for food, only to have some of them hatch into workers. These duly set about working in the captors' nest. Their presence increased the 'fitness' of the colony and these egg-hatching colonies began to dominate, with selection favouring those raiders whose instincts were modified solely for abducting, but not eating, the '*nigra*' menials. Even further along this evolutionary path was a *Polyergus* slave-making ant which was 'abjectly dependent' on its captured menials. This slaver had 'no working neuters, but only warriors, or slave-takers, which have jaws incapable of building a nest' or feeding the young. These masters had lost the instinct to care for the colony: the slaves made the nests, fed the larvae and determined when to migrate. Bizarrely, as Darwin recounted in 'Natural Selection', these turncoat slaves even prevented their masters' marauding expeditions until the pupae of their own target species were ready! So it would appear that the *slaves*' instincts had been altered as well, to serve the captors' ends.[14] That would have been fatal to natural selection. Darwin resolved the conundrum by suggesting that the slaves' instinct to stop its own species from migrating at the wrong time had simply now manifested in preventing the masters from leaving on a raid during the wrong weeks.

While Darwin was naturalizing the slave-making instinct of *Formica sanguinea*, he was decrying the slavery rationale of *Homo sapiens* (white Southern variety). What made his work intriguing was that in the high summer ant-raiding months the family continued on its own abolitionist track. The social issues were unavoidable.

Three weeks before Darwin saw the first *defeat* of a marauding ant party, his eldest child, nineteen-year-old William, studying with a tutor in Norfolk for his Cambridge scholarship exam, posted home a *Times* report on Cuban slavery. No coaching was needed for this son of abolition to know its importance. On 17 June 1858, Bishop Samuel Wilberforce (son of the great abolitionist) brought before the House of Lords a petition from Jamaica calling for their lordships to compel the Spanish government to honour its treaties and end the slave trade with Cuba. 'Love of gain' lay at the root of this traffic, Wilberforce boomed to cries of 'Hear, hear' – unlawful gain 'purchased with . . . blood'. Christian Britain was 'bound by every obligation . . . most sacred' to use all means in her 'power to put a stop to the evils' of the trade.[15] The Bishop's tour de force ended with Aunt Bessy Wedgwood's hero, the octogenarian Lord Brougham, who had led their lordships' charge against West Indian slavery thirty years earlier, rising to lead a chorus of noble assent. It was a 'capital account of the Bishop of Oxford', Darwin responded, thanking William.

So steeped in the anti-slavery literature was Darwin that he slid easily (if not consciously) into 'Domestic Institution' terminology while peering underground. He was unabashed about it. This is how he explained his latest findings in Moor Park to Hooker:

> I have had some fun here in watching a slave-making ant, for I could not help rather doubting the wonderful stories, but I have now seen a defeated marauding party, & I have seen a migration from one nest to another of the slave-makers, carrying their slaves (who are *house* & not field niggers) in their mouths![16]

Smith had noted that the tiny *F. nigra* captives never left the nest, except to migrate. So he too called them 'Household-Slaves', and Darwin would attribute the idea of their being 'strictly household slaves' to Smith. But it is clear that Darwin had earlier fallen in with the language.

One might have expected him to make sure no metaphoric crossover was possible. Slave-making was, after all, 'so extraordinary and odious an instinct', as he admitted. And commentators have indeed seen him acknowledging a practice natural to ants 'while attempting to hold off any naturalization of human slavery'; and of using the South's descriptive vocabulary 'without allowing the behavior of ants to justify the practices of men'.[17] To the contrary, he actually makes no attempt to relate *or* unrelate ant and human behaviour at all – in truth he never mentions humans in the same context. However, because Darwin had always insisted that 'What applies to one animal will apply throughout all time to all animals,' and because he applied terms from America's 'peculiar institution' to ant colonies, an antagonist could argue, prima facie, that he was naturalizing both 'odious' societies.

The absurdity of such a conclusion was highlighted by Erasmus's perennial ant joke, that, said Darwin, 'I shall find some day that they have their Bishops'. That was the gist, of course. There were no abolitionist ants, no booming bishops in an ant Chamber, no looming Formicidae civil war over moral issues. Even if a little 'judgment or reason' did influence 'animals very low in the scale',[18] still ants were largely instinctive, almost automatons. Humans were seen as reasoning moral beings, with faculties allowing huge choice: the difference would have seemed so obvious to Darwin's readers that he simply had no need to flag it.

That choice was apparent now as never before. As he watched his red and black insects, all Britons watched the daily press, wondering when America's racial time bomb would explode, knowing that each tick, each Southern atrocity, each abolitionist attack brought the moment closer. No abolitionist front existed for ending ant atrocities.

Needless to say a new breed of determinists doubted that there was much leeway for choice, even in humans, but the Darwins held to their own views. Self-taught historian Henry Buckle, living, like Darwin, on an inheritance, and like Darwin a stickler for routine, had published his *History of Civilization in England* in 1858. Valiant heroes, glorious victories, slavery and its abolition – all of Albion's greatness was, to Buckle, the statistically certain result of natural laws of progress. Martineau, who had never believed in free will, loved his sense of fate; Erasmus sent her the book. Some deplored the implication that conduct was beyond individual control, others that all science and history was underwritten 'by one glorious principle of universal and undeviating regularity'. But, unlike Knox, there would be no racial primordiality for Buckle, only a natural Prichardian separation of nations to suit clime and time. Indeed his metropolitan liberalism went against the crasser racist dictates of the age (he dismissed talk of an idle 'Celtic race; the simple fact being, that the Irish are unwilling to work, not because they are Celts, but because their work is badly paid').[19] Who better, then, for Fanny and Hensleigh Wedgwood to invite to dine with Erasmus, the Darwins and the Hookers?

It was a mistake. Whatever Buckle's own plodding routine, he was quite unlike Darwin: the domineering Buckle roared all evening, filling the parlour with details of how he had indexed and compiled facts from his 20,000-volume library. Darwin, a fact-indexer himself, eventually read Buckle's *History* twice but doubted whether its 'laws' were 'worth anything'. Everybody considered Buckle a crashing bore. Emma nursed her own nefarious thought: if 'personal character' didn't count in history, how to explain 'the abolition of slavery'? Would 'Buckle . . . say that if Garrison had not arisen' in America, history would have thrown up 'somebody else' of equal stature to purge the sin?[20]

As William's account of the Bishop's speech arrived at Down House, so did a more ominous package. Wallace's manuscript from the Far East turned up on 18 June 1858. Darwin's routine existence was jolted. That day too his daughter Henrietta came down with what was at first thought to be tonsilitis, only for it to be diagnosed as infectious diphtheria. Guests and visiting family – ex-governor Robert Mackintosh included – rushed away for safety. Illness was the family's besetting woe, which Darwin could cope with. What agonized him was Wallace's manuscript.

Lyell had half expected this bombshell. Hadn't it been Wallace's short 1855 paper that had prompted his advice to Darwin to publish, lest he be scooped? Your 'words have come true with a vengeance', conceded Darwin. Wallace's manuscript read like a revelation of the theory he had refined obsessively for twenty years. Wallace might not have mentioned 'selection', he might have denied any analogy between wild and domestic races, but in Malthusian respects of overpopulation, struggle and differential survival, the theory looked similar. Darwin read into it what he feared, a scooping. 'Eleven long chapters' of 'Natural Selection' were in draft – 225,000 words – with about three left to write; the ten pages on sexual selection had been slipped into the manuscript at the start of chapter six, with other additions. After two decades of trepidation, Darwin stood on the brink, only to groan to Lyell, 'All my originality . . . will be smashed'![21]

A flurry of letters led to a gentleman's arrangement. Hooker and Lyell secured Darwin's priority with the joint reading of extracts from his 1844 essay and 1857 letter to Gray, and of Wallace's manuscript, at the Linnean Society of London on 1 July 1858. The papers were published in that order (not that Wallace knew, out in the Far East). This meeting, the world's premiere of natural selection, passed almost unnoticed; 'sexual selection' was not mentioned by name, let alone as an explanation of human racial divergence. Darwin talked in passing of a 'second agency' producing changes, 'namely, the struggle of the males for the females' (there was no clue that he was beginning to swing to female choice) – a struggle by means of battle, beauty or song to win favours. Tussling between males led to sexual differentiation, but there was no elaboration or explanation, and no mention of selection in the formation of racial traits.

Within weeks Darwin was boiling down his huge manuscript for quick publication. At first he thought it would make a 'pamphlet'; then, as the foolscap piled up, a 'popular' book. Lyell floated a proposal to his own publisher, John Murray, with Darwin promising 'my Book is not more *un*-orthodox' than needs be, meaning 'I do not discuss origin of man' or 'bring in any discussion about Genesis'. Murray didn't care for 'Natural Selection' in the title, but Darwin insisted on it because 'selection' was 'constantly used in all works on Breeding' – and this was a book that would begin with fancy pigeons. They settled on 'Natural Selection or the preservation of favoured races', which Murray liked. Everybody knew about race, and not just the pigeon variety. Racial Anglo-Saxonism was sweeping Britain and America, with proselytizers such as Knox and Nott only the extremist tip of a 'manifest destiny' ideology that was digging deep into both cultures. The plight of one enslaved race in America was about to lead to war, and was stirring abolitionist breasts at home, thanks to *Punch* cartoons and a million copies of *Uncle Tom's Cabin*.[22] Racial contact, racial preservation, racial fate were the great arguing points of the age. *On the Origin of Species by Means of Natural Selection, or the Preservation of Favoured Races in the Struggle for Life* may not have discussed the 'origin of man', but a signifier word in Darwin's title, 'races', put the book at the centre of the greatest moral debate of the moment.

It was perhaps as well that Darwin's changing views on sexual selection and the blocking of Blyth's letters left his anti-Agassizian alternative – blacks and whites differentiated by selective mating – for later publication. As an account of creation, the *Origin of Species* would run 'slap counter to Genesis' anyway, so any talk of the human races could further compromise the hearing for natural selection. Yet those who knew Darwin's science would have realized how traditional, in one sense, it was. Muddy-booted breeders, with their hard-won craft knowledge of heredity and 'long lines of descent', took pride of place in the *Origin*. Why shouldn't 'species in a state of nature' be considered 'lineal descendants of other species', he asked, just as 'many of our domestic races have descended from the same parents'?[23] His premise was still that *all* races ultimately stemmed from common parents, including mankind's by implication.

Not every insider caught Darwin's drift as readily as Lyell did. Huxley hated slavery but was prepared to give Nott and Gliddon's *Types of Mankind* a hearing. He was a better hater than most, and for him Agassiz stood alongside Richard Owen and the bishops in his demonic pantheon. Darwin had to agree that Agassiz's 'superficiality & wretched reasoning powers' had reached rock-bottom in his *Essay on Classification* (1857). Having thanked Agassiz for 'this magnificent present', Darwin about-turned and laughed with Huxley at its 'utterly impracticable rubbish'. Imagine a Harvard professor in 1857 claiming (as Darwin scribbled in his copy) that the 'idea of sp[ecies]. proceeding from single pair [was] almost given up by all naturalists!'[24]

But Huxley had his own mental 'rubbish' to dump. He was moving fast, but not fast enough for Darwin. Huxley still doubted life's progression through geological time; he thought that a dinosaur-age man might have speared ancient marsupials 'as the Australian niggers' do now – in short, that mankind might have persisted unchanged for aeons, making the search for origins futile.[25] He could not get to grips with ancient animals being more 'generalized' than their living descendants. Huxley found Darwin difficult. Softly, softly, as always, was Darwin's way, as he Socratically probed Huxley's thought. But it was hard work, as Huxley dismissed Darwin's 'genealogical trees' for animals and plants: 'Your pedigree business', Huxley responded, 'has no more to do with pure Zoology – than human pedigree has with the Census'. This was as bad as Agassiz, totting up all the species in a place rather than tracing their family ties. Darwin was less interested in a zoological headcount; common descent to him was the *real* 'plan on which the Creator . . . worked', so eagerly sought by so many. He shot back with a killer question:

Grant all races of man descended from one race; *grant* that all structure of each race of man were perfectly known – grant that a perfect table of descent of each race was perfectly known – grant all this, & then do you not think that most would prefer as the best classification, a genealogical one [?]

What held for human races held for every race. Dare Huxley deny that all humans were related? 'I sh^{d}. like to hear what you w^{d}. say on this purely theoretical case', wrote Darwin. 'Generally, we may safely presume, that the resemblance of races & their pedigrees would go together.'[26] Darwin's bulldogs required the stiffest canine training.

Lyell himself pressed Huxley hard. If fossils *did* show life 'progressing' after all, he warned, then mankind might turn out to be the descendant of a long series of primates. 'A race of savages . . . with small cranial development' would have appeared first, '& out of this the negro & white races & others . . . extinct or yet to come' would be 'evolved in the same way as permanent varieties are formed'.[27] Where the *Origin* was silent, Lyell spoke – and he was reading the writing on the wall. His worry now was how to revise his own *Principles of Geology* to keep up.

In the rush to publish, Darwin's reading habits changed. He normally got through scores of volumes each year. In these months, his reading lists grew shorter until finally he stopped recording. Much that was 'non-scientific' was now read to him by Emma, and their choice of literature at this racially tense time revealed enduring concerns in their engagement with events in America.

One book haunted Darwin more than any other. Everybody recommended Frederick Law Olmsted's trilogy on his travels through the South. It was a small world. Olmsted was a close friend of the Harvard botanist Asa Gray's in-law Charles Loring Brace (named after the fugitive-slave-defending uncle); and on a walking tour of Britain, Gray had shown Olmsted and Brace around Kew Gardens. Gray would commend *A Journey in the Seaboard Slave States* (part of a trilogy) to Hooker too. Slavery might be more tragedy than treachery to Olmsted, but he tackled Southern society with no holds barred. Aunt Fanny pressed the *Journey* on Emma and Charles, knowing it would 'give you great pleasure'[28] – meaning it would reinforce their convictions.

Darwin started the Texas volume of Olmsted's trilogy while recovering at Moor Park part-way through writing the *Origin*. This taster on slavery's 'disastrous' appearance in one state steeled him for the Slave-States *Journey*, though nothing he had ever read prepared him for the shock. At every turn, slavery's power to corrupt appeared with agonizing clarity. By weaving together press reports and interviews, Olmsted recreated the immediacy of real life.[29]

One understands Darwin's sympathy. Olmsted shared his passion for the smallest detail, his curiosity about human nature – mulattos intrigued him – and his belief in the equal dignity of the races. In Virginia he stood with poor Negroes at a child's graveside, their leader speaking eloquently, the women weeping. Then there were the man, boy and girl led to auction like livestock, on the end of a rope, handcuffed and lashed by a 'sleet storm', their ragged clothes drenched. Olmsted took a train south with slaves bound for market, like cattle, each woman valued less as a labourer than a 'brood-mare'. One farmer complained of losing 'one of his best women ... in child-bed just before harvest', also the infant, and this 'came very hard upon him', as he would not have sold such stock for 'a thousand dollars'. In the more northern Slave States, Olmsted noted, 'as much attention is paid to the breeding and growth of negroes as to that of horses and mules'.[30] The image of goateed Southern gentleman Josiah Nott, fancier of horse stock and slaves, was never far from these pages.

Such were the evils sanctified by polygenism. To a gentleman comfortably ensconced in the English countryside, Olmsted's images were a reminder of the poisonous fruits of Agassiz and Nott's philosophy. Catching 'niggers' was like so much field sport. Olmsted's pages were turned by a gentleman who now had no stomach for hunting even a bird himself. '*No particular* breed of dogs is needed for hunting negroes', Olmsted recorded; 'blood-hounds, fox-hounds, bull-dogs, and curs', all could be 'trained for it'. One advertisement from the *West Tennessee Democrat* offered 'BLOOD-HOUNDS ... They can take the trail TWELVE HOURS after the NEGRO HAS PASSED, and catch him with ease'.

Hounds were Darwin's forte; to prove the breeds had a common descent was to prove slave and master had a common descent, and the upshot would finally end such atrocity, as polygenism fell with its perpetrators. He never forgot the 'horror of his sleepless nights' in early 1859 as he read Olmsted.[31] And by then he was writing the last chapters of the *Origin of Species*.

'Light will be thrown on the origin of man and his history.' One short sentence in the *Origin*'s final pages covered him against accusations of concealing his deeper beliefs. He knew everyone would read 'mankind' into the book. Those twelve words meant that 'Man is in [the] same predicament with other animals'. This would be an open secret, as he admitted to Lyell.[32]

But the full explanation of human racial origins was omitted because Darwin lacked the overwhelming evidence to convince a sceptical world. As it was, he had to plead that he came to natural selection without ulterior motive, seeking only the truth. But no special pleading would help if he published on humankind. A crushing weight of evidence would be needed to show not only sexual selection splitting the races, but how the ancestral humans were descended bodily from beasts – vastly more than he could muster in such haste.

Only a third of 'Natural Selection' had been shoehorned into the *Origin*. It had been his plan for the 'big Book' to cover human racial origins, with one volume's worth, at least, on domestic races, hybrids, mongrels and artificial selection. This seemed critical as slavery propagandists published their own huge books separating the many human species and America plunged towards war. But in the rush to the *Origin* and priority, Darwin's race-interests were necessarily put on ice. Edward Blyth's dried-up well of evidence was partly to blame; for, as Darwin said in the *Origin*, paying handsome tribute, he valued Blyth's opinion on racial ancestry 'more than that of almost any one'.

Darwin slipped another telling sentence into a passage on non-adaptive differences between domestic breeds. 'I may add that some little light can apparently be thrown on the origin of these differences' – including differences between the 'races of man' as domestic breeds – 'chiefly through sexual selection of a particular kind, but without here entering on copious details my reasoning would appear frivolous.' It was understated in that quiet English way: 'some little' meant quite a lot. But the race-making power of sexual selection was to be kept under wraps, even if in another passage Darwin teasingly revealed the kind of sexual choice he had in mind.

We know he believed that human males had seized the power of choice, though he would not say so in the *Origin*; selecting females for *their* beauty, just as 'man can give elegant carriage and beauty to his bantams'. Then he looked beyond the male alligators behaving 'like Indians in a war-dance', or cock birds singing and displaying, to the *females* judging these strength or singing pageants. He now hinted that they were the ones who 'at last choose the most attractive partner'. Females, by 'selecting, during thousands of generations according to their standard of beauty', in effect were responsible for picking male gaudiness.[33] His mechanism for differentiating races from one another was still being thought through, but either way, male or female choice, animals in the *Origin* acted as their own breeders, their own race-makers. Then he clammed up again: 'I cannot here enter on the details necessary to support this view.'

His cousin Francis Galton, African traveller and commentator on Hottentot backsides, wanted to know more about this sexual choice. Darwin confessed himself uncertain 'about the share males & females play' until 'I compare all my notes'. More digging was needed, more evidence. He feared being 'cut up' about males and females, birds and butterflies, before he ever got to the subject of human unity.[34]

Man was untouchable in the *Origin*, the human races too sensitive, save for the two cryptic light-throwing passages. But Lyell knew that the book *was* ultimately about man. He knew that as natural selection was granted, white humanity's fortress would be breached. He admitted it while reading the proofs, telling Darwin: 'It is this which has made me so long hesitate always feeling that the case of Man & his Races & of other animals & that of plants is one & the same & that if a "vera causa" [true cause] be admitted for one . . . all the consequences must follow.' Fearful still, he wanted moral safeguards. Above all, he needed God to guarantee a lofty pedigree by using 'creative power' to underwrite selection in changing 'an orang into a Boscheman . . . to a Newton'. By having 'the White man [supersede] the Negro', God would show that His chosen race was in His mind from the outset.

No such hauteur for Darwin, though. God wasn't involved in creating white superiority, as Lyell's Southern gentry believed. Evolution elevated the human mind by operating on 'the races of man; the less intellectual . . . being exterminated'. This looked like the line he had taken in his post-Malthus notebooks: that some races, notably the white, were more 'intellectual' and would supplant the natives during colonization. But apparently not, for Darwin took exception to Lyell's example of competition between the 'negro & white in Liberia'. Struggle on foreign soil was not primarily between single races or species but among whole 'groups or genera' of invaders taking on the locals. He also said he was thinking of 'the almost certain future extinction of genus Ourang by genus Man, not owing to man being better fitted for climate, but owing to the inherited intellectual inferiority of the Ourang-genus – man-genus by his intellect inventing fire-arms & cutting down forest'. It was as if Darwin was frightened by his own consequences. The 'intellectual' line – if it meant those with guns – had ominous implications for racial contact. What is the point of freeing the slaves only to naturalize their death? Darwin, as always, hastily shut down the subject, declaring again that there was no 'space to discuss this point'.[35] It would be the dilemma for the 1860s.

Lyell wouldn't let go. How then to explain 'the distance between the European, Negro, Hottentot & Australian races'? If climate was discounted, did competing intellects alone account for the diversity? (He still knew nothing of sexual selection.) He thought again of Divine intervention, which raised the spectre of Agassiz's separate racial creations: 'I . . . am told that they probably may have sprung from several indigenous stocks or species settled in remote & isolated regions'. In desperation Lyell was resorting to Agassiz's primordial races as a last ditch attempt to retain some white dignity. Indeed, didn't the *Origin* play into Agassiz's hands by allowing that 'our dogs have descended from several wild stocks'?[36] Didn't this subvert the beautiful proof that fancy pigeons, like people, had a common ancestry?

It was a valid point, and not an easy one for Darwin. 'You overrate [the] importance of multiple origin of dogs', he replied. 'I sh^{d} *infinitely* prefer the theory of single origin in all cases; if facts would' allow, but they didn't. Darwin's unitarist commitment ran deep, and the pigeons were his simple exemplar; dogs complicated his public presentation. He did believe that domestic dogs came from a number of wild species. These wild dogs must have hybridized successfully, or done so after being domesticated. Race differences among their descendants were thus part natural and part artificial. But all canines came *ultimately* from 'some one very ancient species' – it was just further back. Darwin told his sister Caroline, worried by the same problem, as much: the hybridization of dogs was a 'distinct question' from 'whether these wild species have descended from one aboriginal stock as I believe has been the case'.[37]

Darwin sometimes felt like cutting an inflated Lyell down to size, but his old mentor was still an ally and had to be kept sweet. All canines came from one progenitor, and for Darwin 'the only question is whether the whole or only a part of difference between our domestic breeds has arisen since man domesticated them . . . The Races of Man offer great difficulty: I do not think [the] doctrine . . . of Agassiz that there are several species of man, helps us [Darwin was tactfully including Lyell] in the least . . . Much too long a subject for letter.' Darwin was losing patience. 'With respect to the Races', he said, 'I have one good speculative line, but a man must have entire credence in N[atural]. Selection before he will even listen to it.' That ruled Lyell out. He was not to hear how sexual selection made the races because his belief in natural selection was in doubt, even if he had come some way on transmutation.[38] Nor it seems did Darwin tell Hooker and Gray. Spelling out sexual selection prematurely would forearm critics, hurting its chances of success. Potentially, Darwin had solved a problem polarizing science, fuelling racial antagonism and dividing the human family; and no one cared more about these things than he. With the United States marching to war, the time might have been ripe to set out all the implications of sexual selection; but no, he waited. Judging by Lyell the world would have enough trouble swallowing natural selection, and Darwin lacked the copious detail needed to finish off sexual selection.

He was also a sick man. For years he had been regularly, often wretchedly, ill. The closer to 'man' and to publication, the worse he became. Five times while writing the *Origin of Species* he was forced to decamp to a rest home to take the water cure, his nerves wrecked. 'No nigger with lash over him could have worked harder', he explained as he struggled with his prose. But the real cause 'of the main part of the ills to which my flesh is heir', he admitted, was the *Origin*'s inflammatory case for the evolution of life by a chancy natural selection, and the expected uproar over its bestial implications. He dreaded being 'execrated as an atheist'. For a respectable gentlemen, for whom reputation and honour were everything, it was barely endurable. Later, at his spa, sending out copies of the *Origin*, it was 'like living in Hell'.[39]

Just as he was putting the *Origin* to bed in 1859, a beehive oozing honey – literally – arrived at Down House. It was from another trusted informant: 'any-thing and every-thing I have are at your Service', the gentleman from Jamaica wrote. Birds and reptiles blown in by storms, local fish and wild fowl, even captive parrots that wouldn't breed:

Richard Hill knew about them. He had even sent a pamphlet on pigeon breeding. Hill, who wrote about the birds while also convalescing at a seaside spa, was clearly a man after Darwin's heart.[40]

Darwin had begun piecing together Hill's story after Philip Gosse vouched for his good offices. Gosse, naturalist and inventor of the sea aquarium, had lived in Alabama and left precisely because of its torture, floggings and man-hunts. Relocated in Jamaica, he joined Hill in writing the standard work on the island's birds, which Darwin had read. Darwin could tell that Hill and Gosse shared a narrow biblical view. Gosse's bizarre *Omphalos* (1857) explained away the marks of prehistory and fossils in the rocks as illusions created by God – Adam's navel was the emblematic illusion (Adam having no mother, as life had no ancestry). *Omphalos* might have been more tendentious than Hill's *Books of Moses: How Say You, True or Not True?* but the same evangelicalism inspired them both.[41] While Gosse preached to a Moravian congregation in Jamaica and blessed mixed-race marriages, his friend Hill presided over Baptist missionary meetings.

The young Hill had followed his father – a Montego Bay merchant from Lincolnshire – in the anti-slavery cause. As a young man he had brought a petition to Westminster from Jamaica's coloured population and in 1827, after lobbying by the Saints, was allowed to stand 'within the bar' of the Commons at its presentation – which was quite uncommon, because Hill's own mother was of African descent. He worked for several years alongside Thomas Clarkson and other allies of the Darwin–Wedgwoods, then visited Haiti on behalf of the Anti-Slavery Society to report on Toussaint l'Ouverture's young republic. With emancipation, Hill had returned to Jamaica to be appointed the island's first 'coloured' stipendiary magistrate, charged with the task of adjudicating between the former slave-holders and their new 'apprentices'.[42]

That might have been the sum of Darwin's knowledge about Hill but for a coincidence. At the water-cure Darwin now found a fellow invalid, Judge William Wilkinson, on leave from the Supreme Court of Jamaica. Inevitably, the conversation turned to the law. Darwin might have cried off jury service, 'incapable of the fatigue of one trial', but that had not stopped this member of the squirearchy sitting as a county magistrate for the last two years. Indeed, in previous weeks Darwin had passed sentences for poaching, malicious damage, assault and other petty crimes.[43] Comparing notes with the judge, he found they had much in common. Wilkinson had chaired the Jamaican quarter sessions and had sat as judge in Middlesex County where Hill was a magistrate. Darwin now heard how, after emancipation, Hill had settled disputes and enforced justice for the apprentices – in fact, he had investigated everything from the atrocities perpetrated aboard a captured slave ship to the cause of black Baptist 'rioters' coming to blows over a chapel.[44] He had been made a Privy Councillor in 1855, despite the governor Henry Barkly's 'complexional prejudice'. The bigoted Barkly had been forced to honour Hill, the *Anti-Slavery Reporter* said: the preferment was 'not one of grace, but compulsion', 'almost a state necessity', given the honourable Mr Hill's 'superior capacity and great merits'.[45]

Jamaica's first non-white magistrate had achieved what Charles Darwin, JP, would have done in his place – what Emma and the whole family wished done – in righting historic wrongs, dressing the wounds and healing the scars that Hill's own people had suffered. They would never meet, Darwin and Hill, but a quiet bond began to grow, surpassing even the fraternity of the bench. Thanking him for the beehive, Darwin struck a rare note, reserved solely for old friends – he became personal: 'Your letters have excited in me much interest about you'. 'I was quite delighted (if you will not think it impertinent in me to say so)' – the parenthesis came automatically, as when touching on politics or moral matters – 'to hear of all your varied accomplishments and knowledge, and of your higher attributes in the sacred cause of humanity'.

Then Darwin started his typical quizzing: did livestock '*long* bred' in Jamaica 'tend to assume any particular colour, or other character'? And what about people? He was back to the way skin and hair colour gave advantage to a race. Everybody was asked – did it relate to disease resistance? Hill too was taxed on whether those bred outside the West Indies, 'pure *Europeans*', tended to suffer more from 'Yellow Fever or other tropical complaints' if their 'complexion and hair' were light or dark? Hill had no idea why Darwin wanted to know, though he would soon discover. 'My little Book [the *Origin of Species*] . . . shall be sent you. I fear you will not at all approve of the results arrived at, but I hope and believe that you will give me credit for an honest zeal for truth.'[46]

No appeal to 'the sacred cause of humanity' would spare Darwin a backlash: a lashing for giving blacks and whites a common humanity, and a lashing for tainting that humanity with ape blood. Rare humility was needed to see humans 'created from animals' as he did. Losing Adam and Eve and the miraculous days of Genesis might be more than Hill and other anti-slavery traditionalists could bear, but Darwin offered them a deeper unity, a 'more humble' theology and more spectacular evolutionary 'grandeur' in its place.[47] He chose an epigraph for the *Origin* from the highest moral authority he could think of, the Revd William Whewell of Cambridge, who taught that events in the material world are brought about, not by isolated 'interpositions of Divine power . . . but by the establishment of general laws'. Darwin's God was growing more distant. He was not involved personally with each creation; He was delegating.

It took a certain naivety to send a copy of the *Origin* to the Revd Adam Sedgwick (Darwin's old geology professor) in early November 1859, knowing that he was 'diametrically opposed' to transmutation. And to Henslow: 'I fear you will not approve of your pupil', said Darwin, trying to cushion the blow. With others the book would be more welcome, and copies went off to Hooker, Lyell, Huxley and Gray; the family – sisters and cousins, Fanny and Hensleigh, brother Erasmus (who whisked a spare copy to Harriet Martineau), not forgetting radical Aunt Fanny Allen, all eager readers. But the first copy to be sent this November went to Agassiz – the man revolted by blacks and averse to common descent. Darwin's disarming note was hardly going to disarm, but he tried anyway. The book's conclusions about consanguinity and racial unity 'differ so widely from yours', Darwin admitted, that 'you might think . . . I had sent it out of a spirit of defiance or bravado; but I assure you that I act under a wholly different frame of mind'.[48] What that was he did not say.

12

Cannibals and the Confederacy in London

The *Origin of Species* came into an Anglo-Saxon world on the brink of war. The first great republic founded on principles of equal liberty and justice was about to be ripped apart, its violent interiors exposed.

Eight days after the *Origin*'s publication, John Brown – guerrilla abolitionist and Kansas veteran – was hanged for raiding the federal arsenal at Harpers Ferry, Virginia, while attempting to arm a slave rebellion. Ten insurgents died; his black co-conspirators were executed. The mass uprising, secretly funded by abolitionists, never had a chance. Brown went to the scaffold calling for human sacrifice, blood atonement for the nation's sin. In England, Martineau pronounced him a 'martyr'. That same day in his pacifist pulpit William Lloyd Garrison demanded again the 'dissolution of this slavery-cursed Union', for 'What concord hath Christ with Belial?'[1] The sacred cause had never seemed so holy, or its devotees so self-righteous, even as carnage in Kansas and fugitive slave-hunting made life hell on the ground.

This became the backdrop to the *Origin*'s reception and the crucial questions over Darwin's unexpressed beliefs: how did humans fit into the book's argument? Where did Darwin's science stand on race relationships? For Darwin, the latter was unfinished business, but readers were not to know that. He was *still* sending out questionnaires via the army on skin colour and fever resistance. And he still planned to publish the unused part of 'Natural Selection' – his encyclopaedic data on the variation and crossing of domestic animals and cultivated plants – even though this work was on the back burner. But incongruously, he threw himself into a study of orchids after publishing the *Origin*, and then plant fertilization. Onlookers tried shunting him back to the big question. Erasmus told Emma how much Lyell 'wished the Dogs to be published [the wild and domestic dog descent that bore so closely on the human question] & Huxley the same, & that even Hooker grudged Charles employing his time on plants instead of animals'.[2]

Would he ever step out of the shadows on 'man'? He told the Saxon-Supremacist and descendant of slave-owners the Revd Charles Kingsley that the 'grand & almost awful question' of human genealogy was 'not so awful & difficult to me', having become inured to the idea, but he knew that it was to many others. Kingsley had raised the spectre in a private letter. Darwin candidly admitted that he had 'seen a good many Barbarians' in Tierra del Fuego and it was this sight of a 'naked painted, shivering hideous savage' which had sparked the initial insight into his own grovelling ancestry. It was 'as revolting to me, nay more revolting than my present belief that an incomparably more remote ancestor was a hairy beast'.

He recognized the fear that black blood instilled – he was, after all, writing to a man whose grandfather was a West Indies planter and Barbados judge. The family had been ruined by emancipation, and Kingsley felt he 'owed the Negroes nothing because they already "had all I ever possessed"'. It showed. Kingsley himself believed the 'lowly' races were Providentially doomed and that the whites would sweep out all before them to usher in God's Kingdom (and he was happily invoking natural selection in the cause). The sentiment was widespread. Even Darwin agreed to the gruesome prospect: 'It is very true what you say about the higher races of men, when high enough, will have spread & exterminated whole nations.'[3] There was a fatalism to the statement. While slavery demanded one's active participation, racial genocide was now normalized by natural selection and rationalized as *nature's* way of producing 'superior' races. Darwin had ended up calibrating human 'rank' no differently from the rest of his society. After shunning talk of 'high' and 'low' in his youthful evolution notebooks, he had ceased to be unique or interesting on the subject.

He declared to Kingsley that he had 'materials for a curious essay on Human expression, & a little on the relation in mind of man to the lower animals. How I sh[d]. be abused if I were to publish such an essay!' And there lay the crux. There was a nagging feeling that Darwin was looking for someone else to relieve him of the need. 'I hope & rather expect that Sir C. Lyell will enter in his new Book on the relations of men & other animals',[4] he added, meaning that he would be saved from having to expose himself.

The great hope was Lyell, his mentor for thirty years. Darwin's hope never dimmed, even though Lyell had struggled with the idea that the 'gradations of [human] intellectual powers' could result solely from 'continued natural selection of the most intellectual individuals'. Lyell was moving slowly, but the immensities for him were overwhelming. The need for God to introduce some 'principle of improvement' in evolution was overpowering. What prospect of an afterlife if no moral principle had been added in the past? But Darwin 'would give absolutely nothing for [the] theory of nat[ural]. selection, if it require miraculous additions at any one stage of descent'. He had a touching faith that, with Lyell shifting a little towards human evolution, he would 'go the whole orang'. Hence the encouragement, with Lyell now working up his own book on the 'Antiquity of Man' (Lyell was belatedly visiting all the flint-implement sites): 'You used to caution me to be cautious about man, I suspect I shall have to return the caution a hundred fold! Yours will no doubt be a grand discussion; but it will horrify the world at first more than my' *Origin of Species*.

Lyell's book might draw the poison. Others were less sanguine: Martineau's outburst to Erasmus showed her view as America went to war. She wanted 'to know how *the Lyells* take this American business. They were so completely in the midst of the guilty, unpatriotic, virtually treacherous set, – Ticknors, Appletons &c, that their minds seemed to me darkened or twisted on political subjects.'[5] Darwin could only hope that Lyell had sloughed off his Southern sympathies as he wrote on common descent.

America's struggle would not be for the 'preservation of favoured races' so much as the preservation of a nation, a Union, in which the right of one race to hold property in another was disputed. In this 'struggle for life' – to end or defend slavery as a way of life – the abolitionists' 'favoured race' was black. And so it was for Charles Darwin.

Slavers still plied the Southern coasts. Martineau had exposed the traffic, but fresh cargoes continued to land, black hands for turning white cotton into gold. Their masters hauled the slaves west to fill orders from the Northern and English mills. No compromise or outrage could halt this westward spread. This was a 'house divided' against itself that could not stand, Lincoln declared memorably, but he cared less for abolition than for the Union. As the 1860 presidential race got under way, 'irrepressible conflict' was the watchword in London's *Times* as in New York's. Charles Sumner did his best to keep it that way. Back on his feet in the US Senate, he lambasted 'the barbarism of slavery', its violence and injustice. Quoting Scripture and Olmsted's Slave-States *Journey*, he would not dream of voting on 'whether fellow-men shall be bought and sold like cattle'. Morality was not a voting matter.[6] With Martineau backing him in New York's Garrisonian *Anti-Slavery Standard*, he too called for abolition.

The *Origin of Species* uplifted Martineau at this time. To her, Darwin and Sumner both stood for liberation, the freeing of enslaved minds as well as bodies. The *Origin* had too much 'creation' in it for an atheist, she told Fanny Wedgwood – too many nods towards God. But its argument for natural species-change gave her 'unspeakable satisfaction', and she thanked Erasmus for her copy, which let her look again into his brother's upright mind. It had been two decades since their tête-à-têtes in Great Marlborough Street, when she had first glimpsed its 'earnestness and simplicity, its sagacity, its industry'. Erasmus pronounced the *Origin* 'the most interesting book I ever read', full of reasoning so forceful, so necessary, that 'if the facts wont fit in, why so much the worse for the facts'.[7]

Other old Prichardians made their pitch. Carpenter spoke to fellow Unitarians in his *National Review* piece on the *Origin*. From the start he confronted theologians with Darwin's analogy: was it a religious question 'whether our breeds of dog are derived from one or from several ancestral stocks'? Would anyone object 'if it could be shown that the dog is really a derivation from the wolf', and that from something earlier? Traditionalists saw the races trace their ancestry all the way to Adam and Eve. Why then, after admitting so much change, should they fault the *Origin* for allowing more? And even 'orthodoxy (on this side of the Atlantic at least)' wanted *consistent* abolition, not just of slavery but in natural history: 'the abolition of the two-and-twenty species into which man has been divided'. This was one immediatist who got Darwin's anti-Agassiz drift. Carpenter himself had moved much of the way with Darwin: he 'does not go quite as far as I – but quite far enough', Darwin told Lyell; 'for he admits that all Birds from one progenitor; & probably all fishes & reptiles from another parent. But the last mouthful chokes him – he can hardly admit all Vertebrates from one parent'. Still, 'we shall conquer' ultimately, with such a 'great physiologist on our side'.

Even Catholics understood the 'naturalness' of the connection between human genealogy and Darwin's common-ancestry approach. 'These thoughts are not new to us', the *Dublin Review* said. Darwin's unity of descent among species 'furnishes a fair parallel to what has taken place amongst the citizens of the world', the 'unity . . . and variety of form, of our domestic animals' being like 'the differences amongst our fellow-men'.[8]

But Bishop Samuel Wilberforce spoke for most believers. God's goodness was not to be saved at the expense of His providence, nor the unity of the races by making man a beast. The *Origin* held more loss than gain for those with a faith unshaken by Christian complacency over slavery. Moreover Wilberforce – son of the most famous name in anti-slavery – would now try to turn the tables on Darwin over the slave ants.

Ironically, as Darwin reported, 'more notice' was 'taken of Slave Ants in the Origin than of any other passage'. While he did not intrude his anti-slavery views into the *Origin* when describing ant colonies, he privately refused to accept a parallel. Years later Downe's Tory vicar, the Revd John Brodie Innes, said that he had never been persuaded by Darwin's books: 'I hold to the old belief that a man was made a man though developed into niggers who must be made to work and better men able to make them, if those radicals did not interfere with the salutary chastisement needful, neglecting the lesson taught by the black ants slaves to the white'. Darwin issued an outright contradiction: 'my views', he retorted, 'do not lead me to such conclusions about negroes & slavery as yours do: I consider myself a good way ahead of you, as far as this goes'. Even if Darwin ritually washed his hands in the *Origin* by describing ant slavery as 'so extraordinary and odious an instinct', he still carelessly slipped into slave-owing terminology in talking of 'black' 'household slaves' toiling for the British *Formica* slave ants.

It allowed Wilberforce to practically accuse him of believing 'that the tendency of the lighter-coloured races of mankind to prosecute the Negro slave-trade was really a remains . . . of the "extraordinary and odious instinct" which had possessed them before they had been "improved by natural selection" from Formica Polyerges into Homo.' This was the unkindest cut, but Darwin had let himself in for it by not drawing out the distinction in his book. He had not said what he thought: that ant behaviour was honed by a blind natural selection, and this was no morally culpable agent, even if it created 'instincts causing other animals to suffer'.[9] In humans, slavery was not an instinct. That was the difference. There *was* culpability here.

It took the radical Unitarian preacher Theodore Parker to understand Darwin. Parker now lay dying of tuberculosis in Italy. He had been one of the 'Secret Six' who had helped fund the Harpers Ferry raid, and who believed that a slave had a natural right to kill his captors. He was beyond prosecution when news of the *Origin* reached him, and he immediately turned its argument against the slavery-science he knew in Boston. Darwin subverted 'Agassiz's foolish notion' of a divine miracle taking place whenever the earth needed a 'new form of lizard' – or a new human race. A God 'who only works by fits and starts is no God at all.' Darwin and Parker were alike, both physically racked. Darwin's worn countenance actually resembled Parker's, or so thought the Swiss firebrand and Darwin-admirer Karl Vogt, who set their photos side by side.

In Mobile, Nott got Darwin's message too: 'the man is clearly crazy'. But as a parson-skinner Nott at least saw that the *Origin* 'stirs up Creation and much good comes out' of that.[10]

Even before the *Origin* had appeared, Asa Gray, the Harvard botanist who had been briefed by Darwin, had started proselytizing at a private meeting of the American Academy of Arts and Sciences, held in the Boston parlour of his father-in-law, the fugitive-slave lawyer Charles Greely Loring. Among the fellows and Harvard professors sat the cotton mill magnate John Amory Lowell, with Agassiz and his allies. Gray talked on his recent paper on the flora of east Asia and eastern North America. Why did the same or similar temperate species come to inhabit opposite sides of the globe? Agassiz taught that the groups were created separately, each in its place. Gray's view had Darwin's blessing: God didn't create magnolias twice, in Japan and the Carolinas; all had descended from a single stock, with the dispersed variants cut off in their disparate zones.

Everyone knew what was at stake. Agassiz's belief in human plurality was well known from *Types of Mankind* – Lowell and his first cousin Francis had bought six copies between them – and from his articles in the heavyweight *Christian Examiner*. Agassiz and Gray both insisted their science was above politics and unmixed with religion. But on that winter night in 1859 Gray aimed to 'knock out the underpinning of Agassiz's theories about species and their origin – [and] show . . . the high probability of *single* and *local* creation of species'. Everyone in Loring's parlour knew that Gray's target was Agassiz, an amiable demagogue with a divisive racial science. Reading the report of the discussion that followed, Darwin simply repeated: 'What rubbish, Agassiz talks.'[11]

The Gray–Agassiz debate rumbled on intermittently before larger audiences. Agassiz tried to shift the discussion on to his own geological ground, but Gray stuck with plants. *Races of Man* author Charles Pickering showed up to weigh in. Eventually, on 12 May 1859, a notable number assembled in Gray's parlour at the Harvard herbarium to hear him expound natural selection for the first time on Yankee soil. Darwin's theory was the 'only noteworthy attempt' to understand variety and distribution in terms 'of cause and effect'. Agassiz's miraculous Plan of Creation was being supplanted by an evolutionary mechanism. Gray aimed 'maliciously to vex the soul of Agassiz' with a view 'diametrically opposed to all his pet notions', said an onlooker. And Agassiz's usual extempore ebullience failed to sway the audience. Darwin finally saw his carefully cultivated friendship paying off: 'how very well you argue', he complimented Gray. 'Great' was the usual epithet applied to Agassiz, and Darwin agreed: he was 'great in taking a wrong view'.[12]

In 1860, Lowell, the senior member of Harvard's governing Corporation, sided with Agassiz against Gray and Darwin in the *Christian Examiner*. He called the *Origin*'s science bad and its religion worse. The evidence that all fancy pigeons descended from 'one common stock' was 'very slender'. Fancy pigeons are so many separate species, or else hybrids. Lowell knew his *Types of Mankind*, and Agassiz backed him. Everybody knew about hybrids. Massachusetts had repealed its anti-miscegenation law in 1843. The races married freely in Boston, much to Agassiz's chagrin. Gray rebutted Agassiz's hybrid theory – a 'stunner of an answer', Darwin called it – and stole Lowell's theological thunder by insisting that the *Origin* did not touch the question of God.[13] Natural selection was a *scientific* theory, and Agassiz's creation-science the reverse – 'theistic to excess'.

So said Gray in the *American Journal of Science*. His review of the *Origin* was a running comparison with Agassiz, with Darwin appearing as the pillar of tradition: the one rejecting 'the idea of a common descent as the real bond of union', and 'Mr. Darwin' as assuredly holding 'the orthodox view of the descent of all individuals of a species not only from a local birthplace, but from a single ancestor or pair', including – though the *Origin* never said so – the human races. Then venues changed, as Gray tackled Agassiz before his largest audience to date, the 30,000 subscribers of America's literary flagship, the *Atlantic Monthly*. Another Lowell cousin was editor, but James Russell Lowell had a radical past, having started off on abolitionist papers. In the *Anti-Slavery Standard* he had even potted Prichard's ethnology for Garrisonians, ensuring they knew the scientific as well as scriptural case for human unity. With war on the horizon, Lowell was steering the *Atlantic* towards abolitionism, and he ran Gray's huge (anonymous) Darwin review across three issues without altering, he said, 'a single letter'.[14]

Humans or no in the *Origin*, Gray picked up on Darwin's intent. Go back into prehistory and Agassiz's worst nightmare is realized. Those who fought 'contamination' are thwarted; black and white blood flows together:

> Here the lines converge as they recede into the geological ages, and point to conclusions which, upon the theory, are inevitable, but hardly welcome. The very first step backward makes the negro and the Hottentot our blood-relations – not that reason or Scripture objects to that, though pride may.

Gray subtly linked 'Darwin's hypothesis' with a humble racial attitude, more characteristic of anti-slavery, as part of his strategy to '*baptize*' the *Origin*.[15] What Darwin dared not publish, Gray found timely in an America riven by questions of race and kinship.

The *Atlantic* series was perfect to pacify the religious critics at home. Gray's were by '*far* the best Theistic essays' Darwin had ever read and he paid to have them reprinted as a pamphlet with a reassuring title. Five hundred copies of *Natural Selection Not Inconsistent with Natural Theology* were run off by 1861. Half went to clergymen, journals, libraries, clubs and friends old and new. The next edition of the *Origin* even carried a puff for the pamphlet, giving its price and address for purchase. Lyell, who had struggled with the issues, agreed that Gray's articles were the ablest 'on either side of the Atlantic' and told Ticknor so. An Agassiz promoter might not agree, but Lyell put Ticknor on notice: 'opinions can never go back exactly to what they were before Darwin came out'.[16]

*

As Gray took on Agassiz, the Republicans pledged to halt slavery's expansion rather than abolish it, and put up Lincoln for President. With a Republican victory in the presidential race, South Carolina seceded from the Union, followed by six other Southern states. They declared themselves the Confederate States of America and adopted a constitution that, while forbidding foreign slave imports, protected 'the right of property in negro slaves'. Though Lincoln in his inaugural address did not threaten 'the institution of slavery', the South acted. On 12 April 1861, Confederate forces bombarded the federal garrison at Fort Sumter in Charleston harbour. A mile away, on Sullivan's Island, the cannons erupted near Agassiz's old seaside laboratory. After Lincoln called up 75,000 troops, four more states joined the rebel seven. Two governments were at war, one founded on the 'sacred and undeniable' truth that 'all men are created equal' and entitled to 'Life, Liberty and the pursuit of Happiness', the other embracing the 'great truth' enunciated that spring by the Confederate Vice-President: 'that the negro is not equal to the white man; that slavery subordination to the superior race is his natural and normal condition'. To contemporaries, this 'great physical, philosophical, and moral truth' came from science more than religion.[17] For all of their biblical arguments on behalf of slavery in general, many Southern Christians saw *Negro*-specific slavery justified by science.

Agassiz now applied for US citizenship, his adoptive patriotism full of gratitude. To him, 'onward progress of science in the new world' required a scientifically informed public. A creationist America would be a strong America, a 'more perfect Union'. He wrote a decent, if uncomprehending, review of the *Origin* in the *American Journal of Science*, while muttering privately that it was 'very poor!!' As Gray relayed to Downe, Agassiz 'growls over it, like a well cudgelled dog, – is very much annoyed by it – to our great delight'.[18]

Repelled and fascinated by the horror of war, Darwin asked Gray frequently, at times anxiously, for the latest news. Political and military campaigns interested him, whether against the rebel States or Agassiz: and though a formal distinction had to be drawn between the two types of battle, the line could be blurred. Most of Gray's wartime letters are missing or fragmented, but their message is told in Darwin's replies. Providence governed Gray's war. The outcome lay in His hands, at Shiloh or Bull Run just as much as in nature's battlefields, where Gray saw God guiding natural selection to its destiny. 'Lincoln is a trump, a second Washington', Gray boasted to another English friend. When Darwin questioned the North's intentions, Gray reminded him, 'Natural selection quickly crushes out weak nations'. It 'is simply a struggle for existence on our part'. He might have paraphrased their pamphlet as 'Natural Selection Not Inconsistent with Nationalism'. 'We must be strong to be secure and respected', was the message.[19]

Darwin could no more see God's hand in the Civil War than in the war of nature. The men willed the same moral outcome – the end of slavery – but for Darwin the means were all too bloodily human. No doubt Northern industrial might would overwhelm the Deep South (Darwin was deep in Olmsted's vivid *Journey in the Back Country* as events unfolded). But, he asked Gray, 'what is to follow?' Nothing concerned him more. 'Great God how I sh[d] like to see that greatest curse on Earth Slavery abolished'! He wished 'to God . . . the North would proclaim a crusade' against it. 'In the long run, a million horrid deaths would be amply repaid in the cause of humanity'. In wartime, such words are uttered from the heart, not the head. It was William Wilberforce's younger son Henry who said (writing to Martineau, then briefing him for an article on the war): 'the abolition of slavery would be a great gain to the US & to the world, even if all the American negroes were (literally) exterminated'![20] Armchair participants were willing to sacrifice millions – white in one case, black in the other – for the cause.

But Darwin could not persuade himself 'that Slavery would be annihilated'. Lincoln, he complained, did not 'even mention the word in his Address' before Congress on the Fourth of July 1861, calling the nation to arms. Nor were the Republicans fully committed to abolition. Simply 'to stop the spread of Slavery into the Territories' would be a gain, but without abolition Darwin saw no chance. His position had not changed: 'I sometimes wish the contest to grow so desperate that the north would be led to declare freedom as a diversion against the Enemy', he told Gray – meaning declare the slaves free and foment an insurrection in the South. If the North could not win, or did not have the right abolitionist heart, a two-state solution might be better. Radical that he was, like Garrison preaching 'Disunion', Darwin began to see moral superiority in an independent North throwing a cordon sanitaire round the Cotton South.[21]

Britain, officially neutral, was outraged in November 1861 when a Union warship intercepted the Royal Mail steamer *Trent* and forcibly removed two Confederate envoys bound for Europe. Laird's shipbuilders of Birkenhead started construction of swift paddle steamers to run the blockade of Southern ports. In Southampton, Union and Confederate crews brawled on shore and were arrested. Lancashire, starved of cotton, was swinging towards the Confederacy. Were the war to drag Britain in, it would leave 'we two bound, as good patriots, to hate each other', Darwin said to Gray. 'And what a wretched thing it will be, if we fight on [the] side of slavery'.[22]

Events took a turn with Lincoln's 'emancipation proclamation', issued on 22 September 1862 (and effective from 1 January 1863). Agassiz now desired federal regulation to check further interracial crossing, lest the mulattos hold down white attainment of a 'purer morality' and 'higher civilization'. Mixing was 'abhorrent to our better nature': to avoid it meant curtailing black rights, at least until they had shown some responsibility. Agassiz could now be seen around Harvard 'so cross and sore'. He was writing 'very maundering geology & zoology', Gray told Darwin, and was so clearly 'joined to his idols' that he would not 'be of any more direct use in nat[ural]. history' It cheered Darwin no end: 'I must say I enjoy anything which riles Agassiz. He seems to grow bigoted with increasing years'.[23]

But Lincoln's proclamation was hardly the 'crusade' Darwin sought. It came too late, and only 'rebel' States were liberated (Slave States in the Union were exempt). Like so many, Darwin's old Cambridge teacher Adam Sedgwick was appalled at this restriction. Lincoln almost '[makes] the canker of slavery . . . worse than it was, by making it a bonus to such slave-states as will fight under his colours'. It dampened Darwin's enthusiasm, and he doubted that the 'proclamation would have any effect'. Relations between Gray and Darwin were now deteriorating as fast as between Washington and London. Old Sedgwick might have been flabbergasted by the *Origin*, but he spoke for the Darwins on this issue, telling the future Sedgwick sister-in-law of Darwin's eldest son William: 'If it had been a war for the abolition of slavery, and the vindication of those natural rights of mankind' proclaimed in the Constitution, 'all good Englishmen' would have cheered. But the proclamation was a war expedient, partially granted. 'Such are the thoughts of many honest Englishmen, as true lovers of civil freedom, and as sincere haters of slavery.'[24] They certainly were Darwin's.

*

For many of Darwin's intimates, common descent was the starting point of their reviews, being rooted in human experience. Hooker explained the *Origin* to botanists this way. He played on their green-fingered gift for turning a tiny wild strawberry into a prize 'Alpine' or 'British Queen'. Nature was like a supremely skilled gardener. Thus the pedigree of strawberry varieties and the pedigree of life were 'strictly analogous to that of the members of the human or any other family', united by 'a blood relationship'. Nature had not 'mocked us by imitating hereditary descent in her creations'. She worked that way too.

If the Victorians had never read the *Origin*, but only its reviews, they would hardly have guessed that humans were absent. Not when a review could seriously start, 'Mr. Darwin boldly traces out the genealogy of man, and affirms that the monkey is his brother, and the horse his cousin, and the oyster his remote ancestor.' Darwin's kinfolk approach to nature was grasped immediately. Reviewers saw something that wasn't in the book because they too were looking at an enlarged human genealogy. Huxley understood this, and the *Origin*'s import: 'It is true that Mr. Darwin has not, in so many words, applied his views to ethnology; but even he, who "runs and reads" the "Origin of Species" can hardly fail to do so.'[25]

On the other hand, the more religiously minded baulked at 'common descent' evolution doing outright anti-slavery work. Gray's in-law Charles Loring Brace loved the *Origin* and eventually claimed to have read it thirteen times. He was a 'red-hot Abolitionist', and his compassion for slaves and the poor led to him becoming a New York 'city missionary', running refuges for paupers and vagrant children of immigrants. He too stood on Darwin's shoulders and in *The Races of the Old World* (1863) argued for common human descent from language similarities. For him as much as Huxley, humans were paramount in the *Origin*, even though they weren't there. He pictured his *Races* as an argument against Agassiz and *Types of Mankind*. But he kept ethical questions firmly in the religious realm, for

> Slavery is equally wicked and damnable, whether mankind have one parent or twenty parents. The moral Brotherhood of man does not depend on community of descent, but on a common nature, a similar destiny, and a like relation to their common Father – GOD.

In his view, it was a religious prerogative, not the ethics of common origin, that sanctioned abolition; and he agreed with Gray that racial unity should be regarded as a 'purely scientific' question.[26] But he also shared Gray's underlying faith that a 'purely scientific' answer would finally agree with the religious one.

Darwin had expected so much from Lyell's *Antiquity of Man* that a let-down was inevitable. From rumination to publication had taken seven years, and the first signs in February 1863 looked promising. Lyell got Darwin's drift and posed the question: if all human races have 'diverged from one common stock, how shall we resist the arguments of the transmutationist, who contends that all closely allied species of animals and plants have in like manner sprung from a common parentage?' He too was arguing *from* the human races to the rest of creation. The only alternative was Agassiz's polygenist race theory, which Lyell rejected. His unity confession was the starting point for reviews. Critics could thus open: 'Sir Charles Lyell adopts the theory of the unity of the human race which, no doubt, best accords with the hypothesis of the transmutation of species'. Unity was almost as controversial as transmutation, and critics were hard on him for failing to identify the human 'primordial stock', or for assuming that a genealogy of Indo-European languages back to an Aryan root was proof.[27] For many the book was an unsatisfactory, 'cautious' compendium, a belated digest of the flint tool finds in French and English caves.

It was the other end of the argument that upset Darwin. Lyell, never able to go the last mile, let creation back in. Ultimately, in Darwin's view, Lyell lacked the courage of his convictions: 'half-hearted & whole-headed', were Hooker's words. Lyell had topped off the *Antiquity* by adding a 'creational law' to confer 'the moral and intellectual faculties' on humankind. This supreme 'law' had cut in after untold millions of years of nature trundling along unaided: suddenly, it bestowed the first 'soul' on an animal, making it human. Lyell had been anguishing over this geological instant for years, wanting a blinding flash to herald the human arrival. He now effectively lined up with those more religiously inclined, such as Gray and Richard Owen: perhaps the human species, with its poets, prophets and geniuses responsible for the 'revolutions in the moral and intellectual world', did belong to a 'distinct Kingdom of Nature': a moral realm incomprehensible to any 'inferior' creature.

Old Lyell was speaking to an age marked by the static 'Egyptian-tomb' view of the American School. It was now common to hear of racial stagnation and of Negro limitations. Even Darwin's younger disciples were habituated to it: his youngest, John Lubbock – banker, future MP, armchair anthropologist and a neighbour in Downe village – queried Lyell's 'necessary progress of reason' even as he read. Why must the races necessarily progress? 'Look at the Negroes or the Australians! Are they much more civilised now than they were a hundred thousand billion millions of years ago? Or if left to themselves how many geological periods would elapse before they would have analysed the sun & had a great exhibition at Timbuctoo.'[28]

Never quite relinquishing his Southern sympathies, Lyell even used the 'creational law' to explain 'the origin of the *superiority* of certain races of mankind'. By Almighty leaps were they 'successively introduced', and by Almighty sanction were they blessed. That was the final straw for Darwin: the sentiment 'makes me groan'. His old illness returned. He agonized over Lyell's agonies, never really sympathizing. He cancelled the Lyells' visit to Downe and apologized: 'I will first get out what I hate saying, viz that I have been greatly disappointed that you have not given judgment & spoken fairly out what you think about the derivation of Species . . . I had always thought that your judgment would have been an epoch in the subject. All that is over with me.'

Darwin's own heroes weren't able to keep pace. 'It put me into despair', he wrote to Gray; 'I have sometimes almost wished that Lyell had pronounced against me'. Darwin was distraught at the shilly-shallying. His illness became extreme, with long periods of vomiting and depression, leaving him more or less an invalid until the end of 1865.[29] Through it all he could be deeply sympathetic to the bereavements around him, to the death of Hooker's child, and his father, even to FitzRoy's suicide (Darwin's brittle old captain would cut his own throat in 1865). Yet he never showed any sympathy for old Lyell's failure to keep up.

An equal tetchiness marked his dealings with Gray over the war. 'Yankee affairs' had made Gray 'very bumptious' and himself prickly about English attitudes. Gray sent a pamphlet by his father-in-law Charles Greely Loring on Anglo-American relations, which explained to Darwin their 'horror of Disunion' and complained about British support for the cotton states.[30] Darwin's reply was even more tactless. Thinking that the North could not win (he was still reading the 'detestable'

South-supporting *Times*, even though Emma was desperate to give it up) or, worse, facing the 'dreadful' prospect 'that the South, with its accursed Slavery, shd. triumph, & spread the evil', he now thought the North should 'conquer the border states, & all west of Mississippi' and partition the country, leaving the cotton South to wither.[31]

Darwin still feared that Britain might go to war with the North. Laird's and the Clyde shipyards were building 'steam-rams' – iron-clad warships with the latest gun turrets – for the Confederacy. The *Alabama* had already slipped out of the 'notorious' Laird's under false papers and was even now harassing United States merchantmen. Washington was irate. More than ships were going over. The *Origin of Species* might be a 'Whitworth gun' in the armoury of liberalism, as Huxley cleverly put it (meaning it was rearming a new generation of young rationalists stifled by the old orthodoxies), but at this moment real 'Whitworth sharpshooter' rifles were also being run through the Northern blockade to arm the Confederacy. Darwin's was now the moderate voice, echoing the milder family sentiment. Robert Mackintosh (the ex-Governor General of the Leeward Islands) at his sister Fanny Wedgwood's spoke too of dis-Union being the only course and a guide to British policy. More fiery abolitionists agreed. Fanny's friend Francis Newman, campaigner on women's rights, republicanism, vegetarianism (and anti everything else – slavery, animal cruelty, vivisection), exploded at the thought of preserving the Union and thereby saving 'the worst slavery ever heard of in history'.[32]

'I am afraid we shall not like each other for a good while – the nations', that is, Gray told Darwin. For Gray the only solution was to take the South at all costs, make it 'Yankeefied', '& hold the *U.S.* complete'. Those costs were rising. 'We have no children', he told Darwin, and 'I regret only that I have no son to send to the war'. Darwin, forgetting that he himself had visualized sending millions to their deaths to end slavery, was appalled at the sacrifice of one son. He relayed it all to Hooker: 'Did you ever hear the like'? Children were sacrosanct in Darwin's eyes, death a solemnity to be talked of in hushed terms. His ten-year-old daughter Annie – who had tragically died of fever in 1851 – would always occasion a tear. A million could be sacrificed on paper, but not a son. 'What pleasant letters Asa Gray writes. One might as well write to a madman as to him about the war.'[33]

*

It was no coincidence that a new society to study mankind was formed during the Civil War, and even less that the Confederacy had its paid agents inside. The Anthropological Society of London was founded in 1863, within weeks of Lincoln's emancipation proclamation. Hard-headed dissidents had abandoned the Ethnological Society. They disagreed with its creaking 'unity of races' angle, its philanthropic heritage, its cultural (rather than racial) concerns. The whipper-in for the new party, a Hastings-based 30-year-old doctor, James Hunt, a specialist in stammering, favoured the issues critical to Knox and *Types of Mankind*: racial anatomy and ranking – and, with war raging in America, the 'Negro's place' was paramount. The majority of these self-styled 'anthropologists' shared Agassiz's abhorrence of a common descent. The very premise of the society was one of black disparagement: there was no place in their new science for the sentimental notion that the 'Negro is a man and a brother'. Reliable 'data' were needed on his anatomical deviation, and to peg his 'sensual, tyrannical, sullen, indolent' character. Although actually it was pointless looking 'into the mind of the Negro, because he has very little of it, and it is never worth the trouble'. More formally, the society set itself the object to 'record all deviations from the human standard'.[34] That standard was the white man: the bias of the age was simply built in.

Not only did the Anthropologicals put Nott on their Honorary Fellowship rolls from the first, puffing him outrageously as 'the greatest living anthropologist of America',[35] but they had his friend Henry Hotze, a Confederate agent, permanently on the Council (not merely friend – Hotze was the Southern sympathizer who had translated Gobineau's *Moral and Intellectual Diversity of Races* for the American market under Nott's auspices). Hotze and Hunt were of the same stripe, but when it came to deprecation, Hotze was a master: the 'sickening moral degradation' evident in some human races meant that much had to be left out of the Gobineau translation for fear of upsetting Southern sensibilities. This pro-slavery advocate was actually paid to promote the South's benevolent view of their 'peculiar institution'. He had been recruited by the Confederate secret service and was bankrolled by the Confederate government in Richmond. Hotze acted as one might have expected of an *agent provocateur*. Briefed to swing London opinion during the war, he put Nott and Gliddon's *Types of Mankind* and *Indigenous Races* into the Anthropologicals' new library,[36] paid Confederates inside the society (there were three on the Council alone),[37] and swayed articles and pronouncements (he was probably responsible for getting Nott's piece on 'The Negro Race' into the society's *Popular Magazine of Anthropology*).

And that was just the start. Hotze also bought journalists, printed and distributed thousands of sympathetic books and pamphlets, and flyposted Confederate flags (joined to the British ensign) over street walls and railway stations. He ran a propaganda weekly in Fleet Street, *The Index* (1862–5). Between Southern war reports it promoted the Anthropological Society for exposing that 'most dangerous dogma', the 'equality of man', when the Creator had clearly made 'distinct groups' marked for different service. It proudly tied their science of Negro inferiority to the Confederate cause. The Anti-Slavery Society was attacked. Its claim that the slaves' refusal to revolt proved that they were 'wise enough to be trusted' was countered by the comforting thought that they did not have any reason to rise up. As proof of Hunt's value to Dixie, amid battle reports from Chattanooga and the Federals' push on Richmond, the *Index* at the end of 1863 gave huge coverage across three weeks to his revaluation of the Negro's pitiable place in nature. The paper pushed Hunt's conclusions to their limit. Either these small-brained, inferior humans were freed to return to 'savage barbarity', as in Africa, where 'human sacrifices' and unquenchable 'immorality' were the norm, or they continued their 'wonderful progress' under the 'mild, humane, and Christian rule' of the patrician South. *The Index* made Hunt's message political: the Negro's place was in bondage.[38]

The paper praised anthropologists for plumbing the 'invariable and immemorial specific difference' of peoples rather than their unfathomable origin. These were good men who made the 'Negro and European . . . as much distinct species as are horse and ass'. The lowly black was incapable of civilization 'except where he has been made the white man's slave'. This was the Confederate message, its science fixed on the 'congenital' characters of race. It denounced the 'fanatic' spouting equality, and called it 'absurd and misleading' to speak of blacks and whites 'as if they belonged to one stock . . . Both M. de Gobineau and Mr. Hotze were wise enough to avoid the rock on which Mr. Darwin made shipwreck'.[39] The pro-slavery *Index* saw through the *Origin* as well as any reviewer; saw the implications of its 'common stock' imagery for the human races, and hated what it saw.

*

Meanwhile Huxley and Hunt were fighting their own civil war over the issue. Huxley was caught between a rock and a hard place. His brother-in-law, the doctor who had apprenticed him and married his favourite sister Lizzie, had moved to Alabama and was now a Confederate surgeon. His forces had just been routed at Chattanooga. (One can scarcely imagine the horrors seen by their fifteen-year-old son Tom – named after Huxley – who helped his father in the hospital tent.) 'My heart goes with the south, and my head with the north', Huxley wrote to Lizzie. 'I have not the smallest sentimental sympathy with the negro . . . But it is clear to me that slavery means, for the white man, bad political economy; bad social morality; bad internal political organization, and a bad influence upon free labour and freedom all over the world.'[40] That was the line he would hold.

In the liberal *Fortnightly Review* he accepted the now trivial point that there was no proof of races changing in historical time, but that alone did not refute a unified origin. Even granting 'the Polygenist premises' – of separate human species – Darwin's evolutionary conclusions on common descent would still stand. Were Mongols and Mandingos separate genera, their evolutionary genealogy would still go back to a joint stock. But they weren't. Racial differences were so tiny 'that the assumption of more than one primitive stock for all is altogether superfluous. Surely no one can now be found to assert that any two stocks of mankind differ as much as a chimpanzee and an orang do'. It was what Darwin wanted to hear: he wrote on the front cover of Huxley's paper 'To be kept' and circled it, and scored this passage about 'one primitive stock'.[41] Huxley was now fighting Darwin's battles for him.

'Mr. Darwin', said Huxley, 'presents his doctrine as the key to ethnology'. By 1863 Huxley was pushing mankind's common ancestor back deep into time, perhaps even to the age of dinosaurs. Nor did he doubt that the modern races are quite inter-fertile. The thriving population on Pitcairn Island, descendants of Bligh's mutinous sailors from the *Bounty* and their Tahitian women, left that point 'dead against' the Notts and Gliddons. And where was the degradation, anyway, in sharing an ancestry with black people, or even apes? All of us should 'find in the lowly stock whence man has sprung, the best evidence of the splendour of his capacities' and a good 'ground of faith in his attainment of a nobler Future'. That was how Huxley ended his *Man's Place in Nature* in 1863. It was not a faith to appease conservatives. So, snapped the

Athenaeum sardonically, 'Any man can now mount armorial bearings in the shape of the long arms of the gibbon or the gorilla.'[42]

As Agassiz's future had no room for blacks, so Darwin's and Huxley's had none for slavery. The corrosive system was dehumanizing for whites, let alone blacks. But Huxley justified abolition on *economic* grounds. It was impossible to sidestep that 'great problem which is being fought out on the other side of the Atlantic', and whatever the Southern virtues

> I can as little comprehend how any man of clear head can doubt that the South is playing a losing game, and that the North is justified in any expenditure of blood or of money, which shall eradicate a system hopelessly inconsistent with the moral elevation, the political freedom, or the economical progress of the American people.

He dismissed the 'fanatical abolitionists' who thought the Negro an equal. But this 'aberration' was as nothing to 'the preposterous ignorance, exaggeration, and misstatement in which the slave-holding interest indulges' – and he held up Hunt's pamphlet on 'The Negro's Place in Nature' (a deliberate play on Huxley's book title) as deserving 'public condemnation'. He regaled his audiences with passages: 'The hair [of a Negro] is very peculiar – three hairs, springing from different orifices, will unite into one', or Knox's old heresy, that ape-like Negroes had elongated heels and could not stand like men.[43]

Stout support was received from a Civil War veteran who had seen not one black, but a whole regiment. A confirmatory letter came from the Union Colonel of the 1st South Carolina Volunteers, Thomas Wentworth Higginson. As a radical abolitionist, he still bore the cutlass scar on his face acquired while storming the Boston federal courthouse to rescue a fugitive slave. Latterly he had commanded the first federally authorized black regiment to fight in the South. The height of Higginson's praise for the soldiers was matched only by the depth of his scorn for Hunt's absurdities about black feet. Some 'nine tenths' of the command was 'purely negro', he told Huxley, and not one needed odd boots or had feet different from a white man's.[44] (The same old warrior would visit Darwin twice in the 1870s, the second occasion spending the night at Down House: Darwin, fascinated by the 'black regiment', was 'proud' to have received such a man, and the feeling was reciprocated: Higginson called Darwin 'even a greater man than I had thought him'.)[45]

Press headlines now proclaimed 'PROFESSOR HUXLEY AN ABOLITIONIST'. Liberals praised him for avowing 'equal personal rights for all men' and pronouncing 'the system of slavery to be root and branch an abomination'. The Professor had made 'his physiological definition of the Negro's place among men equivalent to an earnest plea for Negro emancipation'.[46] It was enough to have the Ladies London Emancipation Society strike off copies of his anti-Hunt paper as a pamphlet.

The older Darwin was an abolitionist, perhaps even fanatical. For him economic reasons never entered into it: the issue was cruelty. Cruelty to all creatures: Charles and Emma were distributing a pamphlet from later 1863, written by themselves, against the use of gin traps, the dog-toothed steel-sprung jaws so favoured by gamekeepers. These traps, which smashed the leg of any animal that stepped into them, were increasingly in use by the sixties to clear predators on gentlemen's shooting estates. Emma worked with the Royal Society for the Prevention of Cruelty to Animals to institute a prize for the design of a humane replacement. Animals or slaves: it was a cruelty issue set in train by the same Buxton–Wilberforce set decades earlier. For the middle classes, compassion (now increasingly extended to wild animals) had become a mark of gentility. And it was cruelty that really mattered across the Atlantic. As Emma said: 'About America I think the slaves are gradually getting freed & that is what I chiefly care for.' To *The Times*'s anti-emancipation rhetoric, she simply retorted: 'I think all England' needs to 'get up its Uncle Tom again'. Darwin too was cursing the 'old Bloody Times', and still wondering whether, if the middle States joined 'with the South on Slavery', the North shouldn't 'marry Canada, & divorce England & make a grand country, counterbalancing the devilish South'.[47]

So much of Darwin's life and science had been bound to this hatred of slavery: a science that, in its 'brotherhood' aspect, looked impeccable. Huxley saw it. When it came to human unity – a bloodline connecting blacks and whites via a common ancestor – Huxley 'was pleased to be able to show that Mr Darwin was for once on the side of orthodoxy'. And having run his own Union flag up the pole, Huxley's civil war intensified as *The Index* and Confederate anthropologists wheeled out their big guns.

Huxley's 'any expenditure of blood' rhetoric (which matched

Darwin's 'million lives') was dismissed by Hunt as the 'usual fanatical language of the Abolitionists'. The Confederate *Index* printed Hunt's rejoinders as war reportage, while ignoring Huxley's attacks. Huxley made his point plainer. Hunt, amid an 'embarrassing cloud of verbiage', 'gives an "indignant denial" to what he terms my insinuation (which, however, I shall be happy, at any time, to exchange for a broad and direct assertion) that his "views were brought forward in behalf of the slaveholding interest."' It was hard to escape that conclusion. Certainly Hunt sided with Nott's slave lobby by admitting 'that the laws in the Southern States of America against the intermarriage of the negroes and the whites were wise laws'.[48]

Huxley never knew the half of Hunt's incestuous relationship with the Americans. Where Hunt in his 'Negro's Place' said that the planters no more fear rebellion among their 'slaves than they do rebellion amongst their cows and horses' he was parroting. The words came from a pamphlet by the New York publisher John van Evrie with the title *Negroes and Negro 'Slavery'; The First, an Inferior Race – The Latter, Its Normal Condition.* Van Evrie himself then *republished* Hunt's paper in New York for thirty-five cents, stripping the references and adding a timely political preface as the war raged. It was made to look like a 'learned' scientific sanction of the South by the 'President of the Association' in Britain. To ensure the message took, van Evrie ran ads for 'white supremacy' and 'anti-abolition' tracts at the back.[49]

So the 'racial' war around 'Darwinism' (a term just coming into use) was not being fought in far-off Alabama or Agassiz's Boston. Nott had his man in London, the Confederacy their pro-slavery lobby inside the Anthropological Society; and Darwin's supporters were combating them. If ever there was a forum for promoting Knox and 'white supremacy' or a springboard for crushing Darwin's racial union, this society was it. Hunt was Knox's disciple, and he made sure Knox was brought in from the cold. Fêted by the society, Knox in his last act derided Darwinism, denying that 'there is or can be any selection, as Mr. Darwin expresses it', of clever 'negro' varieties into a more intelligent race.[50]

The Anthropological Society was the only London 'learned' body which tolerated long debates on Darwinism. No one doubted that the 'struggle for existence' explained racial extermination: 'Charles Darwin is most sound on this point'. The extirpation of the 'savage' Gauls by

Franks, or Britons 'to make room for the Saxon, were of the greatest benefit to humanity'. Darwin's evolutionary gain in the 'production of the higher' types was being generated by the destruction of 'lower' races. They could see Darwin's ideas spreading 'like the cattle-plague', and some thought his moral movement to bring all races into union had the hallmark of an ugly 'religious revival'.[51] Yet speaker after speaker still granted Darwin his licence.

Dislike of the 'Cannibal Club' (as the Anthropological clique was known) ensured sympathy for the gentler Darwin. Hisses greeted Hunt at the British Association meeting in 1863, when he maintained 'that there is as good reason for classifying the Negro as a distinct species from the European as there is for making the ass a distinct species from the zebra'. His slavery affiliations had preceded him. Hunt's henchman Charles Carter Blake, said the *Evening Star*, 'appears to act the part of Confederate physiologist'. One Negro in the audience stood erect (proving he could) to ask who was the human ass. This was the man the audience had come to hear. Everyone knew that the famous fugitive slave William Craft intended to speak, and they had flocked to the spectacle. Craft reminded Hunt that Caesar had found the woad-covered savages of Britain 'such stupid people that they were not fit to make slaves of in Rome. (Laughter.)' More raucous laughter greeted his observation that God had given black men thick skulls to protect against the scorching sun; had He not, 'their brains would probably have become very much like those of many scientific gentlemen'.[52]

War reportage ran seamlessly into this 'Sambo and the *Savans*' story. The *Leeds Mercury* followed an irate article on the ironsides being built in Britain for the Confederates with an account of Hunt's attempt to smuggle in the ideas of the 'South Carolina pamphleteer'. The *Birmingham Daily Post* ran Craft's 'loudly applauded' rebuttal after its own vitriolic articles on the 'nigger-breeding and woman-flogging' South and British loans to the Confederacy. The *Aberdeen Journal* reported 'shrieks of laughter' greeting Craft's joke about thick skulls, before moving on to French views on the Confederate States and their President Jefferson Davis's circulars to European leaders. Such unprecedented booing and laughter by the well-heeled gives an idea of the tensions as America's war raged. All the papers reported cheers for Craft's claim that black and white men 'were all descended from a common parent'. None of the press corps doubted that the issue of common descent was anything other than highly emotive. And the consensus was clearly against the Anthropologicals' 'Godless manifest-destiny theory'.

Craft did not disappoint: he wryly threw in (referring to the stock comparison of a Shakespeare to a Hottentot) 'that he had been in this country for several years and had met with very few Shakspeares'. When the President, the imperialist Tory Sir Roderick Impey Murchison, stopped Craft talking, the audience became 'rather angry'. Even though two papers were read at the meeting disputing 'the unity of the human race', it was clear from the crowd reaction that neither went down well. 'It needed but the added attendance of an ancient Briton or painted Pict to put the inquiry, "Am I not a man and your ancestor?" to complete the confusion of the small band of ethnological *savans*', said one newspaper.[53] The concatenation of a Civil War and heightened emotions over slavery told in these boisterous meetings, and they boded well for discussions around Darwin's 'common ancestor' *Origin*.

Against Darwin and Huxley, the Anthropologicals threw every Continental authority – not merely threw, but translated, edited, introduced and published under Society auspices. Half a dozen weighty tomes appeared in order to prove immutable difference and cross-racial sterility. Titles such as Georges Pouchet, *The Plurality of the Human Race* (1864), and Paul Broca, *On the Phenomena of Hybridity in the Genus Homo* (1864), told their own story. And each got round the problem: Pouchet had life not only spontaneously emerging at the start, but continuing to do so, providing each white and black line with its own originating point. This was too far-fetched, even for the editor. It might have been acceptable in Europe and used to avoid Darwin's human unity and single origin for all life (it was by the discoverer of the Neanderthal skull, Hermann Schaaffhausen), but no British anthropologist adopted it.[54]

By contrast, another of their translations did provide a rallying point. The Fellows had 'shown unanimity' against the Darwinians' common racial descent, and there could be no advance 'in the application of the Darwinian principles to anthropology until we can free the subject from the unity hypothesis'. But they displayed little fear of transmutation, and even some sympathy for ape extraction. 'No one (except Agassiz and his confreres) will deny the possibility of the descent of man from the ape by some unknown law of development', said the President in his opening address. The way out was provided by Professor Karl Vogt at Geneva in his *Lectures on Man* (translated by Hunt in 1864). Vogt was more pugilistic than Huxley, and more scurrilous – an atheist who delighted in calling certain 'simious' heads 'Apostle skulls' on the grounds that they resembled St Peter's. And he, too, tentatively questioned why 'only one stock [of ape] should possess this privilege' of evolving into humans. Probably in *each* tropical region the local ape had. Vogt's multi-ape ancestry was firmed up in the Anthropological literature. Giving one ape ancestor per human race made him 'a logical Darwinite' in their eyes.[55]

What Down House thought is not known. Correspondents pointed Vogt out, telling Darwin he would 'be amused at his Huxleyan outspokenness'. Darwin, scouring the translation, was made well aware of Vogt's 'origin of Man from distinct Ape-families'. But the Anthropologicals were dressing Vogt's image in Confederate grey (Vogt himself stoutly opposed slavery) and ignoring his 'ifs' and 'buts'. Though enamoured of Darwinism, said the Anthropologicals, Vogt 'entirely repudiates the opinions respecting man's unity of origin, which a section of Darwinians in this country are now endeavouring to promulgate.' This multi-ape theory was the 'sound and philosophical' way forward – the science to counter the 'antiquated doctrines' of blood relations and genealogy that Huxley had inherited from Darwin.[56]

The 'Cannibal Club' became a rival axis of power, vying with Huxley's 'section of Darwinians'. The polygenists gloated in their effrontery. In the window to their meeting hall hung a 'savage's' skeleton, generating complaints from the Christian Union across the road. The mace which brought meetings to order was topped by a negro's head gnawing a human thigh bone. Controversial discussions here could have an electric air, and as a result they had a huge fellowship, some 700 to 800, it was claimed (not exactly truthfully).[57] The polygenists had the power to put their own stamp on 'Man's Place in Nature'. And many now pulled behind Vogt to confront Darwin. They used, in the words of the more liberal Thomas Bendyshe (but still another devil in Huxley's lower circles of Hell), 'more apes than one, to reconcile transmutation with the polygenous theory'.[58]

As devils went, James McGrigor Allan's summoning of Knox, Agassiz and Vogt provided the definitive conclave to prove the 'multiple origin of mankind'. Allan recruited them in 'Ape-Origin of Man' for the *Popular*

Magazine of Anthropology (part of the Anthropologicals' drive to drum up public support). Allan was peculiar, for he appreciated the finer points of many tribes. He had lived in America and was familiar with Nicaragua's Miskito Indians. Nor did he have any illusions about tribal extinctions, it being the custom 'to go to them with a Bible in one hand and a bottle of rum in the other, and to tell them to be like us or disappear'.[59] Yet in his 'Ape-Origin' he too baulked at the bad blood caused by common descent. Compare the 'greatest European heroes and Hottentots living at the Cape of Good Hope': no one could conceive them to be the same species, let alone 'descended from the same primitive stock'. Oddly, while he couldn't stomach this, he could conceive of the transmutation of an ape into an Anglo-Saxon 'at a very remote period'. A wild, free ape in ancient times was preferable to an enslaved black at present. Allan took from Vogt multiple ape ancestors, one in each of Agassiz's 'racial realms', and used Knox's racial antagonisms and 'Mr. Darwin's grand hypothesis' of natural selection to explain not merely why America presents 'a huge battle-field', but the 'antipathy between the races continually at war with one another'. All this made Allan 'a polygenist and a Darwinian', showing that there was no coherency yet to the 'Darwinian' label.[60] It was still up for grabs.

Huxley and a number of elite Darwinians regrouped inside the old Ethnological Society, giving it a new lease of life. Darwin's brother-about-town Erasmus joined the Council in 1862–3. Lubbock, from Downe, became President (to be followed by Huxley). Here Huxley explained his pioneering skull-bisecting techniques as he moved further into anthropology, excavating barrows and midden mounds: the human races were calling him too.[61]

The oddball as always was Wallace, home from Borneo, and revelling in the male-only no-holds-barred atmosphere of the Anthropological. He tried to patch things up with a compromise. The returning prodigal, out of touch, was equally out of synch with the Cannibals. Wallace had lived with Borneo's Dyaks and appreciated their moral world. These were not the Dyaks denigrated by Kingsley as 'beasts, all the more dangerous' because of their 'semi-human cunning', creatures fit to be trampled underfoot by a Teutonic race spreading God's Kingdom. (Interestingly, Kingsley had spent three weeks under Dr Hunt in 1857 for therapy to overcome his stammer, although his philosophy of race long predated this meeting.) And no one in Hunt's clique could possibly have rhapsodized with Wallace, 'The more I see of uncivilised people', the more the 'differences between so called civilised and savage man seem to disappear.'[62]

The Dyaks had reinforced the old socialist's notion of the very oneness of the races. He could be elegiac about native peoples and sad at the urban encroachment on their lives – the very antithesis of Kingsley's desire to eradicate the pests.[63] To Wallace, the Dyaks' fear of infringing others' rights, or their refusal to answer a question lest they tell a lie, evinced a 'wonderfully delicate sense of right and wrong'. He suspected that morality was fundamental to humans and could be traced right back to prehistoric flint-tool users. There was even evidence. In a French grotto in Aurignac (from which villagers had removed seventeen human skeletons), Edward Lartet found arrowheads, flints, extinct cave lion and bear bones and worked reindeer antlers. Apparently some animal joints had been left in the tomb with meat on, as 'food of the dead of the pre-historic race'.[64] To Lyell these funerary feasts and viands had been food for the departed on their journey. And to Wallace too the sepulchre proved that moral and spiritual beliefs were ancient.

This sort of insight became the foundation of Wallace's Darwinian compromise between polygenism and monogenism. It was a virtuoso performance at the Anthropological in 1864. He pushed racial origins far back in time. Humans might have lived for a hundred thousand centuries alongside now-extinct mammals. So it was no surprise that Egyptian tomb walls show recognizably modern Negroes, or that redskins have not changed since the 'very infancy of the human race'. Natural selection takes an inordinately long time. Wallace even believed that *anatomical* changes had ceased in humans. Care, sympathy and altruism even in the 'rudest of tribes' allowed them to escape competition, while technology allowed them to circumvent changing conditions: using furs to keep warm, arrows to kill distant game and fires to prepare food. Therefore the ancestral 'homogeneous' group of widespreading pre-humans had split into races in very remote times, when natural selection was still operating. Possibly even as far back as the Eocene, just after the extinction of the dinosaurs. The races' common ancestor at that time, without speech, without 'moral feelings', was still at the *animal* stage, still an ape under selection's influence. Then all the derived lineages developed larger brains and became human as natural selection continued to operate on the *mind* (somewhat differentially, causing distinct skull shapes). Hence the races have varied aptitudes, language abilities, technological capacities and social attitudes.

Using Darwin's theory – and his – Wallace had got round Morton, taken care of Hunt and kept a converging bloodline, all in a bravura show. He then infuriated the clique by suggesting that, in his socialist vision, natural selection would go on to perfect 'the social state'. It would bring all the 'higher' races up to scratch to displace the 'lower', improving them 'till the world is again inhabited by a single homogeneous race, no individual of which will be inferior to the noblest specimens of existing humanity,'[65] when freedom will prevail and coercive government die away.

The white-supremacists could hardly condone such anarcho-socialist Utopianism. Hunt was incredulous. The Anthropological was the apotheosis of Anglo-Saxonism, Knoxian racial antagonism and the pro-slavery 'niggerology' of a South now at war. The 'mischief' in thinking Darwinism could produce 'from one homogeneous race all the diversity now seen in mankind' was infinitely worse for threatening to re-homogenize it! The idea of a 'homogeneous' race – smacking so much of the reviled amalgamation or miscegenation inflaming relations in America – left Hunt furious. True, Wallace had pushed the original union further back than Huxley had, to a stage when the ancestor had neither speech nor 'moral feelings'. But still these were 'startling assertions', and even more so given Vogt's query, 'why must mankind once have been of one race?' Later Hunt recalled that Wallace 'did not find a single supporter' for his 'eloquent dream'. But that was not entirely true, and the odd listener saw Wallace's paper 'constituting a new era in anthropology'.[66]

Darwin had some misgivings about the paper: he denied, for example, that selection ceased among civilized men. But he was happy to see someone else stand in the spotlight. Sensitive to criticism, valuing his standing among the gentry, he himself suffered qualms about publishing on that inflammatory subject, human transmutation. Accusations of bestialization and worse would fly. But Wallace, the Mechanics' Institute-educated specimen-seller, was fearless. He had nothing to lose professionally. The poor ex-surveyor and tropical traveller had no social standing to forfeit, no children whose good names needed protection (so important in society – the Edinburgh publisher Robert Chambers, asked why he published the evolutionary *Vestiges* anonymously, replied 'I have eleven reasons', and pointed to his children). Wallace knew that Darwin had been deterred, and so ploughed ahead himself. They might have shared (almost) joint billing at the inaugural presentation of natural selection at the Linnean Society, but Darwin scarcely understood his socially distant 'co-author'. A Utopian paradise was hardly where he saw evolution headed.[67] Now he made a bigger miscalculation.

In May 1864 Wallace sent Darwin his 'little contribution to the theory of the origin of man' – the Anthropological Society paper. Darwin's marginalia show his problems with it. Since he (unlike Wallace) had never lived among 'savages', he doubted their altruistic behaviour and therefore saw natural selection working on them *still*. Against Wallace's statement that, whereas animals show no 'mutual assistance', 'the rudest of tribes' sympathize with the vulnerable and ill and protect them from selection's scythe, Darwin jotted: 'Does not act . . . only civilized men!' So *uncivilized* individuals, Darwin believed, were still being played off against each other. They were *not* shielded from selection by their altruistic acts.

Despite this difference he still saw Wallace as a Cinderella among the ugly Anthropological sisters. He was hoping Wallace might be like Gray or Huxley – the propagandists who stood forth to fight a shrinking Darwin's battles for him. Wallace could be another sympathizer who could shoulder the burden, if rightly approached. He replied praising the paper: 'the great leading idea is quite new to me', and it was surely true that competition between the modern races 'depended entirely on intellectual & *moral* qualities' – meaning that the morally and technologically superior whites were vanquishing all others.

Then, out of the blue, Darwin revealed his theory of racial origins. This was his idea that aesthetic choices had created the different physical features of the races. Sexual selection had caused him a mental struggle, and he was proud of his theory. He told Wallace,

> I suspect that a sort of sexual selection has been the most powerful means of changing the races of man. I can show that the diff^{t} races have a widely diff^{t} standard of beauty. Among savages the most powerful men will have the pick of the women & they will generally leave the most descendants. I have collected a

few notes on man but I do not suppose I shall ever use them. Do you intend to follow out your views, & if so would you like at some future time to have my few references & notes? I am sure I hardly know whether they are of any value & they are at present in a state of chaos . . . [P.S.] Our aristocracy is handsomer (more hideous according to a Chinese or Negro) than middle classes from [having the] pick of women . . .[68]

Darwin was hoping Wallace would take over the subject. But regaling a plebeian socialist with talk of 'our' handsome aristocracy was courting disaster. The self-deprecating sales pitch failed. Wallace had no intention of relieving him. Worse was his undiplomatic response.

Wallace did not understand how proprietorial Darwin was about his hard won sexual selection, nor therefore the honour implied by the offer. Wallace emphasized how little natural selection could affect human evolution; how sexual selection would have 'equally uncertain' results. And, sticking up for his order, he doubted 'the often repeated assertion that our aristocracy are more beautiful than the middle classes'. 'Mere physical beauty, – that is, a healthy & regular development of the body & features approaching to the *mean* or *type* of European man, – I believe is quite as frequent in one class of society as the other & much more frequent in rural districts than in cities.' The patrician Darwin hardly wanted to hear that, even less Wallace's seeming snub: should he go further into the question some day, he would accept the notes.

Etiquette was always a sticking point with Darwin, and the earthy Wallace had come unstuck. Darwin withdrew the offer, exclaiming that Wallace was probably right, except 'about sexual selection which I will not give up', adding the snippy, 'I doubt whether my notes w^{d} be of any use to you, & as far as I remember they are chiefly on sexual selection.' First Lyell, now Wallace; the disappointments were mounting.

Only the outcome of the Civil War gave him any satisfaction. 'Jefferson Davis richly deserves to be hung' was Gray's opinion of the Confederate President after the North's victory in April 1865. 'I care for nothing else in the Times', Darwin replied. 'How egregiously wrong we English were in thinking that you could not hold the South after conquering it. How well I remember thinking that Slavery would flourish for centuries in your Southern States!'[69]

*

Two briefly existing Darwinian journals collapsed in 1865. Huxley's Darwinians saw their *Natural History Review* go under as it failed to 'appeal to the masses'. The group had also bought into the more general *Reader* (literally, buying £100 shares), but Huxley's attacks on 'Church Policy' caused the Christian Socialist backers to desert them, and despite a price drop to twopence sales fell off and the magazine was 'bound over to Satan': the Cannibal Club's Thomas Bendyshe bought it in late 1865.[70] Things were going rapidly downhill.

Nor was Wallace helping as Huxley's Darwinian cadre vied with the Cannibals for hegemony. Wallace remained a regular at Anthropological meetings. The bachelor rather revelled in the members' freedom to explore all subjects, even savage sexual taboos – defying the primness of society as much as the delicacy of 'Ladies' Nights' at the rival Ethnological. Wallace actually told Darwin that Bendyshe was 'the most talented man in the [Anthropological] Society'. He would change the *Reader* 'for the better, & I only hope the rumour of that *bète noir* the Anthropological Soc., having any thing to do with it, may not cause our best men of science to withdraw their support'. The freewheeling Wallace liked the open atmosphere, and he asked Darwin (this sounded more pointed than it was intended to be): 'Why are men of science so dreadfully afraid to say what they think & believe?' Hooker was passed Wallace's letter:

> It is all very easy for Wallace to wonder . . . he has all 'the freedom of motion in vacuo' in one sense, had he as many kind & good relations as I have, who would be grieved & pained to hear me say all I think, & had he children who would be placed in predicaments most detrimental to childrens minds by such avowals on my part, he would not wonder so much.[71]

There was Darwin's quandary too – the wealthy squire having cautiously waited twenty years to publish his own destabilizing views, even then defusing the *Origin* by removing the humans. 'I fully agree with your remarks', he replied to Hooker.

Darwin now feared that Wallace was going astray. 'As for the Anthropologists being a bête noir to scientific men', they were indeed for Darwin, who loathed their coarseness and conceit as much as their pro-slavery science. From a man who stood on the proprieties, no indictment was worse. The ungentlemanly Anthropologicals had lost caste and Wallace's fall from grace was one upshot. Now Darwin faced the unappealing prospect of publishing on human transmutation himself – or rather on sexual selection – as his unique way of separating the human races apart from their common stock.[72] No one was going to relieve him.

13
The Descent of the Races

In 1866, a decade after he had removed the 'Races of Man' from his aborted 'Natural Selection' manuscript, Darwin finally plucked up the courage to discuss human racial origins. The bloody crushing of a revolt in Jamaica started a chain of events that left him writing in earnest.

Trouble had been brewing since emancipation in the 1830s. As sugar profits trickled away, the freed slaves proliferated rather than prospered. Black people now outnumbered Europeans by twenty to one; black and mixed-race by thirty to one. In the House of Assembly, 400,000 were governed by a 47-man planter oligarchy chosen by fewer than 2,000 electors. If voting did nothing for the majority, it seemed a revolution would, and as the day of racial reckoning dawned, abolitionist dreams of a peaceful, industrious island faded. The old biblical belief that Africans and whites were members of one family lost ground to prejudice and the new racial-ranking science. Race was becoming the be-all and end-all as events in Jamaica turned violent.

Richard Hill, Darwin's bird man in the capital, Spanish Town, knew about the distress at Morant Bay, the settlement on the east coast of Jamaica that would be the flashpoint. In thirty years he had 'never seen the Colony sunk so low'. Produce was dear and work scarce, with no poor law to help the helpless. His voice rattled the authorities. His warnings about wretched prison conditions so irritated Governor Edward John Eyre that a report was commissioned, which scorned his 'perfectly groundless' assertions as unbecoming of a 'gentleman in Mr. Hill's position'. Even as a magistrate, a mulatto had to mind his step. Hill trod tenderly when addressing black 'ragged school' children: he used scripture subtly to indict 'the proud who treat persons beneath them with contempt'. He was sympathetic but no incendiary and, after the riots, he reported that 'nothing had taken place' in his jurisdiction 'beyond ordinary misdemeanours to excite doubts of the loyalty of the black population or apprehension for the peace and safety of the Island'.[1] Elsewhere it was different.

The cause of emancipated blacks was anything but sacred to Governor Eyre. Eagerly embraced as a Fellow of the Anthropological Society not long after his elevation as viceroy, he had the latest race science behind him to counter those damned by Carlyle as 'rabid Nigger-Philanthropists'.[2] Hard facts showed that the feckless Jamaican majority were irreclaimable savages. Now came the proof.

On 7 October 1865, an unpopular verdict in the Morant Bay courthouse led to a scuffle. Police sent to make arrests were overpowered and an angry crowd, marching into town, was met by a small volunteer militia. Stones were thrown, pikes brandished and the riot act read. The volunteers panicked and fired on the crowd, killing several. Reprisals began – the courthouse was torched, the town occupied, and within days eighteen officials and militia men were dead and thirty-one wounded.

Hearing 'the blacks have risen' confirmed the worst, and Eyre proclaimed martial law. Lurid rumours of heads axed open, eyes scooped out and men disembowelled fuelled his rush to Morant Bay with troops. Meeting no organized resistance, the soldiers rampaged and executed 439 'rebels', flogged more than 600 men and women, and burnt down over 1,000 peasant homes. Eyre had his main political enemy, George William Gordon, an assemblyman and champion of the poor, arrested. The son of a planter and a slave woman, Gordon was a planter himself and a preacher in his native Baptist church. He had sat with Hill as a magistrate, but Eyre had removed him from the Morant Bay bench after he exposed a death in custody. Now he was accused of plotting the insurrection.[3] His court martial was a farce. Convicted of high treason and sedition on the flimsiest evidence, Gordon was hanged from the courthouse ruins.

When the news reached London it revived an old alliance. As a Royal Commission inquired into the events, evangelical philanthropists, liberal reformers and parliamentary radicals combined to make capital from Eyre's conduct. The 'Jamaica Committee' recruited Darwin's class of men, the successors of Britain's first anti-slavery activists – men of principle, pro-Yankee in the Civil War and campaigners for the black freedmen afterwards, all of them convinced (like the Darwins and Wedgwoods in the years before emancipation) that equity abroad was inseparable from equity at home. Committee members, including nineteen MPs, linked the Eyre case with another Reform Bill passing through Parliament which would further extend the middle-class vote. One of them, Charles Buxton, son of Wilberforce's successor, took the chair. A young barrister, James Fitzjames Stephen, grandson of a Clapham Saint and of Wilberforce's sister, was their legal adviser. Merchants and manufacturers, professors, journalists and Dissenting clergy (10 per cent of the original Jamaica Committee membership) made the organizers 300-strong, and the philosopher John Stuart Mill, newly elected Liberal MP on a platform of women's suffrage, was the jewel in their crown. All saw this 'fearful epilogue' to the American Civil War as a test of English moral fibre and judicial principle.[4]

Eyre was dismissed and recalled to England. But in June 1866, the Royal Commission stunned everyone with a report praising his swift action against the rebellion, even while censuring his abuse of the noose, lash and torch. There would be no official criminal proceedings. Buxton's defeatism during the Civil War had infuriated abolitionists, including the Wedgwoods. When Buxton also decided that a private prosecution against Eyre was inexpedient, an outraged Mill took over. Though reckoning that a prosecution for 'abuses of power committed against negroes and mulattoes' would be unpopular with the middle classes, he made Gordon the cause célèbre. Hundreds of blacks had died unspeakable deaths, but Gordon was seen as 'chiefly of white colour'. He had an Irish wife, property, political office and religious allies in England. From July, as Darwin was writing the second half of his book on domestic animals and plants, the committee's tactic became the defence of English constitutional law. Illegality rather than immorality would be the issue. 'Due process' had been flouted and Gordon murdered in cold blood. A 'TEN THOUSAND POUNDS FUND' was started to cover the cost of prosecution.[5]

The hacks on both sides had a field day, naming culprits or shaming hypocrites. Even as damning eye-witness evidence turned up in the press, so India-hands with a memory of the 'Mutiny' leapt to Eyre's defence with gory details of massacres committed by dark-skinned men. 'We anthropologists', proclaimed the Cannibal clique, 'have looked on with intense admiration' at the conduct of an honourable Fellow. Some anthropologists saw 'mercy in the massacre' for such lowly sorts. They joked of the advantages in killing savages as 'a philanthropic principle'.

'The merest novice in the study of race-characteristics ought to know that we English can only successfully rule either Jamaica, New Zealand, the Cape, China, or India, by such men as Governor Eyre.' 'We English' meanwhile were sending Mill 'threats of assassination'.[6]

A hero's welcome awaited Eyre when his steamer docked at Southampton in August 1866. Locals organized a lavish banquet in his honour on the 21st, the Mayor presiding, with a military band to serenade the 100-odd gentlemen guests, and their ladies in the gallery. Outside the Philharmonic Hall protestors massed, summoned by the placards all over town advertising this 'feast of blood' and 'Banquet of Death'.

After the toasts, the Earl of Cardigan, 11th Hussars, a flogger in his own right, who had led the fatal charge of the Light Brigade in the Crimea, claimed that Eyre's troops had conducted themselves in 'the most praiseworthy manner possible'. 'No governor, no matter what might be the cruelties – if they were cruelties – necessarily committed', was 'ever seriously found fault with', much less humiliated as Eyre had been. The Earl of Shrewsbury, a retired admiral and Jamaica property-owner (whose son would be a world-class collector of torture instruments), slammed the Jamaica Committee. Eyre might well have averted the 'extermination of the entire . . . white population'. The Revd Charles Kingsley was at his toadying best. Such an approach had got him Huxley's and Darwin's ear, and he was even quoted in the *Origin of Species*. Now he flattered Eyre as 'so noble, brave, and chivalric a man, so undaunted a servant of the Crown . . . and a saviour of society in the West Indies'. Eyre himself happily tarnished the motives of the rioters. Black men with their blood up were rampant, as everyone knew: 'when the lives and proprieties of those intrusted to his care and the honour of their wives and daughters were in peril' – Eyre was playing to the gallery – he had no choice but to 'intimidate' with 'stringent measures' to prevent 'further risings in other parishes, nearly all of which were equally ripe for revolt'.[7] Killing a few hundred did the job.

As the guests departed, the protestors outside yelled slogans and demanded the 'bloodthirsty tyrant' be handed over. Men rushed forward as the carriages drove off, trying to block the way, and in the jostling some fell beneath the wheels. For the diners, it was a dismal last course, and on 23 August readers pored over *The Times*'s report, checking the guest list for casualties. Southampton's finest had turned out together with many eminent names – Shrewsbury, Cardigan, Kingsley . . . and then 'Darwin'.

It could only be William, his father felt in the pit of his stomach. William Erasmus Darwin, his eldest, whom he had coached in botany as a child and worked with that spring on flower physiology; William who had lived in his father's college rooms at Cambridge and, with his help, had bought into a Southampton bank. William, twenty-seven years old, lending the Darwin name to this celebration of wanton cruelty – the destruction of black families, the taking of black lives.[8] Darwin's worst nightmares teemed with such atrocities. His *Journal of Researches* had confessed as much, all 10,000 copies in circulation. Now *The Times* had his eldest supping with the devil.

The paper spread more gloom. The next day (24 August) it carried a report on the British Association giving 'anthropology' its own platform for the first time at an annual meeting. And who should chair the new subsection but Wallace, welcoming 'all students of man, by whatever name they might call themselves'. Wallace, who doubted the efficacy of sexual selection, who was prepared to meet the polygenists part-way: he was a compromiser, and sure enough, joining him on the platform were the Southern sympathizers Hunt and Carter Blake. Darwin knew that the Anthropologists sought professional recognition, and that Wallace wanted to see the men of science cooperate. But with what – fanaticism, eroticism, polygenism, Confederate values? Darwin skimmed the latest issue of the *Anthropological Review* in disgust. How could Wallace tolerate such 'insolence, conceit, dullness & vulgarity'? 'I fear he will not do what he ought in science', Darwin confided to Hooker.[9]

Worse came on 25 August. Hunt delivered a long attack on an 'illogical Darwinism' which – *The Times* paraphrased – had misled men of science 'to the inference of the original unity of the human species'. A 'more accurate scientific age' would consider 'the polyge[n]ist hypothesis the more probable'. There it was, in the Thunderer's black and white, Hunt gloating while standing on an official platform. He caused such a discussion that it went on 'much beyond the usual' closing hour.

A copy of Hunt's paper came not long afterwards and a sorry Darwin surveyed the damage. It was a hatchet job, an attack on Huxley and through him Darwin. Hunt in exclamatory style publicly fingered Huxley in revealing that 'some Darwinites are Monogenists' and 'in this country are even now teaching . . . that there is, at the present day, but one species of man inhabiting the globe'! It flew in the face of the world's authorities – Agassiz, Morton, Nott – and Broca's French 'know too well the business and the methods of science' to waste time on racial unity. Vogt's Germans were protesting against such 'premature speculations' based on Darwin's theory. Races do not change, nor is there 'a single authenticated example' of one having done so. If 'Unity versus Plurality' were to be settled by properly qualified anthropologists, 'the decision would' go to 'the polygenists'.[10]

Hunt was goading Darwin out of the closet, forcing him to stand up and be counted. Climatic differences could not have produced the skulls and brains of the different races. So, said Hunt, when Huxley investigates how their 'psychological characters' were formed, he will conclude that 'mankind is composed of several species'. This will make him 'a logical disciple of his great master'. Darwin scrawled a double 'X' in the margin. 'Logical' Darwinism would entail separate racial lines! And if climate could not alter race, what could? asked Hunt. Not natural selection (and Darwin had to agree – most *racial* characters were useless in the struggle for existence). The 'unity hypothesis' was 'an article of faith' for Darwinites, a dogma as unfounded as the parsons' belief in Adam and Eve.

'At present', Hunt continued, 'we are quite unable to show the causes which produce the formation of the different races of which the different species of man is composed'. Darwin's pencil slashed twice beside this. He knew he had a solution in sexual selection. Again he marked Hunt's parting gibe: 'I beg to express a wish that, in consideration of the conflicting views held on this subject, Mr. Darwin himself may be induced to come forward, and tell us if the application of his theory leads to unity of origin as contended for by Professor Huxley'.[11] The reclusive Darwin, still standing in the shadows as Huxley fought his public battles on the unity question, was being taunted.

William came home from Southampton for the weekend of 22–24 September 1866. Downe was drenched and it poured non-stop. Darwin's feelings had evidently been unfit for a letter, and William's presence at the banquet was thrashed out face-to-face. An Eyre confrontation was brewing nationally too. The prosecution fund was one-third subscribed. A rival 'Eyre Defence and Aid Fund', set up after the banquet, was running adverts daily in *The Times* showing a growing list of KCBs, FRSs, MPs and military men on its committee (and many subscribers who had once owned West Indian slaves). The Earl of Shrewsbury presided, and beside him was the Darwins' old dining companion, the 'nigger'-hating Carlyle (who would have been happy to see Eyre made '*Dictator* over Jamaica for the next 25 years'). *The Times* portrayed the two committees cancelling each other out, the chairmen Carlyle and Mill occupying opposite 'unphilosophical extremes', their cases proving that 'Mr. Eyre must be either all white or all black.'[12]

William evidently considered the Jamaica Committee's effort to prosecute misguided, even as he apparently denied that he was at the Eyre welcoming banquet. But, in that case, why was the name 'Darwin' in *The Times*? Was the guest list released to the press? It seemed likely, so why was 'Darwin' on that list? No other local bankers were reported present (as William might have noted), so why was he alone invited?[13] Was he *never* involved with Eyre's arrival in Southampton? There was no wriggling free, no more for a Darwin than a Hunt or an Agassiz who endorsed racial wrong. By Monday morning, after the weekend confrontation, Darwin was 'giddy & unwell' for the first time in months.[14]

Darwin had to accept that William was telling the truth, but he wouldn't drop the matter. Something had to be done. A letter to *The Times* would only bring more publicity. But mightn't it be put *quietly* on record that William had been wronged and the family disgraced? A complaint might be made at the highest level. But where?

High on a hillside, a mile and a half from Down House, stood an ancient oak, the pride of the Holwood estate. Darwin would ride out from his garden into Downe Valley, then north to Holwood Farm and uphill on the narrow track by the noble tree. He had passed it countless times on his way to Keston Common, where he dug up insect-eating sundews for his glasshouse, or collected saxifrages (rosy-flowered rock plants). Emma liked to take friends there in the pony carriage.[15] Immensely old, its huge trunk was hollow, with massive roots tumbling out above-ground to serve as picnic seats. Under this tree in 1787 William Wilberforce had heard God's call to abolish the slave trade. Now a fine stone bench, carved with his words, bore witness to that legendary moment – the inscription dated '1862' and 'by permission of Lord Cranworth'.

The Cranworths were occasional visitors, and his lordship raised Emma's low opinion of the nobility. Cranworth sent money to the Downe parish charities each year (via Darwin, who kept the books), and as staunch Liberals they saw eye-to-eye on Jamaica. The events at Morant Bay showed Cranworth that, even thirty years after emancipation, the planters still let slavery's 'Incubus' oppress the island. Surely Cranworth would sympathize with Darwin; perhaps he would have a word in the right ear about *The Times*'s slur. As former Lord Chancellor (until a few months previously) it was in his power. Had the Liberals not left office, the Lord of the 'Wilberforce oak' might have been called to sit in judgement on ex-Governor Eyre. That's how worried Darwin was by the affair – it occasioned a private complaint to the former head of the English judiciary.[16]

Hundreds that autumn responded to the Jamaica Committee's public appeal to make up the £10,000 needed for Eyre's prosecution. The *Anti-Slavery Reporter* printed their circular, calling for 'the rancour of race' to be put aside. But a certain rancour re-entered when an antagonistic *Pall Mall Gazette* noticed Lyell and Huxley's names among the subscribers. It inquired whether their 'peculiar views on the development of species have influenced them in bestowing on the negro that sympathetic recognition which they are willing to extend even to the ape as "a man and a brother".' The suggestion was that the beastly science of evolutionary 'brotherhood' had led to a beastly politics, and Darwin's disciples had shown how low they would stoop to destroy an eminent public servant.

A hard-headed Huxley dismissed the allusion to the famous Wedgwood medallion. His support was not motivated by any particular 'admiration of the negro – still less by any miserable desire to wreak vengeance' on Eyre. The issue was simply: does 'killing a man in the way Mr. Gordon was killed constitute murder in the eye of the law or does it not?' It was a legal question, not a scientific one, nor – and Darwin surely disagreed – a moral one. For Huxley it sufficed that 'English law does not permit good persons, as such, to strangle bad persons as such'.[17]

Friends fell out over this. 'How odd Huxley joining the Eyre prosecution fund – I suppose you approve': Hooker was right, Darwin did. After twenty years, Hooker – the new Director of Kew Gardens in 1865, and a botanist with wide experience of collecting plants in India – knew Darwin better than anyone outside the family. On this issue they disagreed. In a riot situation, legalities 'are fiddle sticks', Hooker insisted.

He spared Darwin what he really thought of 'negroes'; the Eyre defence committee was told instead. Hooker drummed up a sexual threat, as did so many in this racial context. The Jamaican black 'is pestilential'. 'When his blood is up, very cruel acts' follow. Hooker approved of 'those British officers in India who shot their wives before blowing themselves to pieces, rather than allow what they loved' to be raped by the mutinous sepoys. He would have sanctioned the same in Jamaica rather than see white women face such 'unutterable horrors', had not Eyre saved them.[18]

Even Darwin's best friend was on the wrong side. Darwin had never felt easy with Hooker's contempt for democracy and Civil War politics. Just as worrying was his elitist belief that 'Blood, Blunt [money], Brains, Beauty' would evolve a natural aristocracy 'or there is no truth in Darwinism'. Even if the Negro was not 'the equal of the Englishman', the real question was whether he was a moral being, of one blood with white people. Hunt's Confederate science and many Eyre supporters said 'No'. Darwin believed passionately that he was. He put his money where his mouth was and subscribed £10 to the Jamaica Committee on 19 November 1866. His name was published in *The Times* before the month was out, with those of Mill, Huxley, Lyell and a hundred others: 'you are on the right (that is *my*) side', Huxley cheered.[19] Darwin was being forced out of his shell as always – pushed into taking a public stand. One more shove might see him lining up in print with Huxley on the deeper issue of racial origins.

On the 22nd, Darwin and Emma went to London to be with Erasmus. Their sister Susan had died and the family treasures from The Mount (their childhood home in Shrewsbury) were being auctioned. That weekend William too arrived at Erasmus's. Conversation turned to Eyre's prosecution. Darwin's donation to the 'Ten Thousand Pounds Fund' had joined those from 700 others, putting plenty in the pot. Indictments were being prepared using the windfall, black witnesses brought in from Jamaica. The papers were full of the story, and some had turned against Eyre's defenders for their 'ostentatious show of opinion'[20] at the Southampton banquet. William, who opposed the prosecution, made a 'foolish' joke about the Jamaica Committee having enough money left to celebrate Eyre's conviction with their own banquet. William remembered vividly what happened next:

> [T]wo subjects which moved my Father perhaps more deeply than any others were cruelty to animals & slavery – his detestation of both was intense, and his indignation was overwhelming in case of any levity or want of feeling on these matters. With respect to Governor Eyre's Conduct in Jamaica he felt strongly that J. S. Mill was right in prosecuting him[.] I remember one evening at Uncle Eras' we were talking on the subject, and as I happened to think it was too strong a measure to prosecute Governor Eyre for murder, I made some foolish remark about the prosecutors spending the surplus of the fund in a dinner. My father turned on me almost with fury and told me if those were my feelings I had better go back to Southampton . . . Next morning at 7 o'clock or so my Father came up into my bedroom and sat on my bed and said that he had not been able to sleep from the thought that he had been so angry with me [–] he spoke in the most tender & gentle way. So I said that it had served me quite right for my stupid joke; after a few more kind words he left me.[21]

Making merry over a conviction for the mass murder of blacks was too much to take, even from a son.

Eyre was never indicted. Hunt kept up the attack on the 'ignorant, prejudiced Stupid and bigoted Humbugs of the Jamaica Committee who were hunting to the gallows . . . the brave Heroic Man who had saved Jamaica'. After Hunt resigned the Anthropologists' presidency in 1867, Eyre was tipped to replace him. With his 'knowledge of . . . the mixed-breed population of our West India islands' and his 'well known humanity', backers believed he would 'do much to remove the erroneous impression in the minds of the ignorant that the Fellows of the Anthropological Society have an antipathy to the lower species' of humans.[22]

With philanthropy dying, the world seemed increasingly inhospitable. Baited by Hunt, goaded by Hooker, laughed at by his son, Darwin came home to find the Cranworths asking to call with an 'old friend', Kingsley. Eyre's eulogizer was dying to meet the author of the *Origin of Species*. Darwin was probably not in the best mood when the first volume of Lyell's tenth edition of *Principles of Geology* arrived. After Lyell's equivocating *Antiquity of Man*, Darwin had hoped his old mentor would redeem himself.[23] He took one look, shut the book and waited for volume two.

Wallace was now more hindrance than help. His new pamphlet, *The Scientific Aspect of the Supernatural*, must have taken Darwin by surprise. Its fifty-seven pages were replete with clairvoyance, apparitions, hauntings and séances, from dozens of authorities for whom, Wallace believed, there was no question of delusion. He announced that ' "spirit" alone feels, and perceives, and thinks . . . It is the "spirit" of man that is man.'

From old socialists disillusioned by the failures of reform to old ladies spirit-rapping, spiritualism was a fast-growing parlour religion to counter an increasingly materialistic age. The pamphlet had been pulled together from his series in the twopenny *English Leader* for August and September 1866. The radical rag had been founded by the old socialist secularist George Holyoake, and it was well read by the working-class leaders. If politics couldn't provide, the spirit forces would make sure evolution honed a social Utopia.[24] Darwin, his science anchored to competitive individualism, hated socialism. And from Darwin's perspective, the naturally selected hierarchy of intellect that he saw formed by racial competition would be levelled if all men were equally endowed with spirit-minds – and man himself would stand for ever above the animals that had formed him. Wallace seemed again to be cutting the ground from under Darwin's science, and the cumulative impact of these weeks – capped by Wallace's revelation – probably forced Darwin's hand.

Certainly, before Christmas 1866, when his name first appeared in *The Times* among the Jamaica Committee subscribers, Darwin had decided to put 'man' back into the huge book he had planned and then worked at for a decade – *Variation of Animals and Plants under Domestication.* It was all but finished; humans would have to be slipped in quickly, the definitive answer to critics, baiters and backsliders alike.

The details on the way tame animals and garden plants could be changed had become 'an awful confounded pile' – a dry encyclopaedia of cases. A good part of the old 'Natural Selection' manuscript had gone in, and half as much again. By weight of facts, or simply by sheer weight, the two portly volumes would drive home mankind's race-making capacity. It would show how every loft or pen breed, common or exotic, had been shaped through a series of infinitesimally tiny changes by generations of farmers and fanciers. It would clinch the case for the common descent of all wild species from a single ancestor.[25]

He had the whole farmyard in focus, from cattle to cats, and then –

before the fowl, fish, invertebrates and plants – his centrepiece, a hundred pages on pigeons. No other species was treated in such lavish detail. Marshalling a mountain of evidence, he obliterated the notion that the extraordinary strutting, puffed and quiffed pouters, carriers, fantails, tumblers and jacobins had descended from different 'aboriginal stocks'. From Egypt, India, Persia and China, where pigeons had long been kept, the races had diverged 'from the same wild stock', the rock dove, the Adamic 'parent-form'. It was man who had 'created these remarkable modifications', just as man had created the modifications of his own species.

And as such the human races would appear in *Variation*, as self-made: not just in an annexe to a chapter, but having a chapter in their own right to round out the book. 'Man' was to make his debut as a self-selecting '*domesticated* animal'.[26]

Darwin started working up mankind's anatomical heritage. Off went queries about our rudimentary tail-vertebrae, or ear muscles inherited from monkeys. Old notes and recent books were plundered – the male pipefish that incubates its mate's eggs was 'for Man Chapt', he jotted; Lyell on human skulls and Mackintosh on morals the same. Rereading Huxley's *Man's Place in Nature* reminded Darwin to study 'Lubbock–Wallace–Lyell–Prichard–Pickering–Loring'. He began knitting the new material into an old 'rough draft' left over from the 'note on Man' planned for the 'Natural Selection' manuscript ten years earlier. But by Christmas he was 'much perplexed' at trying to fit it all in; a month later, he braced his publisher John Murray: the 'Chapter on man will excite attention & plenty of abuse', which 'is as good as praise for selling a Book'.[27] Yet he had only reached sexual selection, and from that moment on the subject started to balloon. Not surprisingly, for sexual selection was the point of the chapter. It would explain how male competition and female choice produced the human races from one ancestral stock; how men and women picked the desirable traits in their partners, just as breeders picked the traits in their pigeons. The chapter grew too large. In February 1867 he finally decided that his materials on human ancestry and racial divergence by sexual selection were so rich that they would have to make a self-contained 'small volume, "an essay on the origin of mankind" '.[28]

Hooker was the first to hear why the book was so important. 'I have convinced myself of the means by which the Races of man have been mainly formed', Darwin announced, apparently revealing the significance of sexual selection to his friend for the first time. 'I do not expect that I shall convince anyone else,' he added, no doubt recalling Wallace's rejection. That had been 'the heaviest blow possible', but Darwin was dogged when it came to Wallace. He told Wallace that his 'sole reason' for tackling man was to prove that 'sexual selection has played an important part in the promotion of races'.[29] The provocations were riling: not only Hunt's challenge that Darwin should come clean in his views on racial development, but Wallace's snubbing his prize explanation for human diversity. Darwin would put his utmost into sexual selection because the subject intrigued him, no doubt, but also for a deeper reason: the theory vindicated his lifelong commitment to human brotherhood.

Old helpers were revisited one last time. To London he trudged, to tap the Cape expert Andrew Smith on gnus' horns and Hottentots. Three decades since the *Beagle* had docked in South Africa, and Smith had been a never-failing source of information. But Darwin found a invalid recluse, who had abandoned his monumental five-volume study of African peoples after his wife's death. It would never see the light. Smith was now a Catholic convert uninterested in science and poring over the Scriptures. Darwin's 'small volume' would look increasingly anachronistic as old hands from the thirties disappeared. But finishing was imperative given the ragging by the Cannibals. Darwin fired off questions to South America, wanting details 'most about Negros' and their expressions; all, he explained, for a 'little essay on the origin of Mankind', to be published because 'I have been taunted with concealing my opinions'.[30] It would aim at what Lyell and Wallace signally failed to achieve, a totally naturalistic explanation of the 'common descent' of the races from a human ancestor.

Timing was critical. Even the perennially grumbling Blyth, that other standby, and the stock source on Indian races, was suggesting that each human type was of local ape extraction. But Darwin simply would not buy his old correspondent's 'observation on the similarity of the Orang & Malay &c'. Any superficial likeness between local ape and human race 'must be accidental'. Humans were too alike; their common inheritance too obvious. Such men were Darwin's fact-finders, and he admitted that he had 'picked up more facts on sexual characters' from Blyth than

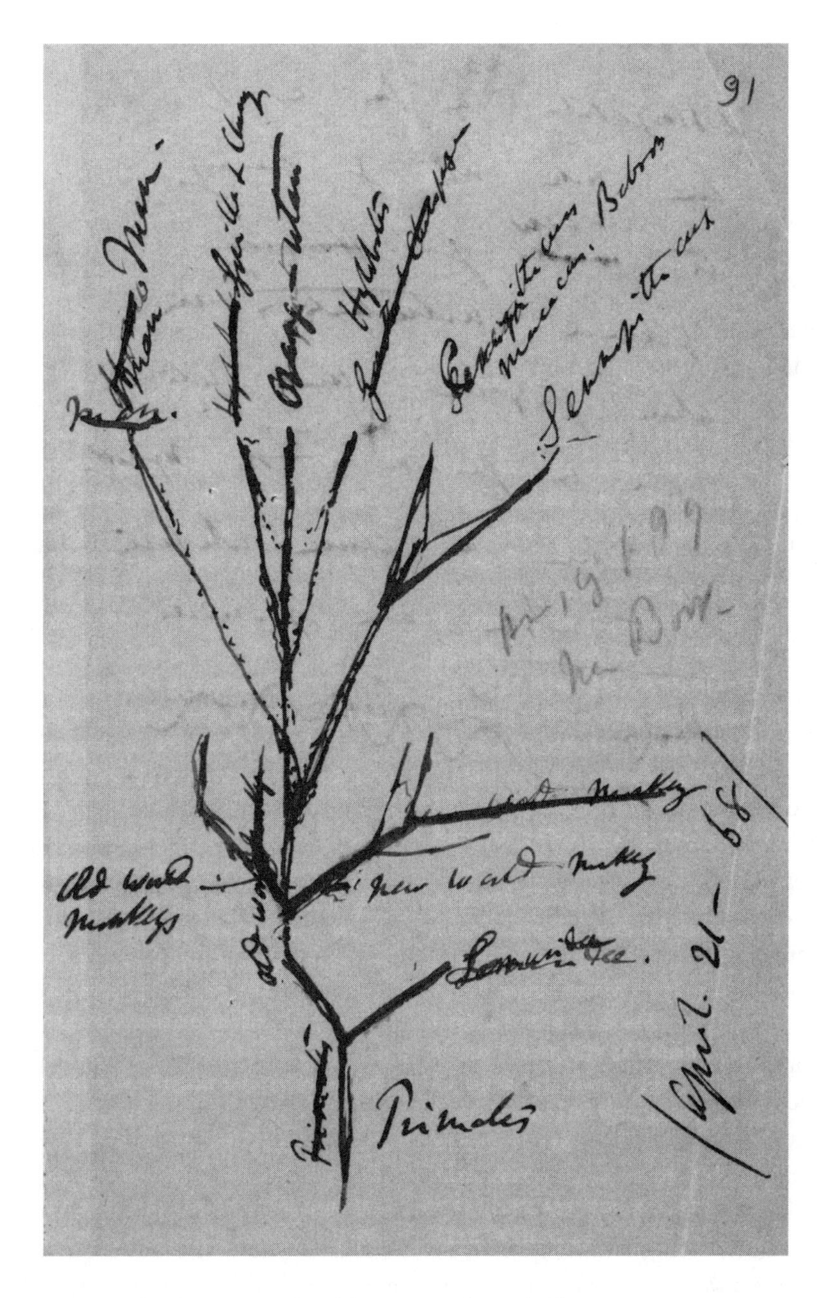

While some 'pluralists' imagined each human race descended from a separate ape species, for Darwin all humans shared a 'common descent', and only our shared human ancestor was descended from the progenitor of the gorilla, chimpanzee, orang, gibbon and Old World Monkeys. He depicted this graphically in a sketch of the Primate family tree, drawn in spring 1868 as he wrote the *Descent of Man*.

from anyone else.[31] But they had grown old waiting while Darwin slowly cogitated the human question.

Darwin remained prima donna-ish about his theories, and on sexual selection particularly. No one was paying much attention. Despite the allusion to it in the *Origin*, most critics had 'shown an utter ignorance of, or disbelief in, the whole matter'. Lyell thought Darwin had exaggerated its importance. Wallace, doubting its worth, was looking to the exceptions, while working up a rival theory of camouflage and coloration. He noted that many tropical butterflies mimicked distasteful species (enough to fool birds and collectors alike). The colours of these had nothing to do with sexual selection. Stranger still, some species flouted convention: the usually drab female, for whom survival was imperative as the exposed egg-layer, could herself be more brightly coloured. In one South American butterfly, *Pleris*, the male was whitish, where the female was a 'rich yellow and buff', precisely mimicking a poisonous species with which she mixed. So it was in numerous insects, Wallace pointed out in print and in person to Darwin.

A sensitive Darwin saw this mimicry explanation undermine the power of his presentation. Wallace's camouflage theory only compounded the problem. While Wallace accepted sexual selection in some groups, notably birds, he explained it very differently. In his view 'the primary action of *sexual selection* is to produce colour pretty equally in *both* sexes, but . . . it is checked in the *females* by the immense importance of *protection* and the danger of conspicuous colouring'. A countervailing natural selection had operated on the female, to ensure that she, sitting unobserved on a nest, had lost any gaudiness. Had she no need of drab camouflage, she too would be gaily coloured. Worse was Wallace's continuing rejection of any application of sexual selection to human races, the very rationale of Darwin's work. Wallace doubted 'if we have a sufficiency of fair & accurate facts to do anything with Man'. There spoke a spiritualist who was looking to the wrong realm.

Edgy, self-protective – or in his own words 'flat' – at the thought of his work 'being almost thrown away', Darwin now virtually shut the door on Wallace, again, announcing that he had collected his notes and intended 'to discuss the whole subject of sexual selection, explaining as I believe it does much with respect to man'.[32]

By March 1867 sexual selection was 'growing into quite a large subject'. It had become *the* cause of human racial divergence, and every corner of the animal kingdom was ransacked in search of supporting proof. Correspondents worldwide were tapped on everything from beetle armour to polygamy in birds and boars; plumage in poultry and pugnacity in partridge cocks – always with an eye, now, to female choice as well as male courtship and combat. Then there were the oddball suggestions, about staining tail feathers or pigeon breasts to test the effects on the paramour. And through it all were the queries about the criteria on which 'savage' women select, their sense of beauty, and so on. May 1868 found him putting it all together, in that encyclopaedic way of his, 'working up from the bottom of the scale'. 'Sexual Selection, which turns out a gigantic subject', was taking on a life of its own. Even he was getting 'sick of the ... everlasting males and females, cocks and hens'.[33]

The more he explored adornments and colours through the animal world, the more lopsided the project became. It was hardly looking like a book on '*man*' at all. By May he had 'as yet only got to fishes' – the bauble-brilliancy of male fish during the breeding season; and Agassiz didn't help by telling him that among Amazonian mouth-breeders, although the male incubates the eggs in his mouth, he is still brightly coloured. Darwin was now 'driven half mad' by these collateral points which required investigation. The title was firming up: he told his young German admirer Ernst Haeckel that the book was 'on the Descent of Man & on sexual selection, which will appear to you an incongruous union'. It was getting more incongruous by the day, as the tail threatened to wag the dog. Humans were being swamped, simply to prove the pervasiveness of sexual selection. This was partly to overwhelm Wallace's rival camouflage theory (which made reef fishes gaudy to match the background corals, an explanation Darwin rejected).[34] Thus Wallace once again takes credit for pushing Darwin, or at least for the topsy-turvy nature of his book.

Huxley came to the rescue. As President of the Ethnological Society, he was taking on Morton's old speciality, America's Indians, but from an opposite, migratory, common-origin perspective. Where Morton had talked of home-grown 'aboriginals', Huxley had waves of humans pouring in through Asia or the Old World, spreading down and intermixing, peoples whose monumental civilizations around the Gulf of Mexico rivalled those of ancient Egypt. Huxley offered to read Darwin's proofs, which raised Darwin's spirits, and he sped on with a new 'feeling of satisfaction instead of vague dread'. They were lifted higher by Haeckel, who was drawing crowds with his lectures on *Darwinismus* and looking himself at the ape origins of mankind: he told Darwin that he was 'awaiting your book on the descent of *man* with great impatience. I, too, believe that the "*Sexual Selection*" plays a very important role in it'. In truth, he meant in producing humans from apes. It seems from his own publications that he had no idea that Darwin was using sexual selection to explain the *races*.[35]

Darwin now had his title: *The Descent of Man*. And the book would enter a world prepared. The expat network so well tapped by Darwin – those colonial wallahs, missionaries and military surgeons, the fingertips of Britannia's reach – served Huxley no less. With Colonial Office help, he started a photographic round-up of Victoria's subject races in 1870. The French (always one jump ahead) had already made a photographic identity parade of the races, nude, in frontal and profile views. Her Majesty's peoples from Cape Town to Australia, Ceylon to Formosa, were now likewise photographed, near-naked, standing beside rules: standardized and ranked, some in convict chains, others dejected. They looked like Agassiz's daguerreotypes of the plantation slaves. The difference was that now the subjects were free, whatever the chains said. Pandering to the sort of voyeurism that manifested at freak shows, the London Stereoscopic and Photographic Company sold photos of naked 'savages' for the new stereo viewers on sale. Nakedness and dejection reinforced the public prejudice of indigenous degradation and childlike innocence. But even as the public became more familiar with them, the races became scarcer. The last native Tasmanian man died in 1869, but his surreptitiously skinned skull never survived the sea crossing to the College of Surgeons in London.[36] The last woman, Truganini, would linger on until 1876.

Darwin's delay meant that books had already started to fill the gap. Disciples had jumped in to explain the evolutionary meaning of these subject races. For his neighbour and protégé, John Lubbock, in *Pre-Historic Times* (1865), as for young E. B. Tylor's *Early History of Mankind* (1865) and *Primitive Culture* (1871), such savages provided living snapshots of the Stone Age. Flint artefacts, like those at the Aurignac grotto with its signs of ritual burial, were mated to aboriginal customs to explain mankind's original state. Humans hadn't fallen from grace, as the Secretary of State for India, the Duke of Argyll, argued (having found time to publish his own *Primeval Man* in 1869). The superstitions of lowly living races were not 'mutilated remnants' of once-higher beliefs; quite the reverse. Tylor took Prichard's 'survivals' – those customs that had made sense in ancestral cultures, and which Prichard worried would be lost with the tribal and racial extinctions – and rethought them as our present-day superstitions. Those irrational acts, from salt over the shoulder to spirit beliefs, were the empty remnants of once meaningful behaviours. They were rudiments, the equivalent of the coccyx bone that Darwin delighted in pointing to as the sign that human ancestors once had tails. Believing, like Darwin, in the 'psychic unity of man', these rising anthropologists – unlike Hunt's Anthropologicals – could use them to reconstruct their ancestors' beliefs.[37]

Given how much Darwin was relying on books published in the sixties, it is not surprising that he accepted their linear human scale. He too now thought of the distance between 'savage' morality and intellect and its highest expression in Europeans, and of the possibility 'that they might pass and be developed into each other' through innumerable gradations. He too could broach the series of stages from fetishism through polytheism to monotheism, as if all cultures passed through them. It suggested that white men had progressed further, leaving their superstitions as 'the remnants of former false religious beliefs'.[38] As he wrote the *Descent*, he adopted the prevailing cultural ladder, with the 'lowest' races on the first rung. It fitted with society's hardening attitudes towards blacks and the view that they could never be made into gentlemen – at least not for a very long time. Tylor explained why. He accepted that the mind was the same in all cultures, whether of Stone or Steam Age. Thus he admitted a Darwinian unity at root. The difference stemmed from accumulated education and technological prowess, which took centuries to acquire. The culture 'improved' at different speeds on its track from 'Stone' to 'Steam'. Those travelling slowly could not jump the huge evolutionary distance to the destination overnight. By falling into line, Darwin was actually backpedalling. He had softened his brilliant view of life's adaptive spread. Back came those 'high' and 'low' calibrations which he had forsworn in the 1830s.

Darwin was cobbling together his anthropology mainly from secondhand sources. But the *Descent*'s rationale was always human sexual selection, and that was being justified by evidence from across the entire zoological spectrum. Darwin himself admitted wearily that it was a 'gigantic subject'. And it was all in aid of explaining the human races, what Tylor named as among the 'most important problems' of anthropology: 'the relation of the bodily characters of the various races, the question of their origin and descent, the development of morals, religion, law'. The 'mind' might have been common across cultures, but subtle variations in the appreciation of beauty – the *beau idéal* – had led to a divergence from the common stock.

Finally, in 1869, one hindrance to the acceptance of his work was removed, with Hunt's premature death. This would also clear the way for Huxley and Lubbock to capture and decapitate the Anthropological Society as the new decade dawned, amalgamating it with the Ethnological as the 'Anthropological Institute'.[39]

Wallace hovered like a spectre to the end, always doubting. Given that the Scottish advocate John M'Lennan devoted a whole book to bride-capture in prehistory – the notion that wives were 'stolen' in raids on neighbouring tribes – one can understand Wallace's doubting that female choice could ever be a primary factor in human race formation. Wife-stealing, he told Darwin, would check it. And in the tribes he had studied 'the women certainly don't choose the men' (there spoke a man whose fiancée had jilted him a few years earlier). Males selected a wife as 'a servant. Beauty is I believe a very small consideration with most savages.'

Darwin knew M'Lennan, and Lubbock had puffed his 'rape of the Sabines' approach in *Primitive Marriage*, which linked the need for bride capture to female infanticide, and even saw in the idea of snatching accomplices – 'groomsmen' – the origin of the modern-day custom of the 'best man'. But it cast the same shadow over sexual selection as did concubine slavery and arranged marriages. The women, as Wallace said, had no choice in the matter. Darwin himself believed that 'barbarians' treated their wives as useful slaves, although he doubted that they were valued solely for this because the women adorned themselves. But the subject was troublesome and he could only suggest that the captors picked 'the handsomest slaves according to their standard of beauty'. It was the final irony: newer studies of slavery were threatening Darwin's theory which itself originated in a reaction to an older slavery science.[40]

Where Wallace was concerned, there was one final twist. Sir Charles Lyell, mentor and tortured Darwin supporter for so long, had 'grown very old' (it was obvious on his visits to Downe, where his stories had started to ramble). He now put Wallace up to review the tenth edition of his *Principles of Geology*, which – as Darwin had feared – still made God's creative impulse drive human evolution. (It needed a 'profoundly scientific' man, Lyell advised the publisher; what he meant was it needed a sympathizer.) Wallace's review in 1869 argued that the 'Power' driving animal evolution to its Utopian conclusion had pre-adapted the huge 'savage' brain, giving an aborigine something far beyond his animal-like 'mental requirements'. People were being readied for the civilizational peak, guided by a spirit-world where Lyell also saw his own and mankind's elevation. Darwin begged to differ, and 'grievously'. On body, brain and race, now, Wallace stood apart and cheerfully said so. He expected to find 'more to differ from' in the *Descent of Man*, he told Darwin, 'than in any of your other books'.[41]

The onus was on Darwin to make human mental evolution plausible. He would present the interbreeding races as close in mental make-up. That was proved by their independent invention of similar artefacts. Stone tools and arrowheads were common to cultures through history and across the globe. He would not countenance Wallace's compromise idea that these mental attainments were reached by the various races *after* they had crossed the human threshold. Darwin was adamant:

> all the races agree in so many unimportant details of structure and in so many mental peculiarities, that these can be accounted for only through inheritance from a common progenitor; and a progenitor thus characterised would probably have deserved to rank as man.[42]

He would never capitulate on this core issue, so constitutive was it of his lifelong 'brotherhood of man' belief. It pushed him perhaps harder into proving mental evolution, as he finished the long journey. Now it was imperative to stump Wallace: to prove not only that the big brain had evolved naturally, but that Malthusian selection had gone on honing the social and moral faculties from 'savage' through civilized tribes.

So strong was his belief in its continuance that Darwin saw natural selection still working in the classes of modern society ('Social Darwinism' this would later be called). His nature and society had always been Malthusian, based on excess mouths and a scramble for limited resources. His governing Whigs in the 1830s had cut the outdoor charities and forced the able-bodied poor to compete; to the same end, Darwin's evolution thrust individual animals and plants in an overpopulated nature into rivalry. But now he incorporated the statistical and eugenic work of his cousin Francis Galton and old friend W. R. Greg (whose conservative migration away from his Edinburgh radicalism was complete). In the 1860s Galton and Greg both reapplied natural selection to society and raised fears of degeneration, because 'we do the utmost' to care for 'the imbecile, the maimed, and the sick'. Darwin echoed their fears about the 'poor and the reckless' breeding early and increasing the number of gutter urchins: 'Thus the reckless, degraded, and often vicious members of society, tend to increase at a quicker rate than the provident and generally virtuous members.' This was the hardening belief of the 1860s' middle classes. As a statistician tabulating working-class mortality, who assumed his own class's intellectual superiority, Galton's data reinforced his cousin's patrician beliefs. Darwin hoped that the 'weaker and inferior' would marry less frequently to help 'check' this enfeeblement of society. But, ever the humanitarian, he still declared that we must bear the consequences of the weak surviving 'without complaining'. Indeed, to curtail 'the aid which we feel impelled to give to the helpless' would no less cause a 'deterioration in the noblest part of our nature'.[43]

Darwin's book also endorsed the ethnic stereotyping so characteristic of mid-Victorians. He quoted Greg on the 'careless, squalid, unaspiring Irishman' from a race which 'multiplies like rabbits'. The Knox-ish xenophobia simply oozed into the text – that Middle England dislike of the Irish Catholic working classes. Darwin even refrained from removing the offensive quotation when an Irishman politely asked him to.[44] So Darwin's transference of Malthusian competition from politics to animal populations gave rise to a piece of supposed 'hard' science that was then reapplied to society in the *Descent of Man*, bolstered by the bigotry of the day.

Darwin knew that, however charming his iridescent birds and jewelled butterflies in the sexually coy selection chapters, he could expect outrage at an evolutionary derivation of mind and morality. Having natural selection account for religious devotion would only add insult to the injury. He tried applying balm by making an 'ennobling belief in God' an evolved virtue among the highest races – a gesture diplomatically tuned to a devout culture, for Darwin had long shed his own Christianity. This was where the worry lay, and before publication he started softening up his critics and readers. Out went a series of self-abasing letters. He warned an old *Beagle* shipmate that the book would 'disgust you'. An old adversary was told that 'I shall meet with universal disapprobation, if not execration'. And Wallace? Darwin thought it 'will quite kill me in your good estimation'. Even from Asa Gray he expected 'a few stabs from your polished stiletto of a pen' for broaching the evolution of morality.[45]

And so it came about. The long-awaited and long-gestated two-volume *Descent of Man, and Selection in Relation to Sex* was published on 24 February 1871. The following days saw 2,500 copies fly off the shelves so fast that an immediate 2,000-copy reprint was called up. The publisher John Murray was 'torn to pieces' by people wanting copies.[46]

The book was received in a 'storm of mingled wrath, wonder, and admiration'. 'Wrath' was reserved for the moral debasement, and 'admiration' for the collation of nature's charming ways of wooing. *The Times* thought the latter was 'one of the most delightful studies in natural history ever written'. Not so the rest of it. Morals had not evolved from brute instinct. Knowing right from wrong was a divine gift which guaranteed salvation. This gift also sanctioned Anglican society's policing of the populace: working-class activists spouting evolution had been jailed for their seditious attack on priestly privilege. Naturalizing morality as the revolutionary Paris Commune was passing laws to recognize bastards only invited *The Times*'s damnation as 'reckless'. Of course, Darwin's and Huxley's expensive, reverent books escaped the ire of the bench.[47] Still, the impeccably correct Darwin, a Justice of the Peace himself, had needed courage to make conscience the better part of troop instinct.

It was easier to broach this heresy now. An evolutionary explanation of religion could rest on Tylor's and Lubbock's work on its primitive prehistory of fetishism, dreams and animism. Here again Darwin was following his disciples in exploiting the aborigines' 'rude state of civilization' as snapshots of our own ancestry. The *Descent of Man* presented a sort of 'pinnacle' view of the past. A male norm was assumed, and Darwin too saw his British class at the apex of civilization. But he also displayed a gentleman's liberal sensitivities ('some savages take a horrid pleasure in cruelty to animals'). And like many, Darwin equated 'savagery' in its 'utter licentiousness' and 'unnatural crimes' with the values of his own under-class (two groups which the socialist Wallace held in higher regard). But by lowering 'savage' morality and raising ape capabilities, Darwin made the continuum towards civilization seem more feasible. For a moment, too, he could mute his rampant Malthusian individualism and replace the goal of individual success and happiness with 'the general good or welfare of the community'. And thereby he extended his explanation to account for the social evolution of what he held to be the highest virtue of civilized Europeans: their humanity towards other peoples and species.[48] It was a humanitarianism that Darwin took pride in. His anti-slavery and anti-cruelty ethic was inviolate. Yet the incongruity of his class holding this ethic sacrosanct while disparaging the 'lower' races (even as colonists displaced or exterminated them) is impossible to comprehend by twenty-first century standards.

However, the evolution of morals was a side engagement. Darwin had focused overwhelmingly on sexual selection to *prove* that all peoples could be extracted from one stock. It left an idiosyncratic book with the oddest centre of gravity. Darwin was his own man, and a rather old-fashioned one at that, reworking an agenda inherited from Prichard's generation, which sought to shore up racial unity as a bulwark against slavery. In the 1860s, the accumulating 'evidence of the descent of man from some lower form', as Darwin said in his introduction, 'appeared to me interesting', and he summarized much of it in the preliminary chapters of *Descent* before tackling the pièce de résistance, sexual selection as explaining man's racial ancestry – which occupied two-thirds of the book.

But this trajectory cannot be understood from the text itself. True, the book contains hints. Even the title is a tell-tale sign: 'Descent' had been Darwin's favoured word from the first, to convey a human-style hereditary ancestry for nature. The newer buzzword 'evolution' was coming into vogue around 1870, but Darwin kept the 'common descent' signification in the title. Few understood; some saw only negative connotations. Even Hooker wished it had been simply titled the *Origin of Man*.[49] But title and content evinced a deeper 1830s' provenance and meaning.

The racial slant left the *Descent of Man* scarcely like a book on 'human evolution' as we think of it today. Nor was it what the public had come to expect. Nothing was said of Lartet's Aurignac grotto, with its proof that flint-making peoples had lived alongside extinct cave bears. Nor those finds from a Brixham cave – notably an extinct bear's arm-bone, severed by the flint knives lying around, and clearly brought into the cave by humans. The Brixham evidence forced a belated acceptance of the hatchet finds of the maligned amateur archaeologist Boucher de Perthes at Abbeville in France. These flint tools had finally placed a long human prehistory beyond question. By 1870, the issue had gone beyond the contemporaneity of men and mammoths to the geological age of mankind.[50] But there was nothing on them in Darwin's book.

Then he tackled mankind's ape descent in an old-fashioned way, familiar to students of comparative anatomy for three decades. He resorted, for instance, to rudimentary organs, those inherited relics, like the coccyx, as pointers to human parentage. Or, with Haeckel's help, he delved into the womb, where the five-month human foetus had a visibly hairy pelt, as a sign that our ancestors were furry. With more ingenuity he studied the anatomy of throwbacks: he relayed disconcerting stories of people still able to wrinkle their scalps like baboons or twitch their ears. He piled on such anecdotal evidence, lost in a normalizing biology. From all of this he developed his infamous image of our ancestor (grist for so many caricatures) as 'covered with hair, both sexes having beards; their ears were pointed and capable of movement; and their bodies were provided with a tail'.[51]

Darwin deliberately forewarned readers that there were 'hardly any original facts' in these earlier human evolution chapters, inviting the readers to skip straight to sexual selection. Not merely nothing new, there was almost nothing *at all* on fossils. His only positive citing was the discovery by Albert Gaudry from the Paris Natural History Museum of the primitive Greek Miocene monkey *Mesopithecus*. Darwin made *Mesopithecus* the ancestor of macaques and langur monkeys (as had Vogt). But he introduced the fossil to illustrate how 'higher groups were once blended together' – how two primate lineages could be traced to a same starting point.[52] Divergence from a 'common stock' was the rule. The fossil was used to illustrate a principle, not trace the primate route.

The strangest omission was of fossil humans. There was barely a word on Neanderthal Man. First found in 1857 in the Neander caves outside Düsseldorf, the beetle-browed fossil man had been made widely known by Huxley's talks – so widely that the Catholic Cardinal Wiseman, one of Huxley's auditors, could issue a Pastoral denouncing anatomists who could weigh 'a solitary cranium . . . in the scale' against Scripture.[53] One could argue that Huxley's *Man's Place in Nature* had covered the subject; or that, since no one knew how old the Neanderthal skull was, Darwin need not discuss its significance. But the omission is interesting.

As is the lack of discussion of the ape brain. Acrimonious exchanges over the uniqueness of the human brain had first alerted polite society to the possibility of human transmutation in the early 1860s. Huxley had categorically denied Owen's claim that the ape's cerebral hemispheres lacked features found in humans (and called him a liar to boot). The two had clashed at the Royal Institution and the British Association, and sometimes weekly in the literary intelligentsia's Saturday paper, the *Athenaeum*. Owen's strategy had been to make bestial transmutation the issue and prejudice readers against Huxley. Darwin himself had come to despise 'the fiend, Owen', but he still put nothing about the proximity of human and ape brains in the *Descent of Man*.[54]

Or about the ecology of wild gorillas, apart from the odd aside on Dr Livingstone's observations. Gorilla behaviour just did not seem to interest Darwin, even though the public could not get enough of it. This huge ape was still a novelty, and one of the first exhibits in the Anthropological Society museum was a 'nearly perfect skin of a large male gorilla'. The gorilla's recent discovery explains its prominence in cartoons lampooning the *Origin of Species*. Mixing in racist slurs, *Punch* now depicted the Irish 'yahoo' as Mr G. O'Rilla. Perhaps one can understand Darwin's shunning of the adventurer Paul du Chaillu's sensationalist reports of the gorilla's ferocity in the 1860s, which made salacious tabloid copy but were seen by the Darwinians as useless. (It hardly helped that du Chaillu had backslapped Owen in London and returned to Africa to catch him a live gorilla in 1864.)[55] So while gorillas and mammoth-hunting humans had coloured reviews of the *Origin of Species*, Darwin ignored them in his book on human descent. What was missing only served to emphasize his single-minded approach.

Instead he moved the manuscript on gingerly to his *idée fixe*, the cause of human racial change. He was diverting all eyes towards sexual selection. And it dominated the text, running above 550 pages in print. He cleared rival explanations of the racial differences first: changing environments, greater exercise of certain parts, even natural selection, because it could only augment 'beneficial variations' and as 'far as we are enabled to judge . . . not one of the external differences between the races of man' was other than aesthetic. For Darwin they were skin-deep and a matter of preference. He presented scenarios for apes standing erect and being counted men, with all the correlated changes in the pelvis, spine, skull, hands and feet; but the most peculiar human attribute, hairlessness, he believed, served an ornamental purpose. Partners were increasingly chosen with thinning pelts. For Darwin pink or black skin had an aesthetic appeal, like hair type and face shape – people simply preferred these slightly divergent looks. And where the antagonistic Argyll in *Primeval Man* refused to believe that competition for the fittest could produce an 'unclothed and unprotected' helpless being (let alone that morality could emerge from troop instincts), Darwin countered that our ancestors' exposed state might have been the impetus for that cohesion so characteristic of human society. Our social-huddling progenitor would have taken strength in the 'sympathy and the love of his fellow-creatures'.[56]

Historical judgements on the *Descent of Man* can be quite skewed. (A late doyen of Darwin studies declared that 'there was no compelling reason that a book-sized manuscript on the subject of Sexual Selection must be combined with a book on the descent of man.') Yet the London press twigged the connection immediately. They encountered the *Descent* in its post-Civil War context where the hottest topic for a generation had been the 'Unity versus Plurality' debate. The *Daily News* pre-empted publication by a day and on 23 February 1871 launched favourably into Darwin's 'courageous, if hazardous, speculation'. His 'minute and painstaking observation' on sexual selection, with natural selection, was 'intimately related' to his explanation of 'the origin of man' by means of 'a pedigree extending rather beyond the time of William the Conqueror'.[57]

Darwin's talk of apes got the larger 'common descent' message across. Man 'is related to the gorilla, not as a grandson or great grandson, but as a grand-nephew or great grand-nephew', the London *Examiner* reported. Human beings, Darwin conceded, ended up with 'a pedigree of prodigious length, but not, it may be said, of noble quality'. To which *The Times* added obligingly – perhaps quelling Lyell's lingering fears – that the nobility 'would seem to be reflected backwards, on the Chinese system, according to which persons who distinguish themselves ennoble, not their children but their ancestors.'[58] This wasn't going to placate many.

But *The Times* also understood the book's racial angle. With natural selection failing to 'explain the differentiation of the various races of man', Darwin had resorted to sexual selection to account for physique and physiognomy. 'He thinks that females have entertained a preference for males, or males for females, possessing certain specialities of form either for use or ornament, and have thus exercised an unconscious but continuous selection in favour of such peculiarities. To support this theory he passes in review the whole animal kingdom.'[59] There was no doubt here about the book's *raison d'être*.

The book sold to a new generation with birds in their hair. The trend for feather bonnets was already leading to dire warnings about declining bird populations. From 1870 some twenty thousand tons of ornamental plumage was shipped into Britain yearly. The clinching proof of sexual selection came in the *Descent*'s four bird chapters (running to no fewer than 200 pages), and women in shimmering feather hats took Darwin's point that an aesthetic appreciation ran right through nature, from humans to hummingbirds. He extended a sense of beauty, thought by many to be uniquely human, to the birds, and he subtly suggested that there was an interchangeable appreciation. 'As women everywhere deck themselves with these plumes, the beauty of such ornaments cannot be disputed.' Darwin expected his readers to empathize on avian choice. But he was also clear that beauty was being appreciated instinctively by the birds, whereas it was cultivated by the highest orders in society. For it was *his* class Darwin was talking to: 'savages' and the 'uneducated' knew what they liked but lacked any higher aesthetic refinement.[60] Again, he was intimating the steps of evolutionary ascent through savagery towards a Victorian gentleperson's taste for the beautiful, whether in landscapes, music or adornment.

This sexual selection running up through nature carried on to prove human unity. To emphasize the point he sieved the old monogenist literature for illustrations. Bachman's evidence on the fertility of mulattoes was roped in, and Darwin's own experience on the *Beagle* of 'mongrel' Brazil and the Indian–Spanish blend on Chiloé island. He had seen the most 'complex crosses between Negroes, Indians, and Europeans', and 'such triple crosses afford the severest test' of fertility, as he knew from his countless botanical and pigeon experiments. But the weightiest argument 'against treating the races of man as distinct species' rested on their natural (non-crossed) gradations. The variability of the races explained why polygenists had been confused on the number of human types – suggesting anything from two to sixty-three species (Agassiz reckoned eight, Morton twenty-two). Anyone who 'had the misfortune' to monograph a zoological class and be cursed by every family's variability ('I speak after experience' said the barnacle expert) would not be surprised at human variation. And in this variation lay the answer to the big question, the one so 'much agitated' of 'late years'. The polygenists had to 'look at species either as separate creations or as in some manner distinct entities'. They were fixed, invariable. But those who accepted evolution, wrought by selection from the variations to which humans are prone, 'will feel no doubt that all the races of man are descended from a single primitive stock'.[61] Human genealogy was more than a metaphor for Darwin's common-descent evolution. It was the prototype explanation.

So many points of racial anatomy spoke of this shared origin, so many shared 'tastes, dispositions and habits'; so much coincidence in a common love of 'dancing, rude music, acting, painting, tattooing'. In short, body, mind and manners could not 'have been independently acquired, they must have been inherited from progenitors' who possessed them. As final proof of this unity, the 62-year-old Darwin took readers all the way back to his experience in Edinburgh, forty-five years earlier. Like the swallowtail-coated and 'civilized' Fuegians on the *Beagle* who, by their 'many little traits of character', showed 'how similar their minds were to ours', so 'it was with a full-blooded negro with whom I happened once to be intimate'.[62] That was the 'blackamoor' John, now a warm, distant memory for Darwin: the ex-slave bird-stuffer who taught the boy week in, week out, during those lonely, frosty days in Edinburgh.

Darwin was approaching the end of his long evolutionary haul, begun in those heightened anti-slavery years of the 1830s. He now paid a telling debt. In the *Descent* he talked of the moral 'scale' and the heights achieved by mankind. The intellectual distance between a savage and 'a Newton or Shakspeare' was huge, but bridged by every conceivable intermediate. As was the 'moral disposition between a barbarian, such as the man described by the old navigator Byron, who dashed his child on the rocks for dropping a basket of sea-urchins, and a Howard or Clarkson'.[63] There was Thomas Clarkson, the family hero, dead these twenty-five years, given his due in Darwin's crowning evolutionary work on the united races. Clarkson, prime mover of the anti-slavery campaign in Britain, whose local lieutenants and backers Darwin had known as a boy, took his place at the moral apex.

So Darwin's journey to the *Descent of Man* ended, on the anti-slavery terrain where it began. Clarkson, who had come to epitomize Darwin's sacred cause, sat at the top of Darwin's pantheon – he was the Newton or Shakespeare of Darwin's moral cosmos.

Darwin's 'common descent' image of evolution was foundational to all of this. It was anathema to the scientific-pluralism-and-slavery lobby because it had originated in an Adam-and-Eve monogenist text. And to a certain extent, they were right: the Christian 'brotherhood of man' was ultimately constitutive of common descent in Darwin's day. Sympathizers knew that his scientific support for racial unity, now detached from its religious roots, was inimical to the pluralist pro-slavery message. As one Darwin and anti-slavery supporter spelled it out, 'Many of our narrow prejudices and false theories in regard to Race – ideas which have been at the base of ancient abuses and long-established institutions of oppression – are removed' by Darwinian ethnology. This was Darwin's dream too. 'Finally', he said in the *Descent of Man*, 'when the principles of evolution are generally accepted . . . the dispute between the monogenists and the polygenists will die a silent and unobserved death.'[64]

Notes

Abbreviations Used in the Notes

See the Bibliography for publication details of titles below and in the Notes.

ANB	*American National Biography*, 24 vols (Oxford University Press, 1999)
APS	American Philosophical Society
AR	*Anthropological Review*
Autobiography	Barlow, ed., *Autobiography of Charles Darwin*
CCD	Burkhardt et al., eds, *The Correspondence of Charles Darwin*
CJ	*Journals of the House of Commons*
CP	Barrett, *Collected Papers of Charles Darwin*
CUL	Cambridge University Library
DAR	Darwin Manuscripts, Cambridge University Library
Descent	C. Darwin, *Descent of Man*, 1871
Diary	R. D. Keynes, ed., *Charles Darwin's 'Beagle' Diary*
DRC	Darwin Reprint Collection, Cambridge University Library
ED	Litchfield, *Emma Darwin*, 1904 (or 1915 edition, as cited)
Foundations	F. Darwin, ed., *Foundations of the 'Origin of Species'*
JASL	*Journal of the Anthropological Society of London*
Journal	C. Darwin, *Journal of Researches*, 1839 (or 1860 reprint of 1845 edition, as cited)
LJ	*Journals of the House of Lords*
LJH	L. Huxley, *Life and Letters of Sir Joseph Dalton Hooker*
LLD	F. Darwin, ed., *Life and Letters of Charles Darwin*
LLL	K. M. Lyell, ed., *Life, Letters, and Journals of Sir Charles Lyell*
LTH	L. Huxley, *Life and Letters of Thomas Henry Huxley*
Marginalia	Di Gregorio and Gill, eds, *Charles Darwin's Marginalia*
MLD	F. Darwin and A. C. Seward, eds, *More Letters of Charles Darwin*
Narrative	FitzRoy, *Narrative of the Surveying Voyages*
Natural Selection	Stauffer, ed., *Charles Darwin's Natural Selection*
Notebooks	Barrett et al., eds, *Charles Darwin's Notebooks*
ODNB	*Oxford Dictionary of National Biography*, 61 vols (Oxford University Press, 2004)
Origin	C. Darwin, *On the Origin of Species*, 1859
PP	*House of Commons Parliamentary Papers*, 1801–
PP, AP	*Parliamentary Papers*, Accounts and Papers
PP, CP	*Parliamentary Papers*, Command Papers
PP, HCP	*Parliamentary Papers*, House of Commons Papers
PP, ST	*Parliamentary Papers*, Slave Trade

TASL	*Transactions of the Anthropological Society of London*
Variation	C. Darwin, *Variation of Animals and Plants under Domestication*, 1868
Wedgwood	B. and H. Wedgwood, *The Wedgwood Circle*
Wedgwood	Wedgwood Archive, Wedgwood Museum
WFP	Alfred Russel Wallace Family Papers, Natural History Museum

INTRODUCTION: UNSHACKLING CREATION

1. *Notebooks*, T79; Napier, *Selection*, 492.
2. Eiseley, 'Intellectual Antecedents', 1; R. Dawkins, 'Darwinism', 234.
3. *Notebooks*, B231.
4. *CCD*, 9:163.
5. *Notebooks*, B232.
6. *Marginalia*, 683.
7. W. P. Garrison to Darwin, 4 Oct. 1879, DAR 165; Darwin to W. P. Garrison, Oct. 1879 and 16 Oct. 1879, quoted in W. P. and F. J. Garrison, *William Lloyd Garrison*, 4:199n. The 16 October letter is in private hands, a copy is with the Darwin correspondence project, CUL. The remaining autograph letter has not been found.
8. Howard Gruber was the first to describe 'compassion for all living things' as a component of Darwin's 'family Weltanschauung' (H. E. Gruber, *Darwin*, chap. 3).

I THE INTIMATE 'BLACKAMOOR'

1. *Journal* (1860), 499, 500; Clarkson, *History*, 1:27.
2. Walvin, 'Rise', 149, 150; Anstey, *Atlantic Slave Trade*, xix; Drescher, 'Whose Abolition?', 165.
3. Anon, *Negro Slavery*, 86; Long, *History*, 1:491–2; Clarkson, *History*, 1:374; Lorimer, *Colour*, 214.
4. King-Hele, *Collected Letters*, 338.
5. King-Hele, *Erasmus Darwin*, 115; Uglow, *Lunar Men*, 259, 410; Shyllon, *Black Slaves*, 184–209.
6. King-Hele, *Essential Writings*, 149 ('Loves of the Plants', III, 441–8, 455–6).
7. E. Darwin, *Temple*, 136 (IV, 73–4).
8. Barclay, *Inquiry*, 148; *Marginalia*, 32.
9. King-Hele, *Essential Writings*, 101; King-Hele, *Erasmus Darwin*, 308; E. Darwin, *Zoonomia*, quoted in R. Porter, 'Erasmus Darwin', 45.
10. Dolan, *Josiah Wedgwood*, 201–2; *Wedgwood*, 38 (cf. King-Hele, *Erasmus Darwin*, 81); Farrer, *Correspondence*, 71; Meteyard, *Life*, 2:40.
11. McNeil, *Under the Banner*, 81; McKendrick, 'Josiah Wedgwood', 42, 46; Dolan, *Josiah Wedgwood*; Watts, *Gender*.
12. Oldfield, *Popular Politics*, 163–6; E. G. Wilson, *Thomas Clarkson*, 48–51; King-Hele, *Collected Letters*, 327–8.
13. Guyatt, 'Wedgwood Slave Medallion'; Oldfield, *Popular Politics*, 155–63; Meteyard, *Life*, 2:566; Compton, 'Josiah Wedgwood'; Bindman, 'Am I Not a Man' for the artistic context.
14. E. Darwin, *Botanic Garden*, 1: 101 (II, 315–16), 110 (II, 421–30); King-Hele, *Collected Letters*, 345; Farrer, *Correspondence*, 88.
15. Farrer, *Correspondence*, 216–17; J. Wedgwood I to O. Equiano, 19 Sep. 1793, Wedgwood E26/18983, also in Farrer, 217–18; Fryer, *Staying Power*, 102ff.; J. Wedgwood I to A. Seward, Feb. 1788, Wedgwood E26/28978, transcribed in J. Wedgwood, *Personal Life*, 246–8; notices from J. F. Garling, 1788–1791, Wedgwood L111/21073–21078.
16. R. I. and S. Wilberforce, *Life*, 1:318; W. Wilberforce to J. Wedgwood I, 6 Apr. 1790,

Wedgwood E36/2770; J. Wedgwood I to W. Wilberforce, 7 Apr. 1790, Wedgwood E36/2771.

17. William Darwin's Recollections, in *Darwin Celebration*, 11–12; R. Keynes, *Annie's Box*, 112–13; Desmond, *Politics*, 186; Clapham Sect, *ODNB*; Howse, *Saints*, chap. 6; Bradley, *Call*, chap. 7.

18. Fryer, *Staying Power*, 110; Gascoigne, *Cambridge*, 223–34, 255–6; Howse, *Saints*, 16–17; Clarkson, *History*, 1:241–58; E. G. Wilson, *Thomas Clarkson*, 21–4; J. D. Walsh, 'Magdalene Evangelicals'.

19. Peckard, *Am I Not a Man?*, 2, 6, 9; Peckard, quoted in J. Walsh and R. Hyam, *Peter Peckard*, 16. The title's slogan had been adopted months earlier by the Abolition Society: Compton, 'Josiah Wedgwood', 52.

20. M. V. Nelson, 'Negro', 206, 211; Curtin, *Image*, 43–4; Horsman, *Race*, 49; R. J. C. Young, *Colonial Desire*, 149; Anon, *Negro Slavery*, 60.

21. Jordan, *White*, 492–3; Greene, 'American Debate', 387; Edward Long, quoted in R. J. C. Young, *Colonial Desire*, 149–51; Lorimer, *Colour*, 24; Malik, *Meaning*, 62.

22. Reported by Henry Cockburn while Darwin was resident in Edinburgh: Cockburn, *Anti-Slavery Monthly Reporter*, no. 9 (26 Feb. 1826), 90; Clarkson, *History*, 1:205, 537–9.

23. T. Clarkson to J. Wedgwood I, 17 June 1793, Wedgwood E32/24742, also Farrer, *Correspondence*, 215–16; E. G. Wilson, *Thomas Clarkson*, 63–5; Thomas Clarkson, *ODNB*. On Sierra Leone: Asiegbu, *Slavery*, chap 1.

24. J. Wedgwood I to T. Clarkson, 18 Jan. 1792, Farrer, *Correspondence*, 187–8; T. Clarkson to J. Wedgwood I, 19 and 20 Jan. 1792, Wedgwood E32/24739, 24740; T. Clarkson to J. Wedgwood I, 25 Aug. and 22 Oct. 1791; 22 Oct. 1792, Wedgwood E32/24738, W/M 1303, E32/24738A; Katherine Plymley diaries, 20–21 Oct. 1791, Shropshire Archives 1066/1, [12–13]; E. G. Wilson, *Thomas Clarkson*, 72–4. On the leading role of women, the 'domestic guardians of morality', in the sugar boycotts: Midgley, 'Slave Sugar Boycotts', 143 and *Women*.

25. T. Clarkson to J. Wedgwood I, 18 Apr. 1794, Wedgwood E32/24743, also Farrar, *Correspondence*, 220–21; T. Clarkson to J. Wedgwood I, 30 Apr. 1794, Wedgwood E32/24744; E. G. Wilson, *Thomas Clarkson*, 74, 80–86.

26. Walvin, 'Rise', 155; Anstey, *Atlantic Slave Trade*, xxii; Drescher, 'Decline Thesis', 3ff.; Jennings, *Business*, 108–12.

27. Farrer, *Correspondence*, 159–60; Clarkson, *History*, 2:341–2; Darwin, 'An Autobiographical Fragment', *CCD*, 4:338–40; *Autobiography*, 21–8; Roberts, 'Josiah Wedgwood'.

28. *ED*, 1:35–8, 66–70; *Autobiography*, 28–9, 55; Drescher, 'Whose Abolition?'.

29. Katherine Plymley, *ODNB*; 'Panton Corbett, Esq.', *Gentlemen's Magazine*, new ser., 45 (Jan. 1856), 87–8; E. G. Wilson, 'Shropshire Lady' and *Thomas Clarkson*, 12; Walvin, *Slavery*, 3; Clarkson, *Strictures*, vi–vii.

30. Browne, *Charles Darwin*, 1:8–10; *Notebooks*, M1, 9, 10, 29, 42, 44, 156; Corbet, *Family; PP*, AP, 1834 (613), xlviii, 217, 228; 1836 (583), xliii, 161, 218–20; Clarkson, *History*, 1:467–8, wrongly identifying the uncle as the archdeacon's father-in-law (cf. Corbet, *Family*, 2:20).

31. J. Wedgwood I to J. Plymley, 2 July 1791, Wedgwood E26/18998, also Farrer, *Correspondence*, 162–3; Clarkson, *History*, 2:341–2; E. G. Wilson, *Thomas Clarkson*, 86; Katherine Plymley diaries, 19 Sep. 1823 and 8 Mar. 1824, Shropshire Archives 1066/130, pp. [35–42] and 1066/132, pp. [12–14]; *CCD*, 1:31. In her diary for 12 Sep. 1825 (Shropshire Archives 1066/134, p. [49]), the Archdeacon's sister Katherine Plymley wrote: 'dined at the Hall, met Joseph, Messrs. Flower, hale [?], Charles Darwin, Henry Johnson, & W. Harrison, Waties's Curate at Corley'. These are probably the members of the day's shooting party, who dined at Longnor Hall. 'Joseph' was Joseph Corbett; 'Flower', 'hale' and 'W. Harrison' have not been identified except the last as a curate. 'Henry Johnson' was a younger contemporary of Darwin's at Shrewsbury School and a fellow student

at Edinburgh. 'Waties' was the Revd Waties Corbett, rector of Corley, Shropshire, the Archdeacon's son.

32. Darwin, 'An Autobiographical Fragment', *CCD*, 2:440; Drescher, 'Slaving Capital'; Costello, *Black Liverpool*; Fryer, *Staying Power*, 58–60; Law, *History*, 11–15; Clarkson, *History*, 1:373.
33. Anon, *Negro Slavery*, 39–42, 53. On the depiction of torture and sexual assault on slave women to garner support for the movement: Paton, 'Decency'.
34. Anon, *Negro Slavery*, 60, 64–5; Walvin, 'Rise', 157–8; Davis, *Inhuman Bondage*, 104.
35. *Journal* (1860), 499, 500; Anon, *Negro Slavery*, 57; Davis, *Inhuman Bondage*, 183, 199; Berger, 'American Slavery', 188.
36. Stange, *British Unitarians*, 35. Shrewsbury householders signed the petition from 8 Mar. 1824; it was received in the Commons on the 16th: *CJ*, 79 (1824), 168; *Salopian Journal*, 10 Mar. 1824; diaries of Katherine Plymley, Shropshire Archives, 1066/132, [12–14]. The Wedgwoods and their relative James Mackintosh contributed £279 13*s*. 0*d*. to the Anti-Slavery Society between 1823 and 1831, of which £230 came from Josiah's sister Sarah. Annual dues were one guinea: Anti-Slavery Society, *Account*, 1823–31. Few supporting families matched this commitment and no woman exceeded Sarah Wedgwood's total contribution.
37. *CCD*, 1:29, 40; 'Darwin in Edinburgh – I', *St James's Gazette*, 16 Feb. 1888, 5–6.
38. *PP*, CP, 1837 (92), xxxv.1, 261–5; A. Grant, *Story*, 2:424–5; 'Medical Faculty of the University of Edinburgh', *Medico-Chirurgical Review*, 20 (1833–4), 315–19.
39. *Autobiography*, 47; *CCD*, 1:2; Colp, *To Be*, 4; 'Darwin in Edinburgh – I', *St James's Gazette*, 16 Feb. 1888, 5–6; J. H. Ashworth, 'Charles Darwin'.
40. *CCD*, 4:36; *Caledonian Mercury*, 12, 23 Jan. 1826; Audubon, *Audubon*, 1:167, 179, 212, 216, 218.
41. Freeman, 'Darwin's Negro Bird Stuffer'; *Autobiography*, 51; *CCD*, 1:28–9; Rice, *Scots Abolitionists*, 23; Myers, 'In Search', 168. We have no other details on John's later years except for an acknowledgment of 'Messrs. FENTON and EDMONDSTON, bird-stuffers, Edinburgh' in MacGillivray's *History of British Birds* (1:v) in 1837, so John seems to have been living in the city a decade later.
42. B. Silliman, *Journal*, 1:209–10; Walvin, *Black*, 189; J. Bachman, *Doctrine*, 105; Myers, 'In Search', 158. Blacks themselves tapped into the mood: placard-bearing Negroes would parade the streets bearing woodcut signs of chained, kneeling slaves with the caption 'Am I not a man and a brother?': Walvin, *Black*, 190; Edwards and Walvin, *Black Personalities*, 46.
43. *Anti-Slavery Monthly Reporter*, no. 9 (26 Feb. 1826), 91–2; Cockburn, *Memorials*, 267–8 on earlier Edinburgh meetings; Edinburgh Anti-Slavery Society, *First Annual Report*; Anti-Slavery Society, *Account* (1823ff.).
44. *CCD*, 1:31, 36–41; *John Bull*, 19 Mar. 1826, 94; 2 Apr. 1826, 108–9; 16 Apr. 1826, 125; on the African Institution, Sarah Elizabeth Wedgwood to her brother J. Wedgwood II, [July 1826], Wedgwood E28/20451; receipts for 1822, 1825 and 1827, Wedgwood E38/28184–28186; Ackerson, *African Institution*.
45. *Descent*, 1:232; Lorimer, *Colour*, 16.
46. William Ferguson, *ODNB*; Pendleton, *Narrative*, 43 lists many others; Lorimer, *Colour*, 31 and 57, 215–17 on other black Edinburgh graduates; Little, *Negroes*, 212; Myers, 'In Search', 171; Shyllon, *Black Slaves*, 160.
47. Waterton, *Wanderings*, 210–15, 242–3; on snake-catching, *Caledonian Mercury*, 28 November 1825; Ishmael, *Guyana Story*; Edgington, *Charles Waterton*, 30, 108, 110–11; Freeman, 'Darwin's Negro Bird-Stuffer'.
48. Carroll, 'Natural History'; S. Smith, *Works*, 2:277, 281; J. Smith, 'Waterton's Wanderings in South America', *Caledonian Mercury*, 6 Jan. 1827; *ED* (1915), 1:91.
49. Irwin, *Letters*, 18; S. Smith, *Works*, 2:294; Grasseni, 'Taxidermy'.
50. Grasseni, 'Taxidermy'; Smith, *Works*, 2:290; Edgington, *Charles Waterton*, 6; for 'vices' see 'Campbell's Travels in South Africa', *The Port-Folio, and New-York Monthly Magazine*, 2 (1822), 265–78 (reprinted from a British review).

51. *Essequebo and Demerary Gazette*, 11 Feb. 1804, 4 May 1805, 31 Jan. 1807; Waterton, *Wanderings*, 113–14; Ishmael, *Guyana Story*, for primary sources on Charles Edmondstone.
52. J. Mackintosh to W. Wilberforce, 27 July 1807, in R. I. and S. Wilberforce, *Life*, 3:302–3; Mrs J. (Bessy) Wedgwood II to her brother Baugh Allen, 26 Dec. 1807, Wedgwood E57/31938; Anstey, *Atlantic Slave Trade*, 401; Northcott, *Slavery's Martyr*.
53. Fanny Allen to her sister Jessie Allen Sismondi, 26 May [1824] (typescript copy), Wedgwood E59/32533; S. Smith, *Works*, 3:143, 145; *ED*, 1:206–7; R. Stewart, *Henry Brougham*, 35; Ishmael, *Guyana Story*.
54. Waterton, *Wanderings*, 10, 17, 116–17; *CCD*, 1:34.
55. *Descent*, 1:232; Edwards and Walvin, *Black Personalities*, 51.

2 RACIAL NUMB-SKULLS

1. Darwin's 1826–7 Edinburgh reading list, DAR 271.5; Freeman, 'Darwin's Negro Bird-Stuffer'; *CCD*, 1:75, 88, 90; Darwin's Edinburgh diary for 1826, DAR 129; Darwin's 'Early notes on guns and shooting', DAR 91:1–3; *Autobiography*, 44; Thompson, *Rise*, 270.
2. *Edinburgh Journal of Natural and Geographical Science*, 1 (1830), 272; Mudie, *Modern Athens*, 221; Audubon, *Audubon*, 1:154; *Autobiography*, 52; *PP*, CP, 1837 (92), xxxv.1, 183–4.
3. Ainsworth, 'Mr. Darwin'; *PP*, CP, 1837, 1:142, 632, and Appendix, 114–18, 'Syllabus of Lectures on Natural History'.
4. *PP*, CP, 1837 (92), xxxv.1, 189, 533, 658, 677, 847; Chitnis, 'University', 87; A. Grant, *Story*, 1:376. Robert Knox published on the platypus, echidna, wombat and cassowary from 1823 to 1828, and Robert Grant on the bandicoot.
5. Darwin read this edition and annotated it: Darwin's 1826–7 Edinburgh reading list, DAR 271.5; *Marginalia*, 173–5.
6. R. Jameson, *Essay*, 334, 378–98, 418, 421, 429–30; Secord, 'Edinburgh Lamarckians', 7–8. Darwin's geological gradualism, strengthened by reading Charles Lyell, had manifested already by 1835, during the *Beagle* voyage: Hodge, 'Darwin', 19–20.
7. R. Jameson, *Essay*, 117–21, 406–7.
8. *Autobiography*, 52; *CCD*, 1:19, 41.
9. Secord, 'Behind the Veil', 166; L. Stephen, *English Utilitarians*, 2: chap 8.
10. Horsman, 'Origins', 390–94.
11. Horsman, 'Origins', 391.
12. Mudie, *Modern Athens*, 253; Audubon, *Audubon*, 1:146, 160, 164, 182–3, 188, 191, 205–7.
13. Shapin, 'Phrenological Knowledge', 224ff.; Calvert, *Illustrations*. By 1831–2 a third of the membership of phrenological societies in London and Edinburgh was medical: Erickson, 'Phrenology'.
14. Audubon, *Audubon*, 1:157, 204; Andrew Boardman's introduction to G. Combe, *Lectures*, v, vi.
15. R. E. Grant, 'An Essay on the Comparative Anatomy of the Brain', bound in 'Essays on Medical Subjects', University College London, MS Add 28, ff. 14–15, 25–26, attacking Gordon, 'Doctrines', 243, 253–4; R. E. Grant, *Dissertatio*, 8, on Erasmus; Jeffrey, 'System', 253, 261–6, 312; *Phrenological Journal*, 1 (1824), iv. On Grant's science: Sloan, 'Darwin's Invertebrate Program'. On his social position and radicalism: Desmond, *Politics*, chap. 2.
16. G. Combe, *Lectures*, 91; *Phrenological Journal*, 1 (1824) 46–55, 54–5; Barclay, *Inquiry*, 380; Lizars, *System*, vii–viii.
17. B. Silliman, 'Phrenology', 68–9; G. Combe, *System*, 45; G. Combe, *Essays*, lv. Monro lecturing on animals: *Lancet*, 1 (1833–4), 97.
18. J. Epps, 'Elements', 100, 118; E. Epps, *Diary*, 61; Sandwith, 'Comparative View', 493.

A copy of Epps' *Internal Evidences of Christianity Deduced from Phrenology* (1827) was sent to the Plinian Society, via Browne (Kirsop, 'W. R. Greg', 388 n17), proving the impossibility of keeping the subject out. The report of Greg's attempt, with Darwin present, to prove that animals possess all the human faculties is in the Plinian Society Minutes M.S.S., 1 (1826–8), f. 51, Edinburgh University Library Dc.2.53.

19. *Edinburgh Journal of Natural and Geographical Science*, 1 (1830), 276; *PP*, CP, 1837 (92), xxxv.1, 674; Chitnis, 'University', 90–91; 'Phrenology and Professor Jameson', *Phrenological Journal*, 1 (1824), 55–8.
20. G. Combe, *Essays*, 382; Shapin, 'Phrenological Knowledge', 224–5.
21. Cox, *Selections*, 169–70; G. Combe, *Essays*, 583.
22. Cox, *Selections*, 160–71; G. Combe, *System*, 171, 567; G. Combe, *Lectures*, 110, 304. On phrenology and race: Stepan, *Idea*, chap. 2; Haller, *Outcasts*, 13ff.
23. G. Combe, *System*, 299, 581, and *passim* on race characteristics; G. Combe, *Life*, 189–90; G. Combe, *Essays*, xlvii; G. Combe, *Constitution*, 180.
24. Shapin, 'Homo Phrenologicus', 57; G. Combe, *Essays*, xlvii, 338.
25. B. Young, 'Lust'; G. Combe, *System*, 129–30, 177, 567; Gibbon, *Life*, 1:160.
26. G. Combe, *System*, 277, 300; murderers figure prominently in Calvert, *Illustrations*. In Joshua Brookes's museum the assassins were in the same cabinet as the 'exotic' (i.e. foreign) crania: Anon, *Museum Brookesianum*, Fifteenth Day's Sale, 86–8.
27. R. Owen, *Catalogue*, 12–14; Lloyd, *Navy*; Gough, 'Sea Power'; G. S. Ritchie, *Admiralty Chart*.
28. Knox, 'Inquiry', 217; Lonsdale, *Sketch*, 24–5; Haller, *Outcasts*, 5 on Blumenbach's Caucasians; Gould, *Mismeasure*, 401–2.
29. *Autobiography*, 47; Rae, *Knox*, 2, 40, 55; Struthers, *Historical Sketch*, 92; 'Pencillings of Eminent Medical Men: Dr. Knox, of Edinburgh', *Medical Times*, 10 (1844), 245–6.
30. Knox, 'Lectures', 283; Lindfors, 'Hottentot'; Lonsdale, *Sketch*, 11–12.
31. Rae, *Knox*, 16; E. Richards, 'Moral Anatomy', 385; Hodgkin, *Catalogue*, unpaginated; Rae, *Knox*, 14–15.
32. Knox, 'Inquiry', 210–12.
33. Darwin's 1826–7 Edinburgh reading list, DAR 271.5; *Autobiography*, 49; Browne, *Charles Darwin*, 1:84; Barclay, *Inquiry*, viii, 143–4, 148; *Marginalia*, 32; Rae, *Knox*, 35–6; Lonsdale, *Sketch*, 39.
34. Rae, *Knox*, 21–2; Ross and Taylor, 'Robert Knox's Catalogue', 273; 'Pencillings of Eminent Medical Men: Dr Knox, of Edinburgh', *Medical Times*, 10 (1844), 245–6.
35. Knox, 'Contributions', 501, 638; E. Richards, 'Moral Anatomy'; Desmond, *Politics*, chap. 2; Rae, *Knox*, 35; Lonsdale, *Sketch*, 36–7.
36. Bell, *Essays*, 4–10, 33–4, 61–2, 68, 147ff.; Browne proposing Darwin on 21 Nov. 1826, Plinian Minutes M.S.S., 1 (1826–8), ff. 34–5 and f. 34, 21 Nov. and 5 Dec. 1826, Edinburgh University Library Dc.2.53, ff. 34–5 for the attacks on Bell; Desmond, *Politics*, 68 on Browne's religious lunatics. It would be over thirty years before Darwin managed 'to upset Sir C. Bell's view' and show the evolutionary pattern to facial expressions: *CCD*, 15:141.
37. Monro, *Morbid Anatomy*, 14–15; 'Pencillings of Eminent Medical Men: Dr Knox, of Edinburgh', *Medical Times*, 10 (1844), 245–6.
38. *Autobiography*, 48; Audubon, *Audubon*, 1:146, 174; Feltoe, *Memorials*, 106, talking of Joshua Brookes; *CCD*, 1:25, 183.
39. G. Wilson and A. Geikie, *Memoir*, 97; Monro, *Morbid Anatomy*, x; H. E. Gruber, *Darwin*, 43; *Notebooks*, M143; Audubon, *Audubon*, 1:176; Monro lectured five days a week for seventy-five minutes for over four and a half months: *PP*, CP, 1837 (92), xxxv.1, 313.
40. Darwin's 1825–7 lecture notes on medicine, DAR 5.16–17; Browne, *Charles Darwin*, 1:60–61; Shapin, 'Politics', 149–53; *Phrenological Journal*, 3 (1826), 166–8, 252–8; vol. 4 (1827), 377–407.
41. W. Hamilton, *Lectures*, 1:648–51, 659.

42. Kirsop, 'W. R. Greg', 379–80, 82. Hudcar Mill was a failing concern under W. R. Greg: M. B. Rose, *Gregs*, 40, 44, 63, 64, 81.
43. Kass and Kass, *Perfecting*, 20–25, 39–42; M. Rose, *Curator*, 22–6, 30–35.
44. Kass and Kass, *Perfecting*, 70–71, 75, 93, 99–100; M. Rose, *Curator*, 41–2. Hodgkin and Morton were in Paris with Knox in 1821–2, lured by the availability of cadavers. Hodgkin was also receiving and probably distributing 'African Committee' literature: T. Hodgkin to S. G. Morton, 30 June 1822, Samuel George Morton Papers, APS.
45. Hodgkin, *Catalogue*, exhibits 93, 171, 983; T. Hodgkin to S. G. Morton, 19 May 1828, 12 May 1830, Samuel George Morton Papers, APS; G. Combe, *Notes*, 2:36–7.
46. T. Hodgkin to S. G. Morton, 17 Dec. 1824; T. Hodgkin to Captain J. Norton, 17 Dec. 1824; T. Hodgkin to S. G. Morton, 28 July 1825, Samuel George Morton Papers, APS.
47. *ED*, 1:130; T. Hodgkin to S. G. Morton, 17 Dec. 1824, Samuel George Morton Papers, APS; Kass and Kass, *Perfecting*, 127. One of Jessie Sismondi's few surviving journals, for 1826, written while Darwin was at Edinburgh, is in DAR 258.2064.
48. Kielstra, *Politics*, 44, 114–15; Thomas, *Slave Trade*, 582–4, 622; R. I. and S. Wilberforce, *Life*, 4:202ff., 238; Audubon, *Audubon*, 1:108.
49. Fanny Allen to her sister Jessie Allen Sismondi, 28 Dec. 1818 (typescript copy), Wedgwood E59/32513; Fanny Allen to her sister Mrs J. (Bessy) Wedgwood II, 2 June [1819], Wedgwood W/M 40; *ED*, 1:165–8; Sismondi, quoted in Lutz, *Economics*, 45; J. Mackintosh to his sister-in-law Fanny Allen, [after 7 Mar. 1831], Wedgwood 59/32699. Cf. J. C. L. Simonde de Sismondi to J. Wedgwood II, 20 May 1833, Wedgwood 11/9847.
50. Audubon, *Audubon*, 1:204; *Phrenological Journal*, 3 (1826), 476–81 for recruiting from the Philadelphia group; Spencer, 'Samuel George Morton's Doctoral Thesis', 324; A. A. Walsh, 'New Science' on the Philadelphia phrenologists.
51. Erickson, 'Phrenology'; Audubon, *Audubon*, 1:191; Spencer, 'Samuel George Morton's Doctoral Thesis', 335–6; G. Combe, *Lectures*, 48, 100.

3 ALL NATIONS OF ONE BLOOD

1. King-Hele, *Erasmus Darwin*, 10–18; J. Pearson, *Exposition*, 522; *Autobiography*, 57.
2. *CCD*, 1:123; Jenyns, *Memoir*, 51.
3. Scholefield, *Memoir*; Overton, *English Church*, 64–5; Jenyns, *Memoir*, 10–12; Walters and Stowe, *Darwin's Mentor*, 7–8. On the African Institution: Sarah Elizabeth Wedgwood to her brother J. Wedgwood II, [July 1826], Wedgwood E28/20451; receipts to J. Wedgwood II, 1822–7, for 3-guinea annual subscriptions, Wedgwood E38/28184–28186; Turley, *Culture*, 55; Howse, *Saints*, 138ff.; receipts to J. S. Henslow, 1822–6, for 1-guinea subscription, and Anti-Slavery Society report, 12 Aug. 1823, Henslow scrapbook, Suffolk County Record Office, Ipswich, HD 654/1; Addenbrookes, *State*, 1826–31. The prize book won by Henslow in 1810 or 1811 that fed his passion for Africa was probably J. Barrow, *Travels*, 1807. Thanks to John Parker, director of the Cambridge University Botanic Garden, for discussing his research on Henslow with us.
4. *Autobiography*, 65; Jenyns, *Memoir*, 52–5.
5. Henslow, *Address*, 7, 8; Bury, *College*, 44–45; Walters and Stow, *Darwin's Mentor*, 115–18; Bourne, *Palmerston*, 61, 234, 239–47.
6. Douglas, *Life*, 126, 136, 158, 183; J. W. Clark and T. M. Hughes, *Life*, 1:335–7; C. H. Cooper, *Annals*, 559; *CP*, 2:72; *Autobiography*, 64; Bourne, *Palmerston*, 243. As bachelors, Sedgwick and Whewell would spend Christmas with the family of Viscount Milton, Whig MP and fervent evangelical – 'Praise God Milton' to his critics – who thought taxes should be withheld until Parliament was reformed: Kriegel, 'Convergence', 425.
7. Whewell, *Philosophy*, 1:iv, xviii, 5–6; *Autobiography*, 66; Snyder, *Reforming Philosophy*; Yeo, *Defining Science*.

8. E. Darwin, *Zoonomia*, quoted in R. Porter, 'Erasmus Darwin', 45; Whewell, *Philosophy*, 1:652, 656, 658, 665, 679, 682, 688–9, 700; Whewell, *History*, 3:476.
9. Whewell, *Philosophy*, 1:657, 662; Augstein, *James Cowles Prichard's Anthropology*, chap. 1; Allen, *Cambridge Apostles*, 3; *CCD*, 1:112.
10. J. B. Sumner, *Treatise*, 1:350, 363, 372, 380; Desmond and Moore, *Darwin*, 48.
11. Moule, *Charles Simeon*, chap. 5; Searby, *History*, 328–9; Carus, *Memoirs*, 658; Zabriskie, 'Charles Simeon', 111; Stock, *History*, 1:141–2; J. Stephen, 'Clapham Sect', 578.
12. Stock, *History*, 1:211–12; Crosby, *Ecological Imperialism*, 236–7; Salisbury, *Border County Worthies*, 196. On Hongi: Jonathan Holmes's gloss on an article by Gerda Morgan in *Cambridge Review*, 2 Dec. 1927, in www.queens.cam.ac.uk/Queens/Record/2001/History/Maoris.html; Carus, *Memoirs*, 217, 235, 258; Samuel Lee and Hongi Hika, *ODNB*. Thanks to Simon Schaffer for bringing Lee to our attention.
13. The debates: Cambridge Union Society, *Laws* 1824–5, and *Laws* 1834 (no record has been found of Oxford's debating slavery until 1829: Oxford Union Society, *Oxford Union Society*, 1831). For the petitions: *Cambridge Chronicle*, 18 Apr. 1823, 13 Feb. 1824, 24 Feb. 1826; *Cambridge Independent*, 25 Feb. 1826; *CJ*, 78 (1823), 238; vol. 79 (1824), 120; vol. 81 (1826), 111; *LJ*, 55 (1822–3), 635; vol. 58 (1826), 70. The University had petitioned Parliament in 1814 over the continuing international slave trade: Cambridge University Archives, CUL CUR 50 (8).
14. Moule, *Charles Simeon*, 53–4, 97; Gray, *Cambridge*, 269–73; Loane, *Cambridge*, 185; Milner, *Life*, 464–5, 663; Searby, *History*, 417; *Cambridge Independent*, 11 Mar. 1826; Anti-Slavery Society, *Account*, 1826.
15. Hilton, *Mad, Bad, and Dangerous People*, 175–7; Sivasundaram, *Nature*, 3–4. The Whig evangelicalism of Darwin's Cambridge is not to be to be confused with the much narrower Tory evangelicalism of the next generation, still less with the liberalism of the Broad Church at mid-century: Brent, *Liberal Anglican Politics*, 126–33, 262–74; and S. F. Cannon, *Science*, chap. 2, which dismisses contemporary evidence of Simeon's continuing impact on Cambridge men with the lofty, 'students often make mistakes' (p. 51).
16. Hilton, *Age*, 42–3, 211; Newsome, *Wilberforces*, 5–12; Hilton, *Mad, Bad, and Dangerous People*, 401–7 (cf. Davis, *Problem*, 288–90); Jenyns, *Memoir*, 132, 136, 142, 145; *CCD*, 1:110. Henslow can hardly be called a theological 'Liberal' (Walters and Stow, *Darwin's Mentor*, chap. 9).
17. Henslow, *Sermon*, 6–7; Jenyns, *Memoir*, 132, 134, 143; Carus, *Memoirs*, 690; Newsome, *Wilberforces*, 11.
18. Not *quite* penitent, in the printed *Sermon* (p. iii) Henslow commended an editorial in the premillennialist organ *Morning Watch* (Anon, 'On the Study') even while seeking assurance from an old Sim, the Revd William Marsh (whose wife was treasurer of the Colchester Ladies' Anti-Slavery Association), that the doctrine of two resurrections was not necessarily unsound (W. Marsh to J. S. Henslow, 20 Mar. 1829, CUL Add. MSS 8176:106; Midgley, *Women*, 80, 82). Watching Irving fail to revive a corpse by prayer finally dispelled Henslow's illusions (Jenyns, *Memoir*, 143). Sedgwick took the whole matter so seriously that he set up a divinity examination with a question about rejecting the Millennium as 'a fable of Jewish dotage' (J. W. Clark and T. M. Hughes, *Life*, 1:340). The phrase, expunged from the Thirty-nine Articles, invited candidates to rethink their attitude to the Jews, whose conversion was expected by evangelicals as a prelude to Christ's Kingdom (Bar-Josef, 'Christian Zionism'). The Clapham-inspired London Society for Promoting Christianity amongst the Jews was bankrolled by Simeon's fanatical friend the Revd Lewis Way (Carus, *Memoirs*, 364–5, 412, 474ff., 511, 550, 574, 635, 659; *The Times*, 14 Jan. 1822, 1; Lewis Way, *ODNB*; Forster, *Marianne Thornton*, 128, 132). Simeon himself sponsored the Cambridge branch, which met as Way's sickly son Albert attended Henslow's soirées – 'poor little Way' to his beetling-companion Darwin (*CCD*, 1:89, 233, 492; Carus, *Memoirs*, 595; Bury, *College*, 18). The 'Jew Society', like the Bible Society, the CMS and the Anti-Slavery Society, was born of evangelical hopes to Christianize the world (Bradley, *Call*, chap. 7). But a persistent problem was the fate of

those never reached by the gospel. Were they to be eternally damned? Henslow's sermon answered with St Paul's picture of 'the whole creation groaning and travailing', the lilies of the field, the fowls of the air, everything that on the sixth day God saw was 'very good'. All these live and die under God's care, Henslow reasoned, so 'are the heathen not better than fowls? . . . Can he forget to redeem?' While the Jews knew of only one resurrection for their own, in St John's post-millennial 'second resurrection', 'the heathen shall be saved "every man according to his works" ', 'all the sons of God', including the Jews (Henslow, *Sermon*, 9–10, 12). Darwin was unconvinced. Fifty years later he would impress Henslow's problem on a young evangelical naturalist with ambitions reminiscent of his own: the problem of 'so small a proportion of mankind having ever heard of Christ' (Romanes, *Thoughts*, 182).

19. Overton, *English Church*, 51, 61; Halévy, *England*, 434–5. Compare Chadwick, *Victorian Church*, 1:396–7 with Hilton, *Mad, Bad, and Dangerous People*, 174–84 and Bebbington, *Evangelicalism*, 271.
20. Quoted in William Paley, *ODNB*; *CCD*, 1:104, 199; Paley, *Principles*, 1:219; *Autobiography*, 65; Clarkson, *History*, 1:465.
21. Paley, *Principles*, 1:96, 237–8.
22. Paley, *Principles*, 1:97; J. M. Herbert's recollections, DAR 112.B64–6; *CCD*, 1:106.
23. Katherine Plymley diaries, 28 Jan. 1826, Shropshire Archives 1066/135, [33–4]; Marshall and Stock, *Ira Aldridge*, chap. 5; Oldfield, *Popular Politics*, 23–38; Ira Aldridge, *ANB*; *CCD*, 1:20, 31. Darwin might have noted the black thespian's visit if only because the famously precocious 'Young Roscius', William Batty, was a Shrewsbury physician's son who studied at Christ's College (Rackham, *Christ's College*, 202–4).
24. *Autobiography*, 55–6. Thanks to Peter Rhodes for the fresh translation.
25. Anti-Slavery Society, *Account*, 1823–30; *ED*, 1:241; Dolan, *Josiah Wedgwood*, 388; *The Times*, 16 Mar. 1825, 3; receipts and invoices with Thomas Allbut and Son, Wedgwood E32/24749–24750; invoices, 17 June and 18 July 1829, Wedgwood E32/24751, 24779; cashbook, Hanley and Shelton Anti-Slavery Society, Wedgwood E32/24784A.
26. *Anti-Slavery Monthly Reporter*, no. 74 (Jan. 1831), 44; bills and receipts to Francis Wedgwood, 1829–31, Wedgwood E32/24747–24784A; the Revd H. Moore to J. Wedgwood II, 10 Mar. 1824, Wedgwood E49/29810; cashbook, Hanley and Shelton Anti-Slavery Society, 1830 expenditures, E32/24874A; J. Wood to J. Wedgwood II, 25 Jan. 1828, Wedgwood E30/22792; J. Wedgwood II to T. Poole, 10 Oct. 1830, Wedgwood E3/2200.
27. *ED*, 1:136, 241; 2:277; *Wedgwood*, 198; J. C. Wedgwood, *History*, 188; Mrs J. (Bessy) Wedgwood II to her sister Emma Allen, 16 Jan. [1827], Wedgwood W/M 39; Anti-Slavery Society, *Account* 1823–31; Sarah Elizabeth Wedgwood to her sister Emma Wedgwood, 19 Dec. 1826, Wedgwood W/M 182; Mrs J. (Bessy) Wedgwood II to her sister Emma Allen, 23 Feb. [1829], Wedgwood W/M 39.
28. Mrs J. (Bessy) Wedgwood II to her sister Fanny Allen, 29 May 1828, Wedgwood W/M 68; Midgley, *Women*, 229 n93; *ED*, 1:242; Anti-Slavery Society, *Ladies' Anti-Slavery Associations*, 5; notebooks of Fanny Wedgwood, 1829–32, Wedgwood W/M 1162; Mrs J. (Bessy) Wedgwood II to her sister Emma Allen, 11 May 1828, Wedgwood W/M 39.
29. Mrs J. (Bessy) Wedgwood II to her sister Emma Allen, 2 Mar. 1830, Wedgwood W/M 39; Midgley, 'Slave Sugar', 155 (cf. Sussman, *Consuming Anxieties*); Mrs J. (Bessy) Wedgwood II to her sister Fanny Allen, 13 Aug. 1828, Wedgwood W/M 68; *ED*, 1:242, 298, 359, 361.
30. R. I. and S. Wilberforce, *Life*, 5:314–15; *ED*, 1:282–3, 295; *Wedgwood*, 201; *Autobiography*, 55.
31. Charles Grant, Robert Grant, *ODNB*; Fanny Mackintosh to her cousin Sarah Elizabeth Wedgwood, [31 July 1830], Wedgwood W/M 167; Fanny Mackintosh to her cousin Sarah Elizabeth Wedgwood, Saturday [1828 or later], Wedgwood W/M 167; Fanny Allen to her sister Jessie Allen Sismondi, 6–7 Mar. [1829], Wedgwood E59/32556.
32. *CCD*, 1:96, 97, 539; Mackintosh, *General View*, 152–3, 230; *ED*, 1:306.

33. Martineau, *Harriet Martineau's Autobiography*, 1:355; Charlotte Wedgwood to her sisters Fanny and Emma Wedgwood, [30 Nov. 1826], Wedgwood W/M 146; Sarah Elizabeth Wedgwood to her sister Emma, 19 Dec. 1826, Wedgwood W/M 182; Mrs J. (Bessy) Wedgwood II to her sister Emma Allen, 26 Jan. 1827, Wedgwood W/M 39; *Anti-Slavery Monthly Reporter*, no. 36 (May 1828), 227; no. 80 (May 1831), 259; *ED*, 1:248; Fanny Allen to her sister Jessie Allen, 8 July [1823], Wedgwood E59/32525.
34. Fanny Mackintosh to her cousin Sarah Elizabeth Wedgwood, Wednesday [after 5 Nov. 1830], Wedgwood W/M 167; G. Stephen, *Antislavery Recollections*, 122, 191; Davis, 'Emergence', 218, 228; Brougham, quoted in Fladeland, *Men*, 196; Mrs J. (Bessy) Wedgwood II to her sister Fanny Allen, 14 June 1831, Wedgwood W/M 68. 'I don't think Papa quite owns to *immediate* abolition', Fanny Mackintosh reported to her cousin Sarah Elizabeth Wedgwood; anyway he wouldn't 'make that a condition' of outlawing slavery (Tuesday [before 19 Apr. 1829], Wedgwood W/M 167; cf. *Anti-Slavery Monthly Reporter*, no. 49 (June 1829), 2–3).
35. Dr Darwin, quoted in *Wedgwood*, 212; *CCD*, 1:111; *Cambridge Chronicle*, 19 Nov. 1830; *CJ*, 86 (1830–31), 157; *LJ*, 63 (1830–31), 125; C. H. Cooper, *Annals*, 567; Bourne, *Palmerston*, 328–9; Mrs J. (Bessy) Wedgwood II to her sister Emma Allen, 26 Jan. 1831, Wedgwood W/M 39; *ED*, 1:322.
36. J. M. Herbert's recollections, 2 June 1882, DAR 112, ff. 66–7; *Autobiography*, 59; Cambridge Union Society, *Laws* (1834), 68; Cradock et al., *Recollections*, 170; *CCD*, 1:117.
37. Whewell, *Elements*, 1:327, 375–7; *The Times*, 22, 23, and 24 Mar. 1831, 4 (Whewell was probably the Trinity College correspondent identified on the next page as 'A Reformer'); C. H. Cooper, *Annals*, 568–9; J. W. Clark and T. M. Hughes, *Life*, 1:375, 394; Sedgwick, *Discourse*, 95–102; Whewell, *Lectures*, chaps 10–11. On Whewell's supposed tolerance of slavery: Donagan, 'Whewell's "Elements of Morality"'; Snyder, *Reforming Philosophy*, 242ff. On his moral philosophy: Fisch and Schaffer, *William Whewell*.
38. *CCD*, 1:122; Bourne, *Palmerston*, 507; Turley, *Culture*, 63–7; *LJ*, 63 (1830–31), 352, 1119–37; *CCD*, 1:121; Halévy, *Triumph*, 33; *Anti-Slavery Reporter*, no. 80 (May 1831), 280; *The Times*, 24 Apr. 1831, 1.
39. J. C. Wedgwood, *Staffordshire Parliamentary History*, 70, 73; election handbills, Wedgwood E4/2966–2986; *The Times*, 4 May 1831, 3; Halévy, *Triumph*, 33.
40. Jenyns, *Memoir*, 30; Innes, 'Memoir'; *CCD*, 1:124.
41. Clowes, *Royal Navy*, 269; O'Byrne, *Naval Biographical Dictionary*; *The Times*, 14 July 1828, 3; 14 Feb. 1829, 4; 20 Apr. 1829, 2; MacGregor, *Fast Sailing Ships*, 14–15; Lyon, *Sailing List*; Ritchie, *Admiralty Chart*; Colledge and Warlow, *Ships*; Muddiman, 'H.M.S. "The Black Joke"'.
42. *CCD*, 1:131; notes of Arthur Gray, Jesus College, Cambridge; *Perth Courier*, 11 Aug. 1831; *Perthshire Advertiser*, 11 Aug. 1831. *Autobiography*, 67 refers to Marmaduke as the 'brother of Sir Alexander Ramsay', not of William. The brothers' father and great-uncle were both Sir Alexanders; Marmaduke's other brother was Edward (Edward Bannerman Ramsay, *ODNB*).

4 LIVING IN SLAVE COUNTRIES

1. Bethell, *Abolition*, 134–42; Lloyd, *Navy*, 275; Gough, 'Sea Power', 27; Conrad, *World*, 16–17; Karasch, *Slave Life*, xv, xxi.
2. 'To the Inhabitant Householders of the Potteries', handbill, Wedgwood E4/2997; speech by Josiah Wedgwood II at a public dinner, Newcastle-under-Lyme town hall, 20 May 1831, Wedgwood E4/2988; J. Wedgwood II to E. Buller, 18 May 1831, Wedgwood E11/9910; E. Buller to J. Wedgwood II, 22 May and 30 June 1831, Wedgwood E11/9911, 9914.

3. Hazlewood, *Savage*, 57–9; *Narrative*, 2; Chapman, *Darwin*, 7–8; Curtis, *Apes*, 123; Lavater, *Physiognomy*, 58; Hartley, *Physiognomy*.
4. *Ipswich Journal*, 30 Apr. 1831, [1], col. 4; R. FitzRoy to F. Beaufort, 10 May 1831, UK Hydrographic Office.
5. *Ipswich Journal*, 7 May 1831, [1–2].
6. *Narrative*, 13–14; Thomson, *HMS Beagle*, 113–16; J. and M. Gribbin, *FitzRoy*, 80, 191; R. FitzRoy to F. Beaufort, 10 May 1831, UK Hydrographic Office; *Ipswich Journal*, 7 May 1831, 4; *The Times*, 5 May 1831, 3. Cf. 4th Duke of Grafton to the 2nd Earl Grey, 1830–32, and C. FitzRoy to 2nd Earl Grey, 6 Dec. 1830, in Grey Papers, Durham University Library, B18/4/3–8 and B14/9/21.
7. *CCD*, 1:134, 136, 139–40, 146; Mrs J. (Bessy) Wedgwood II to her daughters Fanny and Emma Wedgwood, 20 May 1831, Wedgwood W/M 158; Mrs J. (Bessy) Wedgwood II to her sister Fanny Allen, 2 Oct. 1831, Wedgwood W/M 68; *Autobiography*, 172.
8. *CCD*, 1:186; Fanny Allen to her sister Jessie Allen Sismondi, 31 Oct. [1831], Wedgwood E59/32561 (typescript copy); Fanny Mackintosh to 'M', Monday [10 Oct. 1831], Wedgwood W/M 167; Mrs J. (Bessy) Wedgwood II to her sister Emma Allen, 16 Nov. 1831, Wedgwood W/M 39; Mrs J. (Bessy) Wedgwood II to her niece Fanny Mackintosh, 9 Dec. 1831, Wedgwood E57/31970.
9. *The Times*, 26 Dec. 1831, 2, crediting the *Hampshire Telegraph*. A fuller, corrected story appeared in *The Times*, 27 Mar. 1832, 2.
10. *Narrative*, 53; Ship's Log, HMS *Beagle*, 24 Jan. 1832, National Archives ADM 53/236; P. G. King, Journal, Mitchell Library; *PP*, ST, 1842 (561), xliv, 532; Ward, *Royal Navy*, 45–50, 125; Thomas, *Slave Trade*, 334; Conrad, *World*, 70, 72; *Diary*, 23, 26, 28, 30, 32, 33, 71; Muddiman, 'H.M.S. "The Black Joke" '.
11. S. B. Schwartz, *Sugar Plantations*, 431ff., 437, 444–5, 456; Reis, *Slave Rebellion*, 6; Davis, *Inhuman Bondage*, 104; M. V. Nelson, 'Negro', 206; Conrad, *World*, chaps 3–4; Bethell, *Abolition*, 69–72; *PP*, ST, 1833 (007), xliii, 1832 (Class B), 151.
12. M. V. Nelson, 'Negro', 206, 209–10; *Narrative*, 62; *Diary*, 43, 45, 46, 80.
13. R. FitzRoy to his sister Fanny, 8 Apr 1828, CUL Add. 8853.33; R. FitzRoy to his father, 8 Apr. 1828, CUL Add. 8853.34; Ship's Log, HMS *Beagle*, 7 Mar. 1832, National Archives ADM 53/236; Sulivan, *Life*, 14–15; *PP*, ST, 1828 (366), xxvi (Class B), 121.
14. *Autobiography*, 74; R. FitzRoy to F. Beaufort, 5 Mar. 1832, in F. Darwin, 'FitzRoy'; *CCD*, 1:183.
15. *PP*, ST, 1833 (007), xliii, 1832 (Class A), 169–70, 182; *PP*, ST, 1835 (007), li, 1834 (Class B), 225–6; Sulivan, *Life*, ix, xxix; Collister, *Sulivans*, 11, 14, 84, 183; *CCD*, 1:393; Ship's Log, HMS *Samarang*, 31 Aug. 1831 – 15 Mar. 1832, National Archives ADM 53/1318.
16. *Diary*, 45; O'Byrne, *Naval Biographical Dictionary*; Charles Paget, *ODNB*; *The Times*, 13 May 1831, 3; 17 May 1831, 2.
17. Humboldt and Bonpland, *Personal Narrative*, 7:150, 271; *Diary*, 42; cf. Humboldt, *Island*, 1856 and 2001. Darwin's copy of Humboldt is in the Darwin Library, CUL. Volumes 1 and 2, bound together, are inscribed by Henslow to Darwin. FitzRoy told him that he was 'of course welcome to take your Humboldt . . . There will be *plenty* of room for books'. Darwin took all seven volumes and asked for volume eight (never published) to be sent: *Notebooks*, RN24; *CCD*, 1:314.
18. P. G. King, Reminiscences, Mitchell Library; Gough, 'Sea Power', 32–3; *CCD*, 1:219.
19. *Diary*, 49; Gough, 'Sea Power', 29, 31, 32 (cf. Miller, *Britain*, 50); Graham and Humphreys, *Navy*, xxvi; Gosset, *Lost Ships*; Driver and Martins, 'Shipwreck'.
20. Bethell, *Abolition*, 69–76; Conrad, *World*, 15–22.
21. *Journal* (1839), 22; *Diary*, 52, 53, 57, 58, 59; Barlow, *Charles Darwin*, 159, 161.
22. *Diary*, 58; Barlow, *Charles Darwin*, 162; Karasch, *Slave Life*, chap. 2; Conrad, *World*, 51–3.
23. *Diary*, 69; Barlow, *Charles Darwin*, 164; P. G. King, Reminiscences and Autobiography, Mitchell Library; *CCD*, 1:121, 393.

24. *Diary*, 69; Berger, 'American Slavery', 196; J. Campbell, *Negro-Mania*, 459, 461; Stanton, *Leopard's Spots*, 208; Olmsted, *Journey in the Seaboard Slave States*, 312–13; Lorimer, *Colour*, 125; Haller, *Outcasts*, 4; also Malik, *Meaning*, 97; Bolt, *Anti-Slavery Movement*, 3, 122; Curtin, *Image*, 64, 384; Alatas, *Myth*.
25. *The Times*, 27 June 1831; O'Byrne, *Naval Biographical Dictionary*; *CCD*, 1:226, 232; *Narrative*, 74; P. G. King, Autobiography, Mitchell Library; Miller, *Britain*, 41.
26. Bethell, *Abolition*, 67–8, 85 (a Brazilian view is in Rodrigues, *Brazil*, 144ff.); *CCD*, 1:227; *Diary*, 61 (cf. 62, 75, 78); Bindoff et al., *British Diplomatic Representatives*; Miller, *Britain*, 53–4.
27. *CCD*, 1:313, 337; *PP*, ST, 1833 (007), xliii, 1832 (Class A), 102; *PP*, ST, 1834 (471), xliv, 1833 (Class A), 635; Karasch, *Slave Life*, 120.
28. Quoted in Bethell, *Abolition*, 71, 72; *CCD*, 1:313.
29. *CCD*, 1:236, 238; Palmerston, in *PP*, ST, 1834 (471), xliv, 1833 (Class B), 670; R. FitzRoy to F. Beaufort, 28 Apr. 1832, in F. Darwin, 'FitzRoy'; *Diary*, 75; *Autobiography*, 73.
30. *The Times*, 24 Nov. 1828, 3; 6 July 1829, 2; 30 July 1829, 2; 4 Aug. 1829, 2; 14 Sep. 1829, 5; 24 May 1831, 2; 5 Sep. 1831, 4; 3 Oct. 1831, 6; *CCD*, 1:278.
31. *PP*, ST, 1831–2 (010), xlvii, 1831 (Class B), 667–8, 800, 805; *PP*, ST, 1837–8 (533), lii, 161; *The Times*, 5 Sep. 1831, 4; Conrad, *World*, 85.
32. *Narrative*, 95; R. FitzRoy to Capt. G. W. Hamilton, 6–14 Aug. 1832; and T. S. Hood to R. FitzRoy, 5 Aug. 1832, UK Hydrographic Office; Graham and Humphreys, *Navy*, xxiii; Diggs, 'Negro', 296; Castlereagh, quoted in Miller, *Britain*, 35; Gough, 'Sea Power', 30.
33. *Diary*, 89, 90, 91, 93; P. G. King, Journal, Mitchell Library; *Diary*, 89; R. FitzRoy to F. Beaufort, 15 Aug. 1832, UK Hydrographic Office; Ship's Log, HMS *Beagle*, 10 Aug. 1833, National Archives ADM 53/236; *CCD*, 1:250.
34. *PP*, ST, 1834 (471), xliv, 1833 (Class B), 705–6; *Diary*, 148, 161; *The Times*, 4 Feb. 1829, 4; 14 Feb. 1829, 3; 22 Apr. 1829, 2; 22 June, 2; Clowes, Royal Navy, 269. Ironically, Hood himself had once owned the *Adventure*: *Narrative*, 275.
35. R. FitzRoy to F. Beaufort, 16 July 1833, UK Hydrographic Office.
36. *PP*, ST, 1834 (471), xliv (Class A), 640; *PP*, ST, 1835 (007), li, 1834 (Class B), 258; King, *Narrative*, 10, 254, 462; *The Times*, 30 Jan. 1833. Having asked for 'all the gossip', Darwin heard from a Buenos Aires merchant in May 1834 that 'two or three Vessels' had 'arrived at Montevideo with Slaves or African colonists as they term them', and a 'vessel fitting out' in Buenos Aires had been detained by the chargé d'affaires: *Diary*, 191; *CCD*, 1:378, 388.
37. *CCD*, 1:222, 245–7, 259, 276–7, 365; O'Byrne, *Naval Biographical Dictionary*; P. G. King, Journal, Mitchell Library; cf. Ship's Log, HMS *Beagle*, 10–11 Feb. 1832, National Archives ADM 53/236.
38. *CCD*, 1:253, 290, 299, 302, 408; J. C. Wedgwood, *History*, 188; J. C. Wedgwood, *Staffordshire Parliamentary History*, 74–8.
39. *CCD*, 1:287–8, 312–13, 320, 359.
40. Hodge, 'Darwin', 13ff.; *Narrative*, 379; *Journal* (1860), 310; Desmond and Moore, *Darwin*, 160–62; *CCD*, 1:381, 399; *LLL*, 1:268.
41. *Wedgwood*, 215; C. Lyell, *Principles*, 2:62; *Autobiography*, 85; R. N. Hamond to F. Darwin, 19 Sep. 1882, DAR 112.A54–A55; *Narrative*, 115; *CCD*, 1:150, 277, 305. See Brantlinger, 'Missionaries' for the Fijian potentate, also the subject of a hand-coloured etching by William Heath, 'Hokie pokie wankie fum, the King of the Cannibal Islands' ([London]: T. McLean, 22 July 1830).
42. *CCD*, 1:312; Jacoby, 'Slaves'; Hunt, 'On the Physical and Mental Characters', 387; quoting van Evrie, *Negroes*, 23; Davis, *In the Image*, 126–8; *Notebooks*, B231; Rodrigues, *Brazil*, 52–3; *Diary*, 79–80; Zelinsky, 'Historical Geography', 198.
43. R. FitzRoy to F. Beaufort, 16 July 1833, UK Hydrographic Office; Barta, 'Mr Darwin's Shooters', 120–22; *Diary*, 100, 169.
44. *Diary*, 180; Barta, 'Mr Darwin's Shooters', 120; C. Darwin, *Geological Observations*, 78; Knight, *Pictorial Museum*, 1:178.

45. *Diary*, 172, 180–81 (Parodiz, *Darwin* for confirmation); *Beagle* field notebook EH1.11, 16a, Down House.
46. *Diary*, 105, 165, 169, 170, 173–4, 189.
47. *Narrative*, 642, 646; *Beagle* field notebook EH1.14, 142a, Down House; R. FitzRoy to his sister Fanny, 4 Apr. 1834, CUL Add. 8853.43; R. FitzRoy to F. Beaufort, 15 Aug. 1832, UK Hydrographic Office.
48. Bory de Saint-Vincent, 'Orang', 266–7; *CCD*, 1:237, 593; Corsi, *Age*, 218–25; Jacyna, 'Medical Science'; Desmond, *Politics*, 45–6, 289–90. In Darwin's copy of the *Dictionnaire* at Down House, the article on man is lightly marked on several pages in an unknown hand. Thanks to Tori Reeve for sending images.
49. *Narrative*, 641–2, 650; Bory de Saint-Vincent, 'Homme', 277, 281, 313, 325–6.
50. *Diary*, 169; Caldwell, *Thoughts*, iii, vi, 15–16, 35, 37, 40–41, 59ff., 71, 93, 136–8, 140–2, 144, 151.
51. *Narrative*, 152, 154, 167; *Diary*, 218, 221; R. FitzRoy to his sister Fanny, 6 Nov. 1834, CUL Add. 8853.46.
52. *Narrative*, 144, 640–41; *CCD*, 1:97; Cox, *Selections*, 140–43; R. Owen, *Descriptive Catalogue*, 846, no. 5426–7.
53. *Diary*, 143; King, *Narrative*, 397–9; *Narrative*, 176 and App. 16, 142–9; E. Belfour to R. FitzRoy, 31 Jan. 1831, Royal College of Surgeons, London; R. Owen, *Descriptive Catalogue*, 846, no. 5428–40; *CCD*, 1:335.
54. Chapman, *Darwin*, sorts out the tribal groups. Darwin visited the Haush 'foot people' in December 1832, the Yahgan or Yamana 'boat people' in February 1833 and February–March 1834, and the Teheulches or 'foot Patagonians' in January 1834. One of the returning abductees, 'Jemmy Button', was a Yahgan; the others, 'York Minster' and 'Fuegia Basket', were Alakaluf canoe people from whom in 1830 FitzRoy took all four hostages and the remains of the native shot.
55. *Diary*, 124, 125, 134, 135, 137, 139, 222, 223, 224; Chapman, *Darwin*, 46–9; *Beagle* field notebook EH1.12, 21a, 36a, Down House; *Narrative*, 204.
56. *Autobiography*, 126 (cf. 67–8 and *CCD*, 1:305, 311–12); *Diary*, 122, 222, 223, 224, 444; *Narrative*, 138. Though critical of Darwin, a veteran Tierra del Fuego anthropologist states that he 'never sank' to FitzRoy's 'level of contempt': Chapman, *Darwin*, 95.
57. *Diary*, 223, 266, 267, 278, 285; R. D. Keynes, *Charles Darwin's Zoology Notes*, 283; K. V. Smith, 'Darwin's Insects', 43–4. Darwin later pencilled '*species* of' between 'different' and 'parasites' (*CCD*, 3:38n). When it appeared that the lice were different species on the black and white human races, which might support the view that these humans were themselves different species, he produced evidence that different species of parasite can infest the same race and that slight changes in the host body, independent of race, can repel its parasites. See *CCD*, 3:38, 53; 13:359–60; *Descent*, 1:219–20.
58. *Diary*, 290, 366, 384; *Narrative*, 642.
59. *CCD*, 1:354; *Diary*, 138, 376; J. Matthews to D. Coates, 28 Dec. 1835, Birmingham University Library, CMS/B/OMS/ C N O61/; Ellis, *Polynesian Researches*, 1:v.; Sivasundaram, *Nature* on Ellis.
60. *Narrative*, 581, 591; Earle, *Narrative* (1832), 49, 58–9, 69–71, 122–6, chap. 13, 'The Whalers and the Missionaries' and chap. 28, 'A War Expedition and a Cannibal Feast'; *Diary*, 394. While Earle despised the Tory-Anglican Church Missionary Society, he welcomed the humbler Methodist missions: E. H. McCormick, in Earle, *Narrative* (1966), 1–2, 25–6.
61. Confirmed by E. H. McCormick, in Earle, *Narrative* (1966), 23–4; *Diary*, 386, 390.
62. C. Baker to Committee of Correspondence, 9 Jan. 1836, Birmingham University Library CMS/B/OMS/ C N O18/24; *CCD*, 1:254, 284, 466, 471–2, 485; E. H. McCormick, in Earle, *Narrative* (1966), 16–20; *Diary*, 384, 390; Armstrong, 'Darwin's Perception'.
63. *CP*, 1:20, 21, 23, 25–7, 29, 32–3, 34; *CCD*, 1:496, *Diary*, 373. M. W. Graham, 'The Enchanter's Wand' argues that Darwin, 'probably wanting to be part of a public debate', underwent a 'radical change into a diehard supporter of missionaries'. Even if this were

true, FitzRoy's motives and Earle's book are slighted. On the self-presentational conflict between missionaries and 'godless mariners': Sivasundaram, *Nature*, 124–5.

64. Stocking, *Victorian Anthropology*, 278; Brantlinger, *Dark Vanishings*, 125; *Diary*, 408, 411; Hughes, *Fatal Shore*, 414–23.

65. A. Smith, 'Observations', 119–27; *Philosophical Magazine*, 8 (1830), 222–3; P. R. Kirby, *Diary*, 14.

66. Herschel's words: Evans et al., *Herschel*, 42; Qureshi, 'Displaying', 247; Musselman, 'Swords', 424–45. There was a 'stuffed' Hottentot in Brookes's Museum in London: *Brookesian Museum* (London, Gold & Walton, 1828), 94 [Sale Catalogue]: Fifteenth Day's Sale, Friday, 1 August 1828, exhibit 46. In 1848 Knox reported recently seeing one in London: Knox, 'Lectures', 283. On the Boers' efforts at extermination: Brantlinger, *Dark Vanishings*, 75.

67. *Diary*, 424; Armstrong, 'Three Weeks', 13.

68. Quoted in Horsman, *Race*, 185; Morton, *Crania Americana*, 90; Gould, *Mismeasure*, 88; G. Combe, 'Observations', 150; *Diary*, 425; McClintock, *Imperial Leather*, 55; Lindfors, 'Hottentot', 4.

69. P. R. Kirby, *Sir Andrew Smith*, 44, 51, 53, 65, 116; P. R. Kirby, *Diary*, 1:17–18, 41; *Diary*, 424, 426–7; Armstrong, 'Three Weeks', 9; Evans et al., *Herschel*, 42, 225.

70. Armstrong, 'Three Weeks', 13; *Diary*, 426–7; *Autobiography*, 107; *CCD*, 1:498.

71. Crowe, *Calendar*, 153, 3034; Musselman, 'Swords', 429; A. Ross, *John Philip*, 9; Rainger, 'Philanthropy', 706–7; Evans et al., *Herschel*, 43, 152; W. J. Ashworth, 'John Herschel', 172. For a less flattering Afrikaaner perspective on a manipulative Philip: Pretorius, *British Humanitarians*, 30, 159ff.

72. A. Ross, *John Philip*, 94–6; Groves, *Planting*, 1:233–72; FitzRoy and Darwin, 'Letter', in *CP*, 1:19–20.

73. Groves, *Planting*, 251–3, 256, 258–9; Rainger, 'Philanthropy', 706–7; A. Ross, *John Philip*, 3–4. Philip's *Researches in South Africa* (2 vols, 1828) was in the Wedgwoods' library at Maer Hall: *Catalogue*.

74. Kass and Kass, *Perfecting*, 257, 267; Musselman, 'Swords', 420–21, 428; Buxton, *Memoirs*, 369–71; A. Ross, *John Philip*, 142. More Hottentot skulls went off to Epps in London in the 1830s: Bank, 'Of Native Skulls', 390–99. Interestingly, Darwin was still signing Aborigines' Protection Society petitions relating to South Africa (along with Corbett and Buxton's grandsons) as late as 1877: *The Times*, 23 July 1877, 10.

75. Kohn, 'Darwin's Ambiguity', 222; *Notebooks*, RN32; Cannon, 'Impact', 304–11; Evans et al., *Herschel*, 242–3. On the different ways that contemporaries thought new species might be generated: E. Richards, 'Moral Anatomy', 396–406; E. Richards, 'Political Anatomy', 380ff.; E. Richards, 'Question of Property Rights', 133ff.; Desmond, *Archetypes*, 29–37.

76. R. FitzRoy to Lady M. Herschel, 29 June 1836; and R. FitzRoy to Sir J. Herschel, 8 July 1836, Royal Society, Herschel Letters, 7, 245; J. Herschel to R. FitzRoy, 3 Oct. 1836, Royal Society, Herschel Letters, 7, 247. On the *South African Christian Recorder*: Kolbe, 'South African Print Media', 27.

77. Reis, *Slave Rebellion*, xiii; Zelinsky, 'Historical Geography', 168; *Diary*, 434; *PP*, ST, 1836 (006), l, 1835 (Class B), 440–41; *PP*, ST, 1836 (006), l, 1835 (Class B), 444 (cf. 446); *Diary*, 433.

78. *Journal* (1860), 499; *Diary*, 435; *PP*, ST, 1836 (006), l, 1835 (Class B), 445–6.

79. *Diary*, 429, 431, 437; *CCD*, 1:502. On the analogy: Grove, *Green Imperialism*, 343.

80. Van Amringe, *Investigation*, 63; *Narrative*, 644; FitzRoy, 'Outline'; Haynes, *Noah's Curse*; Kenny, 'From the Curse', 370.

81. *Diary*, 45, 441–2; *CCD*, 1:515. It was common shipboard practice to pilfer native skulls: T. H. Huxley in the *Rattlesnake* came home with three from Darnley Island (Desmond, *Huxley*, 127).

5 COMMON DESCENT

1. Hodge, 'Universal Gestation'; Desmond, 'Robert E. Grant's Later Views', 407.
2. *Notebooks*, C138.
3. *Notebooks*, B87–8.
4. *Notebooks*, C204.
5. *Notebooks*, C217.
6. *Notebooks*, B231–2. We may assume that this was aimed at the Cambridge philosopher William Whewell, who for twenty years had refused to extend rights and moral considerations to animals, and attacked Jeremy Bentham's belief in our moral 'duty to regard the pleasures and pains of other animals as those of human beings': compare Mill, 'Whewell's Moral Philosophy' with Whewell, *Elements*, 365–6 and Snyder, *Reforming Philosophy*.
7. Curtin, *Image*, 43; *Notebooks*, B231.
8. *Notebooks*, C154–5.
9. For its effects on science: Desmond, *Politics*; Desmond and Moore, *Darwin*, chaps 16–19.
10. *Notebooks*, C53; B244.
11. *CCD*, 1:345, 367, 469; *Autobiography*, 79.
12. *CCD*, 1:337, 345, 365, 425, 469, 472, 507; 2:7, 11.
13. Cashbook of the Hanley and Shelton Anti-Slavery Society, Wedgwood E32/24784A; *CCD*, 1:372, 504.
14. *CCD*, 1:519–21; Fanny Allen to her sister Jessie Allen Sismondi, 23 Oct. 1833, Wedgwood E57/32076; Emma Wedgwood and her mother Bessy to Fanny Mackintosh Wedgwood, [21–2 Nov. 1836], Wedgwood W/M 233. On Sarah Wedgwood and the role of women in the anti-slavery movement: Midgley, *Women*, 76ff.
15. *CCD*, 1:316, 525; 2:9, 11; J. W. Clark and T. M. Hughes, *Life*, 1:468–69.
16. *LLL*, 2:12; C. Lyell, *Principles*, 2:62; *Autobiography*, 65, 100.
17. *CCD*, 1:532; 2:94, 97, 133–4, 433; *Autobiography*, 126; Booth, *Stranger's Intellectual Guide*, 77–8; Timbs, *Curiosities*, 191; Collini, *Public Moralists*, 16.
18. Colp, *Darwin's Illness*; *Notebooks*, B207, C76–7, 166; M61e. On contemporary debates over these doctrines: Jacyna, 'Immanence'; Desmond, *Politics*, chaps 3–4.
19. *Notebooks*, B3–4; C72.
20. *Notebooks*, B18, 74; C72e; Hodge, 'Darwin', 83–4; Ospovat, *Development*, 211.
21. *Notebooks*, B32–3, 179–82; C140, 215, 228e.
22. *CCD*, 2:8; J. Herschel to R. FitzRoy (draft), 3 Oct. 1836, Royal Society, Herschel Letters.
23. *CCD*, 2:19–21; *Notebooks*, RN 133; *Autobiography*, 119; Cannon, 'Impact', 305, 308.
24. *Notebooks*, B40–43, 147.
25. *Notebooks*, B148.
26. *Notebooks*, C79.
27. 'University College', *The Times*, 2 July 1846, 8; Sen Gupta, 'Soorjo Coomar Goodeve Chuckerbutty'; Lewis, 'Black Letter Day'; Lorimer, *Colour* on other black British graduates; B. Silliman, *Journal*, 1:209–10.
28. G. Combe, *Life*, 303–10; 'Organization of the Brain in the Negro', *Medico-Chirurgical Review*, 28 (1837–8), 249–52; Tiedemann, 'On the Brain', 504; Gould, 'Great Physiologist'; Lorimer, *Colour*, 25, 32, 37.
29. *Notebooks*, B86–7, 119, 142; C233–4.
30. Howse, *Saints*, chap. 8.
31. *CCD*, 1:317–18, 423, 524, 530, 533–4; C. R. Sanders et al., *Collected Letters*, 13:224–6
32. Fanny Mackintosh Wedgwood to her sister-in-law Emma Wedgwood, [28 and 31 Jan. 1837], Wedgwood W/M 199; to her sister-in-law Sarah Elizabeth Wedgwood, [3 May 1837?], Wedgwood W/M 167.
33. *CCD*, 2:86; 4: app. 4.

34. V. Sanders, *Harriet Martineau*, 45; Webb, *Harriet Martineau*, 155–6; Martineau, *Society*, 1:388; Wheatley, *Life*, 156; Stange, *British Unitarians*, 56; Logan, 'Redemption'; Fladeland, *Men*, 229–30.
35. Fanny Mackintosh Wedgwood to her sister-in-law Sarah Elizabeth Wedgwood, [May 1837], Wedgwood W/M 167; Fanny Allen to Patty Smith, 3 June 1837, Wedgwood E57/32107; Fanny Allen to Patty Smith, 27 June 1837, Wedgwood E57/32108; Emma Wedgwood to her sister-in-law Fanny Mackintosh Wedgwood, [23 May 1837], Wedgwood W/M 233; Pierce, *Memoir*, 1:190.
36. *CCD*, 2:80–81, 86.
37. *Notebooks*, M75–7; Martineau, *How to Observe*, 21.
38. Webb, *Harriet Martineau*, 163; Martineau, *How to Observe*, 25; Martineau, *Society*, 2:313, 336.
39. *Notebooks*, D24; M85–7; Fergusson, *Notes*, 203–5; D. Smith, 'Fergusson Papers'.
40. *Notebooks*, T79, E89; Cautley and Falconer, 'On the Remains', 569; Cautley, 'Extract'; Hartwig, '*Protopithecus*', 451–6.
41. *Notebooks*, M138; D137–9.
42. *Notebooks*, M84e, 122–3; N5.
43. *Notebooks*, C244; *ED*, 1:406, 449; *Wedgwood*, 230–32; Arbuckle, *Harriet Martineau's Letters*, 5, 8, 14; *CCD*, 1:359–60; 2:64.
44. Wheatley, *Life*, 202; *Autobiography*, 113; Arbuckle, *Harriet Martineau's Letters*, xix, 5.
45. Norton, *Correspondence*, 1:126–7, 247; Carlyle, 'Characteristics', 381; Horsman, 'Origins', 309–401; Carlyle, 'Signs', 446; Arbuckle, *Harriet Martineau's Letters*, xix.
46. *Autobiography*, 113; Norton, *Correspondence*, 1:126.
47. *ED*, 1:409; *CCD*, 2:95, 431 (cf. 91); Emma Wedgwood to her sister-in-law Fanny Mackintosh Wedgwood, [4 Oct. 1837], Wedgwood W/M 233.
48. Tyrrell, 'Moral Radical Party'; Temperley, *British Antislavery*, 34; Turner, 'British Caribbean'.
49. G. W. Alexander to J. Wedgwood II, 20 [Feb.?] 1838, Wedgwood E32/24780; J. Crisp to Francis Wedgwood, 21 Apr. 1838, E32/24781; G. W. Alexander to Francis Wedgwood, 13 July 1839, E32/24783; J. Crisp to Francis Wedgwood, 17 July 1839, E323/24784; Expenditures, 1838, '. . . carriage of petition to London (19 May) . . .', Cashbook of the Hanley and Shelton Anti-Slavery Society, E32/24784A; *LJ*, 70 (1837), 46.
50. J. Sturge to [Francis Wedgwood], 21 Nov. 1838, Wedgwood E32/24782; Temperley, *British Antislavery*, 123; Fanny Allen to Patty Smith, [1839–40?], Wedgwood E57/32113.
51. Norton, *Correspondence*, 1:126; C. R. Sanders et al., *Collected Letters*, 13:40–43.
52. R. I. and S. Wilberforce, *Life*, 1:318; 3:302–3; 4:212ff., 5:14–15; *ED*, 1:407; C. R. Sanders et al., *Collected Letters*, 13:274–6.
53. J. Wedgwood II to T. Clarkson, [21 Aug.] 1838, Wedgwood W/M 237; Clarkson, *Strictures*, v; Emma Wedgwood Darwin to her aunt Jessie Allen Sismondi, 21 July [1838], Wedgwood W/M 193 (passage omitted in *ED*, 1:409–11); Fanny Allen to her niece Emma Wedgwood Darwin, 12–14 July [1838], Wedgwood W/M 221.
54. Tyrrell, 'Moral Radical', 500; Hochschild, *Bury*, 358.
55. *ED*, 1:419–20, original in Wedgwood W/M 193; *CCD*, 2:114, 116, 128; 7:469.
56. *CCD*, 2:126, 172; *Autobiography*, 95.
57. *The Times*, 29 May 1838, 5; C. R. Sanders et al., *Collected Letters*, 10:246–54.
58. Emma Wedgwood to her aunt Jessie Allen Sismondi, 28 Dec. [1838], Wedgwood W/M 193 (passage omitted in *ED*, 1:431); Fanny Mackintosh to her cousin Sarah Elizabeth Wedgwood, [1828 or later] and [31 July 1830], Wedgwood W/M 167; Fanny Mackintosh to her aunt Mrs J. (Bessy) Wedgwood, [6 Dec. 1831], Wedgwood W/M 210.
59. *CCD*, 2:148; Martineau, *Martyr Age*, 58, 61–2, 64; Webb, *Harriet Martineau*, 191; Martineau, 'Martyr Age'.
60. *CCD*, 1:519; 2:144, 157, 172, 328, 445; 3:326; 4:146, 5:17; *ED*, 2:104.
61. P. James, *Population Malthus*, 323; Waterman, *Revolution*; O'Leary, *Sir James Mackintosh*.

62. *Notebooks*, D135e.
63. *CCD*, 2:19.
64. *Notebooks*, M87.
65. *CCD*, 1:396, 460; *Notebooks*, B161; M32; N26–7; H. E. Gruber, *Darwin*, 188.
66. *Notebooks*, C61, 178; D99; *ED*, 1:412.
67. *Notebooks*, D103, 114e; M149; N64; OUN 8.
68. *Notebooks*, E69; Hodge and Kohn, 'Immediate Origins'; Evans, 'Darwin's Use', 120–26; Durant, 'Ascent'.

6 HYBRIDIZING HUMANS

1. *CCD*, 2:236, 268–9; *ED*, 2:15; Emma Wedgwood Darwin to her sister Sarah Elizabeth Wedgwood, [5 Feb. 1839], Wedgwood W/M 68.
2. T. Hodgkin to S. G. Morton, 2 May 1827, 19 May 1828, 12 May 1830, Samuel George Morton Papers, APS; Wood, *Biographical Memoir*, 7.
3. *CP*, 1:35; *PP*, 1837 (425), vii.1, 45, 81.
4. G. W. Alexander to Francis Wedgwood, 13 July 1839, Wedgwood E32/24783; Rainger, 'Philanthropy', 704–8; Stocking, 'What's in a Name?', 369–72; Kass and Kass, *Perfecting*, 267ff.
5. Morrell and Thackray, *Gentlemen*, 252, 283–5; Fraser, *Power*, 88–9; Royle, *Victorian Infidels*, 50, 62; Desmond, *Politics*, 331.
6. Prichard, 'On the Extinction', 166–70; Anon, 'Varieties', 447; Hodgkin, 'On Inquiries', 53–4; Matthew, *Emigration Fields*, vii, 3, 6, 9; Augstein, *James Cowles Prichard's Anthropology*, 144–6; Brantlinger, *Dark Vanishings*, 36; J. W. Gruber, 'Ethnographic Salvage', 382–3. Only a four-line note of Prichard's paper was published in the BAAS report: 'On the Extinction of the Human Races', *Report of the Ninth Meeting of the British Association for the Advancement of Science; held at Birmingham in August 1839* (London: Murray, 1840), 89.
7. Greg, 'Dr. Arnold'; also in Greg, *Essays*, 1:5–14; Mazrui, 'From Social Darwinism', 71, 75–6.
8. *Notebooks*, D38–9.
9. Darwin's copy of Walker, *Intermarriage*, which he read in June 1839 (*CCD*, 4:457; Darwin Library, CUL), scored a quote from Prichard on page 361, on the benefits of racial intermarriage: between Irish Celt and English settlers; between Russians, Tartars and Mongolians; and in Paraguay, where the mixed breeds were said to be physically superior and more prolific. Historians have discussed blending as containing and restricting variation in Darwin's pre-Malthusian theory, making it a negative force (Kohn, 'Theories', 105–7). But Darwin's reading on racial mixing here pointed up a (short-term) evolutionary benefit.
10. Unreferenced, but it came from Moodie's *Ten Years*, 1:222, quoted in Walker, *Intermarriage*, 362.
11. *PP*, 1837 (425), vii.1, 25, 64–8, 143–51. Philip singled out the Griqua leader, Andreas Waterboer, for his 'integrity and talents' (142). Herschel too was impressed during his Cape sojourn; Waterboer showed more understanding when peering through Herschel's telescope than many a 'civilized' European gawper, as Herschel told Philip: Musselman, 'John Herschel', 42n; Evans et al., *Herschel*, 171.
12. The Griquas appear in Prichard, *Researches* (3rd edn, 1836), 1:147–8; Prichard, *Natural History*, 2nd edn, 1:19–20; J. Bachman, *Doctrine*, 117; Smyth, *Unity*, 197.
13. *Notebooks*, 134e–135e, E9e, 63–4.
14. Finzsch, 'It is scarcely'; Barta, 'Mr Darwin's Shooters', 118.
15. *Diary*, 180–81; Barta, 'Mr Darwin's Shooters', 121.
16. *Journal* (1860), 174, 447; Brantlinger, *Dark Vanishings*, 22, 124–30.
17. *Edinburgh New Philosophical Journal* (1834), 433–4; Pentland, 'On the Ancient Inhabitants'; Prichard, *Researches* (3rd edn, 1836), 1:317.

18. *Notebooks*, E64–5. Pentland had examined hundreds of skulls in the Andes. Some ethnologists would argue that these were not an extinct race, but that the long-heads were produced by applying pressure to the infants' skulls: R. Owen, *Catalogue*, 18–19; Prichard, *Researches* (3rd edn, 1836), 1:319; Martin, *General Introduction*, 206–7; W. B. Carpenter, 'Varieties', 1361. Thus the Titicacans might have been related to other peoples known to do this, or to the present Aymara inhabitants in this region. This better supported the Prichard–Latham unity-and-migration model (Latham, *Natural History*, 458; Latham, *Man*, 59–60). However Morton continued to see this as a unique extinct race: Pentland, 'Ancient Inhabitants', 623–4; Morton, *Crania Americana*, 97; Morton, *Inquiry*, 8.
19. Moore, 'Revolution'. The College in 1831 had three Titicacan skulls, presented by Earl Dudley. They had also acquired some of Pentland's specimens by the time Darwin wrote: Prichard, *Researches* (3rd edn, 1836), 1:316, 318; R. Owen, *Descriptive Catalogue*, 844. For the Darwin–Owen intellectual exchange: Sloan, 'Darwin'. On the new surgeons' museum: Desmond, *Politics*, chap. 6; Rupke, *Richard Owen*, chap. 1.
20. *Notebooks*, T81.
21. *Diary*, 179–80; *Journal* (1839), 120, 520; Barta, 'Mr Darwin's Shooters', 120–21, 127, 129.
22. *Journal*, 520; J. W. Gruber, 'Ethnographic Salvage'.
23. Anon, 'Varieties', 448. Drescher, 'Ending', 432–3, on the different national contexts of French and British abolitionism and ethnology.
24. *Notebooks*, QE1, B244; *Foundations*, 68, 79; Anon, 'Varieties', 448–58; *Athenaeum*, 17 Aug. 1839, 704. Darwin was also asking Dr Andrew Smith questions about 'Savages at Cape' and the ways they selected their cattle and dogs: *Notebooks*, QE16 (11).
25. Of the £3,000 or so dispensed to the committees in 1839, that awarded to Section D, 'For Printing and Circulating a Series of Questions and Suggestions for the Use of travellers and others; with a view to procure Information respecting the different races of Men, and more especially of those which are in an uncivilized state: the Questions to be drawn up by Dr. Prichard, Dr. Hodgkin, Mr. J. Yates, Mr. Gray, Mr. Darwin, Mr. R. Taylor, Dr. Wiseman, and Mr. Yarrell', was the smallest sum, £5: *Report of the Ninth Meeting of the British Association for the Advancement of Science; held at Birmingham in August 1839* (London: Murray, 1840), xxvi; Morrell and Thackray, *Gentlemen*, 285.
26. Caldwell, 'On the Varieties'; Caldwell, *Thoughts*, 74ff. for his views on African–Caucasian anatomical differences, and 88–90 on the black's similarities to an ape; Nott, 'Diversity', 118; Riegel, 'Introduction', 73–8.
27. Caldwell, *Thoughts*, iii–iv, 15–16; Jordan, *White*, 533–4. Caldwell, *Phrenology*, 69ff. denigrates views expressed at the 1837 Liverpool BAAS on the enlightenment of Carib chiefs. On Erasmus Darwin: Warner, *Autobiography*, 52, 170, 295, 297–8.
28. Caldwell, *Thoughts*, v–vii, 35, 37–8, 41, 93, 115, 116, 173, 177; Nott, 'Diversity', 118. Darwin acknowledged Prichard's belief in immutable species: *CCD*, 3:79.
29. Caldwell, *Thoughts*, 42, 55–7, 72–3, 101–2, 114–15, 123–4, 158–60, 164–5.
30. Caldwell, *Thoughts*, 134–8, 141–5, 176; Hodgkin, 'On Inquiries', 54; Erickson, 'Anthropology of Charles Caldwell'; *Notebooks*, B231.
31. Hodgkin, 'On Inquiries', 52–3. Subsequently the questionnaires went to individual contacts all over the globe: Hodgkin, 'Report'.
32. Temperley, *British Antislavery*, 42–55; Thomas, *Slave Trade*, 657.
33. Greg, *Past*, 4, 18, 54, 59–61; M. B. Rose, *Gregs*, 54, 64.
34. Emma Wedgwood Darwin to her sister Sarah Elizabeth Wedgwood, [2 Mar. – 28 May 1841], Wedgwood W/M 168.
35. Society for the Extinction of the Slave Trade, *Proceedings*, 56–9; Fladeland, *Men*, 265–6. In the end Martineau did not attend. On the problems faced by the American women delegates in this male bastion: Sklar, 'Women'.
36. Temperley, *British Antislavery*, 57–61; Kass and Kass, *Perfecting*, 402–11. On those aboard: Hodgkin, 'On Inquiries', 52–3; J. E. Ritchie, *Life*, 2:631; Fyfe, 'Conscientious

Workmen', 206 n40; Edward Blyth's later recollection of Fraser in a memorandum to Darwin, *CCD*, 5:484; C. Darwin, *Structure*, 64n.

37. Bolt, *Anti-Slavery*, 22; Lorimer, *Colour*, 57; Brantlinger, *Dark Vanishings*, 72.
38. J. Campbell, *Negro-Mania*, 39.
39. *CCD*, 3:287; Emma Wedgwood Darwin to her aunt Jessie Allen Sismondi, 2 Apr. [1842], Wedgwood W/M 193; Jardine, 'Description', 187; *ODNB*. J. O. McWilliam Stanger was also a friend of Bachman's: J. Bachman, *Doctrine*, 207.
40. As Hodgkin indicated at the 1841 BAAS meeting: Hodgkin, 'On Inquiries', 54; Stocking, *Victorian Anthropology*, 244; Brantlinger, *Dark Vanishings*, 89.
41. *Notebooks*, C204; Colp, *To Be*, 21; Colp, '*To Be an Invalid* Redux', 214; *CCD*, 4:458. For Darwin at Shrewsbury: CCD, 2:433–4; 4:448; Emma Darwin's diary, 24 Aug. – 2 Oct. 1839, DAR 242.5.
42. *Notebooks*, B147. He carried on, 'May this not be extended to all animals first consider species of cats.— ~~& other tribes~~', showing how human and animal analogies were interchangeable: B148; also B145, C174, QE16 (13).
43. Abstract of Prichard, DAR 71:139–42, f. 140. On the 'Analysis of Contents' page he pencilled another note against this 'Section 6' of Prichard's book: 'Until the action of contagious diseases on close species better known, argument touching men of little value – some difference in predisposition is not denied' (f. 141); Prichard, *Researches* (3rd edn, 1836), 1:152–3.
44. Martin, 'Observations', 3–4; Waterhouse, *Mammalia*, 2:16, in C. Darwin, *Zoology*; *Journal*, 32; Fyfe, 'Conscientious Workmen', 205–7; Desmond, 'Making', 231, 233. On Martin's position: Zoological Society of London, Minutes of Council, 4 (1838), ff. 376, 418–19.
45. *CCD*, 2:2. The new cat had already been correctly reported as '*F. Darwinii*' in the Darwins' other staple daily, the *Morning Chronicle*: 'Zoological Society', *Morning Chronicle*, 12 Jan. 1837. On Gould's position: Desmond, 'Making', 231.
46. For example, on 11 July 1837: 'Zoological Society', *The Times*, 13 July 1837; Martin, 'Monograph'.
47. *Notebooks*, B165, also B209. Orangs and lorises: *Philosophical Magazine*, new ser., 9 (1831), 55–6; 3rd ser., 3 (1833), 61.
48. Martin, *General Introduction*, 167–9, 218, 267, 299–302.
49. He is cited in Nott's appendix in Gobineau, *Moral and Intellectual Diversity*, 475, 503; extensively in Nott et al., *Indigenous Races*; in Nott and Gliddon, *Types of Mankind*; and Nott was heavily indebted to Martin's *History of the Dog* in his parallel pluralist study.
50. Martin, *General Introduction*, 267, 301–3. He was quoted repeatedly in the unity-of-human-species works such as Prichard, *Natural History* (2nd edn, 1845) and J. Bachman, *Doctrine*, where he is described as 'one of the best writers' (p. 93).
51. Meigs, *Memoir*, 20–21, 23, 28, 33; Wood, *Biographical Memoir*, 12–14; Morton, *Crania Americana*, iii; Morton, *Crania Aegyptiaca*, 1; Morton, *Inquiry*, 36.
52. Meigs, *Memoir*, 21, 35–6; Morton, *Crania Americana*, 2–4; Morton, *Inquiry*, 37; 'Certificate of Membership in British & Foreign Aborigines Protection Society, 23 Jan. 1839', Samuel George Morton Papers, APS.
53. Morton, *Inquiry*, 6–8, 13, 16, 19, 37; Morton, *Crania Americana*, iii, 1. Morton distinguished two great families, the Toltecs and the American: they shared an identical racial anatomy, but the latter remained savages where the former – or at least a gifted elite within it – had developed a degree of civilization in the past.
54. Caldwell's review of *Crania Americana*, extracted in his 'Remarks', 208–10, 214–15.
55. *CCD*, 8:171; Kass and Kass, *Perfecting*, 268; Anon, 'Origin', 604, 611.
56. Wyman, 'Morton's Crania Americana', 174–6, 186; Stanton, *Leopard's Spots*, 38; F. and T. Pulszky, *White*, 2:107–8 on the *North American Review*; Morton, *Crania Aegyptiaca*, 158, and Meigs, *Memoir*, 26–7, on 'servants or slaves'; Morton, *Crania Americana*, 88.
57. Morton, *Crania Americana*, 260–61; Gould, *Mismeasure*, 82ff. Whether or not

'Morton's data were completely unsound' (Menand, *Metaphysical Club*, 103; cf. Michael, 'New Look'), we have to understand the shared assumptions and practices of his day which encouraged these results, and treat his pronouncements in context.

58. A. Combe, 'Remarks', 585–9. Not all agreed: William Hamilton re-targeted his anti-phrenological arguments on to Morton, criticizing his failure to distinguish the crania of males and females or to appreciate that millet seeds (the filling medium which determined cranial capacity) gained or lost weight depending on the moisture: W. Hamilton, 'Remarks', 330–33.

59. *American Phrenological Journal*, 3 (1841), 124–6, 191–2, 282–3. Forty pages of extracts and reviews of *Crania Americana* in the *American Phrenological Journal*, 2 (1840), 143–4, 276–82, 385–96, 545–65, testified to its kindred nature; cf. Stanton, *Leopard's Spots*, 37–8. Erickson, 'Phrenology', 92–3, is correct: Morton's commitment to phrenology was stronger than Stanton (or Morton's original biographer, the anti-phrenologist Meigs) allowed. It has been argued (Hume, 'Quantifying Characters') that by firming and fixing racial traits, Morton and his school made a future genetic interpretation easier; but this argument would of course equally be applicable to the hereditarian phrenologists who preceded him.

60. This had begun as early as 1822, with John Bell's preface to the American edition of G. Combe's *Essays*, xlii–xliii. This preface played up the cranial distinctions of human varieties, as indicating inequalities in the mental capacity for improvement through education. Bell also glossed the phrenological factors which would make one race 'the slaves of every invader'. Bell arranged Combe's lectures in Philadelphia in 1839 in the new museum, and had him elected a Corresponding Member of the Academy of Natural Sciences: A. A. Walsh, 'New Science', 403–4; G. Combe, *Notes*, 1:305, 2:374.

61. G. Combe, *System*, 299, 563. Even the expression 'sons and daughters' was effectively a racist phrenological appraisal. For Combe the small skulls of Brazilian Indians indicated a lack of foresight or steadiness of purpose, like a child's, so it was incumbent on Europeans to care for them as children. Had the natives large brains like their conquerors', with enhanced organs of Ideality, Conscientiousness and Causality, 'instead of being their slaves, [the Indians] would become their rivals' (pp. 575–6, 580).

62. G. Combe, *Notes*, 2:21, 48–9, 62–3, 66, 75–8, 84, 86, 112–13; G. Combe, *Lectures*, 306. Of course this rejigging against slavery still left Combe's phrenological system anatomically deterministic with a hierarchical structure and an Anglo-Saxon apotheosis; and when his *Notes* were abstracted in the journals, the racial determinism was still foregrounded, as in G. Combe, 'Observations'.

63. G. Combe, *Notes*, 1:140, 301, 2:157–8, 2:232–3; Morton, *Crania Americana*, iii; G. Combe, 'Phrenological Remarks', 271, 273–5.

64. B. Silliman, 'Phrenology', 65–7, 71; see also 'Lectures on Phrenology', *American Journal of Science*, 38 (1840), 390–91.

65. G. Combe, 'Comparative View', 114–16, 122, 136, a cut-down version of Combe's review in *American Journal of Science*, 38 (1840), 341–75. On the complaints about the de-phrenologizing in the *Edinburgh New Philosophical Journal* version: 'Professor Jameson's Phrenological Prunings', *Phrenological Journal*, 13 (1840), 303–14. Darwin's reading: *CCD*, 4:441, 452, 484.

66. Emma Wedgwood Darwin to her aunt Jessie Allen Sismondi, 9 May [1841], Wedgwood W/M 193 (passage omitted in *ED*, 2:19–20); Greg, *Past*, 97; Temperley, *British Antislavery*, 136–52.

67. Emma Wedgwood Darwin to her aunt Jessie Allen Sismondi, 8 Feb. [1842], Wedgwood W/M 193; Jennings, *Business*, 108; Finkelman and Miller, *Macmillan Encyclopedia*, 229; Rodriguez, *Historical Encyclopedia*, 302; Deyle, *Carry*.

68. Stange, *British Unitarians*, 62; Stanton, *Leopard's Spots*, 64; R. J. C. Young, *Colonial Desire*, 123; Horsman, *Josiah Nott*, 23ff. and chap. 3 on Nott in Mobile. On the different doctoring required by the black slaves – considered as a separate species: Haller, 'Negro', 246–8. Doctors performed many other duties within the 'domestic institution', including

the vetting of slaves at market, and they were well schooled in black anatomy, given the availability of black cadavers in Southern medical schools (including those shipped illicitly from the North in pork barrels): Fisher, 'Physicians', 39, 45–6.

69. Martineau, *Society*, 2:141–2; F. and T. Pulszky, *White*, 2:111–12; Olmsted, *Journey in the Seaboard Slave States*, 565–8.

70. G. Combe, *Notes*, 2:79; Berger, 'American Slavery', 189–90. Combe's American editor John Bell was displeased with *Notes on the United States*, presumably because Combe was so critical of American slavery and back-tracked on black aptitudes: Gibbon, *Life*, 2:124–5. The beauty of women quadroons was a loaded observation not uncommon in anti-slavery literature. On the complexity of the case in an often sexually charged context: Toplin, 'Between Black and White', 190, 194; R. J. C. Young, *Colonial Desire*, 113–14.

71. Nott, 'Mulatto', 252–6; Horsman, *Josiah Nott*, 41, also 18, 86–8; Stanton, *Leopard's Spots*, 65–7; C. Loring Brace, 'Ethnology', 516–17; Toplin, 'Between Black and White', 197–8. That Nott was just giving medical vent to common assumptions in the South about blacks in the North having 'sunk lower': Jenkins, *Pro-Slavery Thought*, 246. The historiography of pro-slavery literature is summarized in Faust, *Ideology*. The South's fears in light of the insurrections are described in Fladeland, *Men*, 190–92. On the effects of Turner's uprising, especially in Alabama: Birney, *James G. Birney*, 72, 85, 104.

7 THIS ODIOUS DEADLY SUBJECT

1. Pierce, *Memoir*, 1:156–7, 160, 190; Dott, 'Lyell'; *CCD*, 2:299.
2. Finkelman and Miller, *Macmillan Encyclopedia*, 563; Drescher and Engerman, *Historical Guide*, 350–51; Menand, *Metaphysical Club*, 10–16; L. G. Wilson, *Lyell*, 38–46, 148; Stange, *British Unitarians*, 100.
3. C. Lyell, *Travels*, 1:49, 169, 183, 186, 193.
4. *LLL*, 2:55, 68–9.
5. L. G. Wilson, *Lyell*, 77–80, 81; Edmund Ravenel, *ANB*; C. Lyell, *Travels*, 1:183, 184.
6. C. Lyell, *Travels*, 1:182, 183, 185, 191, 209, 213–14; *LLL*, 2:66–7.
7. L. G. Wilson, *Lyell*, 81–2; Bellows, 'Study', 517; *LLL*, 1:68.
8. *LLL*, 2:55; C. Lyell, *Travels*, 1:189, 209. Lyell's personal loftiness helps explains why American geologists protested during the trip about his appropriating their fieldwork: R. H. Silliman, 'Hamlet Affair'.
9. Stephens, *Science*, 15, 18, 31–5; Neuffer, *Christopher Happoldt Journal*, 142–4; C. L. Bachman, *John Bachman*, 175; Waddell, 'Bibliography'; *Notebooks*, C251–6, D31–4.
10. J. Bachman, *Doctrine*, 291–2; C. Lyell, *Travels*, 1:172.
11. C. Lyell, *Travels*, 1:178, 261; Dott, 'Lyell', 113; L. G. Wilson, *Lyell*, 106; *LLL*, 2:63.
12. *Notebooks*, C76, N121–84; *Autobiography*, 120; Moore, 'Darwin'.
13. *CCD*, 3:44, 394. Charles asked Emma to offer £400 as a fee for the work involved. This was under a quarter of his £1,748 spending in 1844 (Darwin's account books, CUL, DH/MS* 11:1–17). Assuming that a qualified editor would have been paid a fee commensurate with Darwin's cost of living, and allowing that such an editor, unfamiliar with Darwin's notes, would have needed more time to complete the work than Darwin himself would have taken, it appears that at this time Darwin estimated his essay would have required only a few months' work by him to be ready for publication in that form. Thanks to Randal Keynes for this suggestion.
14. *Foundations*, 68, 79, 92, 93, 115, 241.
15. Cf. Alter, 'Separated'; *CCD*, 4:454, 459, 462, 467; Noll, *Princeton*, 115–24.
16. *Notebooks*, B272; Prichard, *Researches* (1973), 67; Augstein, *James Cowles Prichard's Anthropology*, 52; *CCD*, 13:360. Darwin's only published references to White are in *Variation*, 2:14, 87.
17. C. Lyell, *Principles of Geology* (6th edn, 1840), 1:248–60 in Darwin Library, CUL;

CCD, 2:253; *Beagle* field notebook EH1.3, 4a, Down House. On the platypus: J. W. Gruber, 'Does the Platypus'; Desmond, *Politics*, 279–88.

18. Cf. L. G. Wilson, *Lyell*, 143; *Journal* (1860), 358 (cf. H. E. Gruber, 'Many Voyages'); C. Lyell, *Travels in North America* (1845), 201 in Darwin Library, CUL; *CCD*, 3:233, 234, 242; C. Lyell, *Travels*, 2:36.
19. *CCD*, 3:242 and 243 n8; C. Lyell, *Travels*, 1:184–5. In the first edition of the *Journal* (1839, 27–8), Darwin's remarks on slavery appear at the front, in their chronological place during his sojourn at Rio de Janeiro:

> While staying at this estate [on the Rio Macâe], I was very nearly being an eyewitness to one of those atrocious acts, which can only take place in a slave country. Owing to a quarrel and a lawsuit, the owner was on the point of taking all the women and children from the men, and selling them separately at the public-auction at Rio. Interest, and not any feeling of compassion, prevented this act. Indeed, I do not believe the inhumanity of separating thirty families, who had lived together for many years, even occurred to the person. Yet I will pledge myself, that in humanity and good feeling, he was superior to the common run of men. It may be said there exists no limit to the blindness of interest and selfish habit. I may mention one very trifling anecdote, which at the time struck me more forcibly than any story of cruelty. I was crossing a ferry with a negro, who was uncommonly stupid. In endeavouring to make him understand, I talked loud, and made signs, in doing which I passed my hand near his face. He, I suppose, thought I was in a passion, and was going to strike him; for instantly, with a frightened look and half-shut eyes, he dropped his hands. I shall never forget my feelings of surprise, disgust, and shame, at seeing a great powerful man afraid even to ward off a blow, directed, as he thought, at his face. This man had been trained to a degradation lower than the slavery of the most helpless animal.

20. *Journal* (1860), 499–500; J. and M. Gribbin, *FitzRoy*, 229–35; *CCD*, 3:345.
21. Stange, *British Unitarians*, 51–2; W. B. Carpenter, *Nature*, 9–10; Fryer, *Staying Power*, chap. 3; Thomas, *Slave Trade*, 296, 513–14; Cull, 'Short Biographical Notice'; R. L. Carpenter, *Memoirs*, 271–73.
22. W. B. Carpenter, 'Natural History', 155–67; Jacyna, 'Principles', 59; Desmond, *Politics*, 215.
23. *CCD*, 3:90; Darwin, 'Observations'; W. B. Carpenter, 'Microscopical Structure'; R. D. Keynes, *Charles Darwin's Zoology Notes*, xii–xiii, on *Sagitta*; W. B. Carpenter, *Nature*, 70, quoting Carlyle.
24. O. Dewey, *On American Morals*, 18; M. E. Dewey, *Autobiography*, 79–80, 127–9, 191–2; Stange, *British Unitarians*, 72–3, 110, 119 (and 61ff. for the Carpenters' and Estlins' conversion to evangelical Garrisonianism, that is, radical abolitionism); W. and E. Craft, *Running*, 97.
25. O. Dewey, *On American Morals*, 17–21, 24; Stange, *British Unitarians*, 172–4; Stanton, *Leopard's Spots*, 73–4.
26. W. B. Carpenter, *Zoology*, 1:148–9; W. B. Carpenter, 'Letter', 140–41; *Foundations*, 72–4.
27. W. B. Carpenter, *Zoology*, 1:147, 150–51; W. B. Carpenter, *Principles of Human Physiology*, 67–8; W. B. Carpenter, 'Letter', 139–44; W. B. Carpenter, 'Dr. Carpenter'; W. B. Carpenter, *Nature*, 12, 78–80.
28. W. B. Carpenter, 'Letter', 143; E. G. Wilson, *Thomas Clarkson*, 177, 184–9.
29. Knox, 'Lectures', 97–8, 118–19, 133–4, 231; Knox, *Races*, 20, 565 for the cities toured.
30. Knox, 'Lectures', 97, 148, 299, 332; Biddiss, 'Politics'.
31. Knox, 'Lectures', 233, 299; E. Richards, 'Moral Anatomy', 385, 394–5 for the best revisionist account of Knox.
32. *The Times*, 13 May 1847, 1; Lindfors, 'Hottentot', 10–16, on this San group in London, and Lindfors, *Africans* on the manufacturing of 'racial' types in stage shows.
33. 'Dr. Knox on the Races of Men', *Medical Times*, 18 (1848), 114; Knox, 'Lectures', 97, 363.

34. *Notebooks*, B179–80; *Foundations*, 68–9, 72, 98–9; *CCD*, 2:182–5, 202–5; Kohn, 'Theories', 134ff.; W. B. Carpenter, *Zoology*, 1:150–51.
35. Dickson, 'Letter'; Nott, 'Hybridity of Animals', in Nott and Gliddon, *Types*, 398; L. G. Wilson, *Lyell*, 64; Stanton, *Leopard's Spots*, 73–4; W. B. Carpenter, 'Letter', 142–3.
36. *CCD*, 3:258; L. G. Wilson, *Lyell*, 172; Long, quoted in R. J. C. Young, *Colonial Desire*, 150, 151.
37. C. Lyell, *Second Visit*, 1:366. Darwin marked this statistic in his copy of Lyell's *Second Visit*, Darwin Library, CUL; L. G. Wilson, *Lyell*, 171, 176.
38. C. Lyell, *Second Visit*, 1:304. Darwin marked this passage in his copy of Lyell's *Second Visit*, Darwin Library, CUL; L. G. Wilson, *Lyell*, 177; *Journal* (1860), 378.
39. *LLL*, 2:11, 15, 97, 100; L. G. Wilson, *Lyell*, 181; C. Lyell, *Second Visit*, 1:364; 2:2.
40. Nott, quoted in L. G. Wilson, *Lyell*, 219; *LLL*, 2:42–3; *CCD*, 4:56; Horsman, *Josiah Nott*, 102.
41. Lyell Notebook 134, [16], Kinnordy MSS, by courtesy of Leonard Wilson; Nott, 'Unity', 20–21.
42. Nott, 'Statistics', 279, 280; Nott, quoted in L. G. Wilson, *Lyell*, 219.
43. Lyell Notebook 134, [15–17], Kinnordy MSS, by courtesy of Leonard Wilson.
44. W. T. Hamilton, *Pentateuch*, 314, 317; Nott, *Two Lectures*, 5, 7, 18; Lyell Notebook 134, [19], Kinnordy MSS, by courtesy of Leonard Wilson.
45. W. T. Hamilton, *Pentateuch*, 298; Hamilton considered his Babel solution to the origin of the races 'original' (p. x) and dated it to a lecture in January 1844, though it was not published until 1850 (pp. 303–4n). Nott picked up on and satirized it immediately: 'We might just as well *suppose* that some were changed into Monkeys, while others were changed into negroes. In arguing a question of this kind we want *facts*' (Nott, *Two Lectures*, 28n).
46. *LLL*, 2:99, 100; Lyell Notebook 134, [16], Kinnordy MSS, by courtesy of Leonard Wilson; Fox-Genovese and Genovese, *Mind*, chap. 15; H. S. Smith, *In His Image*, chap. 2.
47. C. Lyell, *Second Visit*, 2:130–31 (cf. L. G. Wilson, *Lyell*, 233); also 2:115, 116.
48. C. Lyell, *Second Visit*, 2:268–9, 345; L. G. Wilson, *Lyell*, 269; *LLL*, 2:101.

8 DOMESTIC ANIMALS AND DOMESTIC INSTITUTIONS

1. Nott, *Two Lectures*, 5, 7, 15–16, 19, 20; Stanton, *Leopard's Spots*, 67.
2. Nott, 'Statistics', 278–9; Haller, 'Negro', 252–3; Horsman, *Josiah Nott*, 88ff.; Jenkins, *Pro-Slavery Thought*, 246.
3. Jenkins, *Pro-Slavery Thought*, 200; C. Loring Brace, 'Ethnology', 518; Messner, 'DeBow's Review', 201–4.
4. Nott, *Two Lectures*, 21–2.
5. *Notebooks* E65; *Origin*, 113.
6. C, 'Unity', 412 (cf. Nott, *Two Lectures*, 17, and 22 on goats and ewes). For rejoinder and riposte: Anon, 'Issue'; Anon, 'Unity'; Ryan, 'Southern Quarterly Review', 184.
7. C, 'Unity', 414, 416, 444–5.
8. *Notebooks*, B228; also on hybridity C30, 34, 135, and C125 on 'Herbert's law' (attributed to the *Amaryllis* hybridizer William Herbert) about adaptive habits and environmental circumstance also affecting fertility.
9. Morton, 'Hybridity in Animals and Plants', 262–4, 269, 275, a reprint of Morton, 'Hybridity in Animals, considered', 39–50, 203–12. Darwin read both: *Natural Selection*, 427, 431, 454; Morton, 'Description', 212 (on his chicken × guineafowl cross), again noted by Darwin, *Natural Selection*, 436n, 454. For Darwin's criticisms: *CCD*, 4:46, 48.
10. Morton, 'Hybridity in Animals and Plants', 277; *CCD*, 4:46.
11. *PP*, 1836 (440), x.i, 200 (a2466); also 72 (a717), 138 (a1507), 157 (a1743–4), 173

(a1996), 189 (a2270); and 210 (a2568) for Temminck's as the largest collection bought by the British Museum. Contemporary accounts of various 'Temminck's' monkeys: Martin, *General Introduction*, 499, 531. The junglefowl *Gallus temminckii* – the colourful cockerel with its spray of drooping tail feathers – was named at the Zoological Society in 1849, although, ironically, Edward Forbes raised the prospect with Darwin that it itself was a hybrid: *CCD*, 6:61. This caused Darwin to float the idea in print (*Variation*, 1:235). Forbes's view has not stood the test of time.

12. J. Bachman, *Doctrine*, 85–9; Smyth, *Unity*, 193; *CCD*, 4:46.
13. This remained Darwin's view of Morton's paper: *CCD*, 4:46, 47–8, 9:52. Darwin also faulted Morton's knowledge of geese: *Natural Selection*, 427.
14. Morton, 'Hybridity in Animals and Plants', 266; C. H. Smith, *Natural History of Horses*, 153–4, also 66–7, 70–71 on horse species hybridizing to form homogeneous domestic breeds. Darwin read *Dogs* and *Horses* shortly after publication, in 1841 and 1842: *CCD*, 4:460, 463, 465; *Marginalia*, 760–66.
15. R. S. Owen, *Life*, 1:182; Swainson, *Preliminary Discourse*, 154–5; Torrens, 'When did the Dinosaur?'
16. C. H. Smith, *Natural History of Dogs*, 2:77–8, cited in *Annals and Magazine of Natural History*, 8 (1841), 137–8; also, 1:104; 179–83 on the *dhole*; 2:79, and 2:81–2 on FitzRoy; *Marginalia*, 763; Morton, 'Hybridity in Animals and Plants', 270–71. Hamilton Smith was well cited in the literature, praised (by pluralists) and refuted (by unitarists, although they recognized his stature). For Smith on dogs: Nott, 'Unity', 4; J. Bachman, *Doctrine*, *passim*, esp. 44ff., and 61ff.; Anon, 'Original Unity', 550; Anon, 'Natural History' (1852), 441–2; Anon, 'On the Unity', 291; Nott and Gliddon, *Types*, 376ff.; Cabell, *Testimony*, 72. The origin of dogs figured heavily in the human-race debate among London's Prichardians: W. B. Carpenter, *Zoology*, 1:36–7; Holland, 'Natural History', 27; W. B. Carpenter, 'Varieties', 1310–11.
17. Kass and Kass, *Perfecting*, 397; Hodgkin, 'On the Dog', 81.
18. In his chapter on 'Hybridity of Animals' in *Types of Mankind* (p. 393), Nott would argue that 'if specific differences among dogs were the result of climate, all the dogs of each separate country should be alike'. (Darwin appreciated this; artificial selection explained why the breeds diverged.) There was so much on domestic animals in *Types* that an antagonistic *Presbyterian Magazine* asked why Nott and Gliddon did not follow up with 'another quarto on the "Types of Dogkind"?' Nott did indeed follow on with a (racially loaded) article, 'A Natural History of Dogs' in the *New Orleans Medical and Surgical Journal*: C. Loring Brace, 'Ethnology', 521; Stanton, *Leopard's Spots*, 181.
19. C. H. Smith, *Natural History of Dogs*, 1:88–9; *Marginalia*, 762; *CCD*, 1:179, 3:126.
20. C. H. Smith, *Natural History of the Human Species*, 111, also x, 92–102, 117–18, 129 (we cite the 1852 edition, which Darwin read); C. H. Smith, *Natural History of Dogs*, 1:86–97, 103.
21. C. H. Smith, *Natural History of the Human Species*, 126–31.
22. Anon, 'Original Unity', 550.
23. J. Bachman, *Doctrine*, 10, 305–6; *Journal* (1860), 145–7.
24. J. Bachman, *Doctrine*, 135–6, also 24, 29, 31, 34, 85, 144–6.
25. J. Bachman, *Doctrine*, 37–41, also 8, 15, 119; 89–92 on Derby's menagerie, and C. L. Bachman, *John Bachman*, 174; Stephens, *Science*, 166, 173, 188; Lurie, 'Louis Agassiz', 231n, quoting Nott's 1850 letter to Lewis R. Gibbes on 'prostitutors' like Bachman.
26. Anon, 'Natural History' (1850), 304 ('re-enslaving'), 327; Horsman, *Race*, 147–8, on this review; B., 'Prichard's Unity', 208, 212 for the 'Parthenon' gibe.
27. C. Loring Brace's view ('Ethnology', 512), that 'in Britain and France the contest was largely decided by the mid-point of the 19th century, with the polygenists left in nearly sole possession of the field', is overstated: elite gentlemen of London science who remained unitarists included Carpenter, Holland, Lyell, Darwin, Owen, Forbes, Prichard, Hodgkin, Latham and others inside the Ethnological Society, for instance, the clinical psychologist Robert Dunn. Nevertheless pluralism was growing stoutly among ethnologists *and breed-*

ers in Britain, and even the Ethnological Society in the later 1850s would change its consensus. On the complexity of Dunn's case – and his compromise with the craniologists: Kenny, 'From the Curse', 378; Livingstone, *Preadamite Theory*, 30–31; Haller, *Outcasts*, 35.

28. J. Bachman, *Doctrine*, 70–75; C. H. Smith, *Natural History of Dogs*, 93–4; Morton, 'Hybridity in Animals and Plants', 273–4.
29. J. M. Herbert's recollections in DAR 112 B57–76; *CCD*, 1:104, 106, 109.
30. *CCD*, 1:414; also 202, 204, 231, 234, 257, 349–51, 389; Freeman, *Charles Darwin*, 138; *Autobiography*, 68.
31. Sulloway, '*Beagle* Collections', 64, 73; C. Darwin, *Zoology*, pt 3, App., 147–56; Steinheimer, 'Charles Darwin's Bird Collection', 313.
32. *Notebooks*, C124–5; B162; *Variation*, 1:74. The 28 February meeting was reported, for instance, in *John Bull*, 5 Mar. 1837, where Eyton is misnamed as 'Mr Acton'.
33. *Magazine of Zoology and Botany*, 1 (1837), 305; *CCD*, 6:206–7, 211; *Variation*, 1:71, for his discussions with Eyton on pig hybrids. The doubting zoologists were Nicholas Vigors and William Yarrell (although he too then conceded that he knew of fertile duck hybrids).
34. Eyton, 'Some Remarks', 358; Eyton, 'Remarks'.
35. *CCD*, 2:181, 183; *Notebooks*, B30, B139; D inside front cover, D23.
36. *Origin*, 253; *Natural Selection*, 430; Darwin, 'Fertility'.
37. *Notebooks*, D23.
38. Dixon, 'Poultry Literature', 324; *CCD*, 4:169.
39. Dixon, *Ornamental and Domestic Poultry*, ix–xiii; *CCD*, 4:222, 476.
40. Secord, 'Nature's Fancy', 167–8; Dixon, 'Poultry Literature' 320, 324.
41. Darwin's marginalia on Dixon, *Ornamental and Domestic Poultry*, xiii, Darwin Library, CUL (*Marginalia*, 202); Dixon, 'Poultry Literature', 337, also 331–2, 336, 345.
42. *CCD*, 5:391; Dixon, 'Poultry Literature,' 347–51; Lorimer, *Colour*, chaps 3–4.
43. From John Charles Hall's introduction to Pickering, *Races of Man*, x–xii (Hall was a Sheffield physician concerned with the welfare of local industrial workers, with a sideline on unitarist publications); Anon, 'Original Unity', 544, 547; *CCD*, 4:299, 478; Stanton, *Leopard's Spots*, 93–6. Darwin annotated the 1850 English edition of Pickering (Darwin Library, CUL), which contained Hall's introduction: *Marginalia*, 667.
44. Carlyle, 'Occasional Discourse', 676, quoted partly in *United States Magazine and Democratic Review*, 26 (1850), 302–4; and in full, and attributed to Carlyle, in *DeBow's Review*, 8 (1850), 527–38.
45. Fielding, 'Froude's Revenge', 86; Erasmus had long known of Carlyle's 'hallucinations' over slavery and his lament that there was 'no chance of the beautiful relations of master & slave being reestablished in Europe' (quoted p. 80). Mackenzie, 'Thomas Carlyle's "The Negro Question"', 219, 222.
46. Stenhouse, 'Imperialism'; Secord, *Victorian Sensation*, 311; Desmond, 'Artisan Resistance'; R. Cooper, *Infidel's Text-Book*, 158–9.
47. Swainson, *On the Habits*, 324–5. This, like Kirby and Spence's similar account, *Introduction*, 2:74–87, was based on the Genevan Pierre Huber's original announcement.
48. Kirby and Spence, *Introduction*, 2:74–87; E. Newman, *Familiar Introduction*, 50–52. Newman's statements on slave ants were singled out with 'amazement' in the *Monthly Review*, as well as in 'Slave Ants', *Chambers's Edinburgh Journal* (Oct. 1841), 320.
49. *CCD*, 2:244, 3:24; Desmond, 'Making', 170–71.
50. J. F. M. Clark, 'Complete Biography', 252–3. Darwin called Smith the 'highest authority': *Natural Selection*, 314; *CP*, 2:139.
51. Swainson, *Treatise*, 2, 14–15. Swainson's position was noted by Van Amringe, *Investigation*, 138, and even Louis Agassiz called Swainson 'learned': Agassiz, *Essay*, 242.
52. Van Amringe, *Investigation*, 311–12.
53. Van Amringe, *Investigation*, 80, 118, 269–70, 308, 422, 423ff., 456–57, 488. Van Amringe's biblical loose ends – squaring separate human species with a single-pair origin

– were tied shortly by the preadamite theory (which assumed that other types of human beings existed before Adam): Livingstone, *Preadamite Theory*, 24ff.

54. Anon, 'Natural History' (1850), 334; Van Amringe, *Investigation*, 639, 654ff. on beauty and race.

55. Van Amringe, *Investigation*, 654ff. His relativism contrasted with Hamilton Smith's universal racial standard of beauty (*Natural History of the Human Species*, 161–2), which was summed up crassly: 'the fair haired white Caucasian woman has been always sought as a wife by every race': J. Campbell, *Negro-mania*, 547. But Gobineau's own universalism was criticized by his Mobile-based translator Henry Hotze, who saw – as did Darwin – each race exhibit its own standards; thus a different appreciation of 'sexual beauty' in the races 'has been instrumental in separating them, and keeping them distinct': Gobineau, *Moral and Intellectual Diversity*, chap. 12, esp. 379–81.

56. Van Amringe, *Investigation*, 308–9.

57. *Notebooks*, QE16, also E183, M127, N111; Browne, *Charles Darwin*, 1:426–7; Colp, *To Be*, 21, 110–11; *CCD*, 4:35, 458; Holland, *Recollections*, 6.

58. Browne, *Charles Darwin*, 1:491–2; Colp, *To Be*, 39, 109; *CCD*, 4:209–10, 384–5. Even at this distressing time Darwin continued to tap Holland on medical matters, specifically reproduction: *Marginalia*, 162; *CCD*, 4:445.

59. Holland, 'Natural History', 5, 16, 20–21, 26, 30. This number of the *Quarterly Review* came out in December 1849, so Holland was probably drafting it in the weeks after Prichard died, when Darwin visited: R. Keynes, *Annie's Box*, 217–18; Augstein, *James Cowles Prichard's Anthropology*, 108–9.

9 OH FOR SHAME AGASSIZ!

1. *CCD*, 3:2, also 3:253, 346; 4:13; Bellon, 'Joseph Hooker', 7–8 on Hooker's 'philosophical botany'; Endersby, *Imperial Nature*, 47 and *passim* on the Darwin–Hooker relationship; Desmond, *Huxley*, 45–6.

2. Rudwick, 'Darwin'; *LLL*, 2:155; *CCD*, 3:346, 350.

3. This assumes that the sketch was written or completed towards the end of Darwin's two-month visit to Shrewsbury and Maer, between 18 June 1842, when he went to Wales in search of glacial evidence, and his return to London on 18 July according to Darwin, or on the 15th in his account books, or on the 16th as per Emma's diary: *CCD*, 2:435, 437; Emma Darwin's 1842 diary, DAR 242.8.

4. *CCD*, 4:148, also 3:333; 4:74.

5. *CCD*, 4:13, 100n; 5:101, 135, 178, 345–6, 357, 359, 363, 372; Browne, *Charles Darwin*, 1:476–7; Love, 'Darwin'.

6. *LLL*, 2:104; L. G. Wilson, *Lyell*, 140; Lurie, *Louis Agassiz*, 119, 122.

7. Story, 'Harvard', 105; Menand, *Metaphysical Club*, 12–13, 98–9; Lurie, *Louis Agassiz*, 138–41, 147; Pierce, *Memoir*, 3:26; *LLL*, 1:457–8, 2:159.

8. *CCD*, 3:79; J. W. Clark and T. M. Hughes, *Life*, 1:447; *LLL*, 2:155 on Forbes; G. Wilson and A. Geikie, *Memoir*, 263, 461; Agassiz, 'Period', 16; Lurie, *Louis Agassiz*, 4–5, 81–8; Rehbock, 'Early Dredgers'.

9. C. Lyell to R. Owen, 19 Oct. 1846, Richard Owen Correspondence 18.136, Natural History Museum; Desmond, *Archetypes*, 58–9; Bartholomew, 'Lyell'; Story, 'Harvard', 104–5.

10. *CCD*, 5:187.

11. L. Agassiz to R. M. Agassiz, 2 Dec. 1846, bMS Am 1419 (66), ff. 13–14, by permission of the Houghton Library, Harvard University (translated from the French); Menand, *Metaphysical Club*, 105; Gould, 'Flaws', 143; Lurie, 'Louis Agassiz', 234; Lurie, *Louis Agassiz*, 143. Morton had only read his papers on hybridity three weeks before the date of this letter, on 4 and 11 Nov. 1846: Morton, 'Hybridity in Animals, considered', 39.

12. Dupree, *Asa Gray*, 153; Moore, 'Geologists'; J. L. Gray, *Letters*, 1:345–7; Stanton, *Leopard's Spots*, 102–4; Lurie, 'Louis Agassiz', 124–5, 233–4.
13. As R. W. Gibbes reported to S. G. Morton: Lurie, 'Louis Agassiz', 234–5; J. Bachman, *Doctrine*, 3; *CCD*, 3:380; *Journal* (1860), 500.
14. *LLL*, 2:128, 155; L. G. Wilson, *Lyell*, 308; *CCD*, 4:239, 246.
15. Dickens, *American Notes*, 159–63; Berger, 'American Slavery'.
16. Anon, 'Notices' (1849), 639. Others merely granted that Lyell was 'more liberal' than his predecessors but still shielded from plantation reality by emancipationist blinkers: W., 'Second Visit', 415–21. But Lyell's criticisms of loudmouth abolitionists, whatever his impertinence in judging Southern 'gentlemen', did go down well: D. J. McCord, 'How', 359–62.
17. Parker, *Sermons*, xxvii; *CCD*, 4:239, 341, 477; C. Lyell, *Second Visit*, 1:184; *LLL*, 2:154; L. G. Wilson, *Lyell*, 311, 321.
18. See Robert Bernasconi's introduction to his edition of Gobineau, *Moral and Intellectual Diversity*.
19. Knox, *Races*, v, 8, 23; Stocking, *Victorian Anthropology*, 64; Malcolm, 'Address', 89; 'Pencillings of Eminent Medical Men: Dr Knox, of Edinburgh', *Medical Times*, 10 (1844), 245–6.
20. L. S. McCord, 'Diversity', 410, 412, 414–15; Knox, *Races*, 244–5, 464; Lounsbury, *Louisa S. McCord*, 2; Fought, *Southern Womanhood*, 3; Malcolm, 'Address', 90; E. Richards, 'Moral Anatomy', on Knox's transcendentalism and imperial idiosyncrasy.
21. Finkelman and Miller, *Macmillan Encyclopedia*, 13–17.
22. Rodriguez, *Chronology*, 317, 346.
23. *CCD*, 4:478–9, 488; Wheatley, *Life*, 183.
24. *CCD*, 4:362. A copy of Mary Howitt's *Our Cousins in Ohio* with Darwin's signature on the front endpaper is in the family; his reading notebook refers to the book as 'Life in Ohio': *CCD*, 4:479; see Oldfield, 'Anti-Slavery Sentiment'.
25. Davis, *In the Image*, 256; Davis, *Inhuman Bondage*, 265; Still, *Underground Rail Road*, 267 for Craft's revolver.
26. Agassiz, 'Geographical Distribution', 181–3, 186, DRC 111. On the *Christian Examiner*: Mott, 'Christian Disciple'; George Edward Ellis, *ANB*. Harvard's then president, Jared Sparks, retired to become Agassiz's personal publicist: Lurie, *Louis Agassiz*, 199.
27. Agassiz, 'Diversity', 110, 112–13, 120, 142–5; Agassiz, 'Geographical Distribution', 200, 204; Lurie, 'Louis Agassiz', 236–8; Gould, *Mismeasure*, 94–9; Stanton, *Leopard's Spots*, 153.
28. *Proceedings of the Academy of Natural Sciences of Philadelphia*, 3 (1846–7), 29, 41–2, 51, 122, 198, 209, 221, etc.; Wallis, 'Black Bodies', 44–6; Lurie, 'Louis Agassiz', and Stanton, *Leopard's Spots*, cite a number of Gibbes' letters; Horsman, *Josiah Nott*, 33 on the preparatory school.
29. Wallis, 'Black Bodies', 40, 44–5; Stephens, *Science*, 173; Stanton, *Leopard's Spots*, 153. Agassiz continued studying slaves during his winter trips to Charleston in 1852 and 1853. He lived on Sullivan's Island outside Charleston harbour and lectured in town three times a week at the medical college (L. Agassiz to J. D. Dana, 26 Jan. 1852, bMS Am 1419 [119], by permission of the Houghton Library, Harvard University). 'I have found here an excellent opportunity of examining the negros', he wrote (L. Agassiz to Mrs [J.] Holbrook, 25 [Oct. 1852], bMS Am 1419 [118], by permission of the Houghton Library, Harvard University). Agassiz was a frequent guest of reptile expert Dr John H. Holbrook (Lurie, *Louis Agassiz*, 143–4) and here, on his verandah in 1852, the 'ever-recurring topic was that of the origin of the human race' (E. C. C. Agassiz, *Louis Agassiz*, 2:497). Holbrook was another who opposed Bachman, backed Nott and Agassiz, and connived with Gibbes and Morton 'to marginalize monogenist' science: Livingstone, 'Science', 397–8.
30. *CCD*, 4:353, also 4:341, 345; C. Darwin, *Monograph*, 253, 457.
31. J. Campbell, *Negro-mania*, 545, also 6, 11; Van Amringe, *Investigation*, 713; R. C. J. Young, *Colonial Desire*, 138; Lurie, 'Louis Agassiz', 238.

32. Morton, *Letter*, 7; J. Bachman, 'Reply'. Thanks to Gene Waddell for sending us a copy of the latter in the *Charleston Medical Journal and Review*.
33. J. Bachman, *Doctrine*, 135; Dunn, 'Some Observations', 58; Agassiz, 'Diversity', 118; Prichard, 'On the Relations', 316. On Ethnological Society membership: *Anthropological Review*, 7 (1869), 119, refers to six members in 1858; Lorimer, *Colour*, 136, for mid-decade numbers; also Stocking, *Victorian Anthropology*, 245ff.; Rainger, 'Philanthropy'.
34. Agassiz, 'Geographical Distribution', 194; Dixon, 'Poultry Literature', 324; Dixon, *Ornamental and Domestic Poultry*, ix; Nott, 'Diversity', 120.
35. *CCD*, 4:344; C. Darwin, *Monograph*, 155; Kohn, 'Darwin's Keystone'; W. A. Newman, 'Darwin', 360ff. While Darwin announced this ubiquitous variation, he struck a deliberate pose as a 'lumper' in his taxonomy, keeping all the variants within known species. This was to prove to Hooker (himself a lumper) that transmutationist beliefs did not lead to chaotic species-splitting (Hooker's main worry): Bellon, 'Joseph Hooker', 22; Endersby, *Imperial Nature*, 156ff.
36. W. B. Carpenter, 'Researches', 213–22; *CCD*, 5:411; Hodgkin, 'Obituary', 189; W. B. Carpenter, *Nature*, 12; W. B. Carpenter, 'Dr Carpenter', 461. At this time, 1855, Darwin was drawing up lists of hybrids. He envied Carpenter, the pre-eminent compiler of text-books, for effortlessly covering the same ground. Not merely effortlessly: Carpenter was explicitly covering it for Prichardian ends. In his discussion of hybrids in *Principles of Comparative Physiology*, 621–8, Carpenter extended the range of varieties and restricted the fertility of hybrids to ensure that the varied humans appeared united. Darwin's difficulty with compiling left him with a 'wonderfully enhanced . . . respect for Carpenter et id genus omne': *CCD*, 5:377.
37. Emma Wedgwood Darwin to her son William, Friday [Feb.? 1852], DAR 219.1.1; Ahlstrom, *Religious History*, 657; Lorimer, *Colour*, 82–3.
38. *CCD*, 5:118, 187; E. C. C. Agassiz, *Louis Agassiz*, 2:486, 538–9; *LLL*, 2:184, 189. For the astonishing roll-call of students inspired by Agassiz: Winsor, *Reading*, 35.
39. *CCD*, 5:167; Agassiz, 'Geographical Distribution', 184, 194.
40. *Notebooks*, QE5 (11); B82, 92, 125, 224, 248; *CCD*, 2:39; 4:49–50; 5:233; Grinnell, 'Rise', 264–71; *Journal*, 541–2; Hooker, 'Reminiscences', 188.
41. *CCD*, 5:237, 241, 263, 299, 305, 308, 321, 328; *LJH*, 1:352; *CP*, 1:254. Taking nothing for granted, Darwin even tried (unsuccessfully) frogspawn and snail eggs: *CCD*, 6:239, 385; snails as listed in Darwin's 'Catalogue of Down Specimens', EH 88202576, Down House (thanks to Randal Keynes for sharing his transcription).
42. *CP*, 1:255–8.
43. 'Ethnographic Map of the World showing the Present Distribution of the Leading Races of Man' and 'Geographical Distribution of Plants according to Humboldt's Statistics', in *Johnston's Physical Atlas; CCD*, 5:279; *CP*, 1:255–8.
44. *CCD*, 5:187–8; 5:363.
45. Bellon, 'Joseph Hooker', 1–2, 14–15; *LJH*, 1:473–5; Endersby, *Imperial Nature*, 280.
46. A. Gray to J. D. Hooker, 21 Feb. 1854, Royal Botanic Gardens, Kew; D. Porter, 'On the Road', 15–17; *LJH*, 1:473–5.
47. *CCD*, 5:186; Dupree, *Asa Gray*, 228.
48. *CCD*, 4:50; 5:68, 338–9, 364–7, 374–5; 6:122, 198, 201, 408; *CP*, 1:257, 261–4; *LJH*, 1:445, 494.
49. *CCD*, 5:326; 6:90, 100, 174, 176, 178, 248, 305, 385.
50. Secord, 'Nature's Fancy', 165; *Notebooks*, D100; QE4; *CCD*, 5:508; C. Darwin, *Expression*, 259.
51. Desmond and Moore, *Darwin*, 327; Moore, 'Darwin', 459–61; 'Darwin Manuscripts and Letters', *Nature*, 150 (1942), 535; A Lady, *Modern Domestic Cookery*, 210.
52. *CCD*, 5:288, 294, 337, 508, 513; Secord, 'Nature's Fancy', 165; *Variation*, 1:207. To date, no studies of Darwin's artificial selection have seen his experiments serving a larger purpose within the human racial context: Sterrett, 'Darwin's Analogy'; Wilner, 'Darwin's Artificial Selection'; Bartley, 'Darwin'.

53. *Variation*, 1:19, 23; Kass and Kass, *Perfecting*, 590 n14; Schomburgk, *History*, viii, ix, 533, 538, 554–5; *Notebooks*, QE inside front cover; *CCD*, 4:111; 5:513.
54. *CCD*, 5:510–11; 6:269; *PP*, AP, 1854–5 (0.3), lvi.1 (Class A), 71ff.; *PP*, AP, 1856 (0.1), lxii.1 (Class A), 101ff.
55. *CCD*, 5:530–31.
56. *CCD*, 5:524–7, 530–31; 6:54.
57. *CCD*, 5:326, 352, 386; 6:152.
58. *Variation*, 1:131, 180, 186; *CCD*, 5:294, 391; Secord, 'Nature's Fancy', 167; Ospovat, *Development*, 156–7; Bartley, 'Darwin', 317–18.
59. *CCD*, 7:204–5, also 5:359; *Variation*, 1:192; *Origin*, 27. A partial list of Darwin's crossed pigeon races (eleven crossings, with typical entries of the type 'Runt Red %, White Trumpeter &') occurs at the back of his 'Catalogue of Down Specimens', EH 88202576, Down House.
60. *CCD*, 4:139; 5:309; Dixon, 'Poultry Literature', 334, 336; *CCD*, 5:xviii and 315 n2 pass over this ethnological comment in Blyth's response.
61. *CCD*, 6:13–14, 62, 238; *Variation*, 1:208–21; Alter, 'Advantages'; Secord, 'Nature's Fancy', 183–4. Extinctions of sheep and bullock breeds since pre-Pharaonic times had already been reported by others: Pickering, *Races*, 315.
62. *Origin*, 19–20; *CCD*, 5:509, 387; 6:236; Secord, 'Nature's Fancy', 172–4.
63. *CCD*, 6:50–51, 57; *Variation*, 1:224; Secord, 'Nature's Fancy', 177–8.
64. *Foundations*, 95.
65. *Origin*, 19.
66. *LLL*, 2:213; L. G. Wilson, *Sir Charles Lyell's Scientific Journals*, 98, 162, 474–7; *CCD*, 5:492: 7:357; 8:368.
67. From Jane Welsh Carlyle's journal, 25 Nov. 1855, in C. R. Sanders, *Collected Letters*, 30:195–262 (November).
68. Godwin, *Finding Aid*, 25; *Wedgwood*, 248–9; Hammett, 'Two Mobs', 859; Blue, 'Poet'; Fanny Allen to her sister Jessie Allen Sismondi, 10 Apr. [1842 or 1843], Wedgwood E57/32120. Molly's brother Tom was a 'rare rebel' (Jaher, 'Nineteenth-Century Elites', 67) against Boston's Brahmins and had established his own link with the Darwin–Wedgwoods, sending Darwin a copy of J. C. Frémont's *Report of the Exploring Expedition to the Rocky Mountains in the Year* (1845), now in Darwin Library, CUL; *Natural Selection*, 491; Darwin to T. G. Appleton, 31 Mar. 1846, Catalogue 92, James Cummins Bookseller (New York, [2003]), 62–4.
69. Mackintosh's governorship can be followed in his reports to the colonial secretary of the day: on proprietors: *PP*, CP, 1849 (1126), xxxiv.1, 1848 (Maps and Plans), 234–5; *PP*, CP, 1854–5 (1919), xxxvi.1, 1853, 161–2; on immigration, *PP*, HCP, 1850 (643), xl.271, 344–5, 649–50; *PP*, CP, 1851 (1421), xxxiv.99, 1850, 215; *PP*, CP, 1859 Session 1 (2452), xvi.1, 444; on population, labour and wages, *PP*, CP, 1850 (1232), xxxvi.35, 1849 (Pt. i), 84–6; *PP*, CP, 1852 (1539), xxx.1, 1851, 138–9; *PP*, CP, 1856 (2050), xlii.1, 1854, 131–5. On violence and the franchise: 'Condition of Affairs in St. Christopher's Island', *New York Times*, 15 Dec. 1853, 2; W. G. S., 'The Leeward Islands. Results of Emancipation in Antigua', *New York Times*, 3 Aug. 1860, 3 and 'The Leeward Islands. Want of Labor in Antigua and Its Origin', *New York Times*, 15 Aug. 1860, 2.
70. *CCD*, 5:509. On the human traffic: *PP*, ST, 1852–3 (.3), ciii pt. ii.1, 1851–2 (Class B), 739; *PP*, ST, 1852–3 (.5), ciii pt. iii. 201, 1852–3 (Class B), 663–5; *PP*, ST, 1854–5 (.4), lvi. 179, Apr 1854 – Mar 1855 (Class B), 783–5.
71. Carlyle, 'Occasional Discourse'; *ED*, 2:68.
72. *ED*, 2:98, 153–4, 177; annual donations lists, *Anti-Slavery Reporter*; *Wedgwood*, 248; Midgley, *Women*, 175; Cobbe, *Life*, 646; Forster, *Marianne Thornton*, 184–5, 193–4, 216, 243, and *passim*. Besides many hundreds of pounds left to groups promoting temperance, female and sailors' welfare, anti-Catholic Irish evangelization and schools, Sarah Wedgwood's will earmarked £200 each for the London Missionary Society and the Baptist Missionary Society, and £500 for the London City Mission (National Archives 11/2242,

268v–270). According to the GDP deflator index, £100 in 1856 equals roughly £10,000 today.

73. *CCD*, 5:309–15, 392–401, 432; Darwin's abstract of Blyth's letter DAR 203.2.4; Brandon-Jones, 'Edward Blyth', on his live-animal trade. Blyth confirmed that a number of India's birds (including yellow-footed pigeons) had geographically contiguous species that 'intergraded' where their ranges overlapped, leaving 'every possible intermediate gradation', suggesting the species were not discrete and 'aboriginal': *CCD*, 6:7–8; Blyth, 'Drafts', 45; *Natural Selection*, 258.
74. *CCD*, 5:440 n54; Darwin's abstract, DAR 203.2.4; Blyth, 'Attempt'.
75. *CCD*, 4:484; 5:444–5; E. Richards, 'Political Anatomy', on Knox's understanding of 'monsters'. The 'Porcupine family' were cited throughout the anthropological literature after Lawrence (*Lectures*, 389–90) floated the notion that, if such people were ostracized to an island, they might separate into a new race.
76. *CCD*, 5:434; Darwin's abstract, DAR 203.5.2; Prichard, *Natural History* (1845), 70, 73.
77. Prichard, *Natural History* (1855) and Darwin's abstract, DAR 71:143–5 (read May 1856; *CCD*, 4:448, 494); Prichard, *Researches* (3rd edn, 1836), in Darwin Library, CUL, and abstract, DAR 71:139–42 (read Sep. 1838; *CCD*, 4:458).
78. Possibly at the September 1855 BAAS meeting in Glasgow, which both attended (*CCD*, 5:538; *LLL*, 2:201; L. G. Wilson, *Sir Charles Lyell's Scientific Journals*, 71 n30), or perhaps when Darwin came to town to attend the Royal Society Council meeting on 25 October (*CCD*, 5:538).
79. Lurie, 'Louis Agassiz', 239; Horsman, *Josiah Nott*, 177; Gilman, *Life*, 323–4; Nott and Gliddon, *Types*, vii; Stanton, *Leopard's Spots*, 163. The relative price of *Types of Mankind* is given using the GDP deflator.
80. J. Bachman, 'Types', 627–30; J. Bachman, 'Examination of a Few', 794; J. Bachman, 'Examination of the Characteristics', 202–3; Nott and Gliddon, *Types*, 326, 592, 603; Waitz, *Introduction*, 92.
81. J. Bachman, 'Examination of Prof. Agassiz's Sketch', 522; J. Bachman, 'Examination of the Characteristics', 213–14, 217; J. Bachman, 'Examination of a Few', 790–95, 805; Agassiz, 'Sketch', lxxv. The 'heads' in Agassiz's 'Tableau' had been traced from the plates in Niccola Rosellini's and Karl Lepsius's books of monument engravings from Egypt and Nubia (which Darwin read). As their explorations had made tomb paintings widely known in Europe and America, so ex-Cairo Consul Gliddon's theatrical tours unwrapping mummies (occasionally with Agassiz in attendance) now whipped up the Egyptian frenzy in America – with its party dresses and fashionable talk – even as it fed the racialization of biology: Nott and Gliddon, *Types*, xiii, xxxvii, 56; Trafton, *Egypt Land*, chap. 1, esp. 42–4.
82. *CCD*, 5:492. Darwin saw that Bachman, too, was wrong in insisting in single origins for all domestic animals. Darwin believed that dogs and pigs *were* bred from multiple species, even if pigeons were not.
83. Darwin actually wrote: 'How false for how distinct S. America & North temperate America'. This and 'what forced reasoning!' refer to that section of Agassiz's essay ('Sketch', lxix–lxx) in which Agassiz lumps a large part of the North and South American fauna together, characterized by the puma which had a range from Canada to Patagonia. Darwin was of course an expert on South America. It caused him to jot on the inside back cover (*Marginalia*, 604):

> Nothing is more odd than similarity of Fuegian & Brazilian. Why Puma shd range continent unvaried & Monkeys differ in every province. – It is great hiatus in knowledge. I may contrast Man with Monkeys, for on my theory, the Monkeys have varied. –

Meaning again that nature was messy and could not be straitjacketed into Creative provinces, all of a piece. Human races were similar where monkeys varied, while the puma had extended its range enormously (*Natural Selection*, 285). Such facts were more amenable to Darwin's dispersionist views than Agassiz's fixed zones.

84. Nott and Gliddon, *Types of Mankind*, Darwin Library, CUL (cf. *Marginalia*, 604, which transcribes this as 'oh fish pudor Agassiz'); Agassiz, 'Sketch', lxxiv, lxxvi; *CCD*, 4:493.
85. *Marginalia*, 603. The greatest number of Darwin's annotations on *Types of Mankind* were reserved for one of Nott's contributions, 'Hybridity of Animals, viewed in Connection with the Natural History of Mankind', 372–410.
86. *CCD*, 4:484; *Marginalia*, 603; Nott and Gliddon, *Types*, 375. Darwin had already read the Pulszkys' *White, Red, Black* (or, as Darwin referred to it, 'Red Black and White') beginning 11 January 1854 (*CCD*, 4:490), another example of the new wave of American-inspired racial literature. The entry on Darwin's reading list, 'Morton in Charlestown Med. Journal' (*CCD*, 4:484) refers to Morton's letter to Bachman, *Charleston Medical Journal and Review*, 5 (3), (May 1850), 328–44. (Its reprint in the *Charleston Daily Courier*, 25 May 1850, provoked Bachman's response: 'For the Courier by J. B.', *Charleston Daily Courier*, 28 May 1850. We are grateful to Gene Waddell for this detail.) Morton struck off a separate pamphlet, and again the title betrayed the content: *Letter to the Rev. John Bachman, D.D., on the Question of Hybridity in Animals, considered in Reference to the Unity of the Human Species*. It was a reassertion of the viability of hybrids in goats, sheep, cats, etc. as well as domestic curassows, fowls and pigeons, in order to sustain the idea that humans were so many hybridizing species. Darwin finally asked Asa Gray for a copy of this in 1861, and received it along with Bachman's rejoinders in the *Charleston Medical Journal*: *CCD*, 9:52, 213.
87. *CCD*, 6:69, 89, 522; L. G. Wilson, *Sir Charles Lyell's Scientific Journals*, 54; *Origin*, 21; *Variation*, 1:194; *Notebooks*, B232.

10 THE CONTAMINATION OF NEGRO BLOOD

1. *The Times*, 7 June 1856, 12, col. E; 11 June, 12, col. B; 17 June, 12, col. C; 20 June, 9, col. E; 25 June, 12, col. C; Pierson, 'All Southern Society'; Sinha, 'Caning'; Pierce, *Memoir*, 1:318–19, 345; 2:19–23, 35, 39, 41–5, 70, etc.; C. Sumner, *Crime*, 9.
2. *CCD*, 4:448, 471; J. C. Frémont, *Report of the Exploring Expedition to the Rocky Mountains* (1845), Darwin Library, CUL; *Natural Selection*, 491; *The Times*, 10 June 1856, 9, col. D.
3. Bernstein, 'Southern Politics'; Wish, 'Revival'; *New York Times*, 4 Nov. 1856; *The Times*, 28 Oct. 1856, 7, col. A.; 10 Nov., 5, col. A; 25 Nov., 8, col. F; Crawford, '*The Times*', 233.
4. Stanton, *Leopard's Spots*, 166 (emphasis added), 193; Lurie, *Louis Agassiz*, 240–41; Fox-Genovese and Genovese, *Mind*, 216; Horsman, *Josiah Nott*, 197–200; H. S. Smith, *In His Image*, 164–5.
5. Quoted in L. G. Wilson, *Lyell*, 377, also 374–9; L. G. Wilson, *Sir Charles Lyell's Scientific Journals*, 101, 123; Bartholomew, 'Lyell'.
6. Horsman, *Josiah Nott*, 177; L. G. Wilson, *Lyell*, 386; *LLL*, 2:185.
7. *CCD*, 6:90, 179; Lyell Notebook 134, [16–17], Kinnordy MSS, by courtesy of Leonard Wilson; *LLL*, 2:331.
8. L. G. Wilson, *Sir Charles Lyell's Scientific Journals*, 87, 88, 106.
9. L. G. Wilson, *Sir Charles Lyell's Scientific Journals*, 94–5, 96, 98, 103.
10. L. G. Wilson, *Sir Charles Lyell's Scientific Journals*, 97, 122, 124.
11. L. G. Wilson, *Sir Charles Lyell's Scientific Journals*, 177; *CCD*, 6:169; 8:28.
12. *CCD*, 2:341, 390; 4:322–4; 5:6; 6:95–7, 100, 115, 147, 169.
13. *CCD*, 2:376; 6:125, also 6:115, 147. Woodward partly sympathized with Waterhouse's quinarian classification, which Darwin had denounced: Woodward, *Manual*, 1:58–9.
14. Harvey, *Sea-Side Book*, 5–6; Woodward, *Manual*, 1:21.
15. *CCD*, 6:125–6; Pickering, *Races*, 314. Some weeks later Darwin reread Pickering: *Marginalia*, 667.
16. *CCD*, 6:184, 189.

17. *CCD*, 5:288; 6:142, 205, 236, 238, 387; 13:390.
18. T. H. Huxley, 'Contemporary Literature', 248, 250–51, 253. Others adopted the primordial regions promoted by Agassiz in *Types of Mankind*. Philip Sclater, a barrister-turning-avian-geographer, accepted that birds were created en bloc across their range, and that 'few philosophical zoologists . . . would now-a-days deny' it. Agassiz's was the 'philosophical view of this subject', and Sclater noted the implication that 'varieties of Man had their origin in the different parts of the world where they are now found': Sclater, 'General Geographical Distribution', 130–31. Others liked the 'philosophical views' of Nott and Gliddon's book for giving science its independence and legitimation: Portlock, 'Anniversary Address', clxii.
19. T. H. Huxley, 'Contemporary Literature', 253.
20. *CCD*, 6:89; L. G. Wilson, *Sir Charles Lyell's Scientific Journals*, 56–7; *LLL*, 2:212.
21. T. H. Huxley, 'Lectures', 482–3; Desmond, *Huxley*, 224–6.
22. *CCD*, 4:265–7, 284–7; 5:79–80; T. H. Huxley, 'Contemporary Literature', 248.
23. Lurie, *Louis Agassiz*, 271; James Dwight Dana, *ANB; CCD*, 4:286, 289; 5:351.
24. Sherwood, 'Genesis'. Lewis wrote 35,000 words as 'Laicus' in the Schenectady papers *Reflector* and *Cabinet* in December and January, 1855–6 (Yetwin, 'Rev. Horace G. Day'; Potter, *Discourses*, 19–20; cf. Noll, *Civil War*, 48). Laicus' identity was known to locals. Dana was born and raised in Utica, just up the Erie Canal from Schenectady, and he continued to visit family living in Utica as late as 1857 (Gilman, *Life*, 181).
25. Gilman, *Life*, 323–4; Agassiz quoted in Stanton, *Leopard's Spots*, 169; Sherwood, 'Genesis', 314.
26. *CCD*, 4:289; 6:180–81, 216; 13:385.
27. *CCD*, 3:2.
28. *CCD*, 6:235–6.
29. *CCD*, 6:236, 237.
30. *CCD*, 6:136, 174, 201, 377.
31. *Notebooks*, D154, 158, 162, 174e, 178.
32. *Foundations*, 92–3, 220–27.
33. Nott and Gliddon, *Types*, 270, 312–13, 405, 415; *CCD*, 4:493.
34. *Marginalia*, 603–4; Knox, *Races*, 159, 198, 356.
35. *CCD*, 6:71, 75; *Marginalia*, 603–4. As Darwin later put it, 'The Lords continually select the most beautiful & charming women out of the lower ranks; so that a good deal of indirect selection improves the Lords': *CCD*, 10:48.
36. Darwin's abstract of Knox's *Races of Men*, DAR 71:65. Cf. Tipton, 'Darwin's Beautiful Notion', which confines analysis of beauty and sexual selection to Darwin's published texts.
37. Nott and Gliddon, *Types*, 54–6 and Darwin's notes in his copy on p. 54 and the back endpaper, Darwin Library, CUL.
38. Prichard, *Researches into the Physical History of Mankind*, 5 vols (3rd and 4th edns, 1841–51), vol. 1, rear notes, Darwin Library, CUL; Augstein, *James Cowles Prichard's Anthropology*, chap. 1; Stocking, 'From Chronology', lxxiv–xc.
39. Prichard, *Researches*, 5 vols (3rd and 4th edns., 1841–51), vol. 1, rear notes; vol. 4, 454, 519, 534, 537, rear notes; vol. 5, rear notes, Darwin Library, CUL. On tufted ducks: Darwin's letters, Dec. 1855 and June 1856, *CCD*, 5:512, 6:135. *Marginalia*, 683 omits to transcribe the verso of a rear note in vol. 1 of Prichard: 'If I ever consider Man look over other & earlier edition'. These words, taken with Darwin's decided 'In my note on Man', on the verso of a rear note in vol. 5, suggest a change of intention during his study of Prichard. In Darwin's copy of the 'other & earlier edition' of Prichard, *Researches* (3rd edn, 1836), Darwin Library, CUL, a rear note to vol. 1 reads, 'March. 1857 I have not looked through all these, but I have gone through the later Edition'; and in vol. 2 another rear note reads, 'March 1857 I have not looked through'.
40. Prosper Lucas, *Traité philosophique et physiologique de l'hérédité naturelle*, 2 vols (1847–50), 2:129, 158, 296, Darwin Library, CUL; *CCD*, 4:495; *Notebooks*, D147e, T37, and p. 649.

41. Desmond and Moore, *Darwin*, 253. Darwin acquired his copy of the 1822 edition, published by the radical William Benbow (who also dealt in pornographic prints), before the *Beagle* voyage (*Marginalia*, 485), possibly during his medical studies at Edinburgh. Darwin wrote notes as early as 1838–9 reminding himself to study the book (*Marginalia*, 161, 162; *CCD*, 2:142n). The only record of his reading it is dated 23 April 1847, when he called it 'poor' (*CCD*, 4:427). Evidently at some time he found *Lectures* more useful because the latter half of his copy is well annotated (Darwin Library, CUL). Among the marked pages are those that would be cited in the *Descent of Man* (1:118; 2:318, 333, 349, 352, 357); these were Darwin's first and only published references to the heretical Lawrence. Lawrence is not mentioned in the extant parts of the 'Natural Selection' manuscript, but Darwin's annotations in *Lectures* show that his later reading was spurred by interests dateable from literature he is known to have studied in 1855–7 while writing the 'big Book', particularly titles in which Lawrence's name appears among a cluster of marked references, *inter alios*, to Blumenbach, Buffon, Meckel and Pallas.
42. Annotations on William Lawrence, *Lectures on Physiology, Zoology, and the Natural History of Man* (1822), 272, 274, 276 (beard), 337 (nose), 354 (ears), 356 (tattoos), 357 (lips), 368 (steatopygia), Darwin Library, CUL. For a sensitive study of Anglo-French attitudes (including Cuvier's) to these Khoisan people in the early nineteenth century, and the 'Hottentot Venus' Sara Baartman's role in the Western trade of imported and exhibited human exotics: Qureshi, 'Displaying'.
43. *Notebooks*, T26, 37, D147e; *CCD*, 6:527; *Natural Selection*, 213; Darwin's copy of Latham, *Man and His Migrations* (1851), 97, front notes, Darwin Library, CUL (read Aug. 1856: *CCD*, 4:495); *CCD*, 6:527. On Latham's decline: Hake, *Memoirs*, 208–10.
44. *CCD*, 6:299–300, 400, 516.
45. *Natural Selection*, 436 n14, also 388–9, 426–7 n1, 431, 436–7, 439, 486, 454. 'Darwin's questions on the breeding of animals in captivity' (*CCD*, 3:404–5) may have led to the report by the Zoological Society on breeding attempts in the Zoological Gardens between 1838 and 1846.
46. *Natural Selection*, 25–6, 441n, 443.
47. *Natural Selection*, 97.
48. Kohn, 'Darwin's Keystone' on the concept of 'divergence', or how sibling groups could split and live alongside one another by adopting a 'division of labour'; Nott et al., *Indigenous Races*, 502–3, 509, 511.
49. Nott et al., *Indigenous Races*, 523–6; L. G. Wilson, *Sir Charles Lyell's Scientific Journals*, 157; Owen's annotated copy of C. Lyell, *Supplement*, 14–15, Palaeontology Library, Natural History Museum, London; Desmond, *Archetypes*, 225 n13.
50. Nott et al, *Indigenous Races*, 428, 510, 614–15, and 402–603 for Gliddon's, 'The Monogenists and the Polygenists'.
51. *CCD*, 6:241–2; 7:346.
52. *CCD*, 6:375, 383; 7:427; *Natural Selection*, 376.
53. *Natural Selection*, 213. The addition on sexual selection to chapter 6 was written, if not inserted, before Darwin referred to sexual selection on f. 76 of chapter 7, which itself was written before 5 June 1857, when he began the passage on equine variation on f. 105: *Natural Selection*, 275, 317–18, 328ff.; cf. *CCD*, 6:409.
54. *CCD*, 6:515, 527. On Darwin's and Wallace's differing collecting techniques: Fagan, 'Wallace'.
55. L. G. Wilson, *Sir Charles Lyell's Scientific Journals*, 137–42; *Natural Selection*, 467–8, 477, 481; *CCD*, 8:28.
56. Owen, 'On the Characters', 19–20; Ulrich, 'Thomas Carlyle', 31–2; Rupke, *Richard Owen*, 268–9; Desmond, *Archetypes*, 74–6. Owen's lecture was worked up with an anti-transmutatory appendix in Owen, *On the Classification*.
57. T. H. Huxley to J. D. Hooker, 5 Sep. 1858, Huxley Papers 2.35, Imperial College; Desmond, *Huxley*, 238–40; *CCD*, 6:367, 419.
58. *CCD*, 7:2–3. Blyth's last known 1856 letter is dated 3 April (*CCD*, 6:67–9); his

letter of 21 April 1857 is known only from Darwin's abstract in the Dibner Collection, Smithsonian Institution, Washington, DC (Burkhardt and Smith, *Calendar*, 2080).
59. *CCD*, 7:358; 8:28.
60. A. Gray to J. D. Hooker, 21 Feb. 1854, Royal Botanic Gardens, Kew; *CCD*, 5:322–3; Dupree, *Asa Gray*, 192; *LJH*, 1:376.
61. J. L. Gray, *Letters*, 1:296, 396; Dupree, *Asa Gray*, 20–21; Marsden, *Evangelical Mind*, chap. 4.
62. Levy, 'Sims' Case', 73, and *passim*; Menand, *Metaphysical Club*, 9; Stone, *Trial*; Dupree, *Asa Gray*, 178–9; F., 'Jane Loring Gray'; Collinson, 'Anti-slavery'; *The Times*, 19 Apr. 1851, 8, col. A.
63. J. L. Gray, *Letters*, 1:346; 2:425, 432, 516; A. Gray to J. D. Hooker, 21 Feb. 1854, Royal Botanic Gardens, Kew (also D. Porter, 'On the Road', 16).
64. *CCD*, 4:128; 5:118; 6:315, 340, 360, 363, 456.
65. *CCD*, 6:236, 387, 416, 432, 433, 437.
66. G. B. Nelson, 'Men'; L. Agassiz to J. Nott and G. Gliddon, 1 Feb. 1857, in Nott et al., *Indigenous Races*, xiii–xv; A. Gray to J. D. Hooker, 21 Feb. 1854, Royal Botanic Gardens, Kew (also D. Porter, 'On the Road', 33); *CCD*, 6:445–6, 492.

11 THE SECRET SCIENCE DRIFTS FROM ITS SACRED CAUSE

1. A. R. Wallace to H. W. Bates, 28 Dec. 1845, WFP. Drawing on Prichard, Chambers also accepted the unity of the human races: *Vestiges*, 281, 283.
2. *CCD*, 5:520; Chambers, *Vestiges*, 296; L. G. Wilson, *Sir Chartes Lyell's Scientific Journals*, 1; Wallace, 'On the Law'; Wallace, *My Life*, 1:341, 354; Marchant, *Alfred Russel Wallace*, 1:53.
3. A. R. Wallace to G. Silk, 30 Nov. 1858, WFP; Wallace's field journal, quoted in J. L. Brooks, *Just before the Origin*, 168; Wallace, 'Letter'; Wallace, *Malay Archipelago*, 19 (cf. Agassiz, 'Sketch', tableau and map), 415; Vetter, 'Wallace's Other Line'.
4. A. R. Wallace to G. Silk, 30 Nov. 1858, WFP; Wallace, *Malay Archipelago*, 20. On Davis: R. C. J. Young, *Colonial Desire*, 74; Stocking, 'What's in a Name?', 374–5; Stocking, *Victorian Anthropology*, 65; Nott and Gliddon, *Types*, 371; Nott et al., *Indigenous Races*, 216; *CCD*, 4:495.
5. Wallace, *My Life*, 1:362, 363; Wallace, *Malay Archipelago*, 92–3, 305, 311, 312, 313–14, 316, 362; Moore, 'Wallace's Malthusian Moment'.
6. Fanny Allen to her niece Emma Wedgwood Darwin, 7 Dec. 1857, Wedgwood W/M 221; *ED*, 2:180–81, 224.
7. *The Times*, 21 Mar. 1857, 5, col. D; 24 Mar., 12, col. A; 31 Mar., 6, col A; 8 Apr., 10, col. E etc.; *Case of Dred Scott*, 9.
8. Wheatley, *Life*, 260; Martineau, 'Slave Trade', 568–9; Webb, *Harriet Martineau*, 16, 315n, 327–8; Arbuckle, *Harriet Martineau's Letters*, xxiii, 143, 156–7, 165, 171n; V. Sanders, *Harriet Martineau*, 156–7.
9. *CCD*, 6:134; *ED*, 2:138; Darwin's comment, reported by Lyell, Feb. 1851, Lyell Notebook 165, 115, Kinnordy MSS, quoted by L. G. Wilson, *Lyell*, 338. Darwin was probably repeating a remark attributed to the English wag Douglas Jerrold: Wheatley, *Life*, 302; Webb, *Harriet Martineau*, 299.
10. V. Sanders, *Harriet Martineau*, 157, 159; Arbuckle, *Harriet Martineau's Letters*, 167; Pierce, *Memoir*, 3:543–4, 550, 567.
11. By now Greg had more articles in the *Edinburgh Review* than in any other quarterly, twenty-two of them written while he was Martineau's mesmerist (V. Sanders, *Harriet Martineau*, 106–8, 124). He stopped editing the Unitarian *National Review* long before Darwin was asked to write for the *Edinburgh* (Drummond and Upton, *Life*, 1:269; *CCD*,

6:205), but anyway Greg would not have approached a friend of Martineau's to write for the *National* when her estranged brother James was behind it. Having landed Martineau as a contributor, Greg's editor at the *Edinburgh* may have asked him to seek the reclusive Darwin's services through Fanny Wedgwood (Arbuckle, *Harriet Martineau's Letters*, 120, 128).

12. Martineau, 'Slave Trade', 542–3, 584.
13. *CCD*, 7:80, 84, 89.
14. *Natural Selection*, 511–12; *Origin*, 219–24; *CCD*, 7:36.
15. *The Times*, 18 June 1858, 6; *CCD*, 7:130.
16. *CCD*, 7:113, 130.
17. Beer, 'Darwin's Reading', 562; J. F. M. Clark, 'Complete Biography', 253–5; *Origin*, 113, 220; *CCD*, 7:287.
18. *Origin*, 208; *CCD*, 9:91.
19. Buckle, *History*, 1:29–30, 49, 68n; 3:482; on slavery, 1:447, 463; Arbuckle, *Harriet Martineau's Letters*, 160–205.
20. Emma Wedgwood Darwin to her sister Sarah Elizabeth Wedgwood, [n.d.], Wedgwood W/M 168; *CCD*, 7:31, 34; *Autobiography*, 109–10.
21. *CCD*, 6:100; 7:62, 107, 115, 117, 513–20.
22. Lorimer, *Colour*, 83–6, 92–6; Maurer, '*Punch*'.
23. *Origin*, 29; *CCD*, 7:270; Secord, 'Nature's Fancy'.
24. L. Agassiz, *Contributions to the Natural History of the United States of North America*, vol. 1, *Essay on Classification*, rear note for p. 166, Darwin Library, CUL; *CCD*, 6:456; 7:26, 262.
25. T. H. Huxley to C. Lyell, 26 June 1859, APS B/D25.L; *LTH*, 1:174; Desmond, *Huxley*, 256 for the context of this exchange with Lyell; Desmond, *Archetypes*, chap. 3.
26. *CCD*, 6:456, 462–3.
27. *CCD*, 7:305; 13:411–12.
28. Fanny Allen to her niece Emma Wedgwood Darwin, [7 Dec. 1857], Wedgwood W/M 221; *LLD*, 1:118–19, 121; J. L. Gray, *Letters*, 2:371, 421–2; Dupree, *Asa Gray*, 292.
29. Olmsted, *Journey through Texas*, xv; *CCD*, 4:496. Olmsted's trilogy was 'well studied' by Darwin, who recommended part three, *A Journey in the Back Country*, to Hooker for its vivid 'picture of man & Slavery': *CCD*, 9:9, 266. On Olmsted's journalism: Schlesinger, 'Editor's Introduction'.
30. Olmsted, *Journey in the Seaboard Slave States*, 25, 30, 55, 83; Toplin, 'Between Black'.
31. *ED* (1915), 2:169; *CCD*, 4:496; 7:504; Olmsted, *Journey in the Seaboard Slave States*, 160–61, 163.
32. *CCD*, 8:28; *Origin*, 488.
33. *Origin*, 18, 88, 89, 199.
34. *CCD*, 7:427; 9:80; Galton, *Narrative*, 87–8.
35. *CCD*, 7:340, 345–6; 13:411–12.
36. *CCD*, 7:384; 13:418–19; *Origin*, 253.
37. *CCD*, 7:386, 392; *Origin*, 253.
38. *CCD*, 7:357; 8:28.
39. *CCD*, 7:247, 296, 362, 392; Colp, *To Be*, 62–8 and *Darwin's Illness*.
40. *CCD*, 6:319, 352–4, 7:233–4; Darwin's annotated copy of Hill, *Week*, DRC G.149. For a partial list of Hill's publications: Griffey, 'Bibliography'.
41. Thwaite, *Glimpses*, 130, 131, 136; *CCD*, 4:488, 494; 6:9, 32–3, 382–3; Gosse, *Letters*, 251–2, 255; Darwin's annotated copies of Gosse's *Letters from Alabama* and *A Naturalist's Sojourn in Jamaica*, with many annotations, Darwin Library, CUL.
42. Cundall, 'Richard Hill' (1896, 1920). See Turner, 'British Caribbean', 313ff.; Curtin, *Two Jamaicas*; and Hill, 'Extraits' (the original English publication has not been found).
43. *Bromley Record*, 1 Apr., 2 May, 3 June 1859 (thanks to Randal Keynes and Richard Milner for pointing out this evidence); *The Times*, 16 June 1859, 8, col. F; *CCD*, 4:103.
44. For Hill's career: *PP*, AP, 1836 (166-i and 166-ii), xlviii–xlix, 104–5, 332–3; 1837 (521),

liii.1, 47–8, 74–5, 233–6, 268–9, 311–12; 1839 (157), l.11, 20–22; 1845 (691-i, 691-ii and 691-iii), xxviii.1, 47–55; 1852–3 (76), lxvii.1, 60–61; 1859 Session 2 (31 and 31-1), xx.1 and xxi.1, 287; and Turner, 'British Caribbean' for the post-emancipation magistracy. For Wilkinson's career: Foster, *Register*, 434; *Jamaica Almanack* (Kingston: printed by Cathcart, 1840, 1843, 1845, 1846); Feurtado manuscript, National Library of Jamaica; *PP*, AP, 1859 Session 1 (239), xvii, 420; and Island Solicitor, *Courts* for background.

45. *Anti-Slavery Reporter*, 3rd ser., 4 (1 Jan. 1856), 11; (1 Apr. 1856), 89; vol. 3 (1 Oct. 1855), 235; also 3rd ser., 3 (1 Sep. 1855), 212 and Macmillan, *Sir Henry Barkly*, chap. 6. Hill was not without his own prejudices: Hall, *Civilising Subjects*, 205.

46. *CCD*, 7:201–2, 322, 346; cf. 369. Hill's battered presentation-copy of the first edition of the *Origin of Species*, signed by one of publisher John Murray's clerks, was offered twice on the second-hand market at Toronto in the 1980s, latterly for 8,000 Canadian dollars: Freeman, *Works* (1986), item 373.

47. *Origin*, 490; *Notebooks*, C197.

48. *CCD*, 7:366, 370, 373, 462; 8:555–7; Arbuckle, *Harriet Martineau's Letters*, 185.

12 CANNIBALS AND THE CONFEDERACY IN LONDON

1. W. L. Garrison, 'On the Death', 300; Webb, *Harriet Martineau*, 325.
2. *CCD*, 11:505, 667.
3. CCD, 10:71–2; Kingsley, quoted in Waller, 'Charles Kingsley', 558; Kingsley, *Charles Kingsley*, 1:4–5.
4. *CCD*, 10:71–2.
5. Arbuckle, *Harriet Martineau's Letters*, 206, 208; *CCD*, 7:345, 347; 8:28, 80, 403; Bynum, 'Charles Lyell's *Antiquity*'; *LLL*, 2:365.
6. Pierce, *Memoir*, 3:607–14; C. Sumner, 'Barbarism', 174–5, 231.
7. *CCD*, 7:390; Arbuckle, *Harriet Martineau's Letters*, 185–6.
8. Morris, 'Darwin', 61; *CCD*, 7:412, 446; 8:21; Carpenter, in Hull, *Darwin*, 93.
9. *Origin*, 220, 475; *CCD*, 9:91, 280; S. Wilberforce, 'On the Origin', 253–4; Stecher, 'Darwin–Innes Letters', 235–7.
10. Nott, quoted in Haller, 'Species Problem', 1323; *CCD*, 15:502; Weiss, *Life*, 2:423; Renehan, *Secret Six*.
11. *CCD*, 7:292; Lurie, *Louis Agassiz*, 276; Nott and Gliddon, *Types*, 735; Farrell, *Elite Families*, 65–7; Dupree, *Asa Gray*, 253.
12. *CCD*, 8:23; Dupree, *Asa Gray*, 255–61.
13. *CCD*, 8:217; J. A. Lowell, 'Darwin's Origin', 449, 451, 453–4, 456; Lurie, *Louis Agassiz*, 291–9; Dupree, *Asa Gray*, 286–7.
14. Dupree, *Asa Gray*, 297; J. R. Lowell, 'Ethnology', in *Anti-Slavery Papers*, 1:25ff.; Farrell, *Elite Families*, 170; James Russell Lowell, *ANB*; A. Gray, *Darwiniana*, 10, 11, 13, 22–3, 41, 50–51.
15. *CCD*, 10:140; A. Gray, *Darwiniana*, 76. Gray argued that both mono- and polygenists had to bend to Darwin's common descent. Traditionalists sworn to Adamic unity had to 'admit an actual diversification into strongly-marked and persistent varieties, and so admit the basis of fact upon which the Darwinian hypothesis is built'. Agassiz and his disciples must concede that human 'species' cannot be 'primordial and supernatural', but must themselves have descended from a common stock: A. Gray, *Darwiniana*, 142, 144.
16. *LLL*, 2:341; Lurie, *Louis Agassiz*, 225; *CCD*, 8:388; 9:393ff.; Peckham, *Origin*, 57.
17. Cleveland, *Alexander H. Stephens*, 721; Preamble to the US Constitution; Noll, *Civil War*, 56–64.
18. Lurie, *Louis Agassiz*, 291, 302, 305–6. On Agassiz's subsequent sniping at Darwin: Morris, 'Louis Agassiz's Additions'; also Ellegård, *Darwin*, 122.

19. *CCD*, 9:384–5; 10:87; J. L. Gray, *Letters*, 2:483.
20. H. W. Wilberforce to H. Martineau, 19 Apr. 1865, Harriet Martineau Papers, HM 1023, Birmingham University Library; *CCD*, 9:266.
21. *CCD*, 9:214–15, 266, 368. To Harriet Martineau, 'ninety-nine in a hundred' Britons believed 'the war is not for the abolition of slavery': Logan, *Collected Letters*, 4:307.
22. *CCD*, 9:368; Bennett, *London*, 226–30.
23. *CCD*, 11:451, 548, 582; E. C. C. Agassiz, *Louis Agassiz*, 2:607. Agassiz's miscegenation fears have been well picked over: Gould, *Mismeasure*, 80; Menand, *Metaphysical Club*, 114–16; Haller, *Outcasts*, 85; R. J. C. Young, *Colonial Desire*, 149.
24. J. W. Clark and T. M. Hughes, *Life*, 2:393; *CCD*, 10:471.
25. T. H. Huxley, *Critiques*, 163; Bowen, 'On the Origin', 475; Hooker, in Hull, *Darwin*, 83–4.
26. C. L. Brace, *Races*, 343; E. Brace, *Life*, 153–4, 175, 179–82, 300; Kalfus, *Frederick Law Olmsted*, 161.
27. 'Lyell on the Geological Evidence of the Antiquity of Man', *AR*, 1 (1863), 129–37; Bynum, 'Charles Lyell's *Antiquity*'; Crawfurd, 'On Sir Charles Lyell's "Antiquity" ', 60–62; C. Lyell, *Geological Evidences*, 387–8. Agassiz is named in C. Lyell, *Principles* (10th edn., 1867–8), 2:475–6.
28. J. Lubbock to C. Lyell, 21 Feb. 1863, Edinburgh University Library, Gen. 113; C. Lyell, *Geological Evidences*, 469, 495, 498; *CCD*, 11:419.
29. *CCD*, 11:206–9, 403; C. Lyell, *Geological Evidences*, 504–5 (our emphasis).
30. *CCD*, 9:214–15, 266, 368; 11:333, 444; Loring and Field, *Correspondence*. For Martineau's sharp response to Loring's presentation copy: Webb, *Harriet Martineau*, 334.
31. *CCD*, 11:166–7; cf. Colp, 'Charles Darwin'. On *The Times*'s support for the South: Crawford, '*The Times*'. Despite its Confederate sympathies, the paper could broach the single versus 'great multiplicity of [racial] stocks' issue and still favour a single origin: *The Times*, 16 Sep. 1861, 6.
32. F. W. Newman to Fanny Mackintosh Wedgwood, 5 Jan. 1863, Wedgwood E58/32357; Sieveking, *Memoir*, 139; Häggman, 'Confederate Imports'; *Leeds Mercury*, 2 Sep. 1863; *CCD*, 11:333.
33. *CCD*, 11:452, 525, 556, 564; 13:208. Gray insisted that the end could only be 'the complete territorial reinstatement of the Union, and the abolition of slavery': J. L. Gray, *Letters*, 2:518.
34. Hunt, 'Introductory Address', 3–4; *JASL*, 2 (1864) xlviii–xlix; Kenny, 'From the Curse', 376; Hunt's details from the meagre biography in *JASL*, 8 (1870), lxxix–lxxxiii.
35. Nott, 'Negro Race', 102n; *TASL*, 1 (1863), i.
36. 'Catalogue of Books in the Library of the Anthropological Society of London, up to December 31st, 1866', bound with *JASL*, 6 (1868); *JASL*, 2 (1864), lxxix for Hotze's £5 donation to the library fund; Gobineau, *Moral and Intellectual Diversity*, 454, also cited in *JASL*, 3 (1865), xcviii; *TASL*, 1 (1863), xxv for Hotze on the Council, which he served through to 1868.
37. Other Confederates in the society were the pro-slavery George McHenry (to whom Hotze gave £300) and Albert Taylor Bledsoe, dispatched by Confederate President Jefferson Davis to Europe in 1863 to work for the Confederacy, and elected to the society on 1 Dec. 1863: *JASL*, 2 (1864), xxiii; Bonner, 'Slavery', 300–304, 309. McHenry put an article in the Anthropologicals' *Popular Magazine of Anthropology* justifying slavery as the most economical way of cultivating the vast cotton crops (McHenry, 'On the Negro'). A subsequent staff reporter on Hotze's pro-Confederate London paper, the *Index*, George Witt, was also on the Anthropological Council with Hotze and donated to the society's library and museum: *TASL*, 1 (1863), xix, xxv; *JASL*, 2 (1864), lxxvi; *AR*, 3 (1865), ii. So there were three known Confederates on the Anthropological Society Council, and four at least who were directly (Hotze, Bledsoe) or indirectly (McHenry, Witt) on Richmond's payroll. On Hotze's expense sheet: J. F. Jameson, 'London Expenditure'. Whether or not this was widely known, the press did not flinch from describing how 'the confederates have wrought

together since [the Anthropological Society's foundation] with amazing success': 'Professor Huxley and the Anthropologists', *National Reformer*, 12 Mar. 1864, 5.

38. 'Dr. Hunt on the Negro's Place in Nature', *Index*, 26 Nov. and 3 Dec. 1863; 'The Negro's Place in Nature', 10 Dec. 1863; also *Index*, 28 May 1863; 'The Natural History of Man', *Index*, 23 July 1863; Bonner, 'Slavery', 302–3; Jameson, 'London Expenditure' for fly-posting.
39. 'The Distinctions of Race', *Index*, 23 Oct. 1862.
40. *LTH*, 1:251–2; Desmond, *Huxley*, 325.
41. T. H. Huxley, 'On the Methods and Results of Ethnology', *Fortnightly Review*, 15 June 1865, DRC G.386; T. H. Huxley, *Critiques*, 163.
42. Leifchild, 'Evidence', 288; T. H. Huxley, *Evidence*, 131, 184; T. H. Huxley, *Critiques*, 138, 157–8, 163; T. H. Huxley, *Professor Huxley*, 7.
43. T. H. Huxley, *Professor Huxley*, 8–9. Coming from a younger generation, Huxley's view of blacks and 'savages' was always more derogatory than Darwin's: Di Gregorio, *T. H. Huxley's Place*, 166.
44. T. W. Higginson to T. H. Huxley, 23 June 1867, Huxley Papers 18.167–8, Imperial College; Conway, *Autobiography*, 1:176. Higginson had been another of the 'Secret Six' who had armed John Brown for the Harpers Ferry raid: Menand, *Metaphysical Club*, 29; Poole, 'Memory', 211–15.
45. Darwin to F. E. Abbot, 2 July 1872, Harvard University Archives; Higginson, *Thomas Wentworth Higginson*, 323–4, 334; *LLD*, 3:176.
46. T. H. Huxley, *Professor Huxley*, 7, 9, 13: 'Professor Huxley's Lectures on "The Structure and Classification of the Mammalia" at the Royal College of Surgeons', *Reader*, 3 (1864), 266–8; 'Professor Huxley an Abolitionist', *Caledonian Mercury*, 7 Mar. 1864; 'Professor Huxley and the Anthropologists', *National Reformer*, 12 Mar. 1864, 5.
47. *CCD*, 11:695, 776; 12:319; Ritvo, *Animal Estate*, 126–48.
48. *The Times*, 29 Aug. 1863, 9, and *Hull Packet*, 4 Sep. 1863 (Hunt's statement was a response to the escaped slave William Craft's attack on the Southern laws banning mixed marriages); 'The Negro's Place in Nature', *Reader*, 3 (1864), 334–5; 'The Negro's Place in Nature', *Index*, 24 Mar. 1864; Huxley paraphrased in Allan, 'On the Ape-Origin', 122; Lorimer, *Colour*, 159.
49. Hunt, *Negro's Place*; van Evrie, *Negroes*, 23 (republished after the war in 1868 as *White Slavery and Negro Subordination*); Hunt, 'Physical and Mental Characters', 387.
50. Knox, 'On the Application', 267; Stocking, *Victorian Anthropology*, 245; Stocking, 'What's in a Name?', 375.
51. *AR*, 6 (1868), 304–5; 'savage': *AR*, 1 (1863), 484. On granting Darwin his licence: S. E. Bouverie-Pusey, A. R. Wallace, Charles Carter Blake (who nonetheless 'objected to having the polygenists and transmutationists confounded together') in *JASL*, 2 (1864), cxxviii, cxxix, cxxx–cxxxi, clxxiii. Even Hunt, shortly before his death, admitted that 'the Origin of Species is one of the most glorious and most praiseworthy publications of the nineteenth century': *AR*, 6 (1868), 78. C. S. Wake, a rare monogenist remaining in the society, was prepared at least to talk of Darwin's theory of a 'common progenitor', though without much approval of Darwin: *JASL*, 5 (1867), cxiv–cxvi. Another, the Revd Mr Macbeth, thought that Darwin 'had, indeed, removed certain objections to the unity of man' and that the common origin of languages supported him: *JASL*, 6 (1868), cxv.
52. *Daily News*, 31 Aug. 1863, 2; *AR*, 1 (1863), 388–9, 391; T. H. Huxley, 'Negro's Place'; Ellegård, *Darwin*, 75; Hunt, 'On the Physical and Mental Characters', 387; Stocking, *Victorian Anthropology*, 251; Driver, *Geography Militant*, 98–9; *Leeds Mercury*, 3 Sep. 1863.
53. *Caledonian Mercury*, 4 Sep. 1863; *Hull Packet*, 4 Sep. 1863; *The Times*, 29 Aug. 1863, 9 for Craft on slavery, and 31 Aug. 1863; *Lloyd's Weekly Newspaper*, 6 Sep. 1863; *Aberdeen Journal*, 9 Sep. 1863; *Birmingham Daily Post*, 1 Sep. 1863; also on Confederate loans, *Belfast News-Letter*, 31 Aug. 1863; *Leeds Mercury*, 2 Sep. 1863.

54. Schaaffhausen, 'Darwinism', cx; Rupke, 'Neither Creation'.
55. Hunt, 'On the Application', 339; Amrein and Nickelsen, 'Gentleman', 243, 255–8; Hunt, 'Introductory Address', 8; Hunt in *AR*, 6 (1868), 78.
56. *AR*, 6 (1868), 311; Vogt, *Lectures*, 378, 461–8; *CCD*, 13:230–31; 15:96–7. Vogt's iffy words actually were: 'But if this plurality of races be a fact . . . if this constancy be another proof for the great antiquity of the various types . . . then all these facts do not lead us to one common fundamental stock, to one intermediate form between man and ape, but to many parallel series' (*Lectures*, 467).
57. E. Richards, 'Moral Anatomy', 422–30; Stocking, 'What's in a Name?', 380–85; Driver, *Geography Militant*, 98–9; Desmond, *Huxley*, 343, 700; J. Crawfurd to T. H. Huxley, 6 Oct. 1866, Huxley Papers 12.335, Imperial College.
58. *JASL*, 2 (1864) cxxxii, cxxxiv. Hunt, Bendyshe and James McGrigor Allan were Vogt's chief advocates. Bendyshe was less extreme than some others; he doubted that Negroes were a separate species, thought African skulls showed a diversity, with many quite European-like, and would not rule out the influence of climate in changing races, even if he too inclined towards the multi-ape theory: *JASL*, 2 (1864), xxxiv–xxxv. He also raised interesting ethical problems. Suppose an ape-man, which 'the Darwinian theory' predicted, exists still in Africa: given the hunting ethos of the day, what should an explorer do? At the moment he 'shoots and presents us with the skin of a Gorilla', but 'when brought face to face with the intermediate creature, how should we act? How should we distinguish between the animal we ought to shoot with triumph, and whose skin should be sent home and stuffed, and the man whose skull and skeleton would be equally interesting: but the manner of whose decease we should not care to know'? With the Anthropologicals blurring the boundaries between man and brute, one senses an ethical seed which had the potential to grow into a troubling bush, whose caring branches might one day cover the gorilla itself: *JASL*, 2 (1864), xxxv.
59. *JASL*, 6 (1868), xvi, cxxvi.
60. James McGrigor Allan, *JASL* 6 (1868) cxv–cxvi and cxxvi–cxxvii; *AR*, 7 (1869), 178; Allan, 'On the Ape-Origin', 124; Moore, 'Deconstructing' and Desmond, *Huxley*, 374ff. on the fortunes of the term 'Darwinism' until about 1870.
61. Desmond, *Huxley*, 333. On the Ethnological Society at this time: E. Richards, 'Moral Anatomy', 421; E. Richards, 'Huxley', 262, 266ff.; Stocking, 'What's in a Name?', 375–9; Stocking, *Victorian Anthropology*, 248–56; Rainger, 'Race'.
62. Marchant, *Alfred Russel Wallace*, 1:54; Moore, 'Wallace's Malthusian Moment', 298; Fichman, *Elusive Victorian*, 31; Chitty, *Beast*, 193; Banton, 'Kingsley's Racial Philosophy'; Horsman, *Race*, 77; Kingsley, *Charles Kingsley*, 1:223.
63. Brantlinger, *Dark Vanishings*, 183; Stepan, *Idea*, 75 on Wallace's belief that the Polynesians might one day produce a higher civilization than Europe's. For all of that, Wallace too recognized that, given white seizure of colonial lands and with native acquisition of civilization a slow organic process, 'sooner or later the lowest races, those we designate as savages, must disappear from the face of the earth': *JASL*, 2 (1864), cx–cxi.
64. Wallace, *AR*, 7 (1869), 42; Jones, *Social Darwinism*, 33; Lartet, 'New Researches', 58–65 on Aurignac.
65. Wallace, 'Origin', clviii, clix, clxiii, clxv–clxvii, clxix. Huxley had been invited to the reading but refused to attend, objecting to the 'Anthropological people': A. R. Wallace to T. H. Huxley, 26 Feb. 1864, Huxley Papers 28.91, Imperial College. On Wallace's mediating science: E. Richards, 'Moral Anatomy', 418ff.; Durant, 'Scientific Naturalism', 40–45; R. Smith, 'Alfred Russel Wallace', 179–80; Kottler, 'Charles Darwin', 388; J. S. Schwartz, 'Darwin', 283–4. On Wallace's socialism: Jones, 'Alfred Russel Wallace'.
66. Sidney Edward Bouverie-Pusey, *JASL*, 2 (1864), clxxiii: Hunt, 'On the Application', 330, 333 and Hunt, 'On the Doctrine', 113 for his attack on Wallace. Bouverie-Pusey also differed from the extremist Anthropologicals in believing that the Negro could only display his full potential outside of slavery: Bouverie-Pusey, 'Negro', cclxxiv.
67. *CCD*, 6:515; Secord, *Victorian Sensation*, 371. On selection continuing in civilized man:

Lubbock, *Pre-Historic Times*, 480; *Descent*, 167. R. A. Richards, 'Darwin', considers Wallace's rejection of Darwin's analogy between domestic and wild species.

68. *CCD*, 12:173, 216–17. Darwin's point about aristocracy originated in his notes on Nott and Gliddon's *Types of Mankind* made in 1855–6 (*Marginalia*, 603).
69. *CCD*, 12:220–22, 248; 13:208, 223.
70. Desmond, *Huxley*, 321, 330, 342–3; M. Foster to T. H. Huxley, 23 Oct. 1865, Huxley Papers 4.153, Imperial College.
71. *CCD*, 13:256, 262–3; E. Richards, 'Huxley', 264–7, and A. R. Wallace to T. H. Huxley, 26 Feb. 1864, Huxley Papers 28.91, Imperial College, on Wallace's enjoyment of the all-male company.
72. *CCD*, 13:278. The longer Darwin delayed in publishing on the human races the more likely that some scooping 'precursor' would be uncovered. So it happened in 1865: Brace told Darwin of an 1818 Royal Society paper that had applied 'most distinctly the principle of N. Selection to the races of man' (*CCD*, 13:279). Darwin added a note to the 'Historical Sketch' in the fourth edition of the *Origin* (see Johnson, 'Preface'), where he created some distance between himself and the old author, William Charles Wells, by explaining that he had applied natural selection 'only to the races of man, and to certain characters alone', rather than to all species: Peckham, *Origin*, 62; *CCD*, 14:283; Wells, 'William Charles Wells'; Erickson, 'Anthropology and Evolution'.

13 THE DESCENT OF THE RACES

1. *PP*, CP, 1866 ([3595 and 3749]), li.507, 791, pp. 544, 618–19, 622, 642; *PP*, CP, 1866 ([3682]), xxx.1, pp. 115–16. Cf. Hall, *Civilising*, 59–60, 197.
2. Carlyle, 'Shooting', 325; Hall et al., *Defining*, 185; *JASL*, 3 (1865), i; Edward John Eyre, *ODNB*.
3. Semmel, *Governor Eyre*, 46–53; Hall, *Civilising*, 23, 57–8; George William Gordon, *ODNB*.
4. Semmel, *Governor Eyre Controversy*, 62; Hall et al., *Defining*, 179–204; Colaiaco, *James Fitzjames Stephen*, 38; Bolt, *Anti-Slavery Movement*, 39–40. Dutton, *Hero*, stands up for Eyre in the light of his earlier achievements as an Australian explorer.
5. Kostal, *Jurisprudence*, 135, 146, 148–9, 159; Mill, *Autobiography*, 298. Buxton 'talks of "Peace" forsooth! If I were a Northerner, I would as soon make peace with a young tiger at my side, as peace with the Southern fanaticism', Francis Newman (now one of the Jamaica Committee) had told Fanny Mackintosh Wedgwood on 5 Jan. 1863, Wedgwood E58/32357.
6. Mill, *Autobiography*, 299; *JASL*, 4 (1866), lxxviii; Stocking, 'What's in a Name?', 379.
7. *The Times*, 23 Aug. 1866, 7; Semmel, *Governor Eyre Controversy*, 88–95; *CCD*, 7:379–80, 407, 409, 411 for Kingsley and Darwin; Peckham, *Origin*, 748 (*183.3:b*). The 20th Earl would amass Europe's largest and most celebrated collection of second-hand torture instruments – shackles, branding irons, thumbscrews, scourges, whips, gags and so on: Ichenhäuser, *Illustrated Catalogue*.
8. *CCD*, 9:302–3; 13:450–51; 'Ex-Governor Eyre at Southampton', *The Times*, 23 Aug. 1866, 7. The only 'Darwin' householder at this time living in Southampton appears to have been William Erasmus Darwin. The post office directory for 1863 lists William E. Darwin of 1 Carlton Terrace; the directories for 1865 and 1867 have W. E. Darwin at 25 High Street (address of the Southampton and Hampshire Bank, of which he was a partner: William Erasmus Darwin, *ODNB*); and the directory for 1869 places him at Ashton Lodge, Bassett, near Southampton. The only 'Darwin' in the local 1871 census is William E. Darwin, Banker and Landowner, of Ashton Lodge, North Stoneham parish. Thanks to Vicky Green for this and other local information.
9. *CCD*, 13:256, 278; *The Times*, 24 Aug. 1866, 6. Wallace did not regard Hunt as 'fit to

be President' of the Anthropological Society: A. R. Wallace to T. H. Huxley, 26 Feb. 1864, Huxley Papers 28.91, Imperial College.

10. Hunt, 'Application', 320–21, 326–7, 337 (cf. Kenny, 'From the Curse', 378); *The Times*, 25 Aug. 1866, 9. Darwin's copy of the offprint (DRC G.387) is inscribed 'John Crawfurd Esq with the best respects of The Author' in, presumably, Hunt's hand, so Hunt had sent Crawfurd this copy. The rear cover is addressed to Darwin at Downe in a different hand, probably Crawfurd's, with a penny stamp stuck on. Therefore Crawfurd was passing on Hunt's complimentary copy.
11. Hunt, 'Application', 329, 330–31, 339–40 and DRC G.387, 9, 19, 21; *Origin*, 199.
12. *The Times*, 14 Sep. 1866, 6; Carlyle, quoted in Kostal, *Jurisprudence*, 184. Of the sixty-eight committee members, a dozen have been identified as former slave-owners or their immediate descendants: Draper, 'Possessing'.
13. The *Hampshire Independent* (22 Aug. 1866), *Hampshire Advertiser* (25 Aug. 1866) and *Southampton Times* (25 Aug. 1866) all have 'Darwin' in the guest list, which may have been supplied to the press, although the papers give slightly different versions. The names of Atherley, Maddison, Pearce and Hankinson, the other banking partners in Southampton between 1865 and 1869, do not appear among the guests, nor in the Eyre Defence and Aid Fund Committee's subscription list dated 11 October 1866: *Eyre Defence*, 25–31.
14. *CCD*, 14:322; Emma Darwin's 1866 diary, DAR 242.30; William Erasmus Darwin's recollections, 4 Jan. 1883, DAR 112.2. In Emma's diaries, Charles is not otherwise noted as being ill between 29 July 1866 and 19 February 1867. William's uncle Harry Wedgwood, an amiable retired barrister, was also at Downe that weekend. Darwin was possibly also agitated by knowing that his sister Susan was desperately ill and would die shortly.
15. Botanical notes in DAR 48:A16, DAR 49:75 and DAR 54:27; *CCD*, 12:188. Thanks to Randal Keynes for these references. Darwin would advise visitors to the neighbourhood to 'cross Holwood Park by footpath – very pretty' (*CCD*, 9:165). See Killingray, 'Beneath the Wilberforce Oak'.
16. William Erasmus Darwin's recollections, 4 Jan. 1883, DAR 112.2; Robert Monsey Rolfe, *ODNB*; *CCD*, 10:575–6, 14:414; R. M. Rolfe, Baron Cranworth, to G. Ticknor, 30 Aug. 1866, Dartmouth College Library. Cranworth corresponded for years with the US Republican senator Charles Sumner who, while recuperating in Europe from his beating on the Senate floor, 'walked in the grounds' at Holwood with Cranworth (Pierce, *Memoir*, 3:550; also 543). On the Civil War: Cranworth to C. Sumner, 13 Mar. 1865, MS Am 1 (1571), by permission of the Houghton Library, Harvard University.
17. T. H. Huxley, 'Mr. Huxley'; *Pall Mall Gazette*, 29 Oct. 1866; *Anti-Slavery Reporter*, 3rd ser. (1866), 278–9, 286.
18. Hooker, quoted in H. Hume, *Life*, 283; *CCD*, 14:373.
19. *LTH*, 1:278ff.; H. Hume, *Life*, 283; *CCD*, 10:127; 14:385, 393.
20. Kostal, *Jurisprudence*, 181; Semmel, *Governor Eyre Controversy*, 143–5; *The Times*, 29 Nov. 1866, 8; Emma Darwin's 1866 diary, DAR 242.30; *CCD*, 14:340–41, 482.
21. William Erasmus Darwin's recollections, 4 Jan. 1883, DAR 112.2. We take the words 'spending the surplus of the fund in a dinner' to imply that William envisaged a successful prosecution as the cause for the celebration: cf. Desmond and Moore, *Darwin*, 541.
22. *AR*, 6 (1868), 461 (that this *was* irony is shown by the follow-on story, about the constabulary's success in shooting 'troublesome' blacks in northern Queensland); Hunt quoted in S. Courtauld to H. Martineau, 29 Dec. 1866 – 2 Jan. 1867, Harriet Martineau Papers HM 270, Birmingham University Library.
23. *CCD*, 14:404, 411 n4. On Kingsley: *CCD*, 14:414, 15:31; cf. Emma Wedgwood Darwin to her daughter Henrietta, [15 Sep. 1866], DAR 219.9:45.
24. Wallace, *Scientific Aspect*, 42–3, 49–50, originally in *English Leader*, 11 Aug. – 29 Sep. 1866; Holyoake, *History*, 517; McCabe, *Life*, 2:6; Holyoake, *Sixty Years*, 2:78; cf. L. Barrow, *Independent Spirits*, 25ff. Darwin may have received the pamphlet from Wallace during his stay in London, 22–9 November: *CCD*, 15:39. Wallace sent Huxley's copy on the 22nd: Marchant, *Alfred Russel Wallace*, 2:187–8. Darwin's unmarked copy

came to Cambridge University Library in 1899, after Emma's death, with other books from her library: Anon, 'List'.

25. *Natural Selection*, 11–14; *CCD*, 14:438. Going beyond the *Origin*'s 'few forms or . . . one' into which life's 'powers' had been 'originally breathed' (p. 490), Darwin now argued confidently by analogy that it was 'probable that all living creatures have descended from a single prototype' (*Variation*, 1:13).

26. *CCD*, 15:141; *Variation*, 1:135, 203–4. As in the 'Natural Selection' manuscript, Darwin uses examples from humans to illustrate general points about animals (*Variation*, 2: chap. 12).

27. *CCD*, 14:430, 437, 439; 15:15, 53, 58, 80; *Marginalia*, 424, 526, 557; *Notebooks*, T26.

28. *CCD*, 15:74; cf. p. 141. Only fragments of the manuscript of Darwin's *Descent of Man* (1871) have been found. The first part, chapters 1–6, existed as a 'rough draft' – also 'my essay' and 'my manuscript' (*Descent*, 1:3–4) – by 6 November 1868, when Darwin received a copy of Ernst Haeckel's *Natürliche Schöpfungschichte*. He then added a dozen references to Haeckel to that draft, all but two of which (beyond the introductory pages) appear in chapters 1–6. Six of the references are to Haeckel's *Generelle Morphologie*, received by Darwin in autumn 1866; it is not known when these references were added. Chapter 7, 'The Races of Man', does not mention Haeckel. Darwin last mooted adding a chapter on man to *Variation* on 27 January 1867 (*CCD*, 15:53). Four 'man chapter' references in Darwin's unpublished papers concern material incorporated into chapters 1–6; one reference, 'All for Man Chapter' (*Marginalia*, 336), is jotted in a book published in 1869. So while the 'rough draft', 'essay' or 'manuscript' to which Darwin added Haeckel references was likely part of, or the basis of, chapters 1–6, Darwin continued to refer to this first part of *Descent* as 'Man Chapter' long after he knew he was writing a book.

29. *CCD*, 15:74, 141.

30. *CCD*, 15:92–3, 179–80; P. R. Kirby, *Sir Andrew Smith*, 339–40. Darwin told Alphonse de Candolle the same in 1868, that he was 'partly led' to publish 'by having been taunted that I concealed my views': *LLD*, 3:93–9.

31. *CCD*, 15:96–7.

32. *CCD*, 15:105–6, 109, 132–3, 137, 141, 237–8, 240, 250–51; Wallace, 'Mimicry', 37–8; Wallace, 'Darwin's "The Descent" ', 177; *LLL*, 2:432. Wallace was extending the mimicry studies of Henry Walter Bates, much admired by Darwin, who had reviewed Bates' work: C. Darwin, 'Contributions'; cf. Kottler, 'Charles Darwin', 417–19; Cronin, *Ant*, chaps 5–6; J. Smith, 'Grant Allen'.

33. *MLD*, 1:303, 316; 2:78.

34. *Descent*, 2:17, 20; *MLD*, 2:78; *LLD*, 3:92; Darwin to E. Haeckel, 6 Feb. 1868, Ernst-Haeckel-Haus, Friedrich-Schiller-Universität Jena.

35. E. Haeckel to Darwin, 23 Mar. 1868, DAR 166:47. Cf. R. J. Richards, *Tragic Sense*, 156–8; Darwin to T. H. Huxley, 21 Feb. [1868], Huxley Papers 5.260, Imperial College; T. H. Huxley, 'On the Ethnology'. In a strategic compliment, Darwin credited Haeckel as 'the sole author' in the 1860s who 'has discussed . . . the subject of sexual selection, and has seen its full importance' (*Descent*, 1:5). But there is nothing in Haeckel's *Natürliche Schöpfungschichte* about sexual selection and human races – Darwin jotted in his copy, received in November 1868, 'Nothing about Sexual Selection' (*Marginalia*, 358) – nor in the passage to which Haeckel pointed Darwin in his *Generelle Morphologie* (2:247, received by Darwin in October or November 1866; E. Haeckel to Darwin, 23 Mar. 1868, DAR 166:47), nor in Haeckel's 1868 article to which Darwin refers in *Descent*, 1:199n.

36. Hughes, *Fatal Shore*, 422–4; for the images, Huxley Manuscripts 1:16, Imperial College, and Desmond, *Huxley*, 397–8; Beer, 'Travelling', 327; Bravo, 'Ethnological Encounters', 344; Staum, 'Nature', 480.

37. Stocking, *Victorian Anthropology*, 126; Augstein, *James Cowles Prichard's Anthropology*, 236; Tylor, *Researches*, 365; Lubbock, *Pre-Historic Times*, 460–64; for Argyll, J. D.

Hooker to T. H. Huxley, [Sep. 1869], Huxley Papers 3.126, Imperial College, and Gillespie, 'Duke', 44ff.

38. *Descent*, 1:35, 68–9, 182; Stringer, 'Rethinking', 542–3 on Tylor; Jones, 'Social History', 11–14; Bowler, 'From "Savage"'; Lorimer, *Colour* on the hardening attitudes against black gentlemanliness; Kuper, 'On Human Nature' on Darwin's 'pre-Darwinian' ethnography; Radick, 'Darwin', 9–10 for another look at Darwin's constant resort to 'superior' and 'inferior' in nature; Alter, 'Race', on the conflict between Darwin's early and late views on the hierarchy of nature.
39. Tylor, *Researches*, 2–3, 5–6; *MLD*, 1:303; Stocking, 'What's in a Name?', 383–4 and Rainger, 'Race', 68 on the politicking; Lorimer, 'Theoretical Racism', 412ff. on the subsequent direction of the Institute and return of updated Prichardian approaches.
40. *Descent*, 1:94, 2:337, 343, 357, 365–8, 383, 404–5; M'lennan, *Primitive Marriage*, 71–2, 85; A. R. Wallace to C. Darwin, 2 Mar. 1869, DAR 82:13, 85:98 on the women; Moore, 'Wallace' on the broken engagement. For M'Lennan's having met Darwin: J. F. M'Lennan to Darwin, 6 Mar. 1871, DAR 171:17.
41. A. R. Wallace to C. Darwin, 20 Oct. 1869, DAR 106.7 (ser. 2): 86–7; Marchant, *Alfred Russel Wallace*, 1:243; Wallace, 'Charles Lyell', 391–2; C. Lyell to J. Murray, [1869?], 3 Feb. 1869, John Murray Archives, National Library of Scotland; Emma Wedgwood Darwin to her sister Sarah Elizabeth Wedgwood, [Mar. 1871], Wedgwood W/M 168.
42. *Descent*, 2:388.
43. *Descent*, 1:168–9.
44. 'Irishman' to C. Darwin, 13 June 1877, DAR 69:12; Moore and Desmond, 'Introduction', xlvi; Greg, quoted in *Descent*, 1:174 (cf. R. J. Richards, *Darwin*, 173, and Helmstadter, 'W. R. Greg'); *Descent*, 1:167–80; Jones, *Social Darwinism*, 102. On the rush of Darwinian satires on the Irish 'apes': Curtis, *Apes*.
45. *LLD*, 3:131; Marchant, *Alfred Russel Wallace*, 1:254; Burkhardt and Smith, *Calendar*, 7171, 7400; *Descent*, 1:65, 106; Moore, 'Of Love', 204–6. On the coyness of the sexual talk in *Descent*: Dawson, *Darwin*, 32–3.
46. Charles Darwin to Francis Darwin, [28 Feb. 1871], DAR 271; De Beer, 'Darwin's Journal', 18. At the trade sale in the Albion Tavern, Murray initially took orders for 2,390 copies (Sales Ledger, 1869–70, John Murray Archives, National Library of Scotland), but the book sold so well that the resellers immediately reordered, and the next month Murray offered Darwin an £800 advance for a second edition (Paston, *At John Murray's*, 232).
47. 'Mr. Darwin on the Descent of Man', *The Times*, 7–8 Apr. 1871; Dawson, *Darwin*, 41–2; W. B. Dawkins, 'Darwin', 195. On the judicial threats to the lower orders even late in the century: Desmond, *Huxley*, 642; Marsh, *Word Crimes*.
48. *Descent*, 1:65–70, 94, 96, 98, 101–2; Ruse, 'Charles Darwin', 626–8 on Darwin's use of group selection to enhance the moral sense; Sober and Wilson, *Unto Others*, 4ff.; Engels, 'Charles Darwin's Moral Sense', 41–6. On anthropology as a 'conversation among gentlemanly peers' which uncritically permits assumptions about class and race: Beer, *Open Fields*, chap. 4. Moore and Desmond's 'Introduction' broaches gender aspects of sexual selection, but we have not pursued the subject here. It will be fully treated in Evelleen Richards' forthcoming *Sexing Selection: Darwin and the Making of Sexual Selection.*
49. Burkhardt and Smith, *Calendar*, 7323; Moore, 'Deconstructing'.
50. Van Riper, *Men*; Bynum, 'Charles Lyell's *Antiquity*'; Grayson, *Establishment*, chap. 8.
51. *Descent*, 1:20, 25, 199, 206.
52. *Descent*, 1:3, 197; Vogt, *Lectures*, 454; Desmond, *Archetypes*, 165.
53. *The Times*, 25 May 1864, 8–9; E. W. Cooke to T. H. Huxley, 8 Feb. 1862, Huxley Papers 12.314, Imperial College, on the Cardinal at Huxley's lectures.
54. Darwin to T. H. Huxley, 5 Dec. [1873], Huxley Papers 5.305, Imperial College. On the ape-brain debate: Gross, 'Hippocampus Minor'; Cosans, 'Anatomy'; C. U. M. Smith, 'Worlds'; L. G. Wilson, 'Gorilla'; Rupke, *Richard Owen*, 270–82; Desmond, *Archetypes*, 75.
55. P. du Chaillu to R. Owen, 19 Aug. 1864, Richard Owen Correspondence, 10.173,

Natural History Museum; McCook, 'It May be Truth' on du Chaillu; *TASL*, 1 (1863), xxv (the skin was presented by Winwood Reade); Burrow, *Evolution*, 119; *Punch*, 18 May 1861 and Curtis, *Apes*, 122 on *Punch*'s simian Irishman. On the literary impact of apes: Hodgson, 'Defining'. Darwin did not even read du Chaillu's *Explorations and Adventures in Equatorial Africa* (1861) when it came out (*CCD*, 9:149, 202).

56. *Descent*, 1:143, 156, 240–48; 2:376; Argyll, *Primeval Man*, 66. While sexual selection was neglected for a century or more (B. Campbell, *Sexual Selection*), lately there has been renewed interest in the possibility that it not only played a part in altering eye and hair colour, but that it might have accentuated sexual dimorphism in European populations, producing thinner waists and fatter hips in women – and interestingly (given Darwin's own suspicion), that it possibly affected skin colour in Europeans. Here it might even have acted dimorphically to produce different complexions, darker in men and lighter in women. For a way into this latest controversial research: cf. P Frost, 'Human Skin-Color Sexual Dimorphism: A Test of the Sexual Selection Hypothesis', *American Journal of Physical Anthropology*, 133 (2007), 779–80; P. Frost, 'European hair and eye color – A Case of Frequency-Dependent Sexual Selection?', *Evolution and Human Behavior*, 27 (2006), 85–103; and K. Aoki, 'Sexual Selection as a Cause of Human Skin Colour Variation: Darwin's Hypothesis Revisited', *Annals of Human Biology*, 29 (2002), 589–608, with L. Madrigal and W. Kelly, 'Human Skin-Color Sexual Dimorphism: A Test of the Sexual Selection Hypothesis', *American Journal of Physical Anthropology*, 132 (2006), 470–82, which disputes the evidence for sexual selection in skin coloration.

57. 'Mr. Darwin on "The Descent of Man" ', *Daily News*, 23 Feb. 1871; Freeman, 'Introduction', 5. Other commentators have had no trouble seeing that sexual selection was about differentiating the races: Tipton, 'Darwin's Beautiful Notion'; Hiraiwa-Hasegawa, 'Sight', 16–17. Ernst Mayr agreed that sexual selection was developed to explain the origin of racial differences, even if, oddly, 'all the examples which [Darwin] quotes on more than 300 pages dealing with sexual selection in animals relate to sexual dimorphism' rather than racial difference: Mayr, 'Descent', 33–4, 44.

58. 'Mr. Darwin on the Descent of Man', *The Times*, 7–8 Apr. 1871; *Examiner* (London), 4 Mar. 1871; *Descent*, 1:213.

59. 'Mr. Darwin on the Descent of Man', *The Times*, 7–8 Apr. 1871.

60. *Descent*, 1:63–4; 2:233; Briggs, *Victorian Things*, 264, 271; Newton, *Zoological Aspect*, 6–7; Tipton, 'Darwin's Beautiful Notion', 123.

61. *Descent*, 1:221, 225–9.

62. *Descent*, 1:34, 232–3. Even St George Mivart (*Essays*, 2:33), the most fervent (Catholic) critic of *Descent*, swallowed his pride to 'thank' Darwin for his 'very distinct and unqualified statements as to the substantial unity of men's mental powers'.

63. *Descent*, 1:35. No one has identified 'Howard'. Quite possibly Darwin was referring to the eighteenth-century humanitarian prison reformer John Howard (1726–90), whose bust stood on Shrewsbury's prison wall, and who was the inspiration for the Howard League for Penal Reform. The Wedgwood women took an interest in penal reform, and Emma's personal library contained Thomas Wrightson's *On the Punishment of Death* (3rd edn, 1837), while Maer's held Basil Montagu, *Opinions of Different Authors on the Punishment of Death* (3 vols, 1809–24), both on the death penalty (Anon, 'List'; *Catalogue*). Before Darwin's birth, his aunt Sarah Wedgwood had written about 'reading Howards account of the prisons &c in England & Europe': 'I suppose there never was such a man born . . .' (Sarah Elizabeth Wedgwood to Jessie Allen, 1808, Wedgwood W/M 73). The only evidence that Darwin was familiar with John Howard is in his biographical sketch of Erasmus Darwin, which quotes his grandfather's poetry as showing how he 'sympathised warmly with Howard's noble work of reforming the state of the prisons throughout Europe' (C. Darwin, 'Preliminary Notice', 47).

In 1871, however, another 'Howard' was familiar to emancipists. Brigadier-General Otis Howard was the Presidentially appointed Commissioner of the Freedmen's Bureau after the Civil War. He had started the massive schooling programme for freed blacks after the conflict. In Mobile, Nott had told Howard he would rather see his medical school 'burned to the ground' than be turned into a Freedmen's school, to no avail. In a letter to Howard, republished by the Anthropological Society, Nott damned this post-war equal education as threatening the progress of the United States, for the races could not 'live together practically on any other terms than that of master and slave'. Undaunted, Howard had travelled through the South, dispensing a million dollars, founding hospitals and over 2,000 schools for the ex-slaves. He laboured under terrible difficulties, with the teachers harassed by that rising 'monster terrible beyond question', the Ku Klux Klan. Nonetheless he issued emergency clothes and rations, fought for ex-slave land ownership and finally established Howard University for black students in 1867 (Nott, 'Negro Race', 105; Haller, *Outcasts*, 81; C. Loring Brace, 'Ethnology', 523–4). General Howard and the Freedmen's Bureau were extensively discussed in *The Times*, and as Darwin told Gray after the war, 'We continue to be deeply interested on American affairs; indeed I care for nothing else in the Times' (*CCD*, 13:222–3). Many British charities merged into the National Freedmen's Aid Union of Great Britain and Ireland (founded 24 April 1866), which was very active, shared personnel with the Jamaica Committee and collected $800,000 for the Freedmen's Bureau (*The Times*, 25 Apr. 1866, 19; Howard, *Autobiography*, 1:196, 271, 292, 369, 375).

64. *Descent*, 1:235; Haller, *Outcasts*, 86; C. L. Brace, *Races*, vi.

Bibliography

MANUSCRIPTS

American Philosophical Society, Philadelphia

Samuel George Morton Papers
T. H. Huxley to C. Lyell, 26 June 1859, B/D25.L

Birmingham University Library, Edgbaston

Letters of Charles Baker, Joseph and Richard Matthews, Church Missionary Society Unofficial Papers, CMS/B/OMS/ C N O18/24 and O61/.
Harriet Martineau Papers

Cambridge University Library

Charles Darwin Collections (Darwin Manuscripts, Darwin Library, Darwin Reprint Collection)
Robert FitzRoy letters to family members, 182?–1834, Add. 8853
University petitions to Parliament on the slave trade, 1814 and 1823, CUR 50(8)
William Marsh to John Stevens Henslow, 20 March 1829, Add. 8176:106

Dartmouth College Library

Ticknor Autograph Collection

Down House, Downe, Kent

Beagle Field Notebooks, EH1.1–18
'Catalogue of Down Specimens', EH 88202576

Durham University Library

Political and Public Papers of 2nd Earl Grey (1764–1845)

Edinburgh University Library

Correspondence of Sir Charles Lyell
Plinian Society Minutes M.S.S. 1 (1826–28), Dc.2.53.

Ernst-Haeckel-Haus, Friedrich-Schiller-Universität Jena

Correspondence of Ernst Haeckel

Gray Herbarium Library, Harvard University

Asa Gray Papers

Harvard Archives, Harvard University Library

Francis Ellingwood Abbot Papers

Houghton Library, Harvard University

Louis Agassiz Papers
Charles Eliot Norton Papers
Charles Sumner Correspondence

Imperial College of Science and Technology, London

Thomas Henry Huxley Archives (Huxley Manuscripts, Huxley Papers)

Jesus College, Cambridge University

Arthur Gray Notes

Mitchell Library, Sydney, Australia

Philip Gidley King the Younger, Journal, 1831–1833; Autobiography (1894); Reminiscences of Charles Darwin, 1831–1836 (1892), FM 4/6900

National Archives, Kew, Richmond, Surrey

Ship's Log: HMS *Beagle*, 4 July 1831 to 17 November 1836. Records of the Admiralty, Ships' Logs, ADM 53/236
Captain's Log: HMS *Samarang*, 1 March 1831 to 31 December 1839. Records of the Admiralty, Captains' Logs, ADM 51/3432
Ship's Log: HMS *Samarang*, 3 June 1831 to 2 June 1832. Records of the Admiralty, Ships' Logs, ADM 53/1318

National Library of Jamaica, Kingston

Feurtado Manuscript

National Library of Scotland

John Murray Archives

Natural History Museum, London

Richard Owen Correspondence
Alfred Russel Wallace Family Papers

Royal Botanic Gardens, Kew, Richmond, Surrey

Asa Gray Letters

Royal College of Surgeons of England

E. Belfour to R. FitzRoy, 31 Jan. 1831

Royal Society of London

Robert FitzRoy correspondence with Sir John and Lady Herschel, Herschel Papers

Shropshire Archives, Shrewsbury

Katherine Plymley Diaries, Corbett of Longnor Papers

Suffolk County Record Office, Ipswich

Henslow scrapbook, correspondence &c, *c.* 1820–1840, HD 654/1

UK Hydrographic Office, Taunton, Somerset

Robert FitzRoy letters to Francis Beaufort, 1831–33

Wedgwood Museum, Barlaston, Staffordshire

Wedgwood Archive

Zoological Society of London

Minutes of Council

PUBLICATIONS

Many anonymous reviews and editorials and other newspaper articles are cited fully in the Notes. Place of publication is London unless stated otherwise.

Abbreviations Used in the Bibliography

AMNH	*Annals and Magazine of Natural History*
ANH	*Archives of Natural History*
AR	*Anthropological Review*
AS	*Annals of Science*
BBMNH	*Bulletin of the British Museum (Natural History), Historical Series*
BJHS	*British Journal for the History of Science*
CMJR	*Charleston Medical Journal and Review*
CUP	Cambridge University Press
ENPJ	*Edinburgh New Philosophical Journal*
ER	*Edinburgh Review*
HPLS	*History and Philosophy of the Life Sciences*
HS	*History of Science*
JASL	*Journal of the Anthropological Society of London*
JESL	*Journal of the Ethnological Society of London*

JHB	*Journal of the History of Biology*
JHI	*Journal of the History of Ideas*
JHMAS	*Journal of the History of Medicine and Allied Sciences*
JNH	*Journal of Negro History*
OUP	Oxford University Press
SA	*Slavery and Abolition*
SHB	*Studies in History of Biology*
SHPBBS	*Studies in History and Philosophy of Biological and Biomedical Sciences*
SHPS	*Studies in the History and Philosophy of Science*
SL	*Slavery and Liberation*
SQR	*Southern Quarterly Review*
VS	*Victorian Studies*
UP	University Press

Ackerson, W., *The African Institution (1807–1827) and the Antislavery Movement in Great Britain* (Lewiston, NY: Mellen, 2005).

Addenbrookes Hospital, *The State of Addenbrookes Hospital in the Town of Cambridge for the Year Ending at Michaelmas 1826 &c* (Cambridge: printed for the Hospital, 1826 &c).

Agassiz, E. C. C., *Louis Agassiz: His Life and Correspondence*, 2 vols (Boston: Houghton, Mifflin, 1893).

A[gassiz], L., 'The Diversity of Origin of the Human Races', *Christian Examiner*, 49 (July 1850), 110–45.

——*Essay on Classification*, ed. E. Lurie (Cambridge, MA: Belknap Press of Harvard UP, 1962; first published 1857).

——'Geographical Distribution of Animals', *Christian Examiner*, 48 (Mar. 1850), 181–204.

——'A Period in the History of our Planet', *ENPJ*, 35 (1843), 1–29.

——'Sketch of the Natural Provinces of the Animal World and their Relation to the Different Types of Man', in Nott and Gliddon, *Types of Mankind*, lviii–lxxviii.

Ahlstrom, S. E., *A Religious History of the American People* (New Haven, CT: Yale UP, 1972).

Ainsworth, W. F., 'Mr. Darwin', *Athenaeum*, no. 2,846 (13 May 1882), 604.

Alatas, S., *The Myth of the Lazy Native* (Cass, 1977).

Allan, J. M., 'Europeans, and their Descendants in North America', *JASL*, 6 (1868), cxxvi–cxlii.

——'On the Ape-Origin of Mankind', *Popular Magazine of Anthropology*, 1 (1866), 121–8.

Allen, P., *The Cambridge Apostles: The Early Years* (Cambridge: CUP, 1978).

Alter, S. G. 'The Advantages of Obscurity: Charles Darwin's Negative Inference from the Histories of Domestic Breeds', *AS*, 64 (2007), 235–50.

——'Race, Language, and Mental Evolution in Darwin's "Descent of Man" ', *Journal of the History of the Behavioral Sciences*, 43 (2007), 239–55.

——'Separated at Birth: The Interlinked Origins of Darwin's Unconscious Selection Concept and the Application of Sexual Selection to Race', *JHB*, 40 (2007), 231–8.

Amrein, M., and K. Nickelsen, 'The Gentleman and the Rogue: The Collaboration Between Charles Darwin and Carl Vogt', *JHB*, 41 (2008), 237–66.

Anon, 'An Issue with the Reviewer of "Nott's Caucasian and Negro Races" ', *SQR*, 8 (1845), 148–90.

Anon, 'List of Donations received during the Year 1899: from the Executors of the Late Mrs. Darwin', *Cambridge University Reporter*, 30 (1900), 1079–80.

Anon, *Museum Brookesianum: A Descriptive and Historical Catalogue of the Remainder of the Anatomical & Zootomical Museum, of Joshua Brookes, Esq. F.R.S., F.L.S., F.Z.S. &c.* (Wheatley & Adlard, 1830).

Anon, 'Natural History of Man', *United States Magazine and Democratic Review*, 26 (1850), 327–45.

Anon, 'Natural History of Man', *United States Magazine and Democratic Review*, 30 (1852), 430–44.

Anon, *Negro Slavery; or, A View of Some of the More Prominent Features of that State of Society, as it exists in the United States of America and in the Colonies of the West Indies, especially in Jamaica* (Hatchard, 1823).

Anon, 'Notices of New Works', *Southern Literary Messenger*, 15 (1849), 638–40.

Anon, 'On the Study of Prophecy', *Morning Watch*, 1 (1829), 1–11.

Anon, 'On the Unity of the Human Race', *SQR*, 10 (1854), 273–304.

Anon, 'The Original Unity of the Human Race', *New Englander*, 8 (1850), 543–84.

Anon, 'Origin and Characteristics of the American Aborigines', *United States Magazine and Democratic Review*, new ser., 11 (1842), 603–21.

Anon, 'Statistics of Population and Trade', *DeBow's Review*, 7 (1849), 167–72.

Anon, 'Unity of the Human Race', *SQR*, 9 (1846), 1–57, 372–91.

Anon, 'Varieties of Human Race', *Report of the Tenth Meeting of the British Association for the Advancement of Science; held at Glasgow in August 1840* (Murray, 1841), 449–58.

Anstey, R., *The Atlantic Slave Trade and British Abolition, 1760–1810* (Macmillan, 1975).

Anti-Slavery Society, *Account of the Receipts and Disbursements of the Anti-Slavery Society for the Years 1823 ... with a List of Subscribers* (printed for the Anti-Slavery Society, 1823 &c).

——*Ladies' Anti-Slavery Associations* (printed for the Anti-Slavery Society, [1828]).

Arbuckle, E. S., ed., *Harriet Martineau's Letters to Fanny Wedgwood* (Stanford, CA: Stanford UP, 1983).

Argyll, The Duke of, *Primeval Man* (Strahan, 1869).

Armstrong, P., 'Charles Darwin's Visit to the Bay of Islands, December 1835', *Auckland Waikato Historical Journal* (Apr. 1992), 10–24.

——'Darwin's Perception of the Bay of Islands, New Zealand, 1835', *New Zealand Geographer*, 49 (1993), 26–9.

——'Three Weeks at the Cape of Good Hope 1836: Charles Darwin's African Interlude', *Indian Ocean Review* (June 1991), 8–9.

Ashworth, J. H., 'Charles Darwin as a Student in Edinburgh, 1825–1827', *Proceedings of the Royal Society of Edinburgh*, 55 (1935), 97–113.

Ashworth, W. J., 'John Herschel, George Airy, and the Roaming Eye of the State', *HS*, 36 (1998), 151–78.

Asiegbu, J. U. J., *Slavery and the Politics of Liberation 1787–1861* (Longmans, 1969).

Audubon, M. R., ed., *Audubon and His Journals*, 2 vols (New York: Chelsea House, 1983).

Augstein, H. F., *James Cowles Prichard's Anthropology: Remaking the Science of Man in Early Nineteenth Century Britain* (Amsterdam: Rodopi, 1999).

B., J. Y., 'Prichard's Unity of the Races', *SQR*, 4 (July 1851), 206–38.

Bachman, C. L., *John Bachman, the Pastor of St. John's Lutheran Church, Charleston* (Charleston, SC: Walker, Evans & Cogswell, 1888).

Bachman, J., 'Additional Observations on Hybridity in Animals, and on Some Collateral Subjects; being a Reply to the Essays of Samuel George Morton, M.D., Penna. and Edinb., President of the Academy of Natural Sciences, Philadelphia', *CMJR*, 6 (May 1851), 383–96.

——*Continuation of the Review of 'Nott and Gliddon's Types of Mankind' ... No. II* (Charleston, SC: James, Williams & Gitsinger, 1855).

——*The Doctrine of the Unity of the Human Race examined on the Principles of Science* (Charleston, SC: Canning, 1850).

——'An Examination of a Few of the Statements of Prof. Agassiz, in his "Sketch of the Natural Provinces of the Animal World, and their Relation to the Different Types of Men" ... [No. II]', *CMJR*, 9 (Nov. 1854), 790–806.

——'An Examination of Prof. Agassiz's Sketch of the Natural Provinces of the Animal World, and their Relation to the Different Types of Man, with a Tableau accompanying the Sketch ... No. IV', *CMJR*, 10 (July 1855), 482–534.

——'An Examination of the Characteristics of Genera and Species as applicable to the Doctrine of the Unity of the Human Race ... No. III', *CMJR*, 10 (Mar. 1855), 201–22.

——'A Reply to the Letter of Samuel George Morton, M.D., on the Question of Hybridity in Animals considered in reference to the Unity of the Human Species', *CMJR*, 5 (May 1850), 466–508.

——'Second Letter to Samuel G. Morton on the Question of Hybridity in Animals, considered in Reference to the Unity of Human Species', *CMJR*, 5 (Sep. 1850), 621–60.

——'Types of Mankind ... [No. I]', *CMJR*, 9 (Sep. 1854), 627–59.

Bank, A., 'Of "Native Skulls" and "Noble Caucasians": Phrenology in Colonial South Africa', *Journal of Southern African Studies*, 22 (1996), 387–403.

Banton, M., 'Kingsley's Racial Philosophy', *Theology*, no. 655 (1975), 22–30.

Barclay, J., *An Inquiry into the Opinions, Ancient and Modern, concerning Life and Organization* (Edinburgh: Bell & Bradfute, 1822).

Bar-Josef, E., 'Christian Zionism and Victorian Culture', *Israel Studies*, 8 (2003), 18–44.

Barlow, N., ed., *The Autobiography of Charles Darwin 1809–1882, with Original Omissions Restored* (Collins, 1958).

——ed., *Charles Darwin and the Voyage of the 'Beagle'* (Pilot, 1945).

Barrett, P. H., ed., *The Collected Papers of Charles Darwin*, 2 vols (Chicago: University of Chicago Press, 1977).

——et al., eds, *Charles Darwin's Notebooks, 1836–1844: Geology, Transmutation of Species, Metaphysical Enquiries* (Cambridge: British Museum (Natural History)/CUP, 1987).

——et al., eds, *A Concordance to Darwin's 'The Descent of Man, and Selection in Relation to Sex'* (Ithaca, NY: Cornell UP, 1987).

Barrow, J., *Travels in Southern Africa ... and in the Interior Districts of Africa by M. Le Vaillant*, abridged by W. Mavor (Minerva Press for Lane et al., 1807).

Barrow, L., *Independent Spirits: Spiritualism and English Plebeians, 1850–1910* (Routledge & Kegan Paul, 1986).

Barta, T., 'Mr Darwin's Shooters: On Natural Selection and the Naturalizing of Genocide', *Patterns of Prejudice*, 39 (2005), 116–37.

Bartholomew, M. J., 'Lyell and Evolution: An Account of Lyell's Response to the Prospect of an Evolutionary Ancestry for Man', *BJHS*, 6 (1973), 261–303.

Bartley, M. M., 'Darwin and Domestication: Studies on Inheritance', *JHB*, 25 (1992), 307–33.

Bebbington, D. W., *Evangelicalism in Modern Britain: A History from the 1730s to the 1980s* (Unwin Hyman, 1989).

Beer, G., 'Darwin's Reading and the Fictions of Development', in Kohn, *Darwinian Heritage*, 543–88.

——*Open Fields: Science in Cultural Encounter* (Oxford: Clarendon, 1996).

——'Travelling the Other Way', in N. Jardine, J. A. Secord and E. C. Spary, eds., *Cultures of Natural History* (Cambridge: CUP, 1996), 322–37.

Bell, C., *Essays on the Anatomy and Philosophy of Expression*, 2nd edn (Murray, 1824).

Bellon, R., 'Joseph Hooker Takes a "Fixed Post": Transmutation and the "Present Unsatisfactory State of Systematic Botany", 1844–1860', *JHB*, 39 (2006), 1–39.

Bellows, D., 'A Study of British Conservative Reaction to the American Civil War', *Journal of Southern History*, 51 (1985), 505–26.

Bennett, A. R., *London and Londoners in the Eighteen-Fifties and Sixties* (Unwin, 1924).

Berger, M., 'American Slavery as seen by British Visitors, 1836–1860', *JNH*, 30 (1945), 181–202.

Bernstein, B. J., 'Southern Politics and Attempts to Reopen the African Slave Trade', *JNH*, 51 (1966), 16–35.

Bethell, L., *The Abolition of the Brazilian Slave Trade: Britain, Brazil and the Slave Trade Question* (Cambridge: CUP, 1970).

Biddiss, M. D., 'The Politics of Anatomy: Dr Robert Knox and Victorian Racism', *Proceedings of the Royal Society of Medicine*, 69 (1976), 245–50.

Bindman, D., 'Am I Not a Man and a Brother? British Art and Slavery in the Eighteenth Century', *Res*, 26 (1994), 67–82.

Bindoff, S. T., E. F. M. Smith and C. K. Webster, eds, *British Diplomatic Representatives, 1789–1852* (Royal Historical Society, 1934).

Birney, W., *James G. Birney and His Times: The Genesis of the Republican Party with Some Account of Abolition Movements in the South before 1828* (New York: Appleton, 1890).

Blue, F. J., 'The Poet and the Reformer: Longfellow, Sumner, and the Bonds of Male Friendship, 1837–1874', *Journal of the Early Republic*, 15 (1995), 273–97.

Blyth, E., 'An Attempt to Classify the "Varieties" of Animals with Observations on the Marked Seasonal and Other Changes which Naturally Take Place in Various British Species, and which Do Not Constitute Varieties', *Magazine of Natural History*, 8 (1835), 40–53.

——'Drafts for a Fauna Indica', *AMNH*, 19 (1847), 41–53.

Bolt, C., *The Anti-Slavery Movement and Reconstruction: A Study of Anglo-American Co-operation, 1833–77* (New York: OUP, 1969).

Bonner, R. E., 'Slavery, Confederate Diplomacy, and the Racialist Mission of Henry Hotze', *Civil War History*, 51 (2005), 288–316.

Booth, A., *The Stranger's Intellectual Guide to London for 1839–40* (Hooper, 1839).

Bory de Saint-Vincent, J. B. G. M., 'Homme', in *Dictionnaire classique d'histoire naturelle*, vol. 8 (Paris: Rey & Gravier, 1825), 269–346

——'Orang', in *Dictionnaire classique d'histoire naturelle*, vol. 12 (Paris: Rey & Gravier, 1827), 261–85.

Bourne, K., *Palmerston: The Early Years, 1784–1841* (Allen Lane, 1982).

Bouverie-Pusey, S. E., 'The Negro in Relation to Civilised Society', *JASL*, 2 (1864), cclxxiv–ccxc.

[Bowen, F.], 'On the Origin of Species by Means of Natural Selection', *North American Review*, 90 (1860), 474–506.

Bowler, P. J., 'From "Savage" to "Primitive": Victorian Evolutionism and the Interpretation of Marginalized Peoples', *Antiquity*, 66 (1992), 721–9.

Brace, C. L., *The Races of the Old World: A Manual of Ethnology* (Murray, 1863).

Brace, C. Loring, 'The "Ethnology" of Josiah Clark Nott', *Bulletin of the New York Academy of Medicine*, 50 (1974), 509–28.

[Brace, E.], *The Life of Charles Loring Brace, Chiefly told in His Own Letters* (New York: Scribners, 1894).

Bradley, I., *The Call to Seriousness: The Evangelical Impact on the Victorians* (Cape, 1976).

Brandon-Jones, C., 'Edward Blyth, Charles Darwin, and the Animal Trade in Nineteenth-Century India and Britain', *JHB*, 30 (1997), 145–78.

Brantlinger, P., *Dark Vanishings: Discourse on the Extinction of Primitive Races, 1800–1930* (Ithaca, NY: Cornell UP, 2003).

——'Missionaries and Cannibals in Nineteenth-Century Fiji', *History and Anthropology*, 17 (2006), 21–38.

Bravo, M. T., 'Ethnological Encounters', in N. Jardine, J. A. Secord and E. C. Spary, eds., *Cultures of Natural History* (Cambridge: CUP, 1996), 338–57.

Brent, R., *Liberal Anglican Politics: Whiggery, Religion and Reform, 1830–41* (Oxford: Clarendon, 1987).

Briggs, A., *Victorian Things* (Penguin, 1990).

Brooks, J. L., *Just before the Origin: Alfred Russel Wallace's Theory of Evolution* (New York: Columbia UP, 1984).

Browne, J., *Charles Darwin*, vol. 1, *Voyaging*; vol. 2, *The Power of Place* (Cape, 1995–2002).

——'Missionaries and the Human Mind: Charles Darwin and Robert FitzRoy', in R. Macleod and P. F. Rehbock, eds, *Darwin's Laboratory: Evolutionary Theory and Natural History in the Pacific* (Honolulu: University of Hawai'i Press, 1994), 263–82.

Buckle, H. T., *The History of Civilization in England*, 3 vols (Longmans et al., 1867).

Burkhardt, F., and S. Smith, eds, *A Calendar of the Correspondence of Charles Darwin, 1821–1882*, rev. edn (Cambridge: CUP, 1994).
——et al., eds, *The Correspondence of Charles Darwin*, 16 vols to date (Cambridge: CUP, 1985–).
Burrow, J. W., *Evolution and Society: A Study in Victorian Social Theory* (Cambridge: CUP, 1966).
Bury, P., *The College of Corpus Christi and of the Blessed Virgin Mary: A History from 1822 to 1952* (Cambridge: printed for the College, 1952).
Buxton, C., ed., *Memoirs of Sir Thomas Fowell Buxton, Baronet, with Selections from His Correspondence* (Murray, 1848).
Bynum, W. F., 'Charles Lyell's *Antiquity of Man* and its Critics', *JHB*, 17 (1984), 153–87.
C, 'Unity of the Races', *SQR*, 7 (1845), 372–448.
Cabell, J. L., *The Testimony of Modern Science to the Unity of Mankind: Being a Summary of the Conclusions announced by the Highest Authorities in the Several Departments of Physiology, Zoology, and Comparative Philology in Favor of the Specific Unity and Common Origin of All the Varieties of Man* (New York: Carter, 1859).
Caldwell, C., 'On the Varieties of the Human Race', *Report of the Eleventh Meeting of the British Association for the Advancement of Science; held at Plymouth in July 1841* (Murray, 1842), 75.
——*Phrenology Vindicated, and Antiphrenology Unmasked* (New York: Colman, 1838).
——'Remarks on the Cerebral Organisation of the American Indians and Ancient Peruvians', *American Phrenological Journal*, 3 (1841), 207–17.
——*Thoughts on the Original Unity of the Human Race* (New York: Bliss, 1830).
Calvert, G. H., ed., *Illustrations of Phrenology; Being a Selection of Articles from the Edinburg Phrenological Journal, and the Transactions of the Edinburg Phrenological Society* (Baltimore: Neal, 1832).
Cambridge Philosophical Society, *Regulations of the Cambridge Philosophical Society . . .* [Cambridge: printed for the Society, 1822 &c].
Cambridge Union Society, *Laws and Transactions of the Union Society . . .* (Cambridge: printed for the Society, 1823 &c).
——*Laws and Transactions of the Union Society, revised and corrected to March, M.DCCC.XXXIV., to which is annexed a List of the Members and Officers, from its Formation in M.DCCC.XV and a List of the Periodical and Other Works taken in by the Society* (Cambridge: printed for the Society, 1834).
Campbell, B., ed., *Sexual Selection and the Descent of Man* (Chicago: Aldine, 1972).
Campbell, J., *Negro-mania: Being An Examination of the Falsely Assumed Equality of the Various Races of Men* (Philadelphia: Campbell & Power, 1851).
Cannon, W. F., 'The Impact of Uniformitarianism: Two Letters from John Herschel to Charles Lyell, 1836–1837', *Proceedings of the American Philosophical Society*, 105 (1961), 301–14.
——[S. F.], *Science in Culture: The Early Victorian Period* (New York: Science History, 1978).
[Carlyle, T.], 'Characteristics', *ER*, 54 (Dec. 1831), 351–83.
[——] 'Occasional Discourse on the Negro Question', *Fraser's Magazine*, 40 (Dec. 1849), 670–79.
——*Occasional Discourse on the Nigger Question* (Bosworth, 1853).
——'Shooting Niagara – And After?', *Macmillan's Magazine*, 16 (Apr. 1867), 319–37.
[——] 'Signs of the Times', *ER*, 49 (June 1829), 439–59.
Carpenter, R. L., *Memoirs of the Life of the Rev. Lant Carpenter, LL.D., with Selections from His Correpondence* (Bristol: Philp & Evans, 1842).
[Carpenter, W. B.], 'Darwin on the Origin of Species', *National Review*, 10 (1860), 188–214.
——'Dr. Carpenter and His Reviewer', *Athenaeum* (4 Apr. 1863), 461.
——'Letter from W. B. Carpenter, M.D.', *Christian Examiner*, 4th ser., 2 (1844), 139–44.
——'Microscopical Structure of Shells', *AMNH*, 13 (1844), 486–7.

——'Natural History of Creation', *British and Foreign Medical Review*, 19 (1845), 155–81.
——*Nature and Man: Essays Scientific and Philosophical* (New York: Appleton, 1889).
——*Principles of Comparative Physiology*, 4th and rev. London edn (Philadelphia: Blanchard & Lea, 1854).
——*Principles of Human Physiology, with Their Chief Applications to Pathology, Hygiene, and Forensic Medicine: Especially Designed for the Use of Students* (Philadelphia: Lea & Blanchard, 1845).
——'Researches on the Foraminifera: Part I. Containing General Introduction, and Monograph of the Genus *Orbitolites*', *Philosophical Transactions of the Royal Society of London*, 146 (1856), 181–236.
——'Varieties of Mankind', in R. B. Todd, ed., *The Cyclopedia of Anatomy and Physiology*, 4 (2) (1849–52), 1294–1367
——*Zoology: A Sketch of the Classification, Structure, Distribution, and Habits, of Animals*, 2 vols (Orr, 1844).
Carroll, V., 'The Natural History of Visiting: Responses to Charles Waterton and Walton Hall', *SHPBBS*, 35 (2004), 31–64.
Carus, W., *Memoirs of the Life of the Rev. Charles Simeon, M.A. . . .* (Hatchard, 1847).
The Case of Dred Scot in the United States Supreme Court: The Full Opinions of Chief Justice Taney and Justice Curtis, and Abstracts of the Opinions of the Other Judges; with an Analysis of the Points Ruled, and Some Concluding Observations (New York: Greeley, 1860).
Catalogue of the Miscellaneous Library of the Late Josiah Wedgwood, Esq. removed from Maer Hall, Staffordshire . . . which will be sold by auction by Messrs. S. Leigh Sotheby & Co. at their house, 3, Wellington Street, Strand, on Monday November 16th, 1846, and five following days [printed for Sotheby, 1846].
Cautley, P., 'An Extract of a Letter', *Proceedings of the Geological Society of London*, 2 (1837), 544–5.
——and H. Falconer, 'On the Remains of a Fossil Monkey', *Proceedings of the Geological Society of London*, 2 (1837), 568–9.
Chadwick, O., *The Victorian Church*, 2 vols (New York: OUP, 1966–70).
[Chambers., R.], *Vestiges of the Natural History of Creation* (Churchill, 1844).
Chapman, A., *Darwin in Tierra del Fuego* (Buenos Aires, Imago Mundi, 2006).
Chitnis, A. C., 'The University of Edinburgh's Natural History Museum and the Huttonian–Wernerian Debate', *AS*, 26 (1970), 85–94.
Chitty, S., *The Beast and the Monk: A Life of Charles Kingsley* (New York: Mason/Charter, 1975).
Clark, J. F. M., ' "The ants were duly visited": Making Sense of John Lubbock, Scientific Naturalism and the Senses of Social Insects', *BJHS*, 30 (1997), 151–76.
——' "The Complete Biography of Every Animal": Ants, Bees, and Humanity in Nineteenth-Century England', *SHPBBS*, 29 (1998), 249–67.
Clark, J. W., and T. M. Hughes, *The Life and Letters of the Reverend Adam Sedgwick . . .* 2 vols (Cambridge: at the UP, 1890).
Clarkson, T., *An Essay on the Slavery and Commerce of the Human Species, particularly the African; translated from a Latin dissertation which was honoured with the First Prize in the University of Cambridge for the year 1785, with additions* (printed by J. Phillips, 1786).
——*The History of the Rise, Progress, and Accomplishment of the Abolition of the African Slave-Trade by the British Parliament*, 2 vols (Longman et al., 1808).
——*Strictures on a Life of William Wilberforce by the Rev. W.*[*sic*] *Wilberforce, and the Rev. S. Wilberforce . . .* (Longman et al., 1838).
Cleveland, H., *Alexander H. Stephens, in Public and Private, with Letters and Speeches, before, during, and since the War* (Philadelphia: National Publishing, 1866).
Clowes, W. L., *The Royal Navy: A History from the Earliest Times to the Present*, vol. 6 (Sampson Low, Marston, 1901).

Cobbe, F. P., *Life of Frances Power Cobbe as told by Herself, with Additions by the Author* (Sonnenschein, 1904).
Cockburn, H., *Memorials of His Time* (New York: Appleton, 1856).
Colaiaco, J. A., *James Fitzjames Stephen and the Crisis of Victorian Thought* (Macmillan, 1983).
Colledge, J. J., and B. Warlow, *Ships of the Royal Navy: The Complete Record of All Fighting Ships of the Royal Navy*, rev. edn (Greenhill, 2003).
Collini, S., *Public Moralists: Political Thought and Intellectual Life in Britain, 1850–1930* (Oxford: Clarendon, 1993).
Collinson, G., 'Anti-slavery, Blacks, and the Boston Elite: Notes on the Reverend Charles Lowell and the West Church', *New England Quarterly*, 61 (1988), 419–29.
Collister, P., *The Sulivans and the Slave Trade* (Rex Collings, 1980).
Colp, R., Jr, 'Charles Darwin, Slavery and the American Civil War', *Harvard Library Bulletin*, 26 (1978), 478–89.
——*Darwin's Illness* (Gainesville: University Press of Florida, 2008).
——'*To Be an Invalid* Redux', *JHB*, 31 (1998), 211–40.
——*To Be an Invalid: The Illness of Charles Darwin* (Chicago: University of Chicago Press, 1977).
Combe, A., 'Remarks on the Fallacy of Professor Tiedemann's Comparison of the Negro Brain and Intellect with those of the European', *British and Foreign Medico-Chirurgical Review*, 5 (1838), 585–9.
[Combe, G.], 'Comparative View of the Skulls of the Various Aboriginal Nations of North and South America. By S. G. Morton, Professor of Anatomy at Philadelphia', *ENPJ*, 29 (1840), 111–39.
——*The Constitution of Man Considered in Relation to External Objects*, 2nd edn (Edinburgh: Anderson, 1835).
[——] ' "Crania Americana . . ." by Samuel George Morton . . .', *American Journal of Science and Arts*, 38 (1840), 341–75.
——*Essays on Phrenology; or An Inquiry into the Principles and Utility of the System of Drs. Gall and Spurzheim, and into the Objections Made Against It* (Philadelphia: Carey & Lea, 1822).
——*Lectures on Phrenology, including Its Application to the Present and Prospective Condition of the United States, with Notes, an Introductory Essay, and an Historical Sketch by Andrew Boardman* (New York: Colman, 1839).
——*The Life and Correspondence of Andrew Combe, M.D.* (Edinburgh: Maclachlan & Stewart, 1850).
——*Notes on the United States of North America, during a Phrenological Visit in 1838–39–40*, 2 vols (Edinburgh: Maclachlan & Stewart, 1841).
——'Observations on the Heads and Mental Qualities of the Negroes and North American Indians', *Phrenological Journal*, 15 (1842), 147–54.
——'Phrenological Remarks on the Relation between the Natural Talents and Dispositions of Nations, and the Developments of their Brains. By George Combe, Esq.', in Morton, *Crania Americana*, 269–91.
——*System of Phrenology*, 3rd edn (Edinburgh: Anderson, 1830).
Compton, L. A., 'Josiah Wedgwood and the Slave Trade: A Wider View', *Northern Ceramic Society Newsletter*, no. 100 (1995), 50–69.
Conrad, R. E., *World of Sorrow: The African Slave Trade to Brazil* (Baton Rouge: Louisiana State UP, 1986).
Conway, M. D., *Autobiography: Memories and Experiences*, 2 vols (Boston: Houghton, Mifflin, 1904).
Cooper, C. H., *Annals of Cambridge*, vol. 4 (Cambridge: printed by Metcalfe & Palmer, 1852).
Cooper, R., *The Infidel's Text-Book* (Hull: Johnson, 1846).
C[orbet], A. E., *The Family of Corbet, Its Life and Times*, 2 vols (St Catherine Press, 1915).
Corbey, R., and B. Theunissen, eds, *Ape, Man, Apeman: Changing Views since 1600:*

Evaluative Proceedings of a Symposium at Leiden, 28 June – 1 July 1993 (Leiden: University of Leiden, Department of Prehistory, 1995).

Corsi, P., *The Age of Lamarck: Evolutionary Theories in France, 1790–1830*, trans. J. Mandelbaum (Berkeley: University of California Press, 1988).

Cosans, C., 'Anatomy, Metaphysics, and Values: The Ape Brain Debate Reconsidered', *Biology and Philosophy*, 9 (1994), 129–65.

Costello, R., *Black Liverpool: The Early History of Britain's Oldest Black Community, 1730–1918* (Portland, ME: Picton, 2001).

Cox, R., ed., *Selections from the Phrenological Journal* (Edinburgh: MacLachlan & Stewart, 1836).

Cradock, P., et al., *Recollections of the Cambridge Union, 1815–1939* (Cambridge: Bowes & Bowes, [1953]).

Craft, W., and E. Craft, *Running a Thousand Miles for Freedom, or, The Escape of William and Ellen Craft from Slavery* (Tweedie, 1860).

Crawford, M., '*The Times* and American Slavery in the 1850s', *SL*, 3 (1982), 228–42.

Crawfurd, J., 'On Sir Charles Lyell's "Antiquity of Man", and on Professor Huxley's "Evidence as to Man's Place in Nature"', *Transactions of the Ethnological Society of London*, 3 (1865), 58–70.

Cronin, H., *The Ant and the Peacock: Altruism and Sexual Selection from Darwin to Today* (Cambridge: CUP, 1991).

Crosby, A. W., *Ecological Imperialism: The Biological Expansion of Europe, 900–1900* (Cambridge: CUP, 1986).

Crowe, M. J., ed., *A Calendar of the Correspondence of Sir John Herschel* (Cambridge: CUP, 1998).

Cull, R., 'Short Biographical Notice of the Author . . .', in Prichard, *Natural History of Man* (1855), 1:xxi–xxiv.

[Cundall, F.], 'Richard Hill', *Journal of the Institute of Jamaica* (July 1896), 223–30.

——'Richard Hill', *JNH*, 5 (1920), 37–44.

Curtin, P. D., *The Image of Africa: British Ideas and Action, 1780–1850* (Madison: University of Wisconsin Press, 1964).

——*Two Jamaicas: The Role of Ideas in a Tropical Colony, 1830–1865* (Cambridge, MA: Harvard UP, 1955).

Curtis, L. P., *Apes and Angels: The Irishman in Victorian Caricature*, rev. edn (Washington, DC: Smithsonian Institution Press, 1997).

[Darwin, C.], 'Contributions to an Insect Fauna of the Amazon Valley. By Henry Walter Bates, Esq.', *Natural History Review*, new ser., 3 (1863), 219–24.

——*The Descent of Man, and Selection in Relation to Sex*, 2 vols (Murray, 1871).

——*The Expression of the Emotions in Man and Animals* (Murray, 1872).

——'Fertility of Hybrids from the Common and Chinese Goose', *Nature*, 21 (1880), 207.

——*Geological Observations on South America: Being the Third Part of the Geology of the Voyage of the Beagle . . .* (Smith Elder, 1846).

——*Journal of Researches into the Geology and Natural History of the Various Countries visited by H.M.S. 'Beagle'* (Colburn, 1839; rev. edn, Murray, 1845; new edn, 1860).

——*A Monograph on the Sub-Class Cirripedia, with Figures of All the Species: The Balanidae, (Or Sessile Cirripedes); The Verrucidae, Etc., Etc., Etc.* (Ray Society, 1854).

——'Observations on the Structure and Propagation of the Genus Sagitta', *AMNH*, 13 (1844), 1–6.

——*On the Origin of Species by Means of Natural Selection, or the Preservation of Favoured Races in the Struggle for Life* (Murray, 1859).

——'Preliminary Notice', in E. Krause, *Erasmus Darwin*, trans. W. S. Dallas (Murray, 1879), 1–127.

——*The Structure and Distribution of Coral Reefs: Being the First Part of the Geology of the Voyage of the Beagle, under the Command of Capt. Fitzroy, R.N. during the years 1832 to 1836* (Smith Elder, 1842).

——*The Variation of Animals and Plants under Domestication*, 2 vols (Murray, 1868).
——ed., *The Zoology of the Voyage of H. M. S. Beagle* ... 5 pts (Smith Elder, 1838).
[Darwin, E.], *The Botanic Garden: A Poem in Two Parts ... with Philosophical Notes*, Pt 1, *The Economy of Vegetation*; Pt 2, *The Loves of the Plants*, 4th edn (J. Johnson, 1799).
——*The Temple of Nature; or, the Origin of Society: A Poem, with Philosophical Notes* (J. Johnson, 1803).
Darwin, F., ed., 'FitzRoy and Darwin, 1831–36', *Nature*, 88 (1912), 547–8.
——ed., *The Foundations of the 'Origin of Species': Two Essays written in 1842 and 1844* (Cambridge: at the UP, 1909).
——ed., *The Life and Letters of Charles Darwin, including an Autobiographical Chapter*, 3 vols (Murray, 1887).
——and A. C. Seward, eds, *More Letters of Charles Darwin: A Record of His Work in a Series of Hitherto Unpublished Letters*, 2 vols (Murray, 1903).
Darwin Celebration, Cambridge, June, 1909: Speeches delivered at the Banquet held on June 23rd (Cambridge: printed by the *Cambridge Daily News*, 1909).
Davis, D. B., 'The Emergence of Immediatism in British and American Antislavery Thought', *Mississippi Valley Historical Review*, 49 (1962), 209–30.
——*In the Image of God: Religion, Moral Values, and Our Heritage of Slavery* (New Haven, CT: Yale UP, 2001).
——*Inhuman Bondage: The Rise and Fall of Slavery in the New World* (New York: OUP, 2006).
——*The Problem of Slavery in the Age of Revolution, 1770–1823* (Ithaca, NY: Cornell UP, 1975).
Dawkins, R., 'Darwinism and Unbelief', in T. Flynn, ed., *The New Encyclopedia of Unbelief* (Amherst, NY: Prometheus, 2007), 230–35.
[Dawkins, W. B.], 'Darwin on the Descent of Man', *ER*, 134 (1871), 195–235.
Dawson, G., *Darwin, Literature and Victorian Respectability* (Cambridge: CUP, 2007).
De Beer, G., 'Darwin's Journal', *BBMNH*, 2 (1959), 1–21.
De Paolo, C. S., 'Of Tribes and Hordes: Coleridge and the Emancipation of the Slaves, 1808', *Theoria*, 60 (1983), 27–43.
Desmond, A., *Archetypes and Ancestors: Paleontology in Victorian London, 1850–1875* (Blond & Briggs, 1982).
——'Artisan Resistance and Evolution in Britain, 1819–1848', *Osiris*, 3 (1987), 77–110.
——*Huxley: From Devil's Disciple to Evolution's High Priest* (Penguin, 1998).
——'The Making of Institutional Zoology in London, 1822–1836', *HS*, 23 (1985), 153–85, 223–50.
——*The Politics of Evolution: Morphology, Medicine and Reform in Radical London* (Chicago: University of Chicago Press, 1989).
——'Robert E. Grant's Later Views on Organic Development: The Swiney Lectures on "Palaeozoology", 1853–1857', *ANH*, 11 (1984), 395–413.
——and J. Moore, *Darwin* (Michael Joseph, 1991).
Dewey, M. E., ed., *Autobiography and Letters of Orville Dewey, D.D.* (Boston: Roberts, 1883).
Dewey, O., *On American Morals and Manners: Reprinted From the 'Christian Examiner and Religious Miscellany'* (Boston: Crosby, 1844).
Deyle, S., *Carry Me Back: The Domestic Slave Trade in American Life* (New York: OUP, 2005).
Dickens, C., *American Notes for General Circulation* (Chapman & Hall, 1850).
Dickson, S. H., 'Letter From S. H. Dickson, M. D.', *Christian Examiner*, 4th ser., 2 (1844), 427–32.
Diggs, I., 'The Negro in the Viceroyalty of the Rio de la Plata', *JNH*, 36 (1951), 281–301.
Di Gregorio, M. A., *T. H. Huxley's Place in Natural Science* (New Haven, CT: Yale UP, 1984).
——and N. W. Gill, eds, *Charles Darwin's Marginalia*, vol. 1 (New York: Garland, 1990).

Dixon, E. S., *Ornamental and Domestic Poultry, Their History and Management* (Gardeners' Chronicle, 1848).
[——] 'Poultry Literature', *Quarterly Review*, 88 (1851), 317–51.
Dolan, B., *Josiah Wedgwood, Entrepreneur to the Enlightenment* (HarperCollins, 2004).
Donagan, A., 'Whewell's "Elements of Morality"', *Journal of Philosophy*, 71 (1974), 724–36.
Dott, R. H., Jr, 'Lyell in America – His Lectures, Field Work, and Mutual Influences, 1841–1853', *Earth Sciences History*, 15 (1996), 101–40.
Douglas, J. M., *The Life and Selections from the Correspondence of William Whewell, D.D., Late Master of Trinity College* (Kegan Paul, 1881).
Draper, N., ' "Possessing Slaves": Ownership, Compensation and Metropolitan Society in Britain at the Time of Emancipation, 1834–40', *History Workshop Journal*, 64 (2007), 74–102.
Drescher, S., 'The Decline Thesis of British Slavery since Econocide', *SL*, 7 (1986), 3–24.
——'The Ending of the Slave Trade and the Evolution of European Scientific Racism', *Social Science History*, 14 (1990), 415–50.
——'The Slaving Capital of the World: Liverpool and National Opinion in the Age of Abolition', *SA*, 9 (1988), 128–43.
——'Whose Abolition? Popular Pressure and the Ending of the British Slave Trade', *Past and Present*, no. 143 (1994), 136–66.
——and S. L. Engerman, eds, *A Historical Guide to World Slavery* (New York: OUP, 1998).
Driver, F., *Geography Militant: Cultures of Exploration and Empire* (Blackwell, 2001).
——and L. Martins, 'Shipwreck and Salvage in the Tropics: The Case of HMS *Thetis*, 1830–1854', *Journal of Historical Geography*, 32 (2006), 539–62.
Drummond, J., and C. B. Upton, *The Life and Letters of James Martineau*, 2 vols (Nisbet, 1902).
Dunn, R., 'Some Observations on the Varying Forms of the Human Cranium, considered in Relation to the Outward Circumstances, Social State, and Intellectual Condition of Man', *JESL*, 4 (1856), 33–54.
Dupree, A. H., *Asa Gray, 1810–1888* (Cambridge, MA: Harvard UP, 1959).
Durant, J. R., 'The Ascent of Nature in Darwin's *Descent of Man*', in Kohn, *Darwinian Heritage*, 283–306.
——'Scientific Naturalism and Social Reform in the Thought of Alfred Russel Wallace', *BJHS*, 12 (1979), 31–58.
Dutton, G., *The Hero as Murderer: The Life of Edward John Eyre, Australian Explorer and Governor of Jamaica, 1815–1901* (Sydney: Collins, [1967]).
Earle, A., *A Narrative of a Nine Months' Residence in New Zealand, in 1827; together with a Journal of a Residence in Tristan d'Acunha* (Longman et al., 1832).
——*A Narrative of a Residence in New Zealand: Journal of a Residence in Tristan da Cunha*, ed. E. H. McCormick (Oxford: Clarendon, 1966).
Edgington, B. W., *Charles Waterton: A Biography* (Cambridge: Lutterworth, 1996).
Edinburgh Anti-Slavery Society, *The First Annual Report of the Edinburgh Society for Promoting the Mitigation and Ultimate Abolition of Negro Slavery* (Edinburgh: printed for the Edinburgh Anti-Slavery Society, 1824).
Edwards, P., and J. Walvin, *Black Personalities in the Era of the Slave Trade* (Macmillan, 1983).
Eiseley, L., 'The Intellectual Antecedents of the *Descent of Man*', in Campbell, *Sexual Selection*, 1–16.
Ellegård, A., *Darwin and the General Reader: The Reception of Darwin's Theory of Evolution in the British Periodical Press, 1859–1872* (Chicago: University of Chicago Press, 1990; first published 1958).
Ellis, W., *Polynesian Researches, during a Residence of Nearly Eight Years in the Society and Sandwich Islands*, 2nd edn, 4 vols (Fisher, 1832).
Endersby, J., *Imperial Nature: Joseph Hooker and the Practices of Victorian Science* (Chicago: University of Chicago Press, 2008).

Engels, E., 'Charles Darwin's Moral Sense – On Darwin's Ethics of Non-violence', *Annals of the History and Philosophy of Biology*, 10 (2005), 31–54.
Epps, E., ed., *Diary of the Late John Epps* (Kent, 1875).
[Epps, J.], 'Elements of Physiology', *Medico-Chirurgical Review*, 9 (1828), 97–120.
Erickson, P. A., 'Anthropology and Evolution: A Comment on Wells', *Isis*, 66 (1975), 96–7.
——'The Anthropology of Charles Caldwell, M.D.', *Isis*, 72 (1981), 252–6.
——'Phrenology and Physical Anthropology: The George Combe Connection', *Current Anthropology*, 18 (1977), 92–3.
Evans, D. S., et al., eds, *Herschel at the Cape: Diaries and Correspondence of Sir John Herschel, 1834–1838* (Austin: University of Texas Press, 1969).
Evans, L. T., 'Darwin's Use of Analogy between Artificial and Natural Selection', *JHB*, 17 (1984), 113–40.
The Eyre Defence and Aid Fund (printed by Pelican, [1866]).
Eyton, T. C., 'Remarks on the Skeletons of the Common Tame Goose, the Chinese Goose, and the Hybrid between the Two', *Magazine of Natural History*, 4 (1840), 90–92.
——'Some Remarks upon the Theory of Hybridity', *Magazine of Natural History*, 1 (1837), 357–9.
F., W. G., 'Jane Loring Gray', *Rhodora*, 12 (1910), 41–2.
Fagan, M. B., 'Wallace, Darwin, and the Practice of Natural History', *JHB*, 40 (2007), 601–35.
Farrell, B. G., *Elite Families: Class and Power in Nineteenth-Century Boston* (Albany: State University of New York Press, 1993).
Farrer, K. E., ed., *Correspondence of Josiah Wedgwood, 1781–1794, with an Appendix containing Some Letters on Canals and Bentley's Pamphlet on Inland Navigation* (printed by Women's Printing Society, 1906).
Faust, D. G., ed., *The Ideology of Slavery: Proslavery Thought in the Antebellum South, 1830–1860* (Baton Rouge: Louisiana State UP, 1981).
Feltoe, C. L., ed., *Memorials of John Flint South* (Murray, 1884).
Fergusson, J., ed., *Notes and Recollections of A Professional Life, by the Late William Fergusson, Esq., M.D. Inspector General of Military Hospitals* (Longman et al., 1846).
Fichman, M., *The Elusive Victorian: The Evolution of Alfred Russel Wallace* (Chicago: University of Chicago Press, 2004).
Fielding, K. J., 'Froude's Revenge, or the Carlyles and Erasmus A. Darwin', in W. W. Robson, ed., *Essays and Studies, 1978 . . .* (Murray, 1978), 75–97.
Finkelman, P., and J. C. Miller, eds, *Macmillan Encyclopedia of World Slavery* (New York: Simon & Schuster, 1998).
Finzsch, N. ' "It is scarcely possible to conceive that human beings could be so hideous and loathsome": Discourses of Genocide in Eighteenth- and Nineteenth-Century America and Australia', *Patterns of Prejudice*, 39 (2005), 97–115.
Fisch, M., and S. Schaffer, eds, *William Whewell: A Composite Portrait* (Oxford: Clarendon, 1991).
Fisher, W., 'Physicians and Slavery in the Antebellum "Southern Medical Journal" ', *JHMAS*, 23 (1968), 36–49.
Fitzroy, R., *Narrative of the Surveying Voyages of His Majesty's Ships 'Adventure' and 'Beagle' between the years 1826 and 1836, describing their Examination of the Southern Shores of South America, and the 'Beagle's' Circumnavigation of the Globe. Proceedings of the Second Expedition, 1831–36, under the Command of Captain Robert Fitz-Roy, R.N.*, [with *Appendix to Vol. II*] (Henry Colburn, 1839).
——'Outline Sketch of the Principal Varieties and Early Migrations of the Human Race', *Transactions of the Ethnological Society of London*, 1 (1861), 1–11.
——and C. Darwin, 'A Letter, containing Remarks on the Moral State of Tahiti, New Zealand, &c.', *South African Christian Recorder*, 2 (Sep. 1836), 221–38.
Fladeland, B., *Men and Brothers: Anglo-American Antislavery Cooperation* (Urbana: University of Illinois Press, 1972).

Forster, E. M., *Marianne Thornton, 1797–1887: A Domestic Biography* (Edward Arnold, 1956).
Foster, J., *The Register of Admissions to Gray's Inn, 1521–1889 . . .* (printed by Hansard, 1889).
Fought, L., *Southern Womanhood and Slavery: A Biography of Louisa S. McCord, 1810–1879* (Columbia: University of Missouri Press, 2003).
Fox-Genovese, E., and E. D. Genovese, *The Mind of the Master Class: History and Faith in the Southern Slaveholders' Worldview* (Cambridge: CUP, 2005).
Fraser, D., *Power and Authority in the Victorian City* (Oxford: Blackwell, 1979).
Freeman, R. B., *Charles Darwin: A Companion* (Folkestone: Dawson).
——'Darwin's Negro Bird-Stuffer', *Notes and Records of the Royal Society of London*, 33 (1978), 83–6.
——'Introduction', in C. Darwin, *The Descent of Man, and Selection in Relation to Sex*, in P. H. Barrett and R. B. Freeman, eds, *The Works of Charles Darwin*, vol. 21 (Pickering, 1989), 5–6.
——*The Works of Charles Darwin: An Annotated Bibliographical Handlist*, 2nd edn (Folkestone: Dawson, 1977; additions and corrections published by the author, 1986).
Froude, J. A., ed., *Letters and Memorials of Jane Welsh Carlyle, prepared for Publication by Thomas Carlyle*, 3 vols (Longmans et al., 1883).
Fryer, P., *Staying Power: The History of Black People in Britain* (Pluto, 1984).
Fyfe, A., 'Conscientious Workmen or Booksellers' Hacks? The Professional Identities of Science Writers in the Mid-Nineteenth Century', *Isis*, 96 (2005), 192–223.
Galton, F., *The Narrative of an Explorer in Tropical South Africa* (Murray, 1853).
Garrison, W. L., 'On the Death of John Brown', in Lewis Copeland, Lawrence W. Lamm and Stephen J. McKenna, eds, *The World's Great Speeches: 292 Speeches from Pericles to Nelson Mandela*, 4th edn (Mineola, NY: Dover, 1999), 299–301.
Garrison, W. P., and F. J. Garrison, *William Lloyd Garrison, 1805–1879: The Story of His Life*, 4 vols (Unwin, 1885–9).
Gascoigne, J., *Cambridge in the Age of Enlightenment: Science, Religion and Politics from the Restoration to the French Revolution* (Cambridge: CUP, 1989).
Gibbon, C., *The Life of George Combe: Author of 'The Constitution of Man'*, 2 vols (Macmillan, 1878).
Gillespie, N. C., 'The Duke of Argyll, Evolutionary Anthropology, and the Art of Scientific Controversy', *Isis*, 68 (1977), 40–54.
Gilman, D. C., *The Life of James Dwight Dana* (New York: Harper, 1899).
Gobineau, J. A. de, *The Moral and Intellectual Diversity of Races, with Particular Reference to Their Respective Influence in the Civil and Political History of Mankind*, trans. and notes by H. Hotz[e] (Philadelphia: Lippincott, 1856).
——*The Moral and Intellectual Diversity of Races*, ed. R. Bernasconi (Bristol: Thoemmes, 2002).
Godwin, D. E. W., et al., *Finding Aid for Appleton Family Papers, 1752–1962 (Bulk Dates 1831–1885): Longfellow National Historic Site, Cambridge, Massachusetts*, 3rd edn, rev. by M. Welch (Charlestown, MA: Northeast Museum Services Center, 2006).
[Gordon, J.], 'The Doctrines of Gall and Spurzheim', *ER*, 25 (1815), 227–68.
Gosse, P. H., *Letters from Alabama (U.S.) Chiefly relating to Natural History* (Morgan & Chase, 1859).
——*A Naturalist's Sojourn in Jamaica* (Longman et al., 1851).
Gosset, W. P., *The Lost Ships of the Royal Navy, 1793–1900* (Mansell, 1986).
Gough, B. M., 'Sea Power and South America: The "Brazils" or South American Station of the Royal Navy, 1808–1837', *American Neptune*, 50 (1990), 26–34.
Gould, S. J., 'Flaws in the Victorian Veil', in *The Panda's Thumb: More Reflections in Natural History* (Penguin, 1980), 140–45.
——'The Great Physiologist of Heidelberg – Friedrich Tiedemann', *Natural History* (July, 1999), 26–9, 62–70.

——*The Mismeasure of Man* (Penguin, 1992; first published 1981).

Graham, G. S., and R. A. Humphreys, eds, *The Navy and South America, 1807–1823: Correspondence of the Commanders-in-Chief on the South American Station* (printed for the Navy Records Society, 1962).

Graham, M. W., ' "The Enchanter's Wand": Charles Darwin, Foreign Missions, and the Voyage of H.M.S. "Beagle" ', *Journal of Religious History*, 31 (2007), 131–50.

Grant, A., *The Story of the University of Edinburgh during Its First Three Hundred Years*, 2 vols (Longmans et al., 1884).

Grant, R. E., *Dissertatio Physiologica Inauguralis, de Circuitu Sanguinis in Foetu* (Edinburgh: Ballantyne, 1814).

Grasseni, C., 'Taxidermy as Rhetoric of Self-Making: Charles Waterton (1782–1865), Wandering Naturalist', *SHPBBS*, 29 (1998), 269–94.

Gray, A., *Cambridge* (Methuen, 1912).

Gray, A., *Darwiniana: Essays and Reviews Pertaining to Darwinism*, ed. A. H. Dupree (Cambridge, MA: Belknap Press of Harvard UP, 1963).

Gray, J. L., *The Letters of Asa Gray*, 2 vols (Boston: Houghton, Mifflin, 1893).

Grayson, D. K., *The Establishment of Human Antiquity* (New York: Academic Press, 1983).

Greene, J. C., 'The American Debate on the Negro's Place in Nature, 1780–1815', *JHI*, 15 (1954), 384–96.

G[reg], W. R., 'Dr. Arnold', *Westminster Review*, 39 (Feb. 1843), 1–33.

——*Essays on Political and Social Science, contributed Chiefly to the Edinburgh Review*, 2 vols (Longman et al., 1853).

——*Past and Present Efforts for the Extinction of the Slave Trade* (Ridgway, 1840).

Gribbin, J., and M. Gribbin, *FitzRoy: The Remarkable Story of Darwin's Captain and the Invention of the Weather Forecast* (Review, 2003).

Griffey, W. A., 'A Bibliography of Richard Hill, Negro, Scholar, Scientist; Native of Spanish Town, Jamaica', *American Book Collector*, 2 (1932), 220–24.

Grinnell, G., 'The Rise and Fall of Darwin's First Theory of Transmutation', *JHB*, (1974), 259–73.

Gross, C. G., 'Hippocampus Minor and Man's Place in Nature', *Hippocampus*, 3 (1993), 403–16.

Grove, R. H., *Green Imperialism: Colonial Expansion, Tropical Island Edens and the Origins of Environmentalism, 1600–1860* (Cambridge: CUP, 1995).

Groves, C. P., *The Planting of Christianity in Africa*, vol. 1, to 1840 (Lutterworth, 1948).

Gruber, H. E., *Darwin on Man: A Psychological Study of Scientific Creativity*, 2nd edn (Chicago: University of Chicago Press, 1981).

——'The Many Voyages of the *Beagle*', in Gruber, *Darwin on Man*, 359–99.

Gruber, J. W., 'Does the Platypus Lay Eggs? The History of an Event in Science', *ANH*, 18 (1991), 51–123.

——'Ethnographic Salvage and the Shaping of Anthropology', *American Anthropologist*, 61 (1959), 379–89.

Guyatt, M., 'The Wedgwood Slave Medallion: Values in Eighteenth-Century Design', *Journal of Design History*, 13 (2000), 93–105.

Haeckel, E., *Generelle Morphologie: Allgemeine Grundzüge der organischen Formen-Wissenschaft, mechanisch begründet durch die von Charles Darwin reformirte Descendenz-Theorie* (Berlin: Reimer, 1866).

Häggman, B., 'Confederate Imports of Whitworth Sharpshooter Rifles from England, 1861–1865', *Crossfire*, no. 66 (Sep. 2001).

Hake, T. G., *Memoirs of Eighty Years* (Bentley, 1892).

Halévy, E., *England in 1815*, rev. edn (Benn, 1949).

——*The Triumph of Reform, 1830–1841*, rev. edn (Benn, 1950).

Hall, C., *Civilising Subjects: Metropole and Colony in the English Imagination, 1830–1867* (Cambridge: Polity, 2002).

——K. McClelland and J. Rendall, *Defining the Victorian Nation: Class, Race, Gender and the Reform Act of 1867* (Cambridge: CUP, 2000).

Haller, J. S., Jr, 'The Negro and the Southern Physician: A Study of Medical and Racial Attitudes, 1800–1860', *Medical History*, 16 (1972), 238–53.

——*Outcasts from Evolution: Scientific Attitudes of Racial Inferiority, 1859–1900* (Urbana: University of Illinois Press, 1971).

——'The Species Problem: Nineteenth-Century Concepts of Racial Inferiority in the Origin of Man Controversy', *American Anthropologist*, 72 (1970), 1319–29.

Hamilton, W., *Lectures on Metaphysics and Logic*, ed. H. L. Mansel and J. Veitch, 2 vols (Boston: Gould & Lincoln, 1859).

——'Remarks on Dr. Morton's Tables on the Size of the Brain', *ENPJ*, 48 (1850), 330–33.

Hamilton, W. T., *The Pentateuch and Its Assailants: A Refutation of the Objections of Modern Scepticism to the Pentateuch* (Edinburgh: Clark, 1852).

Hammett, T. M., 'Two Mobs of Jacksonian Boston: Ideology and Interest', *Journal of American History*, 62 (1976), 845–68.

Hartley, L., *Physiognomy and the Meaning of Expression in Nineteenth-Century Culture* (Cambridge: CUP, 2001).

Hartwig, W. C., '*Protopithecus*: Rediscovering the First Fossil Primate', *HPLS*, 17 (1995), 447–60.

Harvey, W. H., *The Sea-Side Book: Being an Introduction to the Natural History of the British Coasts* (Van Voorst, 1857).

Haynes, S. R., *Noah's Curse: The Biblical Justification of American Slavery* (New York: OUP, 2002).

Hazlewood, N., *Savage: The Life and Times of Jemmy Button* (Hodder & Stoughton, 2000).

Helmstadter, R. J., 'W. R. Greg: A Manchester Creed', in R. J. Helmstadter and B. Lightman, eds, *Victorian Faith in Crisis: Essays on Continuity and Change in Nineteenth-Century Religious Belief* (Basingstoke: Macmillan, 1990), 187–222.

Hennell, J., *Sons of the Prophets: Evangelical Leaders of the Victorian Church* (SPCK, 1979).

Henslow, J. S., *Address to the Reformers of the Town of Cambridge* (Cambridge: printed by W. Metcalfe [1837]).

——*A Sermon on the First and Second Resurrection . . . preached at Great St. Mary's Church on Feb. 15, 1829* (Cambridge: printed by James Hodson for Deighton & Stevenson, 1829).

Higginson, M. T., *Thomas Wentworth Higginson The Story of His Life* (Port Washington, NY: Kennikat, 1971; first published 1914).

Hill, R., *The Books of Moses: How Say You, True or Not True? Being a Consideration of the Critical Objections in Dr. Colenso's Review of the Books of Moses and Joshua* (Kingston, Jamaica: Gall, 1863).

——'Extraits des lettres d'un voyageur à Haiti, pendant les années 1830 et 1831', in Z. Macaulay, *Haiti, ou, Renseignemens authentiques sur l'abolition de l'esclavage et ses résultats à Saint-Dominique et à Guadeloupe, avec des détails sur l'état actuel d'Haiti et des noirs émancipés qui forment sa population, traduit de l'anglais* (Paris: Hachette, 1835).

——*A Week at Port-Royal* (Montego Bay, Jamaica: printed at the *Cornwall Chronicle* Office, 1855).

Hilton, B., *The Age of Atonement: The Influence of Evangelicalism on Social and Economic Thought, 1785–1865* (Oxford: Clarendon, 1988).

——*A Mad, Bad, and Dangerous People: England, 1783–1846* (Oxford: Clarendon, 2006).

Hiraiwa-Hasegawa, M., 'Sight of the Peacock's Tail Makes Me Sick: The Early Arguments on Sexual Selection', *Journal of Biosciences* (Bangalore), 25 (2000), 11–18.

Hochschild, A., *Bury the Chains: Prophets and Rebels in the Fight to Free an Empire's Slaves* (Boston: Houghton Mifflin, 2005).

Hodge, M. J. S., 'Darwin and the Laws of the Animate Part of the Terrestrial System (1835–1837): On the Lyellian Origins of his Zoonomical Explanatory Program', *SHB*, 6 (1983), 1–106.

——'The Universal Gestation of Nature: Chambers' "Vestiges" and "Explanations"', *JHB*, 5 (1972), 127–51.

——and D. Kohn, 'The Immediate Origins of Natural Selection', in Kohn, *Darwinian Heritage*, 185–206.

Hodgkin, T., *A Catalogue of the Preparations in the Anatomical Museum of Guy's Hospital* (Watts, 1829).

——'Obituary of Dr. Prichard', *JESL*, 2 (1850), 182–207.

——'On Inquiries into the Races of Men', *Report of the Eleventh Meeting of the British Association for the Advancement of Science; held at Plymouth in July 1841* (Murray, 1842), 52–5.

——'On the Dog as the Associate of Man', *Report of the Fourteenth Meeting of the British Association for the Advancement of Science; held at York in September 1844* (Murray, 1845), 81.

——'Report of the Committee to Investigate the Varieties of the Human Race', *Report of the Fourteenth Meeting of the British Association for the Advancement of Science; held at York in September 1844* (Murray, 1845), 93.

Hodgson, A., 'Defining the Species: Apes, Savages and Humans in Scientific and Literary Writing of the 1860s', *Journal of Victorian Culture*, 4 (1999), 228–51.

[Holland, H.], 'Natural History of Man', *Quarterly Review*, 86 (1849–50), 1–40.

——*Recollections of Past Life* (Longmans et al., 1872).

Holyoake, G. J., *The History of Co-Operation* (Unwin, 1908).

——*Sixty Years of an Agitator's Life*, 2 vols (Unwin, 1892).

Hooker, J. D., 'Reminiscences of Darwin', *Nature*, 60 (22 June 1899), 187–8.

Horsman, R., *Josiah Nott of Mobile: Southerner, Physician, and Racial Theorist* (Baton Rouge: Louisiana State UP, 1987).

——'Origins of Racial Anglo-Saxonism in Great Britain before 1850', *JHI*, 37 (1976), 387–410.

——*Race and Manifest Destiny: The Origins of American Racial Anglo-Saxonism* (Cambridge, MA: Harvard UP, 1981).

Howard, O. O., *Autobiography of Oliver Otis Howard, Major General, United States Army* (New York: Baker & Taylor, 1907).

Howse, E. M., *Saints in Politics: The 'Clapham Sect' and the Growth of Freedom* (Allen & Unwin, 1971).

Hughes, R., *The Fatal Shore: A History of Transportation of Convicts to Australia, 1787–1868* (Collins Harvill, 1987).

Hull, D. L., *Darwin and His Critics: The Reception of Darwin's Theory of Evolution by the Scientific Community* (Cambridge, MA: Harvard UP, 1973).

Humboldt, A. von, *The Island of Cuba*, trans. J. S. Thrasher (New York: Derby & Jackson, 1856).

——*The Island of Cuba*, ed. L. Martínez-Fernández (Princeton, NJ: Wiener, 2001).

——and A. Bonpland, *Personal Narrative of Travels to the Equinoctial Regions of the New Continent during the Years 1799–1804*, trans. M. H. Williams, 7 vols (Longman et al., 1819–29).

Hume, B. D., 'Quantifying Characters: Polygenist Anthropologists and the Hardening of Heredity', *JHB*, 41 (2008), 119–58.

Hume, H., *The Life of Edward John Eyre, Late Governor of Jamaica* (Bentley, 1867).

Hunt, J., 'Introductory Address on the Study of Anthropology', *AR*, 1 (1863), 1–20.

——*The Negro's Place in Nature* (New York: Van Evrie, Horton, 1864).

——'On the Application of the Principle of Natural Selection to Anthropology', *AR*, 4 (1866), 320–40.

——'On the Doctrine of Continuity applied to Anthropology', *AR*, 5 (1867), 110–20.

——'On the Physical and Mental Characters of the Negro', *AR*, 1 (1863), 386–91.

Huxley, L., *Life and Letters of Sir Joseph Dalton Hooker, O.M., G.C.S.I., based on Materials collected and arranged by Lady Hooker*, 2 vols (Murray, 1918).

——*Life and Letters of Thomas Henry Huxley*, 2 vols (Macmillan, 1900).
[Huxley, T. H.], 'Contemporary Literature – Science', *Westminster Review*, 62 (1854), 242–56.
——*Critiques and Addresses* (New York: Appleton, 1873).
——*Evidence as to Man's Place in Nature* (New York: Appleton, 1863).
——'Lectures on General Natural History: Lecture II', *Medical Times and Gazette*, 17 May 1856, 481–4.
——'Mr. Huxley and Governor Eyre', *Pall Mall Gazette*, 31 Oct. 1866.
——'The Negro's Place in Nature', *Reader* (Mar. 1864), 334–5.
——'On the Ethnology and Archaeology of North America', *JESL*, new ser., 1 (1869), 218–21.
——*Professor Huxley on the Negro Question*, ed. Mrs A. Taylor (Ladies' London Emancipation Society, 1864).
Ichenhäuser, J. D., comp., *Illustrated Catalogue of the Historical and World-Renowned Collection of Torture Instruments, etc. . . . lent for Exhibition by the Right Honourable the Earl of Shrewsbury and Talbot* . . . (New York: Little, 1893).
Innes, C., 'Memoir', in E. B. Ramsay, *Reminiscences of Scottish Life and Character*, 22nd edn (Edinburgh: Edmonston & Douglas, 1874).
Irwin, R. A., ed., *Letters of Charles Waterton of Walton Hall, near Wakefield, Naturalist, Taxidermist and Author of 'Wanderings in South America' and 'Essays on Natural History'* (Rockliff, 1955).
Ishmael, O., *The Guyana Story (From Earliest Times to Independence)* (www.guyana.org/features/guyanastory/guyana_story.html, revised 2006).
An Island Solicitor, *The Courts of Jamaica and Their Jurisdiction*, Pt 1, *The Administration of Criminal Justice* (Smith Elder, 1855).
Jacoby, K., 'Slaves by Nature? Domestic Animals and Human Slaves', *SA*, 15 (1994), 89–97.
Jacyna, L. S., 'Immanence or Transcendence: Theories of Life and Organization in Britain, 1790–1835', *Isis*, 74 (1983), 311–29.
——'Medical Science and Moral Science: The Cultural Relations of Physiology in Restoration France', *HS*, 25 (1987), 111–46.
——'Principles of General Physiology: The Comparative Dimension to British Neuroscience in the 1830s and 1840s', *SHB*, 7 (1984), 47–92.
Jaher, F. C., 'Nineteenth-Century Elites in Boston and New York', *Journal of Social History*, 6 (1972), 32–77.
James, P., *Population Malthus: His Life and Times* (Routledge, 1979).
James, W. M., *The Naval History of Great Britain during the French Revolutionary and Napoleonic Wars*, vol. 6, *1811–1827* (Conway Maritime, 2002).
Jameson, J. F., 'The London Expenditure of the Confederate Secret Service', *American Historical Review*, 35 (1930), 811–24.
Jameson, R., ed., *Essay on the Theory of the Earth, by Baron G. Cuvier . . . with Geological Illustrations*, 5th edn (Edinburgh: Blackwood, 1827).
Jann, R., 'Darwin and the Anthropologists: Sexual Selection and Its Discontents', *VS*, 37 (1994), 287–306.
Jardine, W., 'Description of Some Birds collected during the Last Expedition to the Niger', *AMNH*, 10 (1842), 186–90.
[Jeffrey, F.], 'A System of Phrenology', *ER*, 44 (1826), 253–318.
Jenkins, W. S., *Pro-Slavery Thought in the Old South* (Chapel Hill: University of North Carolina Press, 1935).
Jennings, J., *The Business of Abolishing the British Slave Trade, 1783–1807* (Cass, 1997).
Jenyns, L., *Memoir of the Rev. John Stevens Henslow . . . Late Rector of Hitcham, and Professor of Botany in the University of Cambridge* (Van Voorst, 1862).
Johnson, C. N., 'The Preface to Darwin's *Origin of Species*: The Curious History of the "Historical Sketch" ', *JHB*, 40 (2007), 529–56.
Johnston's Physical Atlas of Natural Phenomena (Blackwood, 1850).

Jones, G., 'Alfred Russel Wallace, Robert Owen and the Theory of Natural Selection', *BJHS*, 35 (2002), 73–96.

——*Social Darwinism and English Thought: The Interaction between Biological and Social Theory* (Brighton: Harvester, 1980).

——'The Social History of Darwin's "Descent of Man" ', *Economy and Society*, 7 (1978), 1–23.

Jordan, W. D., *White over Black: American Attitudes toward the Negro, 1550–1812* (Chapel Hill: University of North Carolina Press, 1968).

Kalfus, M., *Frederick Law Olmsted: The Passion of a Public Artist* (New York: New York UP, 1990).

Karasch, M. C., *Slave Life in Rio de Janeiro, 1808–1850* (Princeton, NJ: Princeton UP, 1987).

Kass, A. M., and E. H. Kass, *Perfecting the World: The Life and Times of Dr. Thomas Hodgkin, 1798–1866* (Boston: Harcourt Brace Jovanovich, 1988).

Kenny, R., 'From the Curse of Ham to the Curse of Nature: The Influence of Natural Selection on the Debate on Human Unity before the Publication of *The Descent of Man*', *BJHS*, 40 (2007), 367–88.

Keynes, R., *Annie's Box: Charles Darwin, His Daughter and Human Evolution* (Fourth Estate, 2001).

Keynes, R. D., ed., *Charles Darwin's 'Beagle' Diary* (Cambridge: CUP, 1988).

——ed., *Charles Darwin's Zoology Notes and Specimen Lists from H.M.S. 'Beagle'* (Cambridge: CUP, 2000).

Kidd, C., *The Forging of Races: Race and Scripture in the Protestant Atlantic World, 1600–2000* (Cambridge: CUP, 2006).

Kielstra, P. M., *The Politics of Slave Trade Suppression in Britain and France, 1814–48: Diplomacy, Morality and Economics* (Basingstoke: Macmillan, 2000).

Killingray, D., 'Beneath the Wilberforce Oak, 1873', *International Bulletin of Missionary Research*, 21 (1997), 11–15.

King, P. P., *Narrative of the Surveying Voyages of His Majesty's Ships 'Adventure' and 'Beagle' between the Years 1826 and 1836, describing their Examination of the Southern Shores of South America, and the 'Beagle's' Circumnavigation of the Globe. Proceedings of the First Expedition, 1826–30, under the Command of Captain P. Parker King, R.N., F.R.S* (Henry Colburn, 1839).

King-Hele, D., ed., *The Collected Letters of Erasmus Darwin* (Cambridge: CUP, 2007).

——*Erasmus Darwin: A Life of Unequalled Achievement* (Giles de la Mare, 1999).

——ed., *The Essential Writings of Erasmus Darwin* (MacGibbon & Kee, 1968).

[Kingsley, F.], *Charles Kingsley: His Letters and Memories of His Life*, 2 vols (King, 1877).

Kirby, P. R., ed., *The Diary of Dr. Andrew Smith, Director of the 'Expedition for Exploring Central Africa', 1834–1836*, 2 vols (Johannesburg: Van Riebeeck Society, 1939).

——*Sir Andrew Smith, M.D., K.C.B., His Life, Letters and Works* (Cape Town: Balkema, 1965).

Kirby, W., and W. Spence, *An Introduction to Entomology; or, Elements of the Natural History of Insects*, 5th edn, 4 vols (Longman et al., 1828).

Kirsop, W., 'W. R. Greg and Charles Darwin in Edinburgh and After – an Antipodean Gloss', *Transactions of the Cambridge Bibliographical Society*, 7 (1979), 376–90.

Knight, C., *The Pictorial Museum of Animated Nature*, 2 vols (Cox, 1844).

Knox, R., 'Contributions to Anatomy and Physiology', *Medical Gazette*, 2 (1843), 463–7, 499–502, 529–32, 537–40, 554–6, 586–9, 860–62.

——'Inquiry into the Origin and Characteristic Differences of the Native Races inhabiting the Extra-tropical Part of Southern Africa', *Memoirs of the Wernerian Natural History Society*, 5 (1824), 206–18.

——'Lectures on the Races of Man', *Medical Times*, 18 (1848), 97–9, 114–15, 117–30, 133–4, 147–8, 163–5, 190, 231–3, 263–4, 283–5, 299–301, 315–16, 331–2, 365–6.

——'On the Application of the Anatomical Method to the Discrimination of Species', *AR*, 1 (1863), 263–70.

——*The Races of Men: A Philosophical Enquiry into the Influence of Race over the Destinies of Nations*, 2nd edn (Renshaw, 1862; first published 1850).

Kohn, D., ed., *The Darwinian Heritage* (Princeton, NJ: Princeton UP, 1985).

——'Darwin's Ambiguity: The Secularization of Biological Meaning', *BJHS*, 22 (1989), 215–39.

——'Darwin's Keystone: The Principle of Divergence', in R. J. Richards and M. Ruse, eds, *The Cambridge Companion to the Origin of Species* (Cambridge: CUP, 2008).

——'Theories to Work By: Rejected Theories, Reproduction, and Darwin's Path to Natural Selection', *SHB*, 4 (1980), 67–170.

Kolbe, H. R., 'The South African Print Media: From Apartheid to Transformation' (Ph.D. thesis, University of Wollongong, 2005).

Kostal, R. W., *A Jurisprudence of Power: Victorian Empire and the Rule of Law* (Oxford: OUP, 2005).

Kottler, M. J., 'Charles Darwin and Alfred Russel Wallace: Two Decades of Debate over Natural Selection', in Kohn, *Darwinian Heritage*, 367–432.

Kriegel, A. D., 'A Convergence of Ethics: Saints and Whigs in British Antislavery', *Journal of British Studies*, 26 (1987), 423–50.

Kuper, A., 'On Human Nature: Darwin and the Anthropologists', in M. Teich, R. Porter and B. Gustaffson, eds, *Nature and Society in Historical Context* (Cambridge: CUP, 1997), 274–90.

A Lady, *Modern Domestic Cookery: Based on the Well-Known Work of Mrs. Rundell . . .* (Murray, 1851).

Lartet, E., 'New Researches Respecting the Co-existence of Man with the Great Fossil Mammals, regarded as Characteristic of the Latest Geological Period', *Natural History Review*, new ser., 2 (1862), 55–71.

Latham, R. G., *Man and His Migrations* (Van Voorst, 1851).

——*The Natural History of the Varieties of Man* (Van Voorst, 1850).

Lavater, J. C., *Physiognomy; or, the Corresponding Analogy between the Conformation of the Features and the Ruling Passions of the Mind* (Tegg, 1866).

Law, I., *A History of Race and Racism in Liverpool, 1660–1950*, ed. J. Henfrey (Liverpool: Merseyside Community Relations Council, 1981).

Lawrence, W., *Lectures on Physiology, Zoology, and the Natural History of Man delivered at the Royal College of Surgeons* (Benbow, 1822).

Lee, D., *Slavery and Romantic Imagination* (Philadelphia: University of Pennsylvania Press, 1982).

[Leifchild, J. R.], 'Evidence as to Man's Place in Nature', *Athenaeum* (28 Feb. 1863), 287.

Levy, L. W., 'Sims' Case: The Fugitive Slave Law in Boston in 1851', *JNH*, 35 (1950), 39–74.

Lewis, A., 'Black Letter Day', *The Bulletin* [*of University College London*], 7 (1989), 18–19.

Lindfors, B., ed., *Africans on Stage: Studies in Ethnological Show Business* (Bloomington: Indiana UP, 1999).

——'Hottentot, Bushman, Kaffir: Taxonomic Tendencies in Nineteenth-Century Racial Iconography', *Nordic Journal of African Studies*, 5 (1996), 1–28.

Litchfield, H. E. [Darwin], *Emma Darwin, Wife of Charles Darwin: A Century of Family Letters*, 2 vols (Cambridge: privately printed at the UP, 1904; Murray, 1915).

Little, K., *Negroes in Britain: A Study of Racial Relations in English Society* (Routledge & Kegan Paul, 1972).

Livingstone, D. N., *Adam's Ancestors: Race, Religion, and the Politics of Human Origins* (Baltimore, MD: Johns Hopkins UP, 2008).

——*The Preadamite Theory and the Marriage of Science and Religion*, 82 (3), *Transactions of the American Philosophical Society* (Philadelphia: American Philosophical Society, 1992).

——'Science, Text and Space: Thoughts on the Geography of Reading', *Transactions of the Institute of British Geographers*, new ser., 30 (2005), 391–401.

Lizars, J., *A System of Anatomical Plates, accompanied with Descriptions, and Physiological, Pathological, and Surgical Observations: Part VII. – The Brain, First Portion. Coloured After Nature* (Edinburgh: Lizars, 1825).

Lloyd, C., *The Navy and the Slave Trade: The Suppression of the African Slave Trade in the Nineteenth Century* (Longman, 1949).

Loane, M. L., *Cambridge and the Evangelical Succession* (Lutterworth, 1952).

Logan, D. A., ed., *The Collected Letters of Harriet Martineau*, 5 vols (Pickering & Chatto, 2007).

——'The Redemption of a Heretic: Harriet Martineau and Anglo-American Abolitionism in Pre-Civil War America' (unpublished paper, Proceedings of the Third Annual Gilder Lehrman Center International Conference at Yale University, Sisterhood and Slavery: Transatlantic Slavery and Women's Rights, October 25–8, 2001).

Long, E., *The History of Jamaica; or, General Survey of the Antient and Modern State of that Island: with Reflections on its Situation, Settlements, Inhabitants, Climate, Products, Commerce, Laws, and Government*, 3 vols (Lowndes, 1774).

Lonsdale, H., *A Sketch of the Life and Writings of Robert Knox, the Anatomist* (Macmillan, 1870).

Lora, R., and W. H. Longton, eds, *The Conservative Press in Eighteenth- and Nineteenth-Century America* (Westport, CT: Greenwood, 1999).

Lorimer, D. A., *Colour, Class and the Victorians: English Attitudes to the Negro in the Mid-Nineteenth Century* (Leicester: Leicester UP, 1978).

——'Role of Anti-Slavery Sentiment in English Reactions to the American Civil War', *Historical Journal*, 19 (1976), 405–20.

——'Science and the Secularization of Victorian Images of Race', in B. Lightman, ed., *Victorian Science in Context* (Chicago: University of Chicago Press, 1997), 212–35.

——'Theoretical Racism in Late Victorian Anthropology, 1870–1900', *VS*, 31 (1988), 405–30.

Loring, C. G., and E. W. Field, *Correspondence on the Present Relations between Great Britain and the United States of America* (Boston: Little, Brown, 1862).

Lounsbury, R. C., ed., *Louisa S. McCord: Political and Social Essays* (Charlottesville: University Press of Virginia, 1995).

Love, A. C., 'Darwin and Cirripedia Prior to 1846: Exploring the Origins of the Barnacle Research', *JHB*, 35 (2002), 251–89.

Lowell, J. A., 'Darwin's Origin of Species', *Christian Examiner*, 68 (May 1860), 449–64.

Lowell, J. R., *The Anti-slavery Papers of James Russell Lowell*, ed. W. B. Parker, 2 vols (Boston: Houghton, Mifflin, 1902).

Lubbock, J., *Pre-Historic Times* (Williams & Norgate, 1865).

Lurie, E., *Louis Agassiz: A Life in Science* (Chicago: University of Chicago Press, 1960).

——'Louis Agassiz and the Races of Man', *Isis*, 45 (1954), 227–42.

Lutz, M. A., *Economics for the Common Good: Two Centuries of Social Economic Thought in the Humanistic Tradition* (Routledge, 1999).

Lyell, C., *The Geological Evidences of the Antiquity of Man, with Remarks on the Theories of the Origin of Species by Variation*, 2nd edn (Murray, 1863).

——*Principles of Geology: Being an Attempt to Explain the Former Changes of the Earth's Surface, by Reference to Causes Now in Operation*, 3 vols (Murray, 1830–33); 10th edn, 2 vols (Murray, 1867–8).

——*A Second Visit to the United States of North America*, 2 vols (Murray, 1849).

——*Supplement to the Fifth Edition of a Manual of Elementary Geology* (Murray, 1859).

——*Travels in North America, with Geological Observations on the United States, Canada, and Nova Scotia*, 2 vols (Murray, 1845).

Lyell, [K. M.], *Life, Letters, and Journals of Sir Charles Lyell, Bart*, 2 vols (Murray, 1881).

Lyon, D., *The Sailing List: All the Ships of the Royal Navy, Built, Purchased and Captured, 1688–1860* (Conway Maritime, 1993).
McCabe, J., ed., *Life and Letters of George Jacob Holyoake*, 2 vols (Watts, 1908).
McClintock, A., *Imperial Leather: Race, Gender and Sexuality in the Colonial Contest* (Routledge, 1995).
McCook, S., ' "It May be Truth, But it is not Evidence": Paul du Chaillu and the Legitimation of Evidence in the Field Sciences', *Osiris*, 11 (1996), 177–97.
McCord, D. J., 'How the South is affected by Her Slave Institutions', *DeBow's Review*, 11 (1851), 349–63.
M[cCord], L. S., 'Diversity of the Races – Its Bearing upon Negro Slavery', *SQR*, 3 (Apr. 1851), 392–419.
Macgillivray, W., *History of British Birds, Indigenous and Migratory*, vol. 1 (Scott et al., 1837).
MacGregor, D. R., *Fast Sailing Ships: Their Design and Construction, 1775–1875* (Lymington: Nautical, 1973).
McHenry, G., 'On the Negro as a Freedman', *Popular Magazine of Anthropology*, 1 (1866), 36–9.
McKendrick, N., 'Josiah Wedgwood and Factory Discipline', *Historical Journal*, 4 (1961), 3–55.
Mackenzie, C. G., 'Thomas Carlyle's "The Negro Question": Black Ireland and the Rhetoric of Famine', *Neohelicon*, 24 (1997), 219–36.
Mackenzie-Grieve, A., *The Last Years of the English Slave Trade: Liverpool, 1750–1807* (Cass, 1968).
Mackintosh, J., *A General View of the Progress of Ethical Philosophy, chiefly during the Seventeenth and Eighteenth Centuries* (Philadelphia: Carey & Lea, 1832).
Macmillan, M., *Sir Henry Barkly, Mediator and Moderator, 1815–1898* (Cape Town: Balkema, 1970).
McNeil, M., *Under the Banner of Science: Erasmus Darwin and His Age* (Manchester: Manchester UP, 1987).
Malcolm, C., 'Address to the Ethnological Society of London, delivered at the Anniversary, 14th May, 1851', *JESL*, 3 (1854), 86–102.
Malik, K., *Man, Beast and Zombie: What Science Can and Cannot Tell Us about Human Nature* (Weidenfeld & Nicolson, 2000).
—— *The Meaning of Race: Race, History and Culture in Western Society* (New York: New York UP, 1996).
Marchant, J., *Alfred Russel Wallace: Letters and Reminiscences*, 2 vols (Cassell, 1916).
Marsden, G. M., *The Evangelical Mind and the New School Presbyterian Experience: A Case Study of Thought and Theology in Nineteenth-Century America* (New Haven, CT: Yale UP, 1970).
Marsh, J., *Word Crimes: Blasphemy, Culture, and Literature in Nineteenth-Century England* (Chicago: University of Chicago Press, 1998).
Marshall, H., and M. Stock, *Ira Aldridge: The Negro Tragedian* (Rockliff, 1958).
Marshall, P., *Bristol and the Abolition of Slavery: The Politics of Emancipation* (Bristol: Historical Association, Bristol Branch, 1975).
Martin, W. C. L., *A General Introduction to the Natural History of Mammiferous Animals, with a Particular View of the Physical History of Man, and the More Closely Allied Genera of the Order Quadrumana, or Monkeys* (Wright, 1841).
—— 'Monograph on Genus Semnopithecus', *Magazine of Natural History*, new ser., 2 (1838), 320–26, 434–41.
—— 'Observations on Three Specimens of the Genus Felis presented to the Society by Charles Darwin, Esq.', *Proceedings of the Zoological Society of London*, 5 (1837), 3–4.
Martineau, H., *Harriet Martineau's Autobiography, with Memorials by Marie Weston Chapman*, 3 vols (Smith Elder, 1877).
—— *How to Observe: Manners and Morals* (Knight, 1838).

[——] 'The Martyr Age of the United States', *London and Westminster Review*, 32 (Dec. 1838), 1–59.
——*The Martyr Age of the United States* (Boston: Weeks, Jordan, 1839).
——*Retrospect of Western Travel*, 3 vols (Saunders & Otley, 1838).
[——] 'The Slave Trade in 1858', *ER*, 108 (Oct. 1858), 541–86.
——*Society in America*, 3 vols (Saunders & Otley, 1837).
Matthew, P., *Emigration Fields: North America, The Cape, Australia, and New Zealand, describing these Countries, and giving a Comparative View of the Advantages they present to British Settlers* (Edinburgh: Black, 1839).
Maurer, O., '*Punch* on Slavery and Civil War in America, 1841–1865', *VS*, 1 (1957), 5–28.
Mayr, E., 'Descent of Man and Sexual Selection', *Atti del Colloquio internazionale sul tema: L'Origine dell'Uomo, indetto in occasione del primo centenario della pubblicazione dell'opera di Darwin, "Descent of Man" (Roma, 28–30 ottobre 1971)* (Rome: Accademia Nazionale dei Lincei, 1973), 33–48.
Mazrui, A. A., 'From Social Darwinism to Current Theories of Modernization: A Tradition of Analysis', *World Politics*, 21 (1968), 69–83.
Meigs, C. D., *A Memoir of Samuel George Morton, M. D., Late President of the Academy of Natural Sciences of Philadelphia* (Philadelphia: Collins, 1851).
Menand, L., *The Metaphysical Club: A Story of Ideas in America* (New York: HarperCollins, 2001).
Messner, W. F., '"DeBow's Review", 1846–1880', in Lora and Longton, *Conservative Press*, 201–10.
Meteyard, E., *The Life of Josiah Wedgwood from His Private Correspondence and Family Papers . . .*, 2 vols (Hurst & Blackett, 1865–6).
Michael, J. S., 'A New Look at Morton's Craniological Research', *Current Anthropology*, 29 (1988), 349–54.
Midgley, C., 'Slave Sugar Boycotts, Female Activism and the Domestic Base of British Anti-Slavery Culture', *SA*, 17 (1996), 137–62.
——*Women against Slavery: The British Campaigns, 1780–1870* (Routledge, 1992).
Mill, J. S., *Autobiography*, 4th edn (Longmans et al., 1874).
——'Whewell's Moral Philosophy', *Westminster Review*, new ser., 2 (Oct. 1852), 349–85.
Miller, R., *Britain and Latin America in the Nineteenth and Twentieth Centuries* (Longman, 1993).
Milner, M., *The Life of Isaac Milner, D.D., F.R.S. . . .* (Parker, 1842).
Mivart, St G. J., *Essays and Criticisms*, 2 vols (Boston: Little, Brown, 1892).
M'Lennan, J. F., *Primitive Marriage: An Inquiry Into the Origin of the Form of Capture in Marriage Ceremonies* (Edinburgh: Black, 1865).
Monro, A., *The Morbid Anatomy of the Brain Vol. I – Hydrocephalus* (Edinburgh: Maclachlan & Stewart, 1827).
Moodie, J. W. D., *Ten Years in South Africa, including a Particular Description of the Wild Sports of that Country*, 2 vols (Bentley, 1835).
Moore, J., 'Darwin of Down: The Evolutionist as Squarson–Naturalist', in Kohn, *Darwinian Heritage*, 435–81.
——'Deconstructing Darwinism: The Politics of Evolution in the 1860s', *JHB*, 24 (1991), 353–408.
——'Geologists and Interpreters of Genesis in the Nineteenth Century', in D. C. Lindberg and R. L. Numbers, eds, *God and Nature: Historical Essays on the Encounter between Christianity and Science* (Berkeley: University of California Press, 1986), 322–50.
——ed., *History, Humanity and Evolution: Essays for John C. Greene* (New York: CUP, 1989).
——'Of Love and Death: Why Darwin "gave up Christianity"', in Moore, *History*, 195–229.
——'Revolution of the Space Invaders: Darwin and Wallace on the Geography of Life', in

D. N. Livingstone and C. W. J. Withers, eds, *Geography and Revolution* (Chicago: University of Chicago Press, 2005), 106–32.

——'Wallace in Wonderland', in C. H. Smith and G. Beccaloni, eds, *Natural Selection and Beyond: The Intellectual Legacy of Alfred Russel Wallace* (New York: OUP, 2008).

——'Wallace's Malthusian Moment: The Common Context Revisited', in B. Lightman, ed., *Victorian Science in Context* (Chicago: University of Chicago Press, 1997), 290–311.

——and A. Desmond, 'Introduction', in Charles Darwin, *The Descent of Man, and Selection in Relation to Sex* (Penguin, 2004), xi–lxvi.

Morrell, J., and A. Thackray, *Gentlemen of Science: Early Years of the British Association for the Advancement of Science* (Oxford: Clarendon, 1981).

[Morris, J.], 'Darwin on the Origin of Species', *Dublin Review*, 48 (1860), 50–81.

Morris, J., 'Louis Agassiz's Additions to the French Translation of His "Essay on Classification" ', *JHB*, 30 (1997), 121–34 .

Morton, S. G., *Crania Aegyptiaca: Observations on Egyptian Ethnography, derived From Anatomy, History and the Monuments* (Philadelphia: Penington, 1844).

——*Crania Americana; or, A Comparative View of the Skulls of Various Aboriginal Nations of North and South America, to which is prefixed an Essay on the Varieties of the Human Species* (Philadelphia: Dobson, 1839).

——'Description of Two Living Hybrid Fowls, between Gallus and Numida', *AMNH*, 19 (1847), 210–12.

——'Hybridity in Animals, considered in Reference to the Question of the Unity of the Human Species', *American Journal of Science and Arts*, 2nd ser., 3 (1847), 39–50, 203–12.

——'Hybridity in Animals and Plants, considered in reference to the Question of the Unity of the Human Species', *ENPJ*, 43 (1847), 262–88.

——*An Inquiry Into the Distinctive Characteristics of the Aboriginal Race of America: Read at the Annual Meeting of the Boston Society of Natural History, Wednesday, April 22, 1842* (Boston: Tuttle & Dennett, 1842).

——*Letter to the Rev. John Bachman, D.D., on the Question of Hybridity in Animals, considered in Reference to the Unity of the Human Species* (Charleston, SC: Walker & James, 1850).

Mott, F. L., 'The "Christian Disciple" and the "Christian Examiner" ', *New England Quarterly*, 1 (1928), 197–207.

Moule, H. C. G., *Charles Simeon* (Inter-Varsity Fellowship, 1965).

The Mount, Shrewsbury: Important Sale of Excellent Household Furniture ... November 19th, 20th, 21st, 22nd, 23rd, 24th, 1866 ... (Shrewsbury: printed by Leake & Evans, [1866]).

Muddiman, J. G., 'H.M.S. "The Black Joke" ', *Notes and Queries*, 172 (1937), 200–201.

Mudie, R., *The Modern Athens: A Dissection and Demonstration of Men and Things in the Scotch Capital* (Knight & Lacey, 1825).

Musselman, E. G., 'Swords into Ploughshares: John Herschel's Progressive View of Astronomical and Imperial Governance', *BJHS*, 31 (1998), 419–35.

Myers, N., 'In Search of the Invisible: British Black Family and the Community, 1780–1830', *SA*, 13 (1992), 156–80.

Napier, M., *Selection of the Correspondence of the Late Macvey Napier* (Macmillan, 1879).

Nelson, G. B., ' "Men before Adam!": American Debates over the Unity and Antiquity of Humanity', in D. C. Lindberg and R. L. Numbers, eds, *When Science and Christianity Meet* (Chicago: University of Chicago Press, 2003), 161–81.

Nelson, M. V., 'The Negro in Brazil as seen through the Chronicles of Travellers, 1800–1868', *JNH*, 30 (1945), 203–18.

Neuffer, C. H., ed., *The Christopher Happoldt Journal: His European Tour with the Rev. John Bachman (June–December, 1838)*, Contributions from the Charleston Museum, 13 (Charleston, SC: Charleston Museum, 1960).

Newman, E., *A Familiar Introduction to the History of Insects: Being a New and Greatly Improved Edition of 'The Grammar of Entomology'* (Van Voorst, 1841).

Newman, W. A., 'Darwin and Cirripedology', *Crustacean Studies*, 7 (1993), 349–434.
Newsome, D., *The Wilberforces and Henry Manning: The Parting of Friends* (Cambridge, MA: Belknap Press of Harvard UP, 1966).
Newton, A., *The Zoological Aspect of Game Laws (British Association Section D, August, 1868)* (Society for the Protection of Birds, no. 13, n.d.).
Noll, M., *The Civil War as a Theological Crisis* (Chapel Hill: University of North Carolina Press, 2006).
——*Princeton and the Republic, 1768–1822: The Search for a Christian Enlightenment in the Era of Samuel Stanhope Smith* (Princeton, NJ: Princeton UP, 1989).
Northcott, C., *Slavery's Martyr: John Smith of Demerara and the Emancipation Movement, 1817–24* (Epworth, 1976).
Norton, C. E., ed., *The Correspondence of Thomas Carlyle and Ralph Waldo Emerson, 1834–1872*, 2 vols (Boston: Ticknor, 1883).
Nott, J. C., 'Ancient and Scriptural Chronology', *SQR*, 2 (1850), 385–426.
——'Diversity of the Human Race', *DeBow's Southern and Western Review*, 10 (1851), 113–32.
——'The Mulatto a Hybrid – Probable Extermination of the Two Races if the Whites and Blacks are allowed to intermarry', *American Journal of Medical Science*, new ser., 6 (July 1843), 252–56.
——'Nature and Destiny of the Negro', *DeBow's Southern and Western Review*, 10 (1851), 329–32.
——'The Negro Race', *Popular Magazine of Anthropology*, 1 (1866), 102–18.
——'Physical History of the Jewish Race', *SQR*, 1 (1850), 426–51.
——'The Problem of the Black Races', *DeBow's Review*, new ser., 1 (Mar 1866), 266–83.
——'Statistics of Southern Slave Population', *DeBow's Review – Agricultural, Commercial, Industrial Progress and Resources*, 4 (1847), 275–89.
——*Two Lectures on the Natural History of the Caucasian and Negro Races* (Mobile, AL: Dade & Thompson, 1844).
——'Unity of the Human Race', *SQR*, 9 (Jan. 1846), 1–57.
——and G. R. Gliddon, *Types of Mankind: or, Ethnological Researches, based upon the Ancient Monuments, Paintings, Sculptures, and Crania of Races, and upon their Natural, Geographical, Philological, and Biblical History; illustrated by Selections from the Inedited Papers of Samuel George Morton, M.D . . . and by Additional Contributions from Prof. L. Agassiz, LL. D.; W. Usher, M.D.; and Prof. H. S. Patterson, M.D.*, 7th edn (Philadelphia: Lippincott, Grambo, 1855; first published 1854).
——G. R. Gliddon, et al., *Indigenous Races of the Earth; or, New Chapters of Ethnological Enquiry: Monographs on Special Departments of Philology, Iconography, Cranioscopy, Paleontology, Pathology, Archeology, Comparative Geography, and Natural History* (Philadelphia: Lippincott, 1857).
O'Byrne, W. R., ed., *A Naval Biographical Dictionary . . .* (Murray, 1849).
Oldfield, J. R., 'Anti-slavery Sentiment in Children's Literature, 1750–1850', *SL*, 10 (1989), 44–59.
——*Popular Politics and British Anti-slavery: The Mobilisation of Public Opinion against the Slave Trade, 1787–1807* (Cass, 1998).
O'Leary, P., *Sir James Mackintosh: The Whig Cicero* (Aberdeen: Aberdeen UP, 1989).
Olmsted, F. L., *A Journey in the Back Country* (New York: Mason, 1860).
——*A Journey in the Seaboard Slave States, with Remarks on Their Economy* (Sampson Low, 1856).
——*A Journey through Texas; or, a Saddle-Trip on the Southwestern Frontier, with a Statistical Appendix* (New York: Dix, Edwards, 1857).
Ospovat, D., *The Development of Darwin's Theory: Natural History, Natural Theology, and Natural Selection, 1838–1859* (Cambridge: CUP, 1981).
Overton, J. H., *The English Church in the Nineteenth Century (1800–1833)* (Longmans et al., 1894).

[Owen, R.], *Catalogue of the Contents of the Museum of the Royal College of Surgeons in London. Part III. Comprehending the Human and Comparative Osteology* (Warr, 1831).

[——] *A Descriptive Catalogue of the Osteological Series contained in the Museum of the Royal College of Surgeons of England*, vol. 2, *Mammalia Placentalia* (printed by Taylor & Francis, 1853).

——'On the Characters, Principles of Division, and Primary Groups of the Class Mammalia', *Journal of the Proceedings of the Linnean Society* (Zoology), 2 (1858), 1–37.

——*On the Classification and Distribution of the Mammalia* (Parker, 1859).

Owen, R. S., ed., *The Life of Richard Owen*, 2 vols (Murray, 1894).

Oxford Union Society, *Oxford Union Society* [*Debates, 18 November 1826 to 9 June 1831*] (Oxford: printed for the Society [1831]).

Paley, W., *The Principles of Moral and Political Philosophy*, 12th edn, 2 vols (printed for R. Faulder, 1799).

Parker, T., *Sermons of Theism, Atheism, and the Popular Theology* (Boston: Ticknor & Fields, 1861).

Parodiz, J. J., *Darwin in the New World* (Leiden: Brill, 1981).

Paston, G., *At John Murray's: Records of a Literary Circle, 1843–1892* (Murray, 1932).

Paton, D., 'Decency, Dependence and the Lash: Gender and the British Debate over Slave Emancipation, 1830–34', *SA*, 17 (1996), 163–84.

Patton, A., *Physicians, Colonial Racism, and Diaspora in West Africa* (Gainesville: University Press of Florida, 1996).

Pearson, J., *An Exposition of the Creed*, ed. W. S. Dobson (Dove, 1832).

[Peckard, P.], *Am I Not a Man? and a Brother? With All Humility addressed to the British Legislature* (Cambridge: printed by J. Archdeacon, printer to the University, 1788).

[——] *The Nature and Extent of Civil and Religious Liberty: A Sermon preached before the University of Cambridge, November the 5th, 1783* (Cambridge: printed by J. Archdeacon, 1783).

Peckham, M., ed., *The Origin of Species by Charles Darwin: A Variorum Text* (Philadelphia: University of Pennsylvania Press, 1959).

Pendleton, L. A., *A Narrative of the Negro* (Washington, DC: Pendleton, 1912).

Pentland, J. B., 'On the Ancient Inhabitants of the Andes', *Report of the Fourth Meeting of the British Association for the Advancement of Science; held at Edinburgh in 1834* (Murray, 1835), 623–4.

Pickering, C., *The Races of Man; and Their Geographical Distribution . . . to which is prefixed an Analytical Synopsis of the Natural History of Man* (Bohn, 1854).

Pierce, E. L., *Memoir and Letters of Charles Sumner*, 4 vols (Sampson Low, 1878–93).

Pierson, M. D., ' "All Southern Society is Assailed by the Foulest Charges": Charles Sumner's "The Crime against Kansas" and the Escalation of Republican Anti-slavery Rhetoric', *New England Quarterly*, 68 (1995), 531–57.

Poole, W. S., 'Memory and the Abolitionist Heritage: Thomas Wentworth Higginson and the Uncertain Meaning of the Civil War', *Civil War History*, 51 (2005), 202–17.

Porter, D., 'On the Road to the *Origin* with Darwin, Hooker, and Gray', *JHB*, 26 (1993), 1–38.

Porter, R., 'Erasmus Darwin: Doctor of Evolution?' in Moore, *History*, 39–69.

——'Science versus Religion?' in *Christ's: A Cambridge College over Five Centuries*, ed. D. Reynolds (Macmillan, 2004), 79–109.

Portlock, J., 'Anniversary Address', *Quarterly Journal of the Geological Society*, 14 (1858), xxiv–clxiii.

Potter, E. N., *Discourses Commemorative of Professor Tayler Lewis, LL.D., L.H.D. . . .* (Albany, NY: printed by J. Munsell, 1878).

Pretorius, J. G., *The British Humanitarians and the Cape Eastern Frontier, 1834–1836* (Pretoria: Government Printer, 1988).

Prichard, J. C., *The Natural History of Man; comprising Inquiries into the Modifying Influence of Physical and Moral Agencies on the Different Tribes of the Human Family*, 2nd edn, 2 vols (Baillière, 1845).

——*The Natural History of Man; comprising Inquiries into the Modifying Influence of Physical and Moral Agencies on the Different Tribes of the Human Family*, 4th edn., ed. E. Norris, 2 vols (Baillière, 1855).
——'On the Extinction of Human Races', *ENPJ*, 28 (1840), 166–70.
——'On the Relations of Ethnology to Other Branches of Knowledge', *JESL*, 1 (1848), 301–29.
——*Researches into the Physical History of Man*, ed. G. W. Stocking (Chicago: University of Chicago Press, 1973; first published 1813).
——*Researches into the Physical History of Mankind*, 3rd edn, 2 vols (Sherwood et al., 1836).
——*Researches into the Physical History of Mankind*, 4th edn, 5 vols (Houlston & Stoneman, 1851; first published 1836–47).
Pulszky, F., and T. Pulszky, *White, Red, Black: Sketches of American Society in the United States during the Visit of Their Guests*, 2 vols (New York: Redfield, 1853).
Qureshi, S., 'Displaying Sara Baartman, the "Hottentot Venus"', *HS*, 42 (2004), 233–57.
Rackham, H., *Christ's College in Former Days: Being Articles reprinted from the College Magazine* (Cambridge: printed at the UP, 1939).
Radick, G., 'Darwin on Language and Selection', *Selection*, 3 (2002), 7–16.
Rae, I., *Knox the Anatomist* (Edinburgh: Oliver & Boyd, 1964).
Rainger, R., 'Philanthropy and Science in the 1830's: The British and Foreign Aborigines Protection Society', *Man*, new ser., 15 (1980), 702–17.
——'Race, Politics, and Science: The Anthropological Society of London in the 1860s', *VS*, 22 (1978), 51–70.
Rehbock, P. F., 'The Early Dredgers: "Naturalizing" in British Seas, 1830–1850', *JHB*, 12 (1979), 293–368.
Reis, J. J., *Slave Rebellion in Brazil: The Muslim Uprising of 1835 in Bahia* (Baltimore, MD: Johns Hopkins UP, 1993).
Renehan, E., *The Secret Six: The True Tale of the Men who conspired with John Brown* (Columbia: University of South Carolina Press, 1996).
Rice, C. D., *The Scots Abolitionists, 1833–1861* (Baton Rouge: Louisiana State UP, 1981).
Richards, E., 'Darwin and the Descent of Woman', in D. R. Oldroyd and I. Langham, eds, *The Wider Domain of Evolutionary Thought* (Dordrecht: Reidel, 1983), 57–111.
——'Huxley and Woman's Place in Science: The "Woman Question" and the Control of Victorian Anthropology', in Moore, *History*, 253–84.
——'The "Moral Anatomy" of Robert Knox: The Interplay Between Biological and Social Thought in Victorian Scientific Naturalism', *JHB*, 22 (1989), 373–436.
——'A Political Anatomy of Monsters, Hopeful and Otherwise', *Isis*, 85 (1994), 377–411.
——'A Question of Property Rights: Richard Owen's Evolutionism Reassessed', *BJHS*, 20 (1987), 129–71.
Richards, R. A., 'Darwin and the Inefficacy of Artificial Selection', *SHPS*, 28 (1997), 75–97.
Richards, R. J., *Darwin and the Emergence of Evolutionary Theories of Mind and Behavior* (Chicago: University of Chicago Press, 1987).
——*The Tragic Sense of Life: Ernst Haeckel and the Struggle over Evolutionary Thought* (Chicago: University of Chicago Press, 2008).
Riegel, R. E., 'The Introduction of Phrenology to the United States', *American Historical Review*, 39 (1933), 73–8.
Ritchie, G. S., *The Admiralty Chart: British Naval Hydrography in the Nineteenth Century*, new edn (Bishop Auckland: Pentland, 1995).
Ritchie, J. E., *The Life and Times of Viscount Palmerston: Embracing the Diplomatic and Domestic History of the British Empire during the Last Half Century*, 2 vols (London Printing & Publishing, 1866–7).
Ritvo, H., *The Animal Estate: The English and Other Creatures in the Victorian Age* (Cambridge, MA: Harvard UP, 1987).
Roberts, G. B., 'Josiah Wedgwood and His Trade Connections with Liverpool', *Proceedings of the Wedgwood Society*, no. 11 (1982), 125–35.

Rodrigues, J. H., *Brazil and Africa* (Berkeley: University of California Press, 1965).
Rodriguez, J. P., ed., *Chronology of World Slavery* (Santa Barbara, CA: ABC-Clio, 1999).
——ed., *The Historical Encyclopedia of World Slavery*, 2 vols (Santa Barbara, CA: ABC-Clio, 1997).
Romanes, G. J., *Thoughts on Religion*, ed. C. Gore (Longmans et al., 1895).
Rose, M., *Curator of the Dead: Thomas Hodgkin (1798–1866)* (Owen, 1981).
Rose, M. B., *The Gregs of Quarry Bank Mill: The Rise and Decline of a Family Firm, 1750–1914* (Cambridge: CUP, 1986).
Ross, A., *John Philip (1775–1851): Missions, Race and Politics in South Africa* (Aberdeen: Aberdeen UP, 1986).
Ross, J. A., and H. W. Y. Taylor, 'Robert Knox's Catalogue', *JHMAS*, 10 (1955), 269–76.
Royle, E., *Victorian Infidels: The Origins of the British Secularist Movement, 1791–1866* (Manchester: Manchester UP, 1974).
Rudwick, M. J. S., 'Darwin and Glen Roy: A "great failure" in Scientific Method?', *SHPS*, 5 (1974), 97–185.
Rupke, N. A., 'Neither Creation Nor Evolution: The Third Way in Mid-Nineteenth Century Thinking about the Origin of Species', *Annals of the History and Philosophy of Biology*, 10 (2005), 143–72.
——*Richard Owen, Victorian Naturalist* (New Haven, CT: Yale UP, 1994).
Ruse, M., 'Charles Darwin and Group Selection', *AS*, 37 (1980), 615–30.
Ryan, F. W., ' "Southern Quarterly Review", 1842–1857', in Lora and Longton, *Conservative Press*, 183–90.
Salisbury, E. G., *Border County Worthies*, 1st and 2nd ser. (Hodder & Stoughton, 1880).
Sanders, C. R., et al., eds, *The Collected Letters of Thomas and Jane Welsh Carlyle*, 34 vols to date (Durham, NC: Duke UP, 1970–).
Sanders, V., ed., *Harriet Martineau: Selected Letters* (Oxford: Clarendon, 1990).
Sandwith, T., 'A Comparative View of the Relations between the Development of the Nervous System and the Functions of Animals', *Phrenological Journal*, 4 (1827) 479–94.
Schaaffhausen, H., 'Darwinism and Anthropology', *JASL*, 6 (1868), cviii–cxi.
Schlesinger, A. M., 'Editor's Introduction', in F. W. Olmsted, *The Cotton Kingdom: A Traveller's Observations on Cotton and Slavery in the American Slave States, based upon Three Former Volumes of Journeys and Investigations by the Same Author* (New York: Knopf, 1953).
[Scholefield, H. C.], *Memoir of the Late Rev. James Scholefield, M.A. ...* (Seeley et al., 1855).
Schomburgk, R. H., *The History of Barbados: Comprising a Geographical and Statistical Description of the Island; a Sketch of the Historical Events Since the Settlement; and an Account of its Geology and Natural Productions* (Longman et al., 1848).
Schwartz, J. S., 'Darwin, Wallace, and the "Descent of Man" ', *JHB*, 17 (1984), 271–89.
Schwartz, S. B., *Sugar Plantations in the Formation of Brazilian Society: Bahia, 1550–1835* (Cambridge: CUP, 1985).
Sclater, P. L., 'On the General Geographical Distribution of the Members of the Class Aves', *Journal of the Proceedings of the Linnean Society* (Zoology), 2 (1858), 130–144.
Searby, P., *A History of the University of Cambridge*, vol. 3, *1750–1870* (Cambridge: CUP, 1997).
Secord, J. A., 'Behind the Veil: Robert Chambers and "Vestiges" ', in Moore, *History*, 165–94.
——'Darwin and the Breeders: A Social History', in Kohn, *Darwinian Heritage*, 519–42.
——'Edinburgh Lamarckians: Robert Jameson and Robert E. Grant', *JHB*, 24 (1991), 1–18.
——'Nature's Fancy: Charles Darwin and the Breeding of Pigeons', *Isis*, 72 (1981), 163–86.
——*Victorian Sensation: The Extraordinary Publication, Reception, and Secret Authorship of 'Vestiges of the Natural History of Creation'* (Chicago: University of Chicago Press, 2000).

Sedgwick, A., *A Discourse on the Studies of the University* (Cambridge: printed at the Pitt Press, 1833).
Semmel, B., *The Governor Eyre Controversy* (Macgibbon & Kee, 1962).
Sen Gupta, P. C., 'Soorjo Coomar Goodeve Chuckerbutty: The First Indian Contributor to Modern Medical Science', *Medical History*, 14 (1970), 183–91.
Shapin, S., 'Homo Phrenologicus: Anthropological Perspectives on an Historical Problem', in B. Barnes and S. Shapin, eds, *Natural Order: Historical Studies of Scientific Culture* (Beverly Hills, CA: Sage, 1979), 41–71.
——'Phrenological Knowledge and the Social Structure of Early Nineteenth-Century Edinburgh', *AS*, 32 (1975), 219–43.
——'The Politics of Observation: Cerebral Anatomy and Social Interests in the Edinburgh Phrenology Disputes', in R. Wallis, ed., *On the Margins of Science: The Social Construction of Rejected Knowledge*, Sociology Review Monograph, no. 27 (Keele, 1979), 139–78.
Sherwood, M. B., 'Genesis, Evolution, and Geology in America before Darwin: The Dana–Lewis Controversy, 1856–1857', in C. J. Schneer, ed., *Toward a History of Geology* . . . (Cambridge, MA: MIT Press, 1967).
Shyllon, F. O., *Black Slaves in Britain* (Oxford: OUP, 1974).
Sieveking, I. G., *Memoir and Letters of Francis W. Newman* . . . (Kegan Paul, 1909).
Silliman, B., *Journal of Travels in England, Holland, and Scotland: Two Passages over the Atlantic, in the Years 1805 and 1806*, 2nd edn, 2 vols (Boston: Wait, 1812).
——'Phrenology', *American Journal of Science and Arts*, 39 (1840), 65–88.
Silliman, R. H., 'The Hamlet Affair: Charles Lyell and the North Americans', *Isis*, 86 (1995), 541–61.
Sinha, M., 'The Caning of Charles Sumner: Slavery, Race, and Ideology in the Age of the Civil War', *Journal of the Early Republic*, 23 (2003), 233–62.
Sivasundaram, S., *Nature and Godly Empire: Science and Evangelical Mission in the Pacific, 1795–1850* (Cambridge: CUP, 2005).
Sklar, K. K., 'Women Who Speak for an Entire Nation: American and British Women compared at the World Anti-Slavery Convention, London, 1840', in J. F. Yellin and J. C. Van Horne, eds, *The Abolitionist Sisterhood: Women's Political Culture in Antebellum America* (Ithaca, NY: Cornell UP, 1994), 301–33.
Sloan, P. R., 'Darwin, Vital Matter, and the Transformism of Species', *JHB*, 19 (1986), 367–95.
——'Darwin's Invertebrate Program, 1826–1836: Preconditions for Transformism', in Kohn, *Darwinian Heritage*, 71–120.
Smith, A., 'Observations Relative to the Origin and History of the Bushmen', *Philosophical Magazine*, new ser., 9 (1831), 119–27, 197–200, 339–42, 419–23.
Smith, C. H., *The Natural History of Horses* (Edinburgh: Lizars, 1845–6).
——*The Natural History of Dogs*, 2 vols (Edinburgh: Lizars, 1839–40).
——*The Natural History of the Human Species, Its Typical Forms, Primaeval Distribution, Filiations, and Migrations* (Bohn, 1852).
Smith, C. U. M., 'Worlds in Collision: Owen and Huxley on the Brain', *Science in Context*, 10 (1997), 343–65.
Smith, D., 'The Fergusson Papers: A Calendar of 92 Manuscript Letters and Documents concerning the Medical History of the Peninsular War (1808–1814)', *JHMAS*, 19 (1964), 267–71.
Smith, H. S., *In His Image, but . . .: Racism in Southern Religion, 1780–1910* (Durham, NC: Duke UP, 1972).
Smith, J., 'Grant Allen, Physiological Aesthetics, and the Dissemination of Darwin's Botany', in G. Cantor and S. Shuttleworth, eds, *Science Serialized: Representations of the Sciences in Nineteenth-Century Periodicals* (Cambridge, MA: MIT Press, 2004).
Smith, K. V., 'Darwin's Insects: Charles Darwin's Entomological Notes, with an Introduction and Comments . . .', *BBMNH*, 14 (1987), 1–143.
Smith, R., 'Alfred Russel Wallace: Philosophy of Nature and Man', *BJHS*, 6 (1972), 177–99.

Smith, S., *The Works of the Rev. Sydney Smith*, 3 vols (Longman et al., 1839).
Smyth, T., *The Unity of the Human Races proved to be the Doctrine of Scripture, Reason, and Science, with a Review of the Present Position and Theory of Professor Agassiz* (New York: Putnam, 1850).
Snyder, L. J., *Reforming Philosophy: A Victorian Debate on Science and Society* (Chicago: University of Chicago Press, 2006).
Sober, E., and D. S. Wilson, *Unto Others: The Evolution and Psychology of Unselfish Behavior* (Cambridge, MA: Harvard UP, 1998).
Society for the Extinction of the Slave Trade, *Proceedings of the First Public Meeting of the Society for the Extinction of the Slave Trade, and for the Civilization of Africa, held at Exeter Hall, on Monday, 1st June, 1840* (printed by Clowes, 1840).
Spencer, F., 'Samuel George Morton's Doctoral Thesis on Bodily Pain: The Probable Source of Morton's Polygenism', *Transactions and Studies of the College of Physicians of Philadelphia*, 5th ser., 5 (1983), 321–38.
Stange, D. C., *British Unitarians against American Slavery, 1833–65* (Cranbury, NJ: Associated University Presses, 1984).
Stanton, W., *The Leopard's Spots: Scientific Attitudes toward Race in America, 1815–59* (Chicago: University of Chicago Press, 1960).
Stauffer, R. C., ed., *Charles Darwin's Natural Selection: Being the Second Part of His Big Species Book written from 1856 to 1858* (Cambridge: CUP, 1975).
Staum, M., 'Nature and Nurture in French Ethnography and Anthropology, 1859–1914', *JHI*, 65 (2004), 475–95.
Stecher, R. M., 'The Darwin–Innes Letters: The Correspondence of an Evolutionist with His Vicar, 1848–1884', *AS*, 17 (1961), 201–58.
Steinheimer, F. D., 'Charles Darwin's Bird Collection and Ornithological Knowledge during the Voyage of H.M.S. *Beagle*, 1831–1836', *Journal of Ornithology*, 145 (2004), 300–320.
Stenhouse, J., 'Imperialism, Atheism, and Race: Charles Southwell, Old Corruption, and the Maori', *Journal of British Studies*, 44 (2005), 754–74.
Stepan, N., *The Idea of Race in Science: Great Britain, 1800–1960* (Macmillan, 1982).
Stephen, G., *Antislavery Recollections, in a Series of Letters addressed to Mrs. Beecher Stowe . . . at Her Request* (Hatchard, 1854).
Stephen, J., 'The Clapham Sect', *Essays in Ecclesiastical Biography*, 5th edn (Longmans et al., 1867), 523–84.
Stephen, L., *The English Utilitarians*, 3 vols (New York: Putnam's, 1900).
Stephens, L. D., *Science, Race, and Religion in the American South: John Bachman and the Charleston Circle of Naturalists, 1815–1895* (Chapel Hill: University of North Carolina Press, 2000).
Sterrett, S. G., 'Darwin's Analogy between Artificial and Natural Selection: How does it go?', *SHPBBS*, 33 (2002), 151–68.
Stewart, R., *Henry Brougham 1778–1868: His Public Career* (Bodley Head, 1985).
Still, William, *The Underground Rail Road: A Record of Facts, Authentic Narratives, Letters, &c., narrating the Hardships, Hair-breadth Escapes and Death Struggles of the Slaves in Their Efforts for Freedom, as related by Themselves and Others, or witnessed by the Author; together with Sketches of Some of the Largest Stockholders, and most Liberal Aiders and Advisers, of the Road* (Medford, NJ: Plexus, 2005; first published 1872).
Stock, E., *The History of the Church Missionary Society: Its Environment, Its Men and Its Work*, 3 vols (Church Missionary Society, 1899).
Stocking, G. W., Jr, 'From Chronology to Ethnology: James Cowles Prichard and British Anthropology, 1800–1850', in Prichard, *Researches into the Physical History of Man* (1973 edn), ix–cxliv.
—— *Victorian Anthropology* (New York: Free Press, 1987).
—— 'What's in A Name? The Origins of the Royal Anthropological Institute (1837–71)', *Man*, 6 (1971), 369–90.
Stokes, J. L., *Discoveries in Australia; with an Account of the Coasts and Rivers explored*

and surveyed during the Voyage of H.M.S. Beagle, in the Years 1837–38–39–40–41–42–43 . . . (T. & W. Boone, 1846).

Stone, J. W., phonographer, *Trial of Thomas Sims, on an Issue of Personal Liberty, on the Claim of James Potter, of Georgia, against Him, as an Alleged Fugitive from Service: Arguments of Robert Rantoul, Jr., and Charles G. Loring, with the Decision of George T. Curtis, Boston, April 7–11, 1851* (Boston: Damrell, 1851).

Story, R., 'Harvard and the Boston Brahmins: A Study in Institutional and Class Development, 1800–1865', *Journal of Social History*, 8 (1975), 94–121.

Stringer, M. D., 'Rethinking Animism: Thoughts from the Infancy of Our Discipline', *Journal of the Royal Anthropological Institute*, 5, (1999), 541–55.

Struthers, J., *Historical Sketch of the Edinburgh Anatomical School* (Edinburgh: Maclachlan & Stewart, 1867).

Sulivan, H. N., ed., *Life and Letters of the Late Admiral Sir Bartholomew James Sulivan, K.C.B., 1810–1890* (Murray, 1896).

Sulloway, F. J., 'The *Beagle* Collections of Darwin's Finches (Geospizinae)', *BBMNH*, 43 (1982), 49–94.

Sumner, C., 'The Barbarism of Slavery', *Charles Sumner: His Complete Works*, vol. 6 (Boston: Lee & Shepherd, 1863), 119–238.

——*The Crime against Kansas. The Apologies for the Crime. The True Remedy: Speech of Hon. Charles Sumner, in the Senate of the United States, 19th and 20th May, 1856* (Boston: Jewett, 1856).

Sumner, J. B., *A Treatise on the Records of the Creation, and on the Moral Attributes of the Creator; with Particular Reference to the Jewish History, and to the Consistency of the Principle of Population with the Wisdom and Goodness of the Deity*, 4th edn, 2 vols (Hatchard, 1825).

Sussman, C., *Consuming Anxieties: Consumer Protest, Gender, and British Slavery, 1713–1833* (Stanford, CA: Stanford UP, 2000).

Swainson, W., *On the Habits and Instincts of Animals* (Longman et al., 1840).

——*A Preliminary Discourse on the Study of Natural History* (Longman et al., 1834).

——*A Treatise on the Geography and Classification of Animals* (Longman et al., 1835).

Temperley, H., *British Antislavery, 1833–70* (Longman, 1972).

Thomas, H., *The Slave Trade: The History of the Atlantic Slave Trade, 1440–1870* (Picador, 1997).

Thompson, F. M. L., *The Rise of Respectable Society: A Social History of Victorian Britain, 1830–1900* (Fontana, 1988).

Thomson, K. S., *HMS Beagle: The Story of Darwin's Ship* (New York: Norton, 1995).

Thwaite, A., *Glimpses of the Wonderful: The Life of Philip Henry Gosse, 1810–1888* (Faber & Faber, 2002).

Tiedemann, F., 'On the Brain of the Negro, compared with that of the European and the Orang-Outang', *Philosophical Transactions of the Royal Society of London*, 126 (1836), 497–527.

Timbs, J., *Curiosities of London: Exhibiting the Most Rare and Remarkable Objects of Interest in the Metropolis* (Bogue, 1855).

Tipton, J. A., 'Darwin's Beautiful Notion: Sexual Selection and the Plurality of Moral Codes', *HPLS*, 21 (1999), 119–35.

Toplin, R. B., 'Between Black and White: Attitudes toward Southern Mulattoes, 1830–1861', *Journal of Southern History*, 45 (1979), 185–200.

Torrens, H., 'When did the Dinosaur get its name?' *New Scientist* (4 Apr. 1992), 40–44.

Trafton, S., *Egypt Land: Race and Nineteenth-Century American Egyptomania* (Durham, NC: Duke UP, 2004).

Turley, D., *The Culture of English Antislavery, 1780–1860* (Routledge, 1991).

Turner, M., 'The British Caribbean, 1823–1838: The Transition from Slave to Free Legal Status', in D. Hay and P. Craven, eds, *Masters, Servants, and Magistrates in Britain and the Empire, 1562–1955* (Chapel Hill: University of North Carolina Press, 2004), 303–22.

Tylor, E. B., *Researches into the Early History of Mankind* (Murray, 1865).
Tyrrell, A., 'The "Moral Radical Party" and the Anglo-Jamaican Campaign for the Abolition of the Negro Apprenticeship System', *English Historical Review*, 99 (1984), 481–502.
Ucelay Da Cal, E., 'The Influence of Animal Breeding on Political Racism', *History of European Ideas*, 15 (1992), 717–25.
Uglow, J., *The Lunar Men: The Friends Who Made the Future, 1730–1810* (Faber & Faber, 2002).
Ulrich, J. M., 'Thomas Carlyle, Richard Owen, and the Paleontological Articulation of the Past', *Journal of Victorian Culture*, 11 (2006), 30–58.
Valone, D. A., 'Hugh James Rose's Anglican Critique of Cambridge: Science, Antirationalism, and Coleridgean Idealism in Late Georgian England', *Albion*, 33 (2001), 218–42.
Van Amringe, W. F., *An Investigation of the Theories of the Natural History of Man, by Lawrence, Prichard, and Others, founded upon Animal Analogies; and an Outline of a New Natural History of Man, founded upon History, Anatomy, Physiology, and Human Analogies* (New York: Baker & Scribner, 1848).
Van Evrie, J. H., *Negroes and Negro 'Slavery': The First, An Inferior Race – The Latter, Its Normal Condition*, 3rd edn (New York: Van Evrie, Horton & Co., 1863; first published 1861).
Van Riper, A. B., *Men among the Mammoths: Victorian Science and the Discovery of Prehistory* (Chicago: University of Chicago Press, 1993).
Vetter, J., 'Wallace's Other Line: Human Biogeography and Field Practice in the Eastern Colonial Tropics', *JHB*, 39 (2006), 89–123.
Vogt, C., *Lectures on Man: His Place in Creation, and in the History of the Earth*, ed. J. Hunt (Longman et al., 1864).
W., C.A., 'A Second Visit to the United States, by Charles Lyell', *SQR*, 1 (July 1850), 406–25.
Waddell, G., 'A Bibliography of John Bachman', *ANH*, 32 (2005), 53–69.
Waitz, T., *Introduction to Anthropology*, trans. J. F. Collingwood (Longman et al., 1863).
Walker, A., *Intermarriage; or, the Mode in which, and the Causes Why, Beauty, Health and Intellect, Result from Certain Unions, and Deformity, Disease and Insanity, from Others . . .* (Churchill, 1838).
[Wallace, A. R.], 'Charles Lyell on Geological Climates and the Origin of Species', *Quarterly Review*, 126 (Apr. 1869), 359–94.
——'Darwin's "The Descent of Man and Selection in Relation to Sex" ', *Academy*, 2 (15 Mar. 1871), 177–83.
——'[Letter dated 21 Aug. 1856]', *Zoologist*, 15 (1857), 5414–16.
——*The Malay Archipelago: The Land of the Orang-Utan and the Bird of Paradise; a Narrative of Travel, with Studies of Man and Nature*, 6th edn (Macmillan, 1877; first published 1869).
[——] 'Mimicry, and Other Protective Resemblances among Animals', *Westminster Review*, new ser., 32 (1 July 1867), 1–43.
——*My Life: A Record of Events and Opinions*, 2 vols (Chapman & Hall, 1905).
——'On the Law which has regulated the Introduction of New Species', *AMNH*, 2nd ser., 16 (Sep. 1855), 184–96.
——'The Origin of Human Races and the Antiquity of Man deduced from the Theory of "Natural Selection" ', *JASL*, 2 (1864), clviii–clxxvi.
——*The Scientific Aspect of the Supernatural: Indicating the Desirableness of an Experimental Enquiry by Men of Science into the Alleged Powers of Clairvoyants and Mediums* (Farrah, 1866).
Waller, J. O., 'Charles Kingsley and the American Civil War', *Studies in Philology*, 60 (1963), 554–68.
Wallis, B., 'Black Bodies, White Science: Louis Agassiz's Slave Daguerreotypes', *American Art*, 9 (1995), 39–61.
Walsh, A. A., 'The "New Science of the Mind" and the Philadelphia Physicians in the Early

1800s', *Transactions and Studies of the College of Physicians of Philadelphia*, 4th ser., 43 (1976), 397–415.

Walsh, J., and R. Hyam, *Peter Peckard: Liberal Churchman and Anti-Slave Trade Campaigner*, Magdalene College Occasional Papers, no. 16 (Cambridge: Magdalene College, 1998).

Walsh, J. D., 'The Magdalene Evangelicals', *Church Quarterly Review*, 159 (1958), 499–511.

Walters, S. M., and E. A. Stow, *Darwin's Mentor: John Stevens Henslow, 1796–1861* (Cambridge: CUP, 2001).

Walvin, J., *Black and White: The Negro and English Society, 1555–1945* (Allen Lane, 1973).

——'The Rise of British Popular Sentiment for Abolition, 1787–1832', in C. Bolt and S. Drescher, eds, *Anti-Slavery, Religion and Reform* (Folkestone: Dawson, 1980), 149–62.

——ed., *Slavery and British Society, 1776–1846* (Longman, 1982).

Ward, W. E. F., *The Royal Navy and the Slavers: The Suppression of the Atlantic Slave Trade* (Allen & Unwin, 1969).

Warner, H. W., ed., *Autobiography of Charles Caldwell, M. D.* (Philadelphia: Lippincott, Grambo, 1855).

Waterhouse, G. R., *Marsupialia, or Pouched Animals* (Edinburgh: Lizars, [1841]).

Waterman, A. M. C., *Revolution, Economics and Religion: Christian Political Economy, 1798–1833* (Cambridge: CUP, 1991).

Waterton, C., *Wanderings in South America, the North-West of the United States, and the Antilles in the Years 1812, 1816, 1820, & 1824*, 2nd edn (Fellowes, 1828).

Watts, R., *Gender, Power and the Unitarians in England, 1760–1860* (Longman, 1998).

Webb, R. K., *Harriet Martineau: A Radical Victorian* (New York: Columbia UP, 1960).

Wedgwood, B., and H. Wedgwood, *The Wedgwood Circle: Four Generations of a Family and Their Friends* (Westfield, NJ: Eastview, 1980).

Wedgwood, J., *The Personal Life of Josiah Wedgwood, the Potter* (Macmillan, 1915).

Wedgwood, J. C., *A History of the Wedgwood Family* (St Catherine Press, 1908).

——ed., *Staffordshire Parliamentary History from the Earliest Times to the Present Day*, vol. 3, *1780 to 1841* (Kendal: Wilson, 1934).

[Wedgwood, S. E.], ed., *British Slavery Described* (Newcastle[-under-Lyme]: printed by J. Mort and sold for the benefit of the North Staffordshire Ladies' Anti-Slavery Society, 1828).

Weiss, J., *Life and Correspondence of Theodore Parker, Minister of the Twenty-eighth Congregational Society, Boston*, 2 vols (New York: Appleton, 1864).

Wells, K. D., 'William Charles Wells and the Races of Man', *Isis*, 64 (1973), 215–25.

Wheatley, V., *The Life and Work of Harriet Martineau* (Secker & Warburg, 1957).

Whewell, W., *The Elements of Morality, including Polity*, 2 vols (New York: Harper, 1845).

——*History of the Inductive Sciences from the Earliest to the Present Time*, 3rd edn, 3 vols (Parker, 1857; first published 1837).

——*Lectures on the History of Moral Philosophy in England* (Parker, 1852).

——*The Philosophy of the Inductive Sciences, founded upon Their History*, 2nd edn, 2 vols (Parker, 1847; first published 1840).

Wilberforce, R. I., and S. Wilberforce, *The Life of William Wilberforce*, 5 vols (Murray, 1838).

[Wilberforce, S.], '*On the Origin of Species, by means of Natural Selection; or the Preservation of Favoured Races in the Struggle for Life*. By Charles Darwin, M.A., F.R.S. London, 1860', *Quarterly Review*, 108 (1860), 225–64.

Wilner, E., 'Darwin's Artificial Selection as an Experiment', *SHPBBS*, 37 (2006), 26–40.

Wilson, E. G., 'A Shropshire Lady in Bath, 1794–1807', *Bath History*, 4 (1992), 95–123.

——*Thomas Clarkson: A Biography* (Basingstoke: Macmillan, 1989).

Wilson, G., and A. Geikie, *Memoir of Edward Forbes, F. R. S.* (Macmillan, 1861).

Wilson, L. G., 'The Gorilla and the Question of Human Origins: The Brain Controversy', *JHMAS*, 51 (1996), 184–207.

——*Lyell in America: Transatlantic Geology, 1841–1853* (Baltimore, MD: Johns Hopkins UP, 1998).

——ed., *Sir Charles Lyell's Scientific Journals on the Species Question* (New Haven, CT: Yale UP, 1970).

Winsor, M., *Reading the Shape of Nature: Comparative Zoology at the Agassiz Museum* (Chicago: University of Chicago Press, 1991).

Wish, H., 'The Revival of the African Slave Trade in the United States, 1856–1860', *Mississippi Valley Historical Review*, 27 (1941), 569–88.

Wood, G. B., *A Biographical Memoir of Samuel George Morton, M.D., prepared by Appointment of the College of Physicians of Philadelphia, and read Before That Body, November 3, 1852* (Philadelphia: Collins, 1853).

Woodward, S. P., *A Manual of the Mollusca; or, A Rudimentary Treatise of Recent and Fossil Shells*, 3 vols (Weale, 1851–6).

[Wyman, J.], 'Morton's Crania Americana', *North American Review*, 51 (1840), 173–86.

Yeo, R., *Defining Science: William Whewell, Natural Knowledge and Public Debate in Early Victorian Britain* (Cambridge: CUP, 1993).

Yetwin, N. B., 'Rev. Horace G. Day, Schenectady's Abolitionist Preacher', *Skenectada*, 1 (Winter 2006), 4–5.

Young, B., '"The Lust of Empire and Religious Hate": Christianity, History, and India, 1790–1820', in S. Collini, R. Whatmore and B. Young, eds, *History, Religion, and Culture: British Intellectual History, 1750–1950* (Cambridge: CUP, 2000), 91–111.

Young, R. J. C., *Colonial Desire: Hybridity in Theory, Culture and Race* (Routledge, 1995).

Zabriskie, A. C., 'Charles Simeon, Anglican Evangelical', *Church History*, 9 (1940), 103–19.

Zelinsky, W., 'The Historical Geography of the Negro Population of Latin America', *JNH*, 34 (1949), 153–221.

Index